RELEVANT PROBLEMS FOR CHEMICAL PRINCIPLES

SECOND EDITION

IAN S. BUTLER • ARTHUR E. GROSSER • *McGill University, Montreal*

RELEVANT PROBLEMS FOR CHEMICAL PRINCIPLES

SECOND EDITION

W. A. BENJAMIN, INC.
Menlo Park, California • Reading, Massachusetts
Don Mills, Ontario • London • Amsterdam • Sydney

RELEVANT PROBLEMS FOR CHEMICAL PRINCIPLES

Second Edition

Library of Congress Catalog Card No. 73-7110

PREFACE

The instructor who regards the plethora of introductory college texts as but another example of increasing entropy has our common sympathy. Moreover, apparently few texts are satisfactory to all instructors. Our experience with the shortcomings of most texts led us to establish two priorities in writing this supplementary problems book: We would provide detailed solutions to *every* problem, and we would include as many problems as possible to show the student that even a limited knowledge of the principles of chemistry would enable him to solve problems concerning real situations encountered not only in chemistry, but in other sciences, and even everyday life. Thus, we have included problems about space science, medicine, geology, dentistry, archaeology, biochemistry, urban pollution and hygiene, engineering, and solar and nuclear energy.

Another of our objectives was to allow students to proceed at their own pace through the book. We have purposely used a multiple-choice format, not to remind the student of the computer age on campus but to construct incorrect answers that correspond to common errors. The reasons for these incorrect answers are often discussed in the solution of the problem. So the student can either work through the solution to find he is correct or to find his error, or he can check only the discussion of the incorrect answer. For the well-prepared student, however, the number corresponding to the correct answer is prominently displayed within a circle in the margin under the solution number so he need not work through an entire solution.

We have taken advantage of the fact that this book is designed to form a part of the complete teaching system accompanying the text *Chemical Principles,* second edition, by Richard E. Dickerson, Harry B. Gray, and Gilbert P. Haight, Jr. Throughout both the text and this book, there are cross references that are intended to indicate to the student the areas in which he might find additional review. The student who continually chooses the wrong answer is referred to the appropriate section in *Programed Reviews of Chemical Principles,* by Jean D. Lassila *et al.* This book, which is another part of the teaching system, offers an introduction to systematic problem-solving that is of use when working the problems in our book.

Although the book closely parallels the text *Chemical Principles,* we have not hesitated to include additional material where we thought it warranted. For example, in Chapter 2, there is a section on molecular beams, and in Chapter 3, there are a number of problems on mass spectrometry.

Not only are students often uncertain of how much they must memorize, but generally they memorize too much. In a "memorization section" in the introduction to each chapter, the student is told exactly what he must memorize to solve the problems. We have tried to keep the memorization to a minimum by showing the student, within a problem, how to proceed quickly from the few memorized facts and equations to the ones necessary to solve the problem. Thus, there are a number of problems that build on each other, in which the results of one are used in the next, and so on. Therefore, we consider some of the problems to be "teaching problems."

The student is invited to evaluate his own progress by working examination questions, which are grouped in Appendix 1. We have not provided detailed solutions but have included answers to all of the examination questions.

We wish to thank our students and colleagues (especially Ferenc Kalos, Garry Stein, Donna Stern, Theodore Waech, Ivor Whorf, and Professor James J. Hogan) who so carefully worked all the problems and criticized our solutions. We would particularly like to thank Professor David N. Harpp, who wrote the problems for Chapter 12, "The Special Role of Carbon." Finally, we alone are, alas, responsible for any errors or inconsistencies throughout the book and would welcome any criticism and suggestions so we may improve it.

IAN S. BUTLER
ARTHUR E. GROSSER

Montreal, Quebec

NOTE TO THE STUDENT

A book of introductory chemistry problems usually seems like a formidable and tiresome epic with neither plot nor characters, and with insufficient narrative. Moreover, the exercises seem to be just that: a chemistry professor's equivalent to a muscle building course.

Our intention is to provide a change from this drudgery. Although we are primarily interested in helping you to master problems in chemistry, we think that it is also important to be learning something about how chemical principles relate to the real world. For this reason, we have tried to think of problems that involve areas such as space science, medicine, geology, archaeology, engineering, biochemistry, and a host of other fields from which you might derive some feeling of involvement.

Our method is to state a problem and to give five possible answers. Then we give a relatively detailed solution (not the *only* way to solve the problem, of course, and not even the shortest; we simply present an understandable way). We have chosen this multiple-choice format so we could deliberately construct incorrect answers that correspond to common errors. The reasons for these errors are often discussed in the solution. Nevertheless, if you are answering the questions correctly, you need not read through the entire solution. The number giving the correct answer is prominently displayed in the margin under the solution number. But if you are having trouble with these problems, we refer you periodically to *Programed Reviews of Chemical Principles*, by Jean D. Lassila *et al.*, for remedial work. This arrangement should allow you to proceed at your own pace. (In addition, we have cross-referenced to the text, *Chemical Principles*, second edition, by Richard E. Dickerson, Harry B. Gray, and Gilbert P. Haight, Jr. We follow the format of this text on a chapter-by-chapter basis.)

A word about memorization. We do not believe that it is important to commit a large quantity of information to memory. (Other scientists appear to agree, for they continually surround themselves with libraries and computers.) But some facts are so basic that it would be cumbersome to look them up time after time. We indicate at the beginning of each chapter what material you should, in our opinion, memorize. We believe that it is minimal.

Finally, we remain aware, even after watching TV commercials, that there are no panaceas. Nature will answer our questions, but only if we know how to ask them. And learning how to ask nature the proper questions is the business of scientists. In learning how to answer these questions, we hope that you learn to ask your own.

CONTENTS

CHAPTER 1 ATOMS, MOLECULES, AND MOLES

This chapter covers the concepts of atoms, gram-atoms, molecules, and moles, and introduces the quantitative aspects of chemical reactions. It is the interrelationship of these concepts with our everyday measures of quantity (weight and volume) that constitutes the basic vocabulary of the chemist.

You should be able to solve problems with a minimum of memorization. But that minimum is not zero. We will suggest what is necessary to memorize at the beginning of each chapter.

Memorize:

$N = 6.0 \times 10^{23}$ atoms g-atom^{-1}
or molecules mole^{-1}
1 liter = 10^3 cm^3
1 kg = 10^3 g

Solutions to problems in Chapter 1 begin on page 12.

PROBLEMS

PROBLEM 1-1

It is estimated that 9.3×10^6 kg (about 20 million lb, avdp) of carbon monoxide (CO) were produced by motor vehicles in Los Angeles in 1967. This is 3.3×10^8 moles of CO. How many gram-atoms of carbon (C) were consumed in producing this much CO?

(at. wt: C = 12, O = 16)

1 3.3×10^8
2 $12 \times 3.3 \times 10^8 = 4.0 \times 10^9$
3 $16 \times 3.3 \times 10^8 = 5.3 \times 10^9$
4 $28 \times 3.3 \times 10^8 = 9.3 \times 10^9$
5 Insufficient information; the chemical reaction to produce CO must be known.

PROBLEM 1-2

A car traveling at 10 miles per hour emits about 0.33 lb avdp of carbon monoxide (CO) gas per mile. How many moles of CO are emitted per mile under these conditions?

(at. wt: C = 12, O = 16; 1 lb avdp = 0.454 kg)

1 5.4×10^{-3} mole
2 1.2×10^{-2} mole
3 5.4 moles
4 9.4 moles
5 12.5 moles

PROBLEM 1-3

During the Apollo 11 moonwalk, a solar wind experiment was deployed. The solar wind collector was an aluminum strip of approximately 3000 cm^2 area (11.5 in. × 40 in.). If the solar wind strikes this foil (and sticks) with an intensity of 1×10^7 H atoms cm^{-2} sec^{-1}, what mass of H atoms was collected during the approximately 100-min experiment?

(at. wt: H = 1.0)

1 1×10^{-13} g
2 5×10^{-12} g
3 3×10^{-10} g
4 1 g
5 1.8×10^{14} g

PROBLEM 1–4

In World War I, 120,000 short tons (1.1×10^8 kg) of poison gas were fired on the Allied forces. (In 1918, half the shells fired by the Germans contained gas.) If the gas is assumed to be phosgene ($COCl_2$), how many *molecules* of the gas correspond to the amount fired on the Allied forces?

(at. wt: C = 12.0, O = 16.0, Cl = 35.5)

1 $1.1 \times 10^9/(6.0 \times 10^{23}) = 1.9 \times 10^{-15}$
2 $1.7 \times 10^9/(6.0 \times 10^{23}) = 2.9 \times 10^{-15}$
3 $6.0 \times 1.1 \times 10^{29} = 6.6 \times 10^{29}$
4 $6.0 \times 1.1 \times 10^{32} = 6.6 \times 10^{32}$
5 $6.0 \times 1.7 \times 10^{32} = 10 \times 10^{32}$

PROBLEM 1–5

The gram molecular weight of hydrogen peroxide (H_2O_2) is 34 g. What are the *units* of molecular weight?

(at. wt: H = 1, O = 16)

1 g
2 mole
3 g $mole^{-1}$
4 mole g^{-1}
5 No units.

PROBLEM 1–6

Which of the following statements concerning an isotope ${}^A_Z$M of an element M is *incorrect?*

(a) Z is the mass number of the element.
(b) A is the mass number of the element.
(c) Z is the number of positive charges on the nucleus.
(d) Z is the atomic number.
(e) A is the sum of the number of protons and the number of neutrons in the nucleus.

1 (a) and (c)
2 (a)
3 (b) and (e)
4 (b) and (d)
5 (d) and (e)

PROBLEM 1-7 Which of the following are pairs of isotopes?

(a) ${}^{2}_{1}H^{+}$ and ${}^{3}_{1}H$
(b) ${}^{3}_{2}He$ and ${}^{4}_{2}He$
(c) ${}^{12}_{6}C$ and ${}^{14}_{7}N^{+}$
(d) ${}^{3}_{1}H$ and ${}^{4}_{2}He^{-}$

1 (b) only
2 (a) and (d)
3 (a) and (c)
4 (c) only
5 (a) and (b)

PROBLEM 1-8 Two thirds of the atoms in the water (H_2O) molecule are hydrogen. What percentage of the weight of a water molecule is the weight of the two hydrogen atoms?

(at. wt: H = 1.008, O = 16.00)

1 5.6%
2 11.2%
3 22.4%
4 33.3%
5 66.7%

PROBLEM 1-9 The atomic composition of the entire universe is approximately as given in Table 1-1.

Table 1-1

Atom	Atom percent (% of the total number of atoms)
H	93
He	7.0

Hydrogen atoms constitute what percentage of the universe by *weight?*

(at. wt: H = 1.0, He = 4.0)

1 20%
2 23%
3 77%
4 87%
5 93%

PROBLEM 1–10

The most abundant element dissolved in sea water is Cl (as chloride ion, Cl^-) at a concentration of 19 grams per kilogram of sea water. The volume of the earth's oceans is 1.4×10^{21} liters (330 million cubic miles). How many gram-atoms of Cl are potentially available from the oceans? (Assume that the density of sea water is $1.0\ g\ cm^{-3}$.)

(at. wt: Cl = 35.5)

1 7.6×10^{20}
2 27×10^{21}
3 27×10^{24}
4 $35.5 \times 27 \times 10^{24} = 96 \times 10^{25}$
5 Insufficient data; the temperature must be known.

PROBLEM 1–11

The chemical composition of a sample of the lunar surface determined by Surveyor 7 is as indicated by Table 1–2.

Table 1–2

Atom	Atom percent
O	58%
Si	18%
Al	9%
Others (average at. wt = 30)	15%

What is the fractional oxygen content by *weight?*

(at. wt: O = 16, Si = 28, Al = 27)

1 $0.58/2 = 0.29$
2 $58/(58 + 18 + 9 + 15) = 0.58$
3 $(58 \times 16)/[(58 \times 16) + (18 \times 28) + (9 \times 27) + (15 \times 30)] = 0.44$
4 $(58 \times 32)/[(58 \times 32) + (18 \times 28) + (9 \times 27) + (15 \times 30)] = 0.61$
5 $(58 + 18 + 9 + 15)/(16 + 28 + 27 + 30) = 100/101 \simeq 1.0$

PROBLEM 1–12

The compound glucose-1-phosphate is of crucial importance in the metabolism of carbohydrates. Its molecular weight is 260 and its density is approximately $1.5\ g\ cm^{-3}$. What is the average volume occupied by one molecule of this compound?

1 $260/(9.0 \times 10^{23})\ cm^3 = 29 \times 10^{-23}\ cm^3$
2 $260/(6.0 \times 10^{23})\ cm^3 = 43 \times 10^{-23}\ cm^3$
3 $(1.0/1.5)\ cm^3 = 0.67\ cm^3$
4 $(10^{23}/6.0)\ cm^3 = 0.17 \times 10^{23}\ cm^3$
5 None of the above.

PROBLEM 1-13

The average density of the universe as a whole is estimated between 1.0×10^{-30} g cm^{-3} and 3.0×10^{-28} g cm^{-3}. If we assume that the entire mass is only hydrogen atoms, what is the average volume of space that contains *one* H atom? (Use an intermediate value for the average density: 1.5×10^{-29} g cm^{-3}.)

(at. wt: H = 1.0)

1 9.0×10^{-6} cm^3
2 55 liters
3 110 liters
4 220 liters
5 110,000 liters

PROBLEM 1-14

Deoxyribonucleic acid (DNA) is thought to be the chemical compound responsible for the process of heredity. We take a DNA sample and find that its density is 1.1 g cm^{-3}, and its molecular weight is estimated to be 6.0×10^8. What is the average volume occupied by one DNA molecule?

1 $\frac{1}{6.0} \times 10^{-23}$ cm$^3 = 0.17 \times 10^{-23}$ cm^3

2 $\frac{1}{1.1} \times 10^{-15}$ cm$^3 = 0.91 \times 10^{-15}$ cm^3

3 1×10^{-15} cm^3

4 $\frac{6.0}{1.1} \times 10^{-15}$ cm$^3 \simeq 5.5 \times 10^{-15}$ cm^3

5 None of the above.

PROBLEM 1-15

Problems 1-15 through 1-20 apply the concepts in Section 1-1 of Chemical Principles *by Dickerson, Gray, and Haight.*

The reaction involved in baking powder (a mixture of cream of tartar and baking soda) is

$$KHC_4H_4O_6 + NaHCO_3 \rightarrow KNaC_4H_4O_6 + H_2O + CO_2$$

$KHC_4H_4O_6$: cream of tartar (potassium hydrogen tartrate); $NaHCO_3$: baking soda

A recipe calls for two level teaspoons (8.0 g) of cream of tartar. How much baking soda must be added for both materials to react completely?

(at. wt: H = 1, C = 12, O = 16, Na = 23, K = 39)

1 $84/188 = 0.45$ g
2 $188/84 = 2.2$ g
3 $8.0 \times 84/210 = 3.2$ g
4 $8.0 \times 84/188 = 3.6$ g
5 $8.0 \times 188/84 = 18$ g

PROBLEM 1–16

Soda-lime glass is used for containers. The glass is prepared by melting sodium carbonate (Na_2CO_3), limestone ($CaCO_3$), and sand (SiO_2). The composition of the glass varies, but the accepted general reaction is

$$Na_2CO_3 + CaCO_3 + 6SiO_2 \rightarrow \underset{\text{soda-lime glass}}{Na_2O \cdot CaO \cdot 6SiO_2} + 2CO_2$$

With this equation, how many kilograms of sand would be required to produce enough glass to make 5000 wine bottles, each weighing 400 g?

(mol wt: $Na_2CO_3 = 106$, $CaCO_3 = 100$, $SiO_2 = 60.0$, $Na_2O \cdot CaO \cdot 6SiO_2 = 478$, $CO_2 = 44.0$)

1 25.0 kg
2 2.52×10^2 kg
3 1.51×10^3 kg
4 7.20×10^5 kg
5 None of the above.

PROBLEM 1–17

The fuel–oxidizer combination used in the Apollo 11 lunar module was Aerozine 50 for fuel—a 50/50 blend by weight of hydrazine, N_2H_4, and unsymmetrical dimethylhydrazine, $(CH_3)_2N_2H_2$—and nitrogen tetroxide, N_2O_4, for oxidizer. Water, nitrogen, hydrogen, carbon monoxide, and carbon dioxide accounted for 95% of the exhaust products. Water was the chief exhaust product; it constituted more than one third of the total weight of the exhaust gases. Two of the reactions that led to the formation of water in the exhaust are

$$2N_2H_4 + N_2O_4 \rightarrow 3N_2 + 4H_2O$$
$$(CH_3)_2N_2H_2 + 2N_2O_4 \rightarrow 2CO_2 + 3N_2 + 4H_2O$$

If we assume that these two reactions were the only ones responsible for the production of water, how many kilograms of water were formed during the ascent of the lunar module from the surface of the moon if about 2200 kg (almost 5000 lb) of Aerozine 50 were consumed?

[mol wt: $N_2H_4 = 32.0$, $(CH_3)_2N_2H_2 = 60.0$, $H_2O = 18.0$, $N_2O_4 = 92.0$]

1 1.24×10^3 kg
2 1.32×10^3 kg
3 2.56×10^3 kg
4 22.0×10^3 kg
5 None of the above.

PROBLEM 1–18

In 1938, the British Interplanetary Society suggested the use of NaOH to remove CO_2 from the cabin atmosphere of a lunar expedition. The average human body discharges about 924 g of CO_2 per day. What minimum weight of NaOH must be carried on board for a 10-day voyage by one astronaut?

$2NaOH + CO_2 \rightarrow Na_2CO_3 + H_2O$

(at. wt: H = 1.0, C = 12.0, O = 16.0, Na = 23.0)

1 1.68 kg
2 8.40 kg
3 9.24 kg
4 16.8 kg
5 None of the above.

PROBLEM 1–19

Two of the reactions that occur in the production of Fe are

$$2C(s) + O_2(g) \rightarrow 2CO(g)$$
$$Fe_2O_3(s) + 3CO(g) \rightarrow 2Fe(s) + 3CO_2(g)$$

Note that the carbon monoxide (CO) formed in the first reaction is consumed in the second reaction. If these reactions are the ones primarily responsible for the production of iron, how many kilograms of O_2 are required to produce 100 kg of Fe?

(at. wt: Fe = 56, C = 12, O = 16)

1 $\frac{3}{7} \times 10^2$ kg

2 $\frac{16}{21} \times 10^2$ kg

3 $\frac{8}{7} \times 10^2$ kg

4 $\frac{12}{7} \times 10^2$ kg

5 2×10^2 kg

PROBLEM 1–20

An astronaut who drinks 2 liters of H_2O per day discharges 2.4 liters of liquid H_2O per day. (The additional amount is produced by metabolism of food.) The use of Li_2O in a spaceship to absorb this discharged water has been suggested. What minimum weight of Li_2O must be carried on board for this purpose for a 10-day voyage by one astronaut?

$Li_2O + H_2O \rightarrow 2LiOH$

(mol wt: $H_2O = 18$, $Li_2O = 30$, LiOH = 24;
D = density of water = 1.0 g cm^{-3})

1 30 g
2 2.4 kg
3 4.0 kg
4 40 kg
5 720 kg

PROBLEM 1–21

An analysis of nicotine (a poisonous compound in tobacco leaves) gave 74.0% C, 8.65% H, and 17.3% N. What is the empirical formula of nicotine?

(at. wt: H = 1.01, C = 12.0, N = 14.0)

1 C_5H_7N
2 C_5H_8N
3 C_6H_7N
4 $C_{10}H_{12}N$
5 $C_{10}H_{14}N$

PROBLEM 1–22

Cortisone (a hormone formed in the adrenal gland, sometimes used in the treatment of rheumatoid arthritis) is found to have the following elemental composition by weight: 69.96% C, 7.83% H, and 22.21% O. It is known from other experimental measurements that the compound's molecular weight is about 360. What is its molecular formula?

(at. wt: H = 1.01, C = 12.0, O = 16.0)

1 $C_{35}H_5O_{11}$
2 $C_{15}H_{20}O_{10}$
3 $C_{21}H_{28}O_5$
4 $C_{20}H_{40}O_5$
5 $C_{20}H_{56}O_4$

PROBLEM 1–23

The hemoglobin from the red corpuscles of most mammals contains approximately 0.33% iron by weight. If physical measurements indicate a molecular weight of 68,000, how many iron atoms are there in each molecule of hemoglobin?

(at. wt: Fe = 56)

1 1
2 2
3 3
4 4
5 5

PROBLEM 1-24 Liquid benzene (C_6H_6) burns in oxygen according to

$$2C_6H_6 + 15O_2 \rightarrow 12CO_2 + 6H_2O$$

If the density of liquid benzene at 20°C is 0.88 g cm^{-3}, how many liters of O_2 gas at STP are needed to complete the combustion of 39 cm^3 of liquid benzene? (Molar volume of a gas at STP $\simeq$ 22.4 liters. STP means "standard conditions" of T and P.)

(mol wt: $O_2 = 32$, $C_6H_6 = 78$)

1 11.2 liters
2 22.4 liters
3 37 liters
4 74 liters
5 84 liters

PROBLEM 1-25 When nitroglycerin, $C_3H_5(NO_3)_3$, explodes, no solid products are produced:

$$4C_3H_5(NO_3)_3(l) \rightarrow 12CO_2(g) + 6N_2(g) + O_2(g) + 10H_2O(g)$$

What is the total volume of the gases produced at STP when a glass vial containing 454 g of nitroglycerin explodes? (Molar volume of a gas at STP $\simeq$ 22.4 liters.)

[mol wt: $C_3H_5(NO_3)_3 = 227$]

1 22.4 liters
2 44.8 liters
3 163 liters
4 325 liters
5 650 liters

PROBLEM 1–26

Many minerals contain sulfur in the form of sulfates [e.g., gypsum (calcium sulfate, $CaSO_4$)] and sulfides [e.g., zinc blende (zinc sulfide, ZnS)]. However, the principal source of commercial supply is deposits of free sulfur, which occur mainly in volcanically active regions. The sulfur was formed during chemical reactions between volcanic vapors, particularly sulfur dioxide and hydrogen sulfide:

$$SO_2(g) + 2H_2S(g) \rightarrow 2H_2O(l) + 3S(s)$$

How many liters of gas (both SO_2 and H_2S) at STP were needed to produce a sulfur deposit of 5.0×10^6 kg found on the slopes of Mount Etna in Sicily? [Molar volume of a gas at STP $\simeq$ 22.4 liters.]

(mol wt: $SO_2 = 64$, $H_2S = 34$, $H_2O = 18$, $S = 32$)

1 1.6×10^5 liters
2 1.6×10^8 liters
3 1.8×10^9 liters
4 3.5×10^9 liters
5 6.2×10^9 liters

SOLUTIONS

SOLUTION 1–1

(1)

We will agree that one molecule of CO contains only one atom of C, so 6.0×10^{23} molecules of CO contain 6.0×10^{23} atoms of C. But 6.0×10^{23} molecules are 1 mole, and 6.0×10^{23} atoms are 1 g-atom. Therefore, in 1 mole of CO there is 1 g-atom of C, and in 3.3×10^8 moles of CO there are 3.3×10^8 g-atoms of C.

Answer **1** is correct. Answer **5** is wrong because the chemical reaction is not necessary; we need to know only the molecular formula.
Answer **2:** 4.0×10^9 g is the *weight* of C;
Answer **3:** 5.3×10^9 g is the *weight* of O;
Answer **4** is the weight of CO produced, just as the problem tells us.

SOLUTION 1–2

(3)

$$\begin{aligned} 0.33 \text{ lb avdp} &= 0.33 \text{ lb avdp} \times 0.454 \text{ kg (lb avdp)}^{-1} \\ &= 0.15 \text{ kg} \\ &= 150 \text{ g} \end{aligned}$$

$$\text{mol wt CO} = 12 + 16 = 28 \text{ g mole}^{-1}$$

$$\begin{aligned} n = \text{moles CO} &= \frac{\text{g CO}}{\text{mol wt CO}} \\ &= \frac{150 \text{ g}}{28 \text{ g mole}^{-1}} \\ &= 5.4 \text{ moles} \end{aligned}$$

3 is correct.
1 You did not convert from kilograms to grams.
2 You did not convert from pounds avoirdupois to grams.
4 You used the atomic weight of O rather than the molecular weight of CO.
5 Same as **4,** but you used the atomic weight of C.

SOLUTION 1–3

(3)

If you have trouble manipulating exponents, see Appendix 2.

The first thing that we want to do is to calculate the number of H atoms that strikes the foil per second:

$$\begin{aligned} \text{H atoms sec}^{-1} &= 1 \times 10^7 \text{ atoms sec}^{-1} \text{ cm}^{-2} \times 3 \times 10^3 \text{ cm}^2 \\ &= 3 \times 10^{10} \text{ atoms sec}^{-1} \end{aligned}$$

The number that strikes the foil during the 100 min is

$$\begin{aligned} \text{H atoms} &= 3 \times 10^{10} \text{ atoms sec}^{-1} \times 10^2 \text{ min} \times 60 \text{ sec min}^{-1} \\ &= 18 \times 10^{13} \text{ atoms} \end{aligned}$$

To calculate the *weight* of hydrogen we should convert this number of atoms to gram-atoms, and then use the atomic weight to obtain the

If you could not solve this problem, you may want to study Section 1-1 of Programed Reviews of Chemical Principles *by Lassila* et al.

number of grams:

$$\text{g-atoms H} = \frac{18 \times 10^{13} \text{ atoms}}{6 \times 10^{23} \text{ atoms g-atom}^{-1}}$$

$$= 3 \times 10^{-10} \text{ g-atom}$$

and since the atomic weight of hydrogen *atoms* is 1.0,

$$\text{g of H} = 3 \times 10^{-10} \text{ g-atom} \times 1.0 \text{ g g-atom}^{-1}$$
$$= 3 \times 10^{-10} \text{ g}$$

3 is correct.
1 The aluminum area was not included.
2 You forgot to convert minutes to seconds.
4 This is, of course, just the atomic weight of hydrogen.
5 You forgot to divide by *N*.

SOLUTION 1-4

(4)

For a review of the concept of moles, see Lassila's book, Section 1-2.

The molecular weight of $COCl_2$ is

$12 + 16 + 2 \times 35.5 = 99 \text{ g mole}^{-1}$

In 1.1×10^8 kg there are 1.1×10^{11} g and

$$\frac{1.1 \times 10^{11} \text{ g}}{99 \text{ g mole}^{-1}} \simeq 1.1 \times 10^9 \text{ moles}$$

[*Note:* $\simeq$ means "approximately equal to."] Since each mole contains 6.0×10^{23} molecules, the total number of molecules is

$1.1 \times 10^9 \text{ moles} \times 6.0 \times 10^{23} \text{ molecules mole}^{-1}$

or

$6.0 \times 1.1 \times 10^{32} \text{ molecules} = 6.6 \times 10^{32} \text{ molecules}$

4 is correct.
1 Division by *N*. This could have been avoided if you had checked units.
5 Wrong molecular weight. The atomic weight for Cl must be multiplied by two.
2 Both the above mistakes.
3 You forgot to change from kilograms to grams.

SOLUTION 1-5

(3)

Gram molecular weight is the weight in grams of a mole of the substance and has the units of g mole^{-1}. We have stressed the importance of this approach to our calculations in the previous problems, and we will continue to do so. The continued use of units in solving problems will take some time, but will prevent many errors. [*Note:* Chemical engineers sometimes use units such as pound mole and ton mole. In this book, we use the term "molecular weight" to mean "gram molecular weight."]

SOLUTION 1–6

(2)

Consider the general term for an isotope, ${}^{A}_{Z}M$. A is the mass number of the element and is the sum of the number of protons and the number of neutrons in the nucleus. Z is the atomic number of the element and is equal to the number of protons (positively charged particles) in the nucleus. Consequently, only statement (a) is incorrect, and the required answer is **2**.

SOLUTION 1–7

(5)

The atomic number determines the element. Different atomic numbers correspond to different elements.

An element has only one atomic number, but usually several atomic masses. Different atomic masses for the same atomic number correspond to different isotopes of the element. (c) and (d) pair different *elements*. (a) and (b) pair the same element but different atomic masses; therefore, they are pairs of isotopes. Answer **5** is correct.

The charge on the element doesn't enter into these considerations, since isotopes involve just the numbers of protons and neutrons in the nucleus. For instance, the pair ${}^{2}_{1}H^{+}$ and ${}^{3}_{1}H$ are isotopes of H in which ${}^{2}_{1}H^{+}$ has had its only electron removed. If you thought the positive charge was obtained by adding a proton to the nucleus, remember that this would change the chemical species of the substance.

SOLUTION 1–8

(2)

If you had difficulty with this problem, see Lassila's book, Section 1–3, Items 1 and 2.

There are usually two ways (at least) to solve composition problems. One is to consider a certain weight, for example, 100 g. But in this case the second way should be easier, namely, to consider 1 mole of H_2O. The molecular weight is 18.016: 18.016 g per 6×10^{23} molecules. In the sample we have chosen, there are 2 g-atoms (12×10^{23} atoms) of H. This amount corresponds to a weight of 2×1.008, or 2.016 g. The percentage of the total weight due to hydrogen is $2.016/18.016 \times 100$ or approximately 11.2% for 6×10^{23} water molecules. And this must *also* be the weight percentage for one molecule.

2 is correct.
1 You forgot that there are 2 g-atoms of H in every mole of H_2O.
3 You probably did the calculation correctly and then multiplied the answer by two because there are two atoms of H per molecule. But you already had taken that fact into account.
4 This is just the O atom percent.
5 This is just the H atom percent.

SOLUTION 1–9

(3)

If you do not understand the relationship between amu and atomic weight, refer to Lassila's book, Section 1–1, Items 17–24.

The percentage by weight will be given by the weight of H atoms divided by the total weight. If we choose to work with a sample of 100 atoms, 93 of them will be H [atomic mass units (amu) = 1] and 7 will be He (amu = 4).

$$\frac{\text{wt of H}}{\text{total wt}} = \frac{93 \text{ atoms} \times 1.0 \text{ amu atom}^{-1}}{93 \text{ atoms} \times 1.0 \text{ amu atom}^{-1} + 7 \text{ atoms} \times 4.0 \text{ amu atom}^{-1}}$$

$$= \frac{93}{93 + 28} = 0.77$$

Weight percentage equals 77%.

3 is correct.
1 is the percentage if the number of H and He atoms are equal.
2 is the weight percentage of He.
4 is the result of using the atomic numbers rather than the atomic weights.
5 is giving up.

SOLUTION 1–10

(1)

The total number of gram-atoms is the total Cl weight divided by the atomic weight. The total weight of the oceans is

$$1.4 \times 10^{21} \text{ liters} \times 10^{3} \text{ cm}^{3} \text{ liter}^{-1} \times 1.0 \text{ g cm}^{-3} = 1.4 \times 10^{24} \text{ g water}$$

The Cl concentration can be expressed as

$$19 \text{ g kg}^{-1} = \frac{19 \text{ g Cl}}{1000 \text{ g water}} = 19 \times 10^{-3} \frac{\text{g Cl}}{\text{g water}}$$

And the total Cl weight is

$$1.4 \times 10^{24} \text{ g water} \times 19 \times 10^{-3} \frac{\text{g Cl}}{\text{g water}} = 2.7 \times 10^{22} \text{ g Cl}$$

Therefore, the number of gram-atoms is

$$\frac{2.7 \times 10^{22} \text{ g Cl}}{35.5 \text{ g g-atom}^{-1}} = 0.076 \times 10^{22} \text{ g-atoms}$$
$$= 7.6 \times 10^{20} \text{ g-atoms}$$

1 is correct.
2 is the number of grams, not gram-atoms.
3 is the weight and an error involving kilograms.
4 is the result of forgetting the meaning of gram-atom or atomic weight. And the units do not work out either.

This amount of Cl is equal to 3×10^{16} tons. We may compare this to the annual rate of commercial extraction of Cl (in the form of NaCl) from sea water, which is 2×10^{7} tons of Cl. (This is 29% of the total world production of Cl by all methods.) We can conclude that the supply is, for our purposes, "limitless."

SOLUTION 1–11

(3)

The fractional oxygen content by weight is the weight of oxygen divided by the total weight. This is simple enough if we remember to use atomic weights with these atom percents. Consider a sample containing Avogadro's number, N, of atoms.

Table 1–3

Atom	Gram-atom	at. wt = g g-atom^{-1}
O	0.58	16
Si	0.18	28
Al	0.09	27
Other	0.15	30

$$\text{wt of O} = 0.58 \text{ g-atom} \times 16 \text{ g g-atom}^{-1}$$
$$= 0.58 \times 16 \text{ g}$$
$$\text{total wt} = 0.58 \times 16 + 0.18 \times 28 + 0.09 \times 27 + 0.15 \times 30 \text{ g}$$

fractional O content by weight

$$= \frac{0.58 \times 16}{(0.58 \times 16) + (0.18 \times 28) + (0.09 \times 27) + (0.15 \times 30)}$$

$$= \frac{58 \times 16}{(58 \times 16) + (18 \times 28) + (9 \times 27) + (15 \times 30)} = 0.44$$

Answer **3** is correct. Answer **2** is the atomic fraction. Answer **4** used the *molecular* weight of O_2 and not the atomic weight of O.

SOLUTION 1–12

(1)

Just find the volume of a mole of glucose-1-phosphate (the volume occupied by N molecules) and divide by N:

$$\frac{\text{volume}}{\text{mole}} = \frac{260 \text{ g mole}^{-1}}{1.5 \text{ g cm}^{-3}} = \frac{260 \text{ cm}^3}{1.5 \text{ moles}}$$

$$\frac{\text{volume}}{\text{molecule}} = \frac{260 \text{ cm}^3}{1.5 \text{ moles}} \times \frac{1}{6.0 \times 10^{23} \text{ molecules mole}^{-1}}$$

$$= \frac{260 \text{ cm}^3}{9.0 \times 10^{23} \text{ molecules}}$$

$$\text{volume} = 29 \times 10^{-23} \text{ cm}^3 \text{ molecule}^{-1}$$

Answer **1** is correct. (To give you an approximate idea of this magnitude: The volume occupied by the molecule is to a volume of a small alarm clock as the volume of the clock is to the volume of the earth's moon.) Answer **3** is the volume occupied by 1 g. Answer **2** is the weight per molecule labeled with the wrong units.

SOLUTION 1–13

(3)

If we can find the number of H atoms per cubic centimeter, we can invert that number to obtain the volume per H atom:

$$\frac{1.5 \times 10^{-29}\text{ g cm}^{-3}}{1.0\text{ g g-atom}^{-1}} = 1.5 \times 10^{-29}\text{ g-atom cm}^{-3}$$

$$1.5 \times 10^{-29}\text{ g-atom cm}^{-3} \times 6.0 \times 10^{23}\text{ atoms g-atom}^{-1} = 9.0 \times 10^{-6}\text{ atom cm}^{-3}$$

We have accomplished our first step. Now let us invert:

$$\frac{1}{9.0 \times 10^{-6}\text{ atom cm}^{-3}} = 0.11 \times 10^{6}\text{ cm}^3\text{ atom}^{-1} = 110 \times 10^{3}\text{ cm}^3\text{ atom}^{-1} = 110\text{ liters atom}^{-1}$$

[*Note:* The universe is about 93% H atoms, so our assumption for this rough calculation is acceptable.]

3 is correct.
1 is the reciprocal of the right answer.
2 and **4** probably involve errors associated with using the molecular weight of H_2 rather than the atomic weight of H.
5 would be right if the number were labeled cubic centimeters instead of liters.

SOLUTION 1–14

(2)

The volume per molecule is

$$\frac{\text{cm}^3}{\text{molecule}} = \frac{6.0 \times 10^{8}\text{ g mole}^{-1}}{1.1\text{ g cm}^{-3}} \times \frac{1}{6.0 \times 10^{23}\text{ molecules mole}^{-1}}$$

$$= \frac{1.0 \times 10^{-15}\text{ cm}^3}{1.1\text{ molecules}} = 0.91 \times 10^{-15}\text{ cm}^3\text{ molecule}^{-1}$$

(Compare this with the volume occupied by a glucose-1-phosphate molecule, which you calculated in Problem 1–12.)

Answer **2** is correct. Answer **1** is just N^{-1}. Answer **3** has left out the density.

SOLUTION 1–15

(4)

The number of moles of $KHC_4H_4O_6$ is

$$\frac{8.0\text{ g}}{\text{mol wt KHC}_4\text{H}_4\text{O}_6} = \frac{8.0\text{ g}}{188\text{ g mole}^{-1}}$$

This is also the number of moles of baking soda that must be added since their stoichiometric coefficients are the same, namely, one.

$$\text{weight of NaHCO}_3 = \frac{8.0}{188}\text{ mole} \times (\text{mol wt NaHCO}_3)$$

$$= \frac{8.0}{188}\text{ mole} \times (84\text{ g mole}^{-1})$$

$$= \frac{8.0 \times 84}{188}\text{ g} = 3.6\text{ g}$$

Answer **4** is correct. This is about one level teaspoon.

SOLUTION 1–16 (3)

Moles of glass to be produced

$$= \frac{5000 \times 400 \text{ g}}{478 \text{ g mole}^{-1}}$$

$$= \frac{2.00 \times 10^6}{478} \text{ moles}$$

From the equation, 6 moles of sand (SiO_2) are required to produce 1 mole of glass. So the moles of sand required

$$= \frac{6 \times 2.00 \times 10^6}{478} \text{ moles}$$

and the grams of SiO_2 needed

$$= \frac{6 \times 2.00 \times 10^6 \text{ moles} \times 60.0 \text{ g mole}^{-1}}{478}$$

$$= \frac{720 \times 10^6 \text{ g}}{478}$$

$$= 1.51 \times 10^6 \text{ g}$$

$$= 1.51 \times 10^3 \text{ kg}$$

3 is the correct answer.
2 corresponds to forgetting the stoichiometric relationship between SiO_2 and soda-lime glass.
1 occurs if you neglected to multiply by the molecular weight of SiO_2.
4 results if you forgot to divide by the molecular weight of soda-lime glass.

SOLUTION 1–17 (3)

The 2200 kg of fuel contained 1100 kg of N_2H_4 and 1100 kg of $(CH_3)_2N_2H_2$. Remembering that each mole of N_2H_4 gives 2 moles of H_2O but each mole of $(CH_3)_2N_2H_2$ gives 4 moles of H_2O, we can arrange this problem in tabular form.

Table 1–4

	N_2H_4	$(CH_3)_2N_2H_2$
Weight	1100×10^3 g	1100×10^3 g
Moles of reactant	$\frac{1.1 \times 10^6 \text{ g}}{32 \text{ g mole}^{-1}}$	$\frac{1.1 \times 10^6 \text{ g}}{60 \text{ g mole}^{-1}}$
Moles of H_2O	$\frac{2 \times 1.1 \times 10^6}{32}$ moles	$\frac{4 \times 1.1 \times 10^6}{60}$ moles
Weight of H_2O	$\frac{18 \times 2 \times 1.1 \times 10^6}{32}$ g	$\frac{18 \times 4 \times 1.1 \times 10^6}{60}$ g

The total weight of H_2O is the sum of these last two terms:

$$\text{wt of } H_2O = \frac{18 \times 2 \times 1.1 \times 10^6}{32} + \frac{18 \times 4 \times 1.1 \times 10^6}{60} \text{ g}$$

$$\simeq 2.56 \times 10^6 \text{ g} = 2.56 \times 10^3 \text{ kg (about 5650 lb) of } H_2O$$

Answer **3** is correct. Answers **1** and **2** are the separate terms in our penultimate equation.

SOLUTION 1–18

(4)

The total weight of CO_2 to be removed during the 10 days is 9240 g. The number of moles of NaOH necessary is

$$\frac{9240 \text{ g}}{\text{mol wt } CO_2} \times 2$$

We have multiplied by two because 2 moles of NaOH are required for each mole of CO_2. The weight of NaOH is

$$\frac{9240 \text{ g}}{44.0 \text{ g mole}^{-1}} \times 2 \times 40.0 \text{ g mole}^{-1} = 16{,}800 \text{ g} = 16.8 \text{ kg}$$

4 is correct.
1 is the amount necessary for one day.
2 neglected the stoichiometric ratio.
3 is just the amount of CO_2 to be removed.

SOLUTION 1–19

(1)

The trick is to be very careful with the stoichiometry. For every 2 g-atoms of Fe, 3 moles of CO are needed (second equation). But to obtain 3 moles of CO we need $1\frac{1}{2}$ moles of O_2 (first equation). So $1\frac{1}{2}$ moles of O_2 give 2 g-atoms of Fe, and $\frac{3}{4}$ mole of O_2 gives 1 g-atom of Fe.

$$100 \text{ kg of Fe are } \frac{1 \times 10^5 \text{ g}}{56 \text{ g g-atom}^{-1}}$$

or

$$\frac{1 \times 10^5}{56} \text{ g-atoms of Fe}$$

which, by the above arguments, requires

$$\frac{3}{4} \times \frac{1 \times 10^5}{56} \text{ moles of } O_2$$

or

$$\frac{3}{4} \times \frac{1 \times 10^5}{56} \times 32 \text{ g of } O_2 = \frac{3}{7} \times 10^5 \text{ g of } O_2$$

$$= \frac{3}{7} \times 10^2 \text{ kg of } O_2$$

Answer **1** is correct. [*Note:* You can also solve this problem by adding the two equations so that CO cancels to obtain an equation for the overall reaction. One and a half times the first reaction is

	$3C(s) + \frac{3}{2}O_2(g) \rightarrow 3CO(g)$
Adding the second equation	$Fe_2O_3(s) + 3CO(g) \rightarrow 2Fe(s) + 3CO_2(g)$
The sum is	$3C(s) + Fe_2O_3(s) + \frac{3}{2}O_2(g) \rightarrow 2Fe(s) + 3CO_2(g)$

You can see that this also gives the correct stoichiometric ratio.]

SOLUTION 1–20

(4)

A *total* volume of 24 liters of H_2O must be removed.

$$24 \text{ liters} = 24 \times 10^3 \text{ cm}^3 = 24 \times 10^3 \text{ g}$$

(since $D = 1.0$ g cm^{-3}). The number of moles of H_2O is

$$\frac{24 \times 10^3 \text{ g}}{18 \text{ g mole}^{-1}}$$

And since 1 mole of Li_2O removes 1 mole of H_2O, the number of moles of Li_2O required is

$$\frac{24 \times 10^3}{18} \text{ moles}$$

and the weight required is

$$\frac{24 \times 10^3}{18} \text{ moles} \times 30 \text{ g mole}^{-1}$$

or

$$40 \times 10^3 \text{ g or } 40 \text{ kg}$$

Answer **4** is the required weight. Answer **3** is almost right. You forgot that the voyage lasts 10 days. The astronaut would be unhappy with the consequences of such a design.

SOLUTION 1–21

(1)

If you obtained a different answer, refer to Lassila's book, Section 1–3, Items 13–27.

This is a common problem in chemistry. The following steps will indicate how empirical formulas can be obtained from percentage composition data.

Step 1. Assume that you have 100 g of the compound. This means that 100 g of nicotine contain 74.0 g C, 8.65 g H, and 17.3 g N.

Step 2. Determine the number of gram-atoms of each element present.

g-atoms of C $= 74.0 \text{ g}/12.0 \text{ g g-atom}^{-1} \simeq 6.16$ g-atoms of C
g-atoms of H $= 8.65 \text{ g}/1.01 \text{ g g-atom}^{-1} = 8.56$ g-atoms of H
g-atoms of N $= 17.3 \text{ g}/14.0 \text{ g g-atom}^{-1} \simeq 1.23$ g-atoms of N

Step 3. Determine the largest common divisor for the numbers of gram-atoms present. In this case the largest common divisor is about 1.23. This step may involve some trial and error. Try the smallest number of gram-atoms first.

Step 4. Divide each of the numbers of gram-atoms present by this common divisor. The quotients should be rounded off to the nearest whole number. For C, 6.16/1.23 ≃ 5; for H, 8.56/1.23 ≃ 7; for N, 1.23/1.23 = 1.

Step 5. Write the whole numbers from Step 4 as subscripts of the various elements present to give the *empirical formula:* C_5H_7N. This is Answer **1**.

SOLUTION 1–22

(3)

We will proceed as in Problem 1–21.

Step 1. 69.96 g of C, 7.83 g of H, 22.21 g of O

Step 2.

$$\text{g-atoms of C} = \frac{69.96 \text{ g}}{12.0 \text{ g g-atom}^{-1}} = 5.83 \text{ g-atoms of C}$$

$$\text{g-atoms of H} = \frac{7.83 \text{ g}}{1.01 \text{ g g-atom}^{-1}} = 7.75 \text{ g-atoms of H}$$

$$\text{g-atoms of O} = \frac{22.21 \text{ g}}{16.0 \text{ g g-atom}^{-1}} = 1.39 \text{ g-atoms of O}$$

Step 3. Choose to divide by 1.39.

Step 4. $\frac{C}{O} = \frac{5.83}{1.39} = 4.19 \simeq 4.2,$

$$\frac{H}{O} = \frac{7.75}{1.39} = 5.58 \simeq 5.6.$$

Step 5. $C_{4.2}H_{5.6}O_1$

Multiply the subscripts by 5 to obtain whole numbers: $C_{21}H_{28}O_5$ (mol wt = 360.3).

Answer **3** is correct. If you chose Answer **1**, you put the weight in the subscripts and divided by two. In addition, the atomic weights in that formula do not sum to 360.

SOLUTION 1–23

(4)

The hemoglobin contains 0.33% Fe and has a molecular weight of 68,000. In 1 mole of hemoglobin there are (0.33/100) × 68,000 g = 224 g Fe. Since 1 g-atom of iron weighs 56 g, there must be

$$\frac{224 \text{ g}}{56 \text{ g g-atom}^{-1}} = 4 \text{ g-atoms of Fe in 1 mole of hemoglobin and 4 atoms of Fe in one molecule of hemoglobin}$$

Answer **4** is correct.

SOLUTION 1–24

(4)

The weight of benzene is

$39\ \text{cm}^3 \times 0.88\ \text{g cm}^{-3}$

and the number of moles is

$$\frac{39 \times 0.88\ \text{g}}{78\ \text{g mole}^{-1}}$$

For every 2 moles of benzene, 15 moles of O_2 are required; for every 1 mole of benzene, 15/2 moles of O_2 are required. The number of moles of O_2 necessary is

$$\frac{15}{2} \times \frac{39 \times 0.88}{78}\ \text{moles}$$

And since each mole occupies 22.4 liters, the volume of O_2 is

$$22.4 \times \frac{15}{2} \times \frac{39 \times 0.88}{78}\ \text{liters} = 74\ \text{liters}$$

Answer **4** is correct. Answer **5** neglected the density.

SOLUTION 1–25

(4)

By inspection, we have 2 moles of nitroglycerin. This amount will give rise to 6 moles of CO_2, 3 moles of N_2, $\frac{1}{2}$ mole of O_2, 5 moles of H_2O, or a total of 14.5 moles of gas. Each mole of gas, irrespective of its nature, will occupy 22.4 liters at STP. Hence, the total volume of gas produced under those conditions is

$14.5 \times 22.4 = 325$ liters

Answer **4** is correct.

SOLUTION 1–26

(4)

For every 3 g-atoms of S produced, 3 moles of gas reactants are required (1 mole of SO_2 and 2 moles of H_2S). The number of liters of reactant gas is

$$22.4\ \text{liters mole}^{-1} \times \underbrace{\frac{3}{3}}_{\substack{\text{stoichiometric}\\ \text{ratio}}} \times \frac{5.0 \times 10^6\ \text{kg} \times 10^3\ \text{g kg}^{-1}}{\underbrace{32\ \text{g g-atom}^{-1}}_{\substack{\text{g-atom}\\ \text{of S}}}}$$

or 3.5×10^9 liters.

4 is correct.
2 is just the number of gram-atoms of S.
3 employs the molecular weight of SO_2 instead of the atomic weight of S.
5 substitutes the molecular weight of H_2O for the atomic weight of S.

CHAPTER 2 THE GAS LAWS AND THE KINETIC THEORY

The relationship between the gas laws and the kinetic molecular theory is an example of the scientific method. These problems will test your capacity to solve problems in these experimental and theoretical areas and your ability to combine the two. As you work these problems, try to make the connection between gas behavior and the kinetic theory.

To do the problems in this chapter, you need to know the following relationships.

Gas Laws

$PV = nRT$
D = density = wt/V
T = °K = °C + 273°
1 atm = 760 mm Hg = 760 torr
STP implies 273°K, 1 atm

Partial Pressures

$P_{\text{total}} = p_A + p_B + \cdots + p_N$

Kinetic Theory

$\epsilon = \frac{1}{2}mv^2$ = kinetic energy per molecule
mv = momentum
$\overline{E}_k = \frac{3}{2}RT$ = average kinetic energy per mole

In the following problems, assume that the gas is ideal unless you are told otherwise.

Solutions to problems in Chapter 2 begin on page 40.

PROBLEMS

PROBLEM 2–1 Consider Boyle's J-tube apparatus in Figure 2–1. When $h = 40$ mm, which one of the following statements applies to the gas trapped in the volume, V_{gas}? (The temperature is constant.)

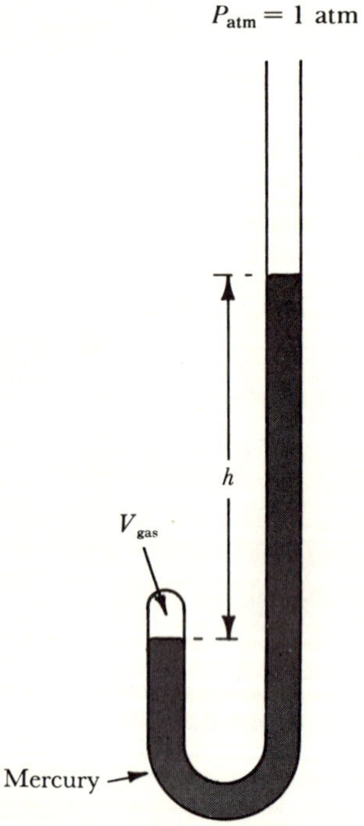

Figure 2–1

1 The gas pressure is 0 torr.
2 The gas pressure is 40 torr.
3 The gas pressure is 720 torr.
4 The gas pressure is 760 torr.
5 The gas pressure is 800 torr.

PROBLEM 2-2 Consider the gas burette in Figure 2-2. Initially, h equals 380 mm. If the mercury reservoir is lowered until h equals 190 mm, which one of the following statements about the volume of the gas trapped in the burette is correct?

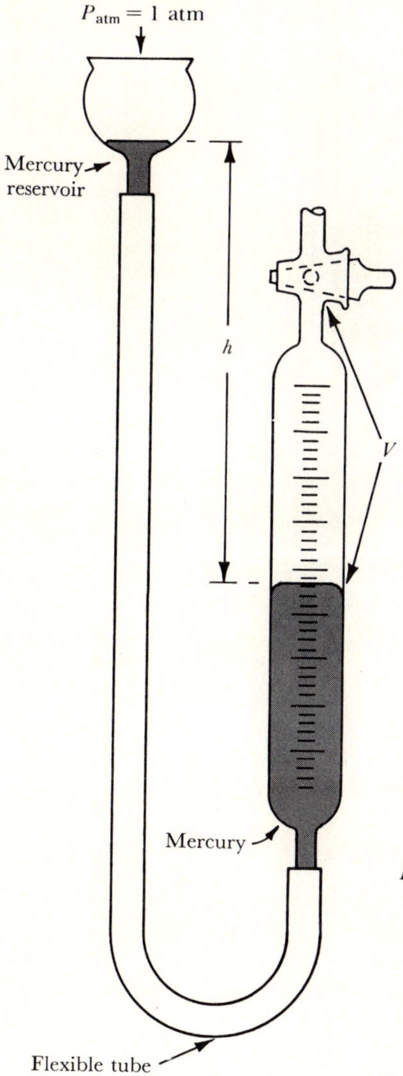

Figure 2-2

1 When the reservoir is lowered, V is halved.
2 When the reservoir is lowered, V remains the same.
3 When the reservoir is lowered, V is 20% larger.
4 When the reservoir is lowered, V is 50% larger.
5 When the reservoir is lowered, V is doubled.

PROBLEM 2–3

Consider the apparatus in Figure 2–3. If the temperature of the oil is raised from 68°C to 136°C, which one of the following statements about the volume of air trapped in the capillary tube is correct? (Ignore the small effect of pressure change; i.e., assume that the pressure remains constant.)

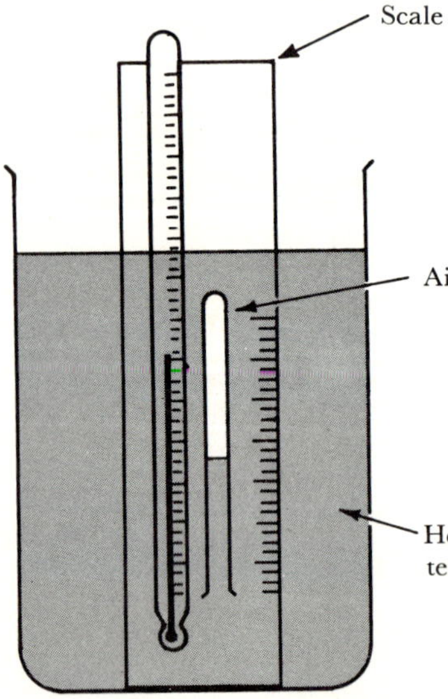

Figure 2–3

1 The volume is reduced by 20%.
2 The volume is increased by 14%.
3 The volume is increased by 20%.
4 The volume is increased by 50%.
5 The volume is doubled.

PROBLEM 2–4

The gauge pressure in an automobile tire when measured during the winter at 32°F was 30 pounds per square inch (psi). The gauge reads the *difference* between the tire pressure and atmospheric pressure (15 psi). In other words, the tire pressure is the gauge reading (psi) plus 15(psi). The same tire was used during the summer, and its temperature rose to 122°F. If we assume that the volume of the tire did not change, and no air leaked out between winter and summer, what is the new pressure as measured on the gauge?

$[°\text{C} = \frac{5}{9}(°\text{F} - 32)]$

1 35 psi
2 38 psi
3 105 psi
4 157 psi
5 The same, namely, 30 psi.

PROBLEM 2-5

Figure 2-4 represents the behavior of an ideal gas if which of the following conditions is met?

For a discussion of ideal gases, see Section 2-5 of Chemical Principles *by Dickerson, Gray, and Haight.*

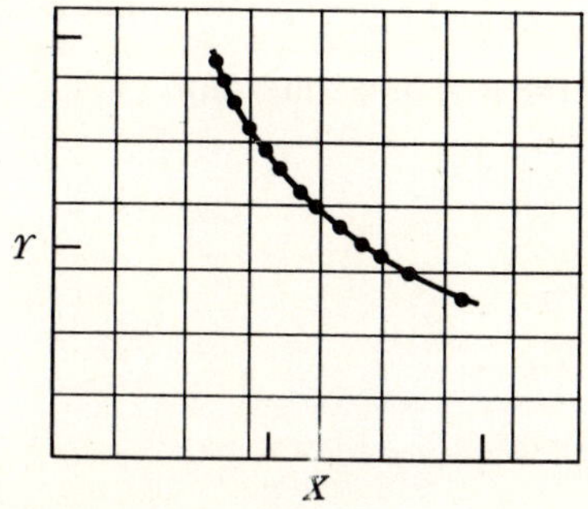

Figure 2-4

1 $X = P$, $Y = T$, and n, V = constants.
2 $X = P$, $Y = V$, and n, T = constants.
3 $X = V$, $Y = T$, and n, P = constants.
4 $X = n$, $Y = P$, and T, V = constants.
5 None of the above.

PROBLEM 2-6

The product PV has the same units as which one of the following? [*Hint:* P = force area^{-1}.]

1 Force area^{-1}
2 Force
3 Pressure
4 Energy
5 Pressure × area

Instead of using ratios and varying only two quantities, let us allow more variables into our problems so we must make full use of the ideal gas equation.

PROBLEM 2-7

A sample of pure gas at 27°C and 380 torr occupied a volume of 492 cm^3. What was the number of moles of gas in this sample?

($R = 0.082$ liter atm mole^{-1} deg^{-1})

1 0.010
2 1/9
3 7.6
4 10
5 6×10^{21}

PROBLEM 2-8

The supersonic transport (SST) already has become a pollution concern. It is estimated that for every hour of flight it will produce about 73 short tons (6.6×10^4 kg) of carbon dioxide (CO_2). This amount of CO_2 gas at 4.1×10^{-1} atm and 627°C corresponds to what volume?

(at. wt: C = 12, O = 16; $R = 0.082$ liter atm mole^{-1} deg^{-1})

1 1.5×10^6 liters
2 1.9×10^8 liters
3 2.7×10^8 liters
4 4.2×10^8 liters
5 2.1×10^{11} liters

PROBLEM 2-9

What is the *number density* of an ideal gas at STP? In other words, what is its density in molecules per cubic centimeter?

($R = 82$ cm^3 atm mole^{-1} deg^{-1})

STP and molar volumes are discussed in Section 2-5 of Chemical Principles.

1 $1/(22.4 \times 6.0 \times 10^{23}) = 7.5 \times 10^{-26}$
2 $1/22{,}400 = 4.5 \times 10^{-5}$
3 $6.0 \times 10^{23}/22{,}400 = 2.7 \times 10^{19}$
4 6.0×10^{23}
5 $22.4 \times 6.0 \times 10^{23} = 13 \times 10^{24}$

PROBLEM 2-10

The molecular weight of a gas, 0.30 g of which occupies a volume of 82 cm^3 at 3.0 atm pressure at 27°C, is ________.

($R = 0.082$ liter atm mole^{-1} deg^{-1})

1 3.0×10^{-2}
2 2.7
3 30
4 1/30
5 Insufficient data.

PROBLEM 2-11

The *exosphere* is defined as the region of the atmosphere in which collisions are so infrequent that the average mean free path is at least the radius of the earth. The region begins at an altitude of about 700 kilometers (km) or 435 miles [mean earth diameter = 12,742 km (7913 miles)] at which height the temperature is about 1200°K. The density of molecules is 1.0×10^6 molecules cm^{-3}. What is the equivalent pressure in atmospheres?

($R = 82$ cm^3 atm $mole^{-1}$ deg^{-1})

1 1.6×10^{-16} atm
2 1.6×10^{-13} atm
3 1.6×10^{-10} atm
4 1.2×10^{-10} atm
5 9.8×10^{10} atm

PROBLEM 2-12

An amount of 1.00 g of a gaseous compound of boron and hydrogen occupies 0.820 liter at 1.00 atm and +3°C. The compound is ________.

($R = 0.0820$ liter atm $mole^{-1}$ $°K^{-1}$;
at. wt: H = 1.0, B = 10.8)

1 BH_3
2 B_4H_{10}
3 B_2H_6
4 B_3H_{12}
5 B_5H_{14}

PROBLEM 2-13

Under what conditions will a pure sample of an ideal gas not only exhibit a pressure of 1 atm but also a concentration of 1 mole $liter^{-1}$?

($R = 0.082$ liter atm $mole^{-1}$ deg^{-1})

1 At STP.
2 When $V = 22.4$ liters.
3 When the gas constant, R, is dimensionless; that is, it has no units.
4 When $T = 12°K$.
5 Impossible to satisfy under any conditions.

PROBLEM 2-14

The change in momentum when a molecule (mass m) of an ideal gas, traveling in the x direction with velocity v_x, hits the container wall and rebounds elastically is ________.

1 mv_x
2 $2mv_x$
3 $-mv_x$
4 m^2v_x
5 Zero.

PROBLEM 2–15

A molecule in a cube of length l has a velocity v_x in the x direction. The number of collisions with one face (YZ wall) per second for this molecule will be given by which one of the following (see Figure 2–5)?

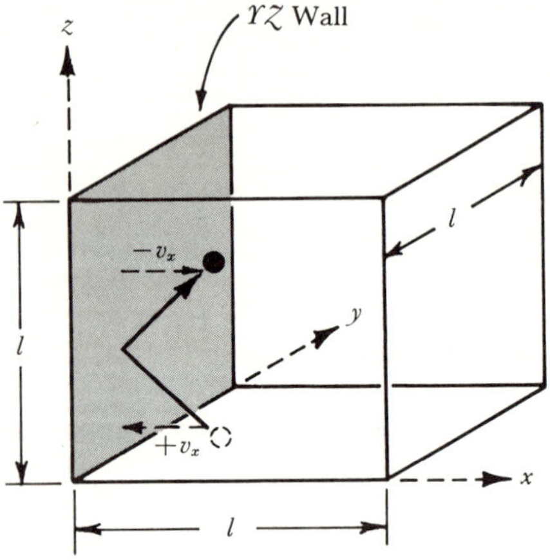

Figure 2–5

1 v_x/l
2 $v_x/2l$
3 $\frac{1}{2}l$
4 $Nmv^2/3l$
5 $\frac{3}{2}RT$

PROBLEM 2–16

At the same T and P, which of the following gases will have the highest average kinetic energy per mole?

(at. wt: H = 1, C = 12, O = 16, S = 32, F = 19)

1 H_2
2 O_2
3 CH_4
4 SF_6
5 All the same.

PROBLEM 2–17

Root-mean-square speed of gas molecules is discussed in Section 2–7 of Chemical Principles.

The root-mean-square speed (v_{rms}) of the molecules of a gas is determined by which one of the following?

[*Note:* v usually denotes a vector velocity (magnitude *and* direction), whereas c usually denotes a scalar speed (magnitude only). However, common usage *in notation* does not conform exactly to this system, so we will often encounter v_{rms}, which is an average quantity without direction and therefore not a vector. Even the terms "velocity" and "speed" are sometimes used interchangeably, and v_{rms} is often called the rms velocity.]

[*Hint:* $v_{rms} = \sqrt{\overline{v^2}}$, in which $\overline{v^2}$ means "the average of v^2."]

1 Solely the molecular weight.
2 The molecular weight and temperature.
3 Solely the temperature.
4 The vapor pressure.
5 The gas density.

PROBLEM 2–18

Attempts to produce an atomic bomb during World War II involved the problem of separating the rare ^{235}U isotope from the ^{238}U species. In Germany, one attempt at separation was made by utilizing the technique of molecular beams. The species effused out of a molecular-beam oven together and then passed through a mechanical velocity selector set at an appropriate speed to pass the ^{235}U species preferentially. The ratio of rms speeds, $v_{rms}(^{235}U)/v_{rms}(^{238}U)$, is given most closely by:

1 $\sqrt{235/238}$
2 1
3 $\sqrt{238/235}$
4 238/235
5 $(238 - 235)/238 \times 235$

PROBLEM 2–19

If equal weights of oxygen and nitrogen are placed in separate containers of equal volume at the same temperature, which one of the following statements is true?

(mol wt: $N_2 = 28$, $O_2 = 32$)

1 Both flasks contain the same number of molecules.
2 The pressure in the nitrogen flask is greater than the one in the oxygen flask.
3 More molecules are present in the oxygen flask.
4 Molecules in the oxygen flask are moving faster on the average than the ones in the nitrogen flask.
5 The nitrogen has a greater average kinetic energy per mole.

PROBLEM 2–20

Suppose that we change the rms speed, v_{rms}, of the gas molecules in a closed container of *fixed* volume from 5×10^4 cm sec^{-1} to 10×10^4 cm sec^{-1}. Which one of the following statements might correctly explain how this change was accomplished?

1 By heating the gas, we doubled the temperature.
2 By pumping out 75% of the gas at constant temperature, we decreased the pressure to one quarter its original value.
3 By heating the gas, we quadrupled the pressure.
4 By pumping in more gas at constant temperature, we quadrupled the pressure.
5 None of the above.

PROBLEM 2–21

The curves in Figure 2–6 represent two different Maxwell-Boltzmann velocity distributions for the *same* gas sample under two different conditions, A and B. Which of the following statements about conditions A and B is true?

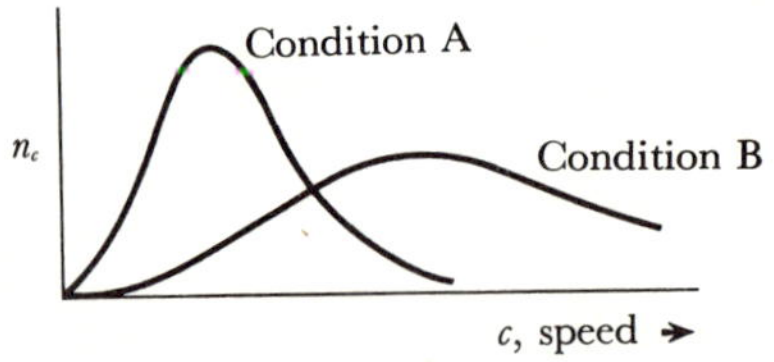

Figure 2–6

1 The temperature for B is greater than for A.
2 The average kinetic energy for A is greater than for B.
3 The intermolecular attractive forces are greater for A than for B.
4 A and B correspond to the same temperature but different volumes.
5 A and B correspond to the same volumes, but the pressure for A is greater than the pressure for B.

PROBLEM 2–22

Mean free path and collision frequency of gases are discussed in Section 2–7 of Chemical Principles.

The rate of collision (as given by a simple kinetic molecular theory) of a gas molecule with others in the gas is proportional to which one of the following?

[*Note:* d = molecular diameter, D^* = density of the gas in molecules cm^{-3} (number density), v = velocity of the molecule, and λ = mean free path.]

1 $D^* \lambda d^2$
2 d^2/v
3 $D^* dv$
4 $D^* d^2/v$
5 $D^* d^2 v$

PROBLEM 2–23

The *escape velocity,* the speed at which molecules can escape the earth's gravitational field, is approximately 1.1×10^6 cm sec^{-1} (about 7 miles sec^{-1}). At what temperature, approximately, will the v_{rms} value for H_2 be equal to the escape velocity?

($R = 8.3 \times 10^7$ ergs mole^{-1} deg^{-1};
at. wt: H = 1)

1 4×10^{-7} °K
2 5×10^3 °K
3 1.5×10^4 °K
4 1×10^4 °K
5 1×10^{13} °K

Having solved the previous problems, you should feel secure about the ideal gas law and a simple kinetic molecular theory of gases. These are our first approximations to the true state of nature, as we appreciate it on a macroscopic and think about it on a molecular level. Now let us branch into the far more usual situation of mixtures of gases.

PROBLEM 2–24

A cyclopropane–oxygen mixture can be used as an anaesthetic. If the partial pressures of cyclopropane

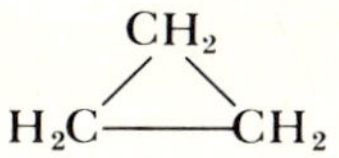

or C_3H_6

and oxygen are 170 torr and 570 torr, respectively, in a gas cylinder, what is the ratio of the number of moles of cyclopropane to the number of moles of oxygen (i.e., $n_{C_3H_6}/n_{O_2}$)?

(mol wt: $C_3H_6 = 42$, $O_2 = 32$)

1 $\frac{170}{42} \Big/ \left(\frac{170}{42} + \frac{570}{32}\right) = 0.19$

2 $170/740 = 0.23$
3 $170/570 = 0.30$
4 $(170 \times 42)/(570 \times 32) = 0.39$

5 $2 \times \frac{170}{42} \Big/ \frac{570}{32} = 0.46$

PROBLEM 2-25 The cylinders in a particular automobile engine have a total volume of 6.15 liters (about 380 cu in.). Assume that air fills this volume, at 1.00 atm and 27°C, and is one fifth oxygen by volume. What weight of pure octane (C_8H_{18}, mol wt = 114) is necessary to combine exactly with the oxygen (mol wt = 32)?

$$2C_8H_{18} + 25O_2 \rightarrow 16CO_2 + 18H_2O$$

($R = 0.082$ liter atm mole^{-1} deg^{-1})

1 0.092 g
2 0.13 g
3 0.46 g
4 5.6 g
5 72 g

PROBLEM 2-26 The stopcock between a 3-liter bulb containing oxygen at 195 torr and a 2-liter bulb containing nitrogen at 530 torr is opened. After equilibration the pressure is which one of the following? (Assume that T is constant.)

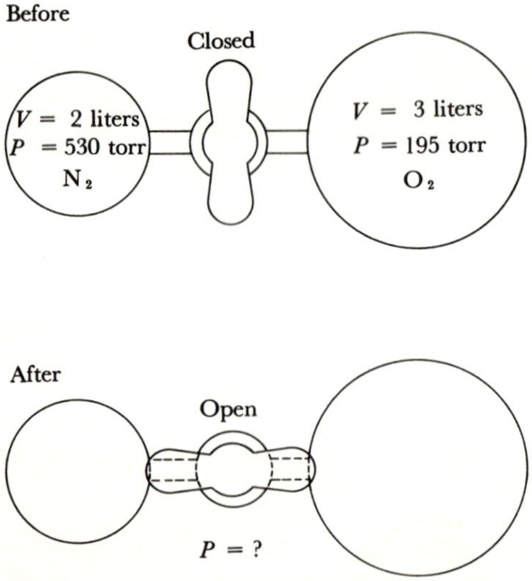

Figure 2–7

1 329 torr
2 363 torr
3 396 torr
4 725 torr
5 Insufficient data; temperature must be known.

PROBLEM 2–27

The human body discharges about 960 g of CO_2 per day. If the cabin for an astronaut has a volume of 7600 liters (about 270 cu ft) and the partial pressure of CO_2 (mol wt = 44) must be maintained at no more than 4.1 torr at 27°C cabin temperature, what weight of CO_2 must be removed on the first day of the voyage? (Assume that the initial partial pressure of CO_2 is zero.)

1 73 g
2 145 g
3 815 g
4 887 g
5 960 g

PROBLEM 2–28

An idling, untuned car engine can produce 600 liters min^{-1} (about 21 cu ft min^{-1}) of exhaust at 27° C that contains 4.1% by volume of carbon monoxide (CO). How many moles of CO does the exhaust deliver per minute?

($R = 0.082$ liter atm mole^{-1} deg^{-1})

1 0.035
2 0.10
3 1.0
4 11
5 Cannot be calculated.

PROBLEM 2–29

A human who has not been conditioned to withstand the gas will succumb to CO_2 poisoning if the concentration of that gas in the atmosphere rises to 10% by volume. A laboratory technician is working with solid CO_2 (Dry Ice) in a closed fume hood whose volume is 6000 liters (about 212 cu ft). If the ventilation stops, what minimum weight of steadily sublimating (vaporizing) CO_2 constitutes a hazard? Assume that the pressure and temperature remain constant at 1.00 atm and 27°C and that there is no CO_2 present initially. The density of solid CO_2 is 1.53 g cm^{-3}.

(mol wt: $CO_2 = 44$)

1 0.69 kg
2 1.1 kg
3 1.6 kg
4 11 kg
5 12 kg

PROBLEM 2-30

A concentration of carbon monoxide of 0.4% by volume in air will result in human death in a short time. An idling, untuned car engine with a 300-cu in. displacement can easily produce as much as 0.60 mole of carbon monoxide per minute as a component of its exhaust fumes. If a garage at 27°C has an air volume of 4.1×10^4 liters (≃1450 cu ft), how long will it take for the sealed garage to reach a lethal concentration of CO? (Neglect the minor effect of total pressure increase; i.e., assume that the pressure remains constant at 1.00 atm and that there is no CO present initially.)

($R = 0.082$ liter atm $mole^{-1}$ deg^{-1})

1 11 sec
2 66 sec
3 5.5 min
4 11 min
5 110 min

PROBLEM 2-31

You are told to prepare a closed experimental environment (a box) for student mice. The box volume will be 294 liters (about 10 ft^3) and the entire air volume will be changed every minute. The relative humidity of the air entering the box is to be controlled at 40% at 21°C. What weight of H_2O must be added to the flow of dry air per minute? (Equilibrium vapor pressure for H_2O at 21°C ≃ 19 torr.)

($R = 0.082$ liter atm $mole^{-1}$ deg^{-1};
mol wt: $H_2O = 18$)

1 0.075 g
2 0.12 g
3 2.2 g
4 5.5 g
5 14 g

PROBLEM 2-32

If this problem is difficult for you, refer to Section 2–8 of Chemical Principles, *which deals with the nonideal behavior of gases and van der Waals' equation of state.*

In the van der Waals equation of state for 1 mole of a nonideal gas,

$$(P + a/V^2)(V - b) = RT$$

the effect of intermolecular forces is accounted for by which of the following?

1 $P + a/V^2$
2 $V - b$
3 RT
4 A different value for R.
5 It is not accounted for in the van der Waals equation.

PROBLEM 2–33 The van der Waals equation of state for a nonideal gas can be rearranged to give

$$\frac{PV}{RT} = \frac{V}{V-b} - \frac{a}{VRT}$$

for 1 mole of gas. The constants a and b are positive numbers. When applied to H_2 at 80°K, the equation gives the curve in Figure 2–8. Which one of the following statements is correct?

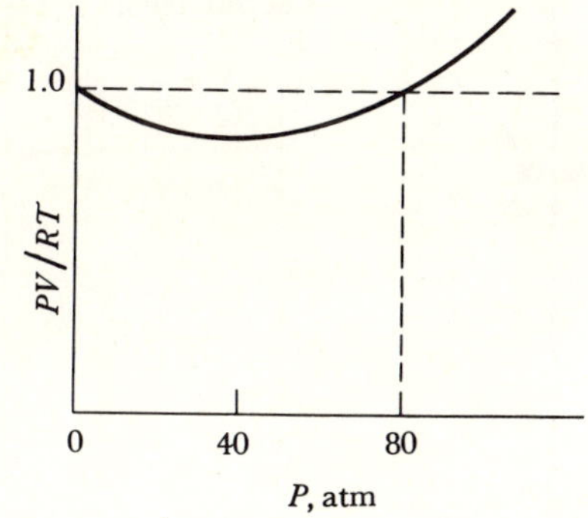

Figure 2–8

1 At 40 atm the two terms, $V/(V-b)$ and a/VRT are equal.
2 At 0 atm the temperature is 0 °K.
3 At 80 atm the two terms $V/(V-b)$ and a/VRT are equal.
4 At 80 atm the gas has condensed.
5 At a pressure greater than 80 atm, the term $V/(V-b)$ is greater than a/VRT.

PROBLEM 2-34 Consider the molecular-beam experiment in Figure 2-9. Such an experiment proves that ________.

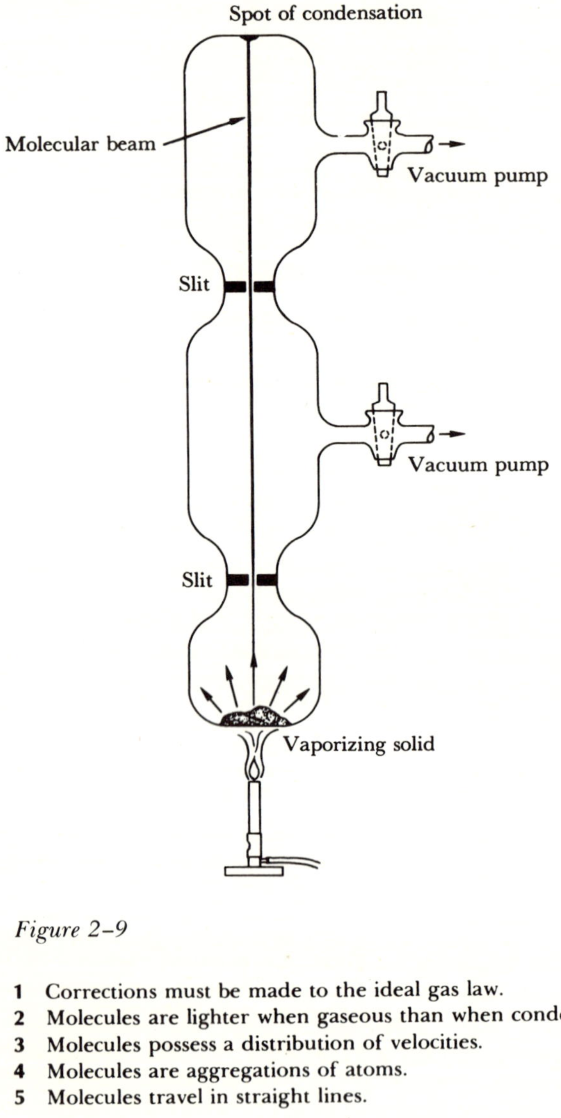

Figure 2-9

1 Corrections must be made to the ideal gas law.
2 Molecules are lighter when gaseous than when condensed as a liquid.
3 Molecules possess a distribution of velocities.
4 Molecules are aggregations of atoms.
5 Molecules travel in straight lines.

PROBLEM 2–35 Examine the experiment in Figure 2–10. Here is an apparatus for determining the "time of flight" of a molecular beam. The spinning disk allows a short "burst" of molecules to pass through the slot, one every revolution. If we have a sensitive detector that can respond to fast variations in intensity, what will the intensity, I, from one burst look like, *at the detector,* as a function of time?
[Data: distance from disk to detector = 100 cm;
$v_{rms} = 5 \times 10^4$ cm sec^{-1}; "burst" lasts 0.05 msec (5×10^{-5} sec)]

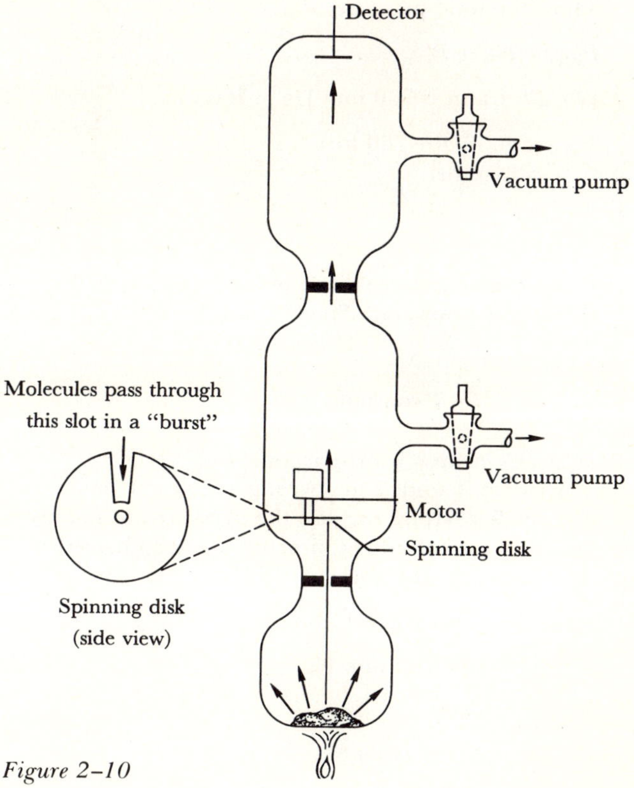

Figure 2–10

1

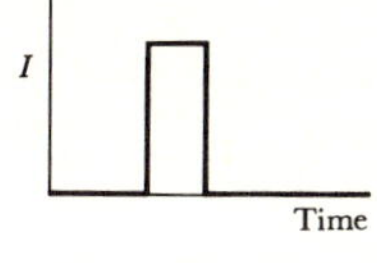

2

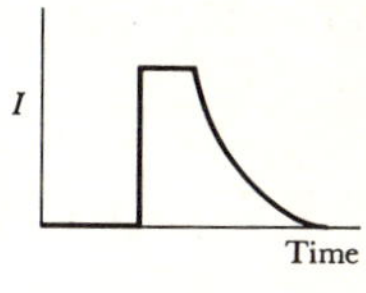

3

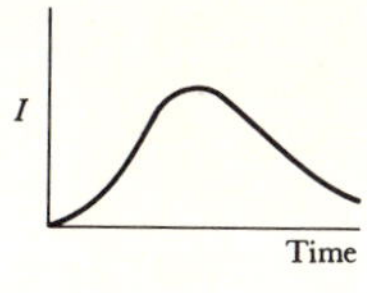

4

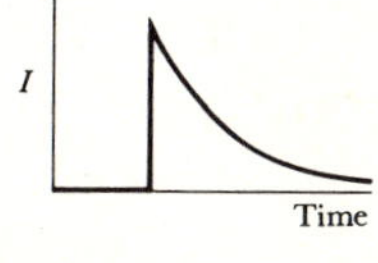

5

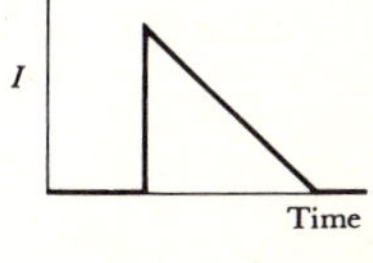

SOLUTION 2-1

(5)

We do not have to know any gas law to solve this. All we have to know is how to think about pressure itself. We must realize that the pressure exerted on the gas in V is due not only to the weight of the column of mercury, but also to the weight of the air pressing on the mercury. Thus, the total pressure, P_{total}, is

$$P_{total} = P_{Hg} + P_{atm}$$

[*Recall:* 1 atm = 760 mm Hg = 760 torr.]

$$\begin{aligned} P_{total} &= 40 \text{ torr} + 760 \text{ torr} \\ &= 800 \text{ torr} \end{aligned}$$

SOLUTION 2-2

(3)

If this problem is difficult for you, refer to Lassila et al. Programed Reviews of Chemical Principles, *Section 2-1, Items 1-14.*

When the height of the column of mercury is decreased, the pressure on the gas is lowered. Since n and T are fixed, V must increase. Recall that

$$P \propto \frac{1}{V} \qquad (n, T \text{ constant})$$

[*Note:* $\propto$ means "is proportional to."]

Answers **1** and **2** are wrong because V must increase. Moreover, Answer **5** is wrong because the pressure has not been halved; only the height of the column of mercury has been halved. The *total* pressure is the height of the mercury column plus the atmospheric pressure.

$$P_{init} = 380 \text{ mm Hg} + 1 \text{ atm}$$

Since 1 atm = 760 mm Hg,

$$P_{init} = 1.5 \text{ atm}$$

$$\begin{aligned} P_{final} &= 190 \text{ mm Hg} + 1 \text{ atm} \\ &= 1.25 \text{ atm} \end{aligned}$$

From our proportionality above,

$$\frac{V_{final}}{V_{init}} = \frac{P_{init}}{P_{final}}$$

$$\begin{aligned} V_{final} &= V_{init}\frac{P_{init}}{P_{final}} = V_{init}\frac{1.50}{1.25} \\ &= 1.2\ V_{init} \end{aligned}$$

Thus, the final volume is 20% larger than the initial, and Answer **3** is correct.

SOLUTION 2-3

(3)

If you couldn't solve this problem, refer to Lassila's book, Section 2-1, Items 15-34.

Since P and n are constant,

$V \propto T \qquad (n, P \text{ constant})$

The temperature doubles so V is doubled. Is Answer **5** correct then? No! The proportionality between V and T is only valid for absolute temperature:

$$\frac{V_{\text{final}}}{V_{\text{init}}} = \frac{T_{\text{final}}}{T_{\text{init}}} = \frac{136 + 273°\text{K}}{68 + 273°\text{K}} = \frac{409}{341}$$

$$\simeq 1.2$$

The final volume is 20% larger than the initial volume; Answer **3** is correct. Answer **2** results if you took the fractional increase in temperature to be 68/409 instead of 68/341.

SOLUTION 2-4

(2)

P and T vary, but the mass of the gas (and therefore, n) and the volume are fixed. From $PV = nRT$

$P = (nR/V)T$

and

$P \propto T \qquad (V, n \text{ constant})$

There are two stumbling blocks to be avoided:

(A) T must be an absolute temperature. Therefore, let us change to °K:

$$°\text{C} = \tfrac{5}{9}(°\text{F} - 32)$$
$$32°\text{F} = 0°\text{C} = 273°\text{K}$$
$$122°\text{F} = 50°\text{C} = 323°\text{K}$$

(B) P must be the total pressure that the gas in the tire exerts, not just its excess over 1 atm as read on the gauge:

$$P = 1 \text{ atm} + 30 \text{ psi} = 15 \text{ psi} + 30 \text{ psi} = 45 \text{ psi}$$

Now we can combine our P and T information in terms of the above proportion:

$$\frac{P_{\text{final}}}{P_{\text{init}}} = \frac{T_{\text{final}}}{T_{\text{init}}}$$

$$\frac{P_{\text{final}}}{45 \text{ psi}} = \frac{323}{273} \simeq 1.18$$

$$P_{\text{final}} \simeq 53 \text{ psi}$$

The *gauge* will read 53 psi − 15 psi ≃ 38 psi. Answer **2** is correct.

1 You did not use the total pressure.
4 You used °F instead of °K.
3 You made both the above mistakes.
5 Only if the gauge is stuck.

SOLUTION 2-5 (2)

What is clear immediately is that Y becomes smaller as X becomes larger. Moreover, the line is not straight. This might be one problem in which we can eliminate a lot of incorrect answers in a short time. An incorrect answer would be one in which the quantities associated with X and Y increase together: Answers **1**, **3**, and **4** pair P with T, V with T, and n with P. For these pairs one increases when the other does (if the remaining quantities are fixed). So these are wrong answers. We are left with Answers **2** and **5**. $PV = nRT =$ constant. When n and T are fixed,

$$P = \frac{\text{constant}}{V} \qquad (n, T \text{ constant})$$

which is the equation of a hyperbola. V decreases as P increases, and their product is constant. Answer **2** is correct.

[*Note:* You don't have to memorize Boyle's law. Derive it each time from the ideal gas law, if you have to. The derivation is only two steps. Also note that if the line were straight instead of curved, the relationship would be $Y = -aX + b$, and not $Y = \text{constant}/X$.]

SOLUTION 2-6 (4)

Our dimensional equation for PV is

$$PV = \frac{\text{force}}{(\text{distance})^2} \times (\text{distance})^3 = \text{force} \times \text{distance}$$

We can make a good guess by eliminating incorrectly dimensioned answers: PV certainly won't have units of pressure (Answer **3**), or pressure × area (Answer **5**); and force × distance certainly won't have units of force area^{-1} (Answer **1**) or just force (Answer **2**). All that remains is Answer **4**. Force × distance must be equal to energy, so PV has units of energy.

Alternatively, we could recall from elementary physics that force × distance = work = energy. Or finally, PV must have the same units as nRT (Appendix 1 in *Chemical Principles* gives R as cal deg^{-1} mole^{-1}, erg deg^{-1} mole^{-1}, etc.), and calories and ergs are energy terms.

SOLUTION 2-7 (1)

Study Items 52–76 in Section 2-2 of Lassila's book if you had difficulty with this problem.

Because we can use the ideal gas law, the problem becomes merely an exercise in choosing the correct units.

$$PV = nRT$$

$$T = 27°\text{C} = 300°\text{K}$$

$$P = 380 \text{ torr} = \frac{380 \text{ torr}}{760 \text{ torr atm}^{-1}} = 0.500 \text{ atm}$$

$$V = 492 \text{ cm}^3 = 0.492 \text{ liter}$$

$$n = ?$$

$$n = PV/RT$$

$$= \frac{0.500 \text{ atm} \times 0.492 \text{ liter}}{0.082 \text{ liter atm mole}^{-1} \text{ deg}^{-1} \times 300°\text{K}}$$

Note that the units in the above fraction cancel to give "mole."

$$n = \frac{0.246}{24.6} = 0.010 \text{ mole}$$

1 is correct.
2 would be the result if you used °C rather than °K. The ideal gas equation is only valid for T in °K.[1]
3 means that you did not make the units of P and R agree. If R has units of liter atm mole^{-1} deg^{-1}, P must also have units of atmospheres.
4 means that you probably did not make the units of R and V agree. If R is in liter atm mole^{-1} deg^{-1}, V must be in liters as well.
5 is the number of molecules, not the number of moles. Divide by N to find the correct answer.

SOLUTION 2–8

(3)

This problem is almost as easy as Problem 2–7, but we have one more step, the calculation of n from the weight of CO_2.

$$PV = nRT$$

$$V = \frac{nRT}{P}$$

$$n = \frac{6.6 \times 10^4 \text{ kg}}{\text{mol wt } CO_2}$$

$$= \frac{6.6 \times 10^7 \text{ g}}{44 \text{ g mole}^{-1}}$$

$$= 1.5 \times 10^6 \text{ moles}$$

$$T = 627°\text{C} + 273° = 900°\text{K}$$

$$V = \frac{1.5 \times 10^6 \text{ moles} \times 0.082 \text{ liter atm mole}^{-1} \text{ deg}^{-1} \times 900°\text{K}}{0.41 \text{ atm}}$$

$$= 1.5 \times 10^6 \times 0.20 \times 9.0 \times 10^2 \text{ liters}$$

$$= 2.7 \times 10^8 \text{ liters}$$

Answer **3** is correct.
2 1.9×10^8 liters if you used 627°C rather than 900°K.
1 1.5×10^6 is n, not V.
4 4.2×10^8 if you used the molecular weight of CO = 28.
5 2.1×10^{11} liters if you divided 0.41 atm by 760 torr.

[1] If you really wanted to, you could find a value for R such that the ideal gas equation would be valid for an absolute temperature scale whose degree intervals are the same size as °F. Not many people want to do this, though.

SOLUTION 2–9

(3)

We have to think a few moments about this special density. We can relate it to the usual density ($D = g/V$), the weight per unit volume. But it is easier to realize that if we have a sample of n moles of gas in a volume, V (cm^3), the number of moles per cubic centimeter is n/V and the number of molecules per cubic centimeter is this quantity multiplied by Avogadro's number: D^*, the number density, equals Nn/V. Can we rearrange the ideal gas equation in such a way that we can solve it for D^*? Of course!

$$PV = nRT$$

$$P = \frac{n}{V} RT$$

$$\frac{n}{V} = \frac{P}{RT}$$

Then multiplying both sides by Avogadro's number, N, we obtain

$$\frac{Nn}{V} = D^* = \frac{PN}{RT}$$

All that's left is to plug in the values on the right-hand side. Use the correct units. (*Recall:* STP implies $T = 273°K$, and $P = 1$ atm.)

$$D^* = \frac{1 \text{ atm} \times 6.0 \times 10^{23} \text{ molecules mole}^{-1}}{82 \text{ cm}^3 \text{ atm mole}^{-1} \text{ deg}^{-1} \times 273°K}$$

$$= \frac{6.0 \times 10^{23} \text{ molecules}}{82 \times 273 \text{ cm}^3} = \frac{6.0 \times 10^{23} \text{ molecules}}{22{,}400 \text{ cm}^3}$$

It would have been faster if we recalled that, *at STP*, 1 mole of ideal gas occupies 22.4 liters, so 6.0×10^{23} molecules occupy 22.4 liters or 22,400 cm^3.

$$D^* = \frac{6.0 \times 10^{23} \text{ molecules}}{22{,}400 \text{ cm}^3}$$

$$= 2.7 \times 10^{19} \text{ molecules cm}^{-3}$$

SOLUTION 2–10

(3)

$$PV = nRT = \frac{g}{\text{mol wt}} RT$$

Solve for molecular weight since it is what we are looking for.

$$\text{mol wt} = \frac{gRT}{PV} = \frac{0.30 \text{ g} \times 0.082 \text{ liter atm mole}^{-1} \text{ deg}^{-1} \times 300°K}{3.0 \text{ atm} \times 82 \text{ cm}^3}$$

will give an incorrect answer (**1**) because the units of V (cm^3) do not match the units of R (liter atm $mole^{-1}$ deg^{-1}). So correct V to units of liter:

$$\text{mol wt} = \frac{0.30 \times 0.082 \times 300 \text{ g liter atm mole}^{-1}}{3.0 \text{ atm} \times 0.082 \text{ liter}}$$

$$= 0.30 \times 100 = 30 \text{ g mole}^{-1}$$

(Note correct units for molecular weight.) Answer **2** arises if you used the wrong T (°C instead of °K). Answer **4** means that your algebra gave you an inverted relationship for molecular weight.

SOLUTION 2-11

(2)

There are many extraneous data given. This is a problem involving the ideal gas law; we will not use the mean free path.

$$PV = nRT$$

$$P = \frac{n}{V} RT$$

$$\frac{n}{V} = \frac{\text{mole}}{\text{volume}}$$

but we are given

$$D^* = \frac{\text{molecules}}{\text{volume}} \qquad \text{(the number density)}$$

$$= \frac{Nn}{V} \qquad \text{(see Problem 2-9)}$$

in which N is Avogadro's number.

$$\frac{D^*}{N} = \frac{n}{V}$$

$$P = \frac{D^*}{N} RT$$

$$= \frac{1.0 \times 10^6 \text{ molecules cm}^{-3} \times 82 \text{ cm}^3 \text{ atm mole}^{-1} \text{ deg}^{-1} \times 1200°\text{K}}{6.0 \times 10^{23} \text{ molecules mole}^{-1}}$$

$$= \frac{82 \times 12 \times 10^8}{6.0 \times 10^{23}} = 82 \times 2.0 \times 10^{-15} \text{ atm}$$

$$= 164 \times 10^{-15} \simeq 1.6 \times 10^{-13} \text{ atm}$$

Answer **2** is correct. If you converted R to 0.082 liter atm mole^{-1} deg^{-1} and consequently had improperly matched units, Answer **1** is calculated. If you converted D^* to 10^9 molecules liter^{-1} but did not also change R to 0.082 liter atm mole^{-1} deg^{-1}, then Answer **3** results. If you did everything right and then converted to torr, the answer is 1.2×10^{-10} *torr* not atm. Answer **4** is therefore wrong. Answer **5** occurs if you forgot to divide by N.

SOLUTION 2-12

(3)

The first task is to determine the molecular weight by using the ideal gas law:

$$PV = nRT = \frac{g}{\text{mol wt}} RT$$

$$\text{mol wt} = \frac{gRT}{PV}$$

$$\begin{aligned} g &= 1.00 \text{ g} \\ R &= 0.0820 \text{ liter atm mole}^{-1} \text{ °K}^{-1} \\ T &= 276\text{°K} \\ P &= 1.00 \text{ atm} \\ V &= 0.820 \text{ liter} \end{aligned}$$

$$\text{mol wt} = \frac{1.00 \times 0.0820 \times 276}{1.00 \times 0.820}$$

$$= 27.6$$

The compound is B_aH_b. We must determine a and b. The mathematical relationship between the atomic weights of B and H (10.8 and 1.0) and the molecular weight (27.6) is

$$a \times 10.8 + b \times 1.0 = 27.6$$

a must be one or two. If $a = 1$, $b = 16.8$ (and not an integer), and BH_{17} is not one of the choices (and does not exist). If $a = 2$, then $b = 6$, which implies B_2H_6, a highly reactive compound called diborane.

SOLUTION 2-13

(4)

If you need help to understand gas law problems, see Chapter 2 of A Study Guide to Chemical Principles *by Hutton.*

We can specify the concentrations in terms of the variables used on the ideal gas equation:

$$P = 1 \text{ atm}$$

$$c = 1 \text{ mole liter}^{-1} = \frac{n}{V}$$

(c = concentration)

The second equation is the crucial one. [We must recognize that if there is a gas sample of 5 moles of N_2 (n) contained in a 10-liter volume (V), its concentration is 0.5 mole liter^{-1} (n/V or c).] All we need to do is to insert these two conditions into an appropriately rearranged ideal gas law:

$$PV = nRT$$

$$P = \frac{n}{V} RT = cRT$$

$$1 \text{ atm} = \frac{1 \text{ mole}}{\text{liter}} RT$$

Choose the units of R correctly!

$$R = 0.082 \frac{\text{liter atm}}{\text{mole deg}}$$

$$1 \text{ atm} = 1 \frac{\cancel{\text{mole}}}{\cancel{\text{liter}}}\left(0.82 \frac{\cancel{\text{liter}} \text{ atm}}{\cancel{\text{mole}} \text{ deg}}\right)T$$

$$T = \frac{1}{0.082} = 12°\text{K}$$

In other words, an ideal gas will satisfy the two conditions only when the absolute temperature is 12°K.

If you have been able to solve the previous problems (or can solve them now), you should feel confident about being able to handle the ideal gas law for pure gases. Your only difficulty in the future should be deciding how the ideal gas law is applied to other (new) situations.

The next 10 problems will give you an opportunity to test your understanding of the kinetic molecular theory and to apply those principles.

SOLUTION 2-14

(2)

We must remember the definition of momentum:

initial momentum $= mv_x$
final momentum $= -mv_x$ (v_x is reversed and changes sign)
change in momentum = initial momentum − final momentum
$= mv_x - (-mv_x) = 2mv_x$

Answer **2** is correct. Answers **1** and **3** are just the initial and final momenta separately. Answer **4** has the wrong units. If you chose Answer **5** you forgot that the final momentum has a sign opposite to that of the initial momentum.

SOLUTION 2–15

(2)

If you are confused by abstract problems, you may find it helpful to reduce it to a concrete example. Suppose that the box is 2 cm long and the velocity is 8 cm sec^{-1}. We have drawn an idealized trajectory for 1 sec (Figure 2–11). It has traversed the 2-cm length of the box four times and has had two collisions with the *YZ* wall: two collisions per second. The correct answer must satisfy this example and must have units of sec^{-1}. Answers **3**, **4**, and **5** can be eliminated on this basis. Answer **2** satisfies the example. Answer **1** is the number of collisions with the *YZ* wall *and* its opposite wall per second.

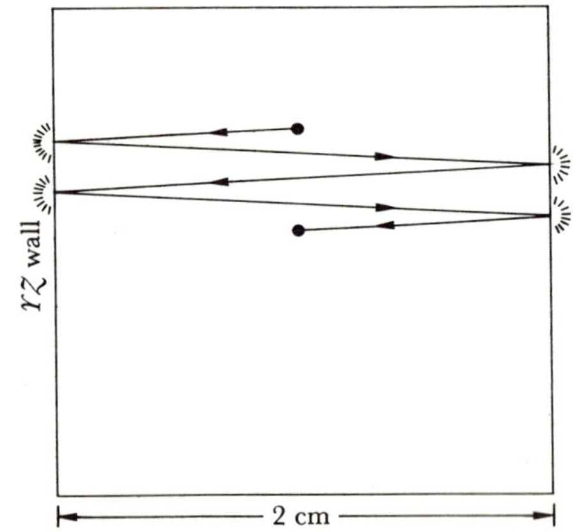

Figure 2–11

SOLUTION 2–16

(5)

$$\overline{E_k} = \tfrac{3}{2}RT$$

(average kinetic energy per mole) = (depends only on temperature and not molecular weight)

The heavier (large m) molecules are moving more slowly (small v) than the lighter ones (small m, large v) in such a way that $N\ (\tfrac{1}{2}m\overline{v^2})$ is the same for *all* gases at the same temperature. Answer **5** is correct.

SOLUTION 2–17

(2)

All we have to remember are two relationships:

(A) The result of the kinetic theory derivation,

$$\overline{E_k} = \tfrac{3}{2}RT = \text{average kinetic energy mole}^{-1}$$

(B) The definition of kinetic energy for a molecule,

$$\epsilon = \tfrac{1}{2}mv^2$$

The average kinetic energy for one molecule, $\bar{\epsilon}$, is just the average of the quantity on the right, namely

$$\bar{\epsilon} = \tfrac{1}{2}m\overline{v^2}$$

(m need not be included in the average since it is constant.) And the average kinetic energy for a mole is just $\bar{\epsilon}$ times N, Avogadro's number:

$$\overline{E_k} = N\bar{\epsilon} = \tfrac{1}{2}Nm\overline{v^2}$$

Since m is the mass of one molecule, Nm is the molecular weight:

$$\overline{E_k} = \tfrac{1}{2}(\text{mol wt})\overline{v^2} = \tfrac{3}{2}RT$$

$$\overline{v^2} = \frac{3RT}{\text{mol wt}}$$

And taking the square root, we find that

$$\sqrt{\overline{v^2}} = v_{\text{rms}} = \sqrt{\frac{3RT}{\text{mol wt}}}$$

which is determined only by T and mol wt. Answer **2** is correct.

A helpful device for remembering the definition of v_{rms} is to write the algebraic symbols in the order in which they appear. Root: $\sqrt{}$, mean: $\sqrt{\overline{}}$, square: $\sqrt{\overline{^2}}$, and then insert v: $\sqrt{\overline{v^2}}$.

SOLUTION 2–18

(3)

$$v_{\text{rms}} = \sqrt{\frac{3RT}{\text{mol wt}}}$$

See Problem 2–17 to find out how to avoid memorizing this relationship.

$$\frac{v_{\text{rms}}(^{235}\text{U})}{v_{\text{rms}}(^{238}\text{U})} = \frac{\sqrt{3RT/235}}{\sqrt{3RT/238}} = \sqrt{238/235}$$

because the T's are the same.[2]

Answer **3** is correct. If **4** was your answer you forgot the square root. If **1** was your answer you had v_{rms} proportional, rather than *inversely* proportional, to the square root of the molecular weight. If you selected **2**, you forgot that at the same T, the average kinetic energy per mole is the same for the two isotopes, but not their speeds.

The German apparatus, built with Russian internee labor, employed the gaseous compound uranium hexafluoride (UF_6) rather than uranium metal.

[2] Although the speed distribution in a molecular beam is slightly different from that given by the Maxwell–Boltzmann equation, this result is still valid.

SOLUTION 2–19

(2)

$$\begin{array}{ccc} \boxed{O_2} & = & \boxed{N_2} \\ V & = & V \\ T & = & T \\ g_{O_2} & = & g_{N_2} \\ \text{mol wt} = 32 & & \text{mol wt} = 28 \end{array} \qquad \begin{array}{c} \\ \\ \\ \text{(A)} \\ \text{(B)} \end{array}$$

Dividing (A) by (B), we find

$$n_{O_2} \quad < \quad n_{N_2}$$

Thus, we have more moles (and molecules) of N_2 than O_2, so Answers **1** and **3** are incorrect. Since $P = nRT/V$ and T/V is constant, the N_2 pressure is higher (Answer **2** is correct). Note that Answer **5** must be incorrect because the average kinetic energy per mole is given by the temperature alone ($\frac{3}{2}RT$) and is the same for both gases. To compare velocities (Answer **4**), we need only recall

$$\overline{E_k} = \tfrac{3}{2}RT = (\tfrac{1}{2}m\overline{v^2})N$$

$$\overline{v^2} = \frac{3RT}{Nm} = \frac{3RT}{\text{mol wt}}$$

$$\sqrt{\overline{v^2}} = v_{\text{rms}} = \sqrt{\frac{3RT}{\text{mol wt}}}$$

(Refer to Problem 2–17.) So the heavier O_2 molecule is, on the average, *slower* than N_2 at the same temperature.

SOLUTION 2-20

(3)

The average velocity is proportional to $[T/(\text{mol wt})]^{1/2}$, but none of the choices involved changing the kind of gas (and therefore molecular weight). Thus, we can say that $v \propto T^{1/2}$ since the molecular weight is constant:

$$\frac{v_{\text{final}}}{v_{\text{init}}} = 2 = \left(\frac{T_{\text{final}}}{T_{\text{init}}}\right)^{1/2}$$

$$\frac{T_{\text{final}}}{T_{\text{init}}} = 4$$

We have almost answered the problem. The temperature has been quadrupled. Thus, Answer **1** is wrong by a factor of two. Answer **2** holds T constant and is wrong. Answer **3** does not include T. Answer **4** is the same as Answer **2**. Is Answer **5** the correct choice? No. Even though Answer **3** does not include T explicitly, to quadruple the pressure in a container of fixed volume and gas sample means that the temperature had to be quadrupled. Answer **3** is correct.

SOLUTION 2–21 The average speeds for B are clearly greater than for A.

(1)

1 True. Higher temperatures imply higher speeds.
2 False. Lower speeds (for the same molecular weight) ensure lower energies.
3 False. Red herring.
4 False. Speed distributions depend only on temperature and molecular weight, not volume.
5 False, but very tricky. Since the gas sample is the same, the number of moles is constant. If V is also constant,

$$P = \frac{nR}{V} T = (\text{constant}) \times T$$

Thus, a higher T implies a higher P. Fortunately, $T_B > T_A$ implies $P_B > P_A$, so the last part is wrong.

SOLUTION 2–22

(5)

1 The rate should not be proportional to λ; the greater the mean free path between collisions, the fewer collisions there are.
2 The rate *should* be proportional to velocity: The faster the molecule is traveling, the more molecules it will encounter per second. Answer **4** is wrong for the same reason.
3 and **5** The rate should be proportional not only to v but to D^* as well. The greater the density, the more difficult it is to miss a molecule. We must decide whether the rate is proportional to d or d^2. We could do this, unthinkingly, on a purely dimensional basis, as follows.

The rate of collisions of one molecule will be the number of collisions per second and will have units of no./sec. Answer **3** has units of

$$\frac{\text{no. molecules}}{\text{cm}^3} \times \text{cm} \times \frac{\text{cm}}{\text{sec}} = \frac{\text{no. molecules}}{\text{cm sec}}$$

whereas Answer **5** has the units

$$\frac{\text{no. molecules}}{\text{cm}^3} \times \text{cm}^2 \times \frac{\text{cm}}{\text{sec}} = \frac{\text{no. molecules}}{\text{sec}}$$

and is correct. Alternatively, we could realize physically that collisions will not occur just at the diameter but on any part of the cross-sectional area that a molecule presents to the world.

SOLUTION 2–23

(4)

$$v_{rms} = \left(\frac{3RT}{\text{mol wt}}\right)^{1/2} \quad \text{(A)}$$

Now square Equation A:

$$v_{rms}^2 = \frac{3RT}{\text{mol wt}}$$

and solve for our unknown, T:

$$T = \frac{v_{rms}^2(\text{mol wt})}{3R}$$

Let us try to insert on the right-hand side:

$$T = \frac{(1.1 \times 10^6)^2 \text{ cm}^2 \text{ sec}^{-2} \times 2 \text{ g mole}^{-1}}{3 \times 8.3 \times 10^7 \text{ ergs mole}^{-1} \text{ deg}^{-1}} \quad \text{(B)}$$

$$T = \frac{2.4 \times 10^{12} \text{ g cm}^2 \text{ sec}^{-2}}{24.9 \times 10^7 \text{ ergs deg}^{-1}}$$

Are these units correct? Erg is an energy unit in the cgs (centimeter, gram, second) system. One form of energy that we know is kinetic energy,

$\frac{1}{2}mv^2$

In cgs units, $m = \text{g}$, $v = \text{cm sec}^{-1}$, $v^2 = \text{cm}^2 \text{ sec}^{-2}$, and $\frac{1}{2}mv^2 = \text{g cm}^2 \text{ sec}^{-2} =$ erg. So the units of Equation B cancel to give degrees (Kelvin, as usual):

$$T = \frac{2.4}{24.9} \times 10^5 \text{ °K} \simeq 1 \times 10^4 \text{ °K}$$

4 is correct.
1 You used miles sec^{-1}!
2 Wrong *molecular* weight for H_2.
3 Probably used the expression for most probable velocity.
5 Tried $R = 0.082$ liter atm $\text{mole}^{-1} \text{ deg}^{-1}$?

[*Note:* The escape velocity does not depend on the mass of the *escaping* body. For the earth's moon, the escape velocity is about 2.3×10^5 cm sec^{-1}. Why is there no atmosphere there?]

SOLUTION 2–24

(3)

Refer to Lassila's book, Section 2–2, Items 80–83 for a review of Dalton's law of partial pressures, if you could not solve this problem.

$$p_1 = n_1 \frac{RT}{V}$$

$$p_2 = n_2 \frac{RT}{V}$$

In a given volume at a temperature T,

$$\frac{p_1}{p_2} = \frac{n_1}{n_2}$$

The ratio of partial pressures is also the mole ratio:

$$\frac{n_{C_3H_6}}{n_{O_2}} = \frac{170}{570}$$

Answer **3** is correct. Answer **2** is what chemists call the "mole fraction": the ratio of moles of a substance to the total number of moles of all substances.

$$\frac{n_{C_3H_6}}{n_{C_3H_6} + n_{O_2}} = \frac{170}{570 + 170} = \frac{170}{740}$$

But that was not what we were looking for. Answer **4** is the ratio of *weights* of cyclopropane to oxygen. Answers **1** and **5** are just wrong.

In the 1850's, when anaesthetics were first being used routinely in surgery, some church groups strongly opposed this practice. One of the innovators, Dr. James Y. Simpson, was able to counter their attack by pointing out that God, "the first surgeon," had used a form of anaesthesia when He put Adam to sleep before performing the first thoracoplasty.

SOLUTION 2–25

(3)

The crucial step is to clear up our thinking about problems that specify gas composition in volume percent. It is preferable to translate these problems into the more common concept of partial pressure. Twenty percent of the volume at 1.00 atm is equivalent to (translates to) 0.20 atm in the total volume, or a partial pressure of 0.20 atm. We use this partial pressure terminology because we know that the gas molecules are not confining themselves to particular parts of the container. If we apply such a translation to this problem,

$$p_{O_2} = 0.20 \text{ atm}$$
$$T = 300°\text{K}$$
$$V = 6.15 \text{ liters}$$

$$n_{O_2} = p_{O_2}\frac{V}{RT} = \frac{0.20 \times 6.15 \text{ liters atm}}{0.082 \times 300 \text{ liter atm mole}^{-1}}$$

$$= 0.050 \text{ mole}$$

For every 25 moles of O_2, we use 2 moles of octane:

$$n_{\text{octane}} = \frac{2}{25} \times 0.050 = \frac{0.10}{25} \text{ mole}$$

$$g_{\text{octane}} = n_{\text{octane}} \times 114 = \frac{0.10}{25} \times 114 = \frac{11.4}{25} \text{ g} = 0.46 \text{ g}$$

The correct answer is **3**.

1 You divided both the pressure *and* the volume by five.
2 Wrong molecular weight (32 rather than 114).
4 You did not use any stoichiometric ratio.
5 You inverted the stoichiometric ratio.

SOLUTION 2-26

1

$V_{final} = 5$ liters
$P_{final} = ?$

But by Dalton's law of partial pressures

$$P_{final} = (n_{O_2} + n_{N_2}) \frac{RT}{V_{final}} \tag{A}$$

Yet we do not know the temperature. However, we do have enough information to calculate the n's. Perhaps the T's will cancel.

Original conditions:

$$n_{O_2} = \frac{3 \text{ liters} \times 195 \text{ torr}}{RT}$$

$$n_{N_2} = \frac{2 \text{ liters} \times 530 \text{ torr}}{RT}$$

Multiply both equations by RT and add:

$$(n_{O_2} + n_{N_2})RT = (195 \times 3 + 530 \times 2) \text{ torr liter}$$

But if we divide by V_{final} this becomes the quantity that we are looking for (see Equation A):

$$(n_{O_2} + n_{N_2}) \frac{RT}{V} = 195 \times \frac{3}{5} + 530 \times \frac{2}{5} = P_{final}$$

$$P_{final} = (117 + 212) \text{ torr} = 329 \text{ torr}$$

SOLUTION 2-27

4

$$p_{CO_2} = \frac{n_{CO_2}RT}{V}$$

$$n_{CO_2} = p_{CO_2}V/RT$$
$$g_{CO_2} = n_{CO_2} \times \text{mol wt } CO_2$$
$$g_{CO_2} = p_{CO_2}V \times \text{mol wt } CO_2/RT$$

Put p_{CO_2} in proper units for the given R:

$$p_{CO_2} = \frac{4.1 \text{ torr}}{760 \text{ torr atm}^{-1}}$$

$$g_{CO_2} = \frac{(4.1)}{760} \text{ atm} \times \frac{7600 \text{ liters} \times 44 \text{ g mole}^{-1}}{0.082 \text{ liter atm mole}^{-1} \text{ deg}^{-1} \times 300°\text{K}}$$

$$= \frac{4.1}{0.082} \times 10 \times \frac{44}{300} \text{ g}$$

$$= \frac{440}{0.02 \times 300} = \frac{440}{6.0} \simeq 73 \text{ g}$$

But this is not the answer. It is the weight of CO_2 that corresponds to the partial pressure of 4.1 torr in the cabin. The amount that has to be removed is 960 − 73, or 887 g. Answer **4** is correct. If you made that old mistake of forgetting to convert °C to °K, g_{CO_2} would be 815 g (Answer **3**), and the difference would be 145 g (Answer **2**), neither of which is correct. (We should note that the usual CO_2 concentration in the atmosphere is not zero but about 0.3%. If we included this in our calculation, it would make a difference of about 6 g.)

SOLUTION 2–28

(3)

We can use the ideal gas equation for this flow system by making a simple modification. The relationship

$$n = \frac{PV}{RT}$$

holds for a nonflowing system. If there is a flow of a volume of gas (which has a *P* and *T*), in liters per minute, then this corresponds to the flow of a mass of gas, or the number of moles of gas in moles per minute. The relationship

$$\dot{n} = \frac{\text{moles}}{\text{min}} = \frac{P(\text{atm})\ \dot{V}(\text{liter min}^{-1})}{R(\text{atm liter mole}^{-1}\ \text{deg}^{-1})T}$$

holds for a flowing system. Note that the units balance; the quantities with overdots signify flow quantities. We are ready to solve this problem once we translate from "4.1% by volume" to our partial pressure concept:

$$p_{CO} = 0.041 \text{ atm}$$

$$\dot{n} = \frac{\text{moles produced}}{\text{min}} = \frac{0.041 \text{ atm} \times 600 \text{ liter min}^{-1}}{0.082(\text{liter atm mole}^{-1}\ \text{deg}^{-1}) \times 300 \text{ deg}}$$

$$= \frac{1.0 \text{ mole}}{\text{min}}$$

Answer **3** is correct. Answer **1** used cubic feet instead of liters; Answer **4** did not convert to °K. Answer **2**: Have you slipped a decimal in percent conversion?

SOLUTION 2–29 (2)

We can translate "percent by volume" into partial pressure language: 10% by volume → partial pressure that is 10% of the total pressure; so p_{CO_2}(lethal) = 0.10 atm is the condition. All we have to do is to find the weight of CO_2, which, when gaseous, will have this partial pressure and will satisfy

$$p_iV = n_iRT$$

(subscript i means that the equations are valid for each individual gas species in the mixture).

$$n_i = p_iV/RT = g_i/(\text{mol wt})_i$$

$$g_i = (\text{mol wt})_i p_iV/RT$$

$$= \frac{44 \text{ g mole}^{-1} \times 0.10 \text{ atm} \times 6000 \text{ liters}}{0.082 \text{ liter atm mole}^{-1} \text{ deg}^{-1} \times 300°\text{K}}$$

$$= \frac{44}{0.082} \times 2.0 = \frac{88}{0.082} = 1070 \text{ g} \simeq 1100 \text{ g}$$

$$\simeq 1.1 \text{ kg } (\simeq 2.3 \text{ lb})$$

Answer **2** is correct. Answers **1** and **3** arise if you divide or multiply the correct answer by the density. But the density does not enter into the calculation at all. The units of these answers are also a mess. Answer **1** has units of cm^3 and Answer **3** has $g^2\ cm^{-3}$. Answer **4** happens if you slipped a decimal point. Answer **5** occurs when you use °C instead of °K.

SOLUTION 2–30 (4)

How many moles of CO correspond to the lethal atmosphere?

$$n_{CO} = \frac{p_{CO}V}{RT}$$

$$p_{CO} = 0.004 \text{ atm}$$

When we have translated from volume percent, we have

$$V = 4.1 \times 10^4 \text{ liters}$$

$$T = 300°\text{K}$$

$$n_{CO} = \frac{4.1 \times 10^4 \times 4 \times 10^{-3}}{8.2 \times 10^{-2} \times 300} = \frac{1}{2} \times \frac{4}{3} \times 10^1$$

$$= 6.7 \text{ moles}$$

If we add 6.7 moles of CO, the lethal condition exists. This will happen in

$$\frac{6.7 \text{ moles}}{0.60 \text{ mole min}^{-1}} \simeq 11 \text{ min}$$

(We should note that a concentration of 0.15–0.20% is dangerous for an exposure of 1 hour.)

SOLUTION 2-31

(3)

The only new point is relative humidity, and we must realize that it gives the H_2O partial pressure in terms of the maximum H_2O partial pressure, namely, the equilibrium vapor pressure at that temperature. In this problem

$$p_{H_2O} = 19 \text{ torr} \times 0.40$$

$$= \frac{19 \times 0.40}{760} \text{ atm} = \frac{0.40}{40}$$

$$= 1.0 \times 10^{-2} \text{ atm}$$

$$n_{H_2O} = p_{H_2O}V/RT$$

$$g_{H_2O} = \text{mol wt } H_2O \times p_{H_2O}V/RT$$

$$= \frac{18 \text{ g mole}^{-1} \times 1.0 \times 10^{-2} \text{ atm} \times 294 \text{ liters}}{0.082 \text{ liter atm mole}^{-1} \text{ deg}^{-1} \times 294°\text{K}}$$

$$= \frac{0.18}{0.082} \text{ g} \simeq 2.2 \text{ g}$$

3 is correct.
4 arises if you forgot to multiply by 0.40, so this corresponds to 100% humidity.
5 arises if you divided by 0.40 instead of multiplied.
2 is the number of *moles* of H_2O.
1 used the wrong units for V.

SOLUTION 2-32

(1)

The quantity $P + a/V^2$ may be thought of as involving intermolecular forces, whereas the term $V - b$ involves molecular size and "unavailable volume." Answer **1** is correct. Of course, R is a constant whose value cannot change.

SOLUTION 2-33

(5)

1 If the two terms are equal, $PV/RT = 0$, which is not true at 40 atm.
2 At $P = 0$ atm you might think that T must be 0°K, so that PV/RT can be 1. Remember that division by zero is not accepted practice. We can extrapolate our measurements of PV/RT to low pressure by using large volumes at reasonable temperatures, in this case 80°K, as specified. As P goes to 0, V must become large, and $V/(V - b)$ approaches 1. Also, a/VRT approaches 0, and so (by our equation) PV/RT approaches 1. This is our ideal gas condition for 1 mole and indicates that real gases approach ideality at low pressure.
3 At 80 atm, $PV/RT = 1$, and the two terms cannot be equal. See Answer **1**.
4 Not necessarily. If it did condense, we would see a precipitous decrease in PV/RT as the volume decreases from gas to liquid or solid.
5 Correct. At $P > 80$ atm, $PV/RT > 1$ and $V/(V - b) > a/VRT$.

SOLUTION 2–34

(5)

All of the statements, except Answer **2**, are correct. But only one of these four is proved by *this* experiment; that statement is **5**: Molecules travel in straight lines. If they did not, there would not be a sharp spot of condensed beam. The molecules could have curved in flight to deposit on the entire surface of the end bulb.

No determination of velocity distribution was made, so this experiment says nothing about Answer **3**. Similarly, Answer **4** has nothing to do with breaking molecules into atoms. Consider Answer **1**. If anything, this experiment supports one of the main concepts of the kinetic theory that agrees with the ideal gas law: Molecules travel in linear trajectories between collisions. As for Answer **2**, each molecule preserves its weight, whether found as a gas or as a liquid.

SOLUTION 2–35

(3)

This experiment is analogous to a horse race. The "gate" opens and the molecules run out. The important principle is the distribution of speeds in the gas. The gas molecules (horses) are not equal: Some are speedy and others are ready for the glue factory. If the track (the disk-to-detector distance) is long enough, there will be a difference in the time that the molecules arrive at the detector, with the fast molecules arriving there almost immediately. (Remember that a Maxwell–Boltzmann distribution contains a small but finite number of *very* fast molecules.) Only Answer **3** has such a wide distribution of speeds.

We can ask if the numbers we have been given are reasonable; that is, is the track long enough? The transit time, τ, for the molecules to cover the disk-to-detector distance, d, is

$$\tau = d/v$$

and for molecules with v_{rms} speed

$$\tau = \frac{100 \text{ cm}}{5 \times 10^4 \text{ cm sec}^{-1}} = 2 \times 10^{-3} \text{ sec}$$

The ratio of this time to the burst time is

$$\frac{2 \times 10^{-3} \text{ sec}}{5 \times 10^{-5} \text{ sec}} = 0.4 \times 10^2 = 40$$

So the length of time spent on the track by molecules with speeds of v_{rms} is 40 times the burst time. And there is more than enough time to effect the separation.

CHAPTER 3 MATTER WITH A CHARGE

The electron is the fundamental constituent of matter that is responsible for the chemical behavior of atoms and molecules. When a neutral atom or molecule loses an electron a cation is formed; when an electron is gained an anion is formed. The problems in this chapter deal with the nature of ions in solutions, gases, and solids.

To answer the questions in this chapter, you will need to memorize the following relationships:

1 coulomb = 1 A sec (A = ampere)
e^- = electron
1 faraday ($\mathscr{F}$) = $N \times$ charge on one electron
1 angstrom (Å) = 10^{-8} cm

Solutions to problems in Chapter 3 begin on page 76

PROBLEMS

PROBLEM 3-1

The same quantity of electricity was passed through two separate electrolytic cells containing solutions of copper sulfate ($CuSO_4$) and silver nitrate ($AgNO_3$), respectively. The reactions at the cathodes in the two cells are

First cell: $Cu^{2+} + 2e^- \rightarrow Cu(s)$
Second cell: $Ag^+ + e^- \rightarrow Ag(s)$

If 3.18 g of Cu were deposited in the first cell, how many grams of Ag were deposited in the second cell?

(at. wt: Cu = 63.6, Ag = 108)

1 0.936 g
2 2.70 g
3 3.75 g
4 5.40 g
5 10.8 g

PROBLEM 3-2

Faraday's laws of electrolysis are discussed in Section 3–1 of Chemical Principles *by Dickerson, Gray, and Haight.*

The same quantity of electricity that caused the deposition of 10.0 g of silver from a $AgNO_3$ solution was passed through a solution of a gold salt containing gold cations of unknown charge (Au^{x+}). An amount of 6.08 g of gold was deposited. What is the charge on the gold cations? [*Note:* The cathode reaction in the $AgNO_3$ cell is $Ag^+ + e^- \rightarrow Ag(s)$.]

(at. wt: Ag = 108, Au = 197)

1 1−
2 1+
3 2+
4 3+
5 4+

PROBLEM 3-3

Electrogalvanizing is the process whereby iron or steel objects such as garbage cans and siding for buildings are coated electrolytically with a thin layer of zinc metal to protect them against rusting. In a typical case, 3.27 g of zinc metal is deposited on a small steel garbage can when a current of 5.36 A is passed for 30 min through an electrolytic cell containing a solution of a zinc salt. What is the combining weight of zinc in the zinc salt?

(1 $\mathscr{F}$ = 96,500 coulombs;
at. wt: Zn = 65.4)

1 0.609 g
2 17.5 g
3 32.7 g
4 1960 g
5 Insufficient data provided; the cell voltage must be given.

PROBLEM 3–4

The most commonly used dry cell nowadays is the flashlight battery, or the zinc and manganese dioxide (MnO_2) dry cell (Figure 3–1). Since it cannot be recharged, a flashlight battery cannot be used after its chemicals are exhausted. The cathode reaction in a flashlight battery can be written as

$$2MnO_2(s) + Zn^{2+} + 2e^- \rightarrow ZnMn_2O_4(s)$$

If the cathode in a typical flashlight battery contains 4.35 g MnO_2, how long could such a battery deliver a current of 2.0 milliamperes (mA) before all its chemicals are exhausted?

(mol wt: $MnO_2 = 87$, $ZnMn_2O_4 = 239$;
$1\ \mathcal{F} = 96{,}500$ coulombs;
$1\ A = 10^3\ mA$)

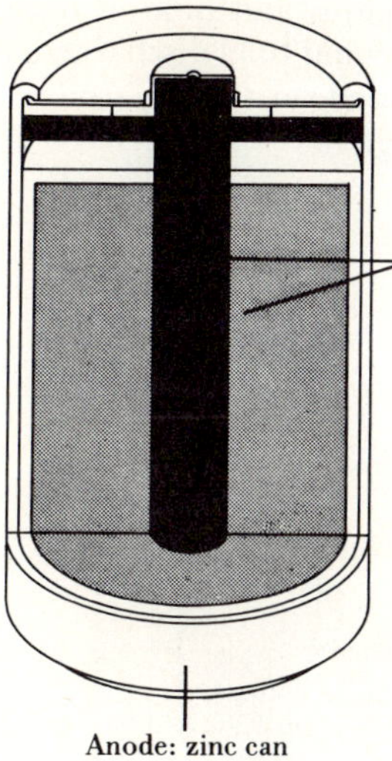

Figure 3–1

1 25 sec
2 2.4×10^3 sec
3 8.7×10^5 sec
4 2.4×10^6 sec
5 4.8×10^6 sec

PROBLEM 3–5

Electrical cells and batteries are examined in Section 17–2 of Chemical Principles.

The lead–acid automobile battery is an example of a storage battery. A storage battery, unlike the flashlight battery described in Problem 3–4, can be reused many times because it can be recharged electrically. The lead–acid automobile battery usually consists of six cells connected in series. The current that flows out of the first cell passes into the second, and so on, until it has passed through all six cells. One of these cells is shown in Figure 3–2.

When the battery is discharging current to the various electrical devices in the automobile, the following reactions occur in the anodes and cathodes, respectively:

Anode reaction

$$Pb(s) + HSO_4^- \rightarrow PbSO_4(s) + H^+ + 2e^-$$

Cathode reaction

$$PbO_2(s) + HSO_4^- + 3H^+ + 2e^- \rightarrow PbSO_4(s) + 2H_2O$$

Lead sulfate is formed at *all* the electrodes and actually adheres to them. The battery is completely discharged when the lead sulfate covering the electrodes is so thick that the chemical reactions cease.

Typical automobile batteries consisting of six cells are rated at "100 ampere hours," which means that they have the chemical capacity to deliver a current of 100 A for 1.00 h, 1.00 A for 100 h, or any other combination of current in amperes and time in hours such that the product equals 100 before becoming fully discharged. How many kilograms of $PbSO_4$ will be deposited on the electrodes of one of these batteries when it is fully discharged?

(mol wt: $PbSO_4 = 303$;
$1\ \mathcal{F} = 96{,}500$ coulombs)

Figure 3–2

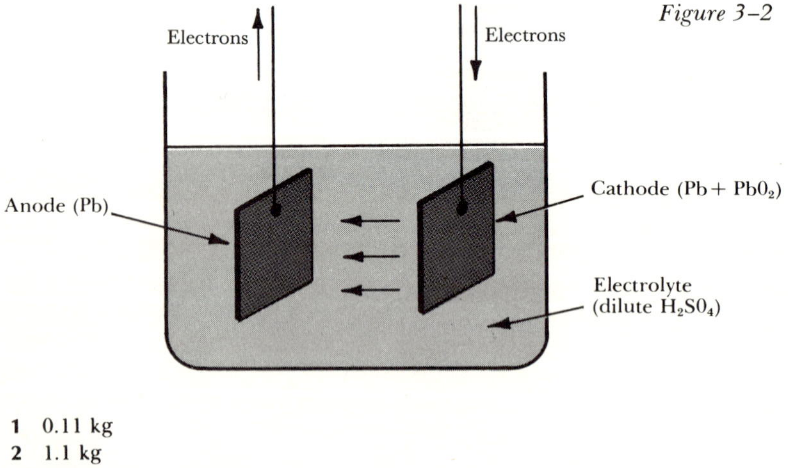

1 0.11 kg
2 1.1 kg
3 3.4 kg
4 6.8 kg
5 14 kg

PROBLEM 3–6

Figure 3–3 illustrates the silver-plating of a table fork. The fork and the silver rod are used as electrodes, and a direct current from a battery is passed through the silver cyanide (AgCN) solution. The fork acts as the cathode, whereas the silver rod acts as the anode. When the current is passed, Ag^+ ions in the electrolyte are attracted toward the fork, where they are deposited as silver atoms. The Ag^+ ions that are lost from the electrolyte are replaced by Ag^+ ions produced from the silver anode by the reaction $Ag(s) \rightarrow Ag^+ + e^-$.

If a current of 1.0 A is passed for 1.0 h through a silver-plating system such as the one described above, how thick a coating will be deposited on a table fork of surface area 30 cm^2? The density of silver is about 10 g cm^{-3}. Assume a current efficiency of 89% (i.e., only 89% of the available current leads to silver production).

(at wt: Ag = 108)

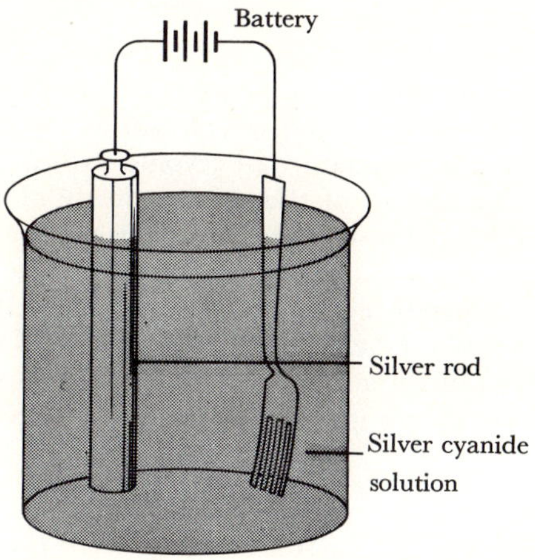

Figure 3–3

1 3.3×10^{-6} cm
2 2.0×10^{-4} cm
3 0.012 cm
4 0.013 cm
5 Insufficient data; the concentration of the silver cyanide solution is needed for the calculation.

PROBLEM 3-7

When a six-cell lead–acid storage battery is used to start an automobile, the battery delivers a certain current for 1.20 sec. Under these typical conditions, the total weight of lead from the anodes converted into lead sulfate is exactly 1.242 g. If tne anode reaction is

$$Pb(s) + HSO_4^- \rightarrow PbSO_4(s) + H^+ + 2e^-$$

what current does the battery deliver?

(at. wt: Pb = 207;
mol wt: $PbSO_4 = 303$;
1 $\mathcal{F}$ = 96,500 coulombs)

1 2.68 A
2 80.5 A
3 110 A
4 161 A
5 966 A

PROBLEM 3-8

All commercial aluminum is made by the Hall–Héroult electrolytic process. A carbon-lined iron pot, which acts as the cathode, contains the electrolyte—a molten mixture of cryolite (Na_3AlF_6) and aluminum oxide (Al_2O_3). The anodes of the cell are made of carbon. The cell operates at a 5.00-V potential difference between the electrodes.

One of the largest aluminum plants in the world is at Arvida, Quebec, in Canada. It can produce 3.40×10^8 kg (3.75×10^5 short tons) of aluminum a year. If it takes 20.0 kilowatt hours (20.0 kW for 1.00 h) to produce 1.00 kg of aluminum, how many coulombs of electricity are used in one year of aluminum production by using cells of the type described above?

(1 kW = 10^3 W;
1 W = 1 coulomb V sec^{-1} = 1 A sec^{-1})

1 1.36×10^{12} coulombs
2 2.45×10^{13} coulombs
3 8.17×10^{13} coulombs
4 4.90×10^{15} coulombs
5 1.23×10^{17} coulombs

Now that you can solve problems involving ions in solution, we will go on to study ions in the gas phase. As you will see, we will be particularly interested in the m/e ratios of gaseous ions, that is, the ratios of their masses to their charges. These ratios are of extreme importance in the identification of chemical compounds.

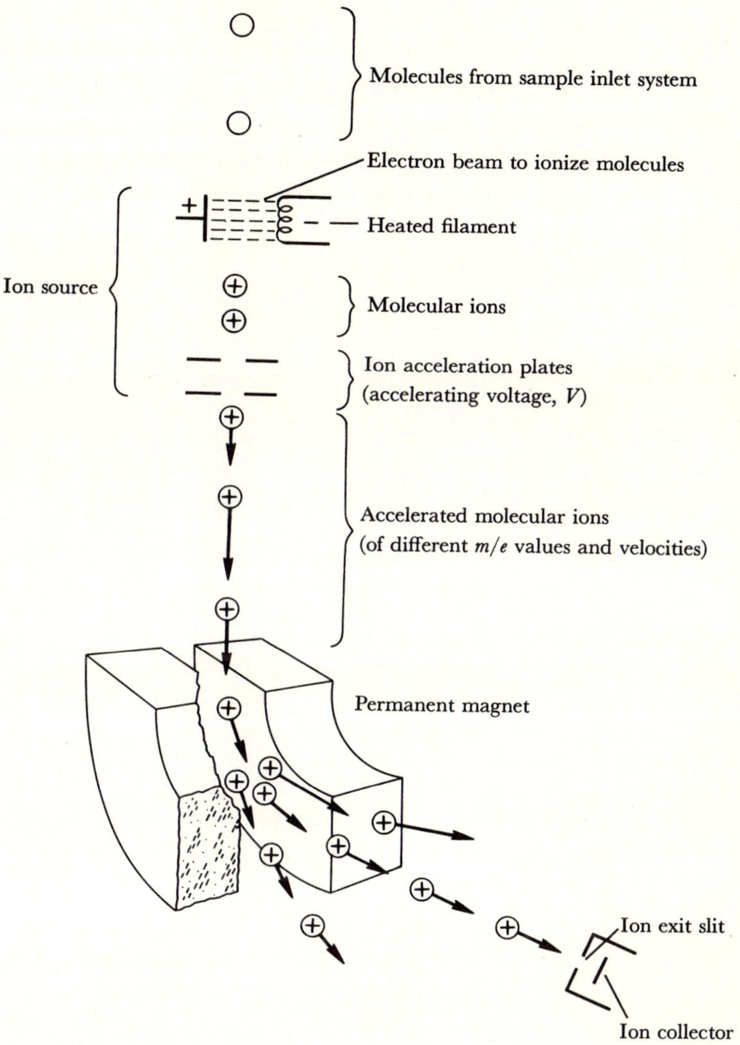

Figure 3–4. The mass spectrometer. Arrows of different lengths correspond to ions of different m/e values. Ions of large m/e (short arrows) are actually traveling slower than ions of small m/e (long arrows).

Figure 3–4 illustrates the mass spectrometer. This is an instrument that measures the m/e ratios of gaseous ions. Let us assume that we have managed to introduce a gaseous sample of a compound into the high vacuum system ($P \simeq 10^{-6}$ torr) of the spectrometer. In the ion source, the vaporized sample molecules are bombarded with a beam of electrons from a heated metal filament. This *electron bombardment results in the ionization of the molecules. A bombarding electron knocks an electron off the molecule.*

$$M + e^- \rightarrow M^+ + 2e^-$$

For a fraction of the molecules, some of the chemical bonds are broken. This results in the formation of positive fragment ions of lower m/e ratios than that of the parent molecular ion, M^+. For instance,

$$BF_3 + e^- \rightarrow (BF_3)^+ + 2e^-$$

parent molecular ion

$$BF_3 + e^- \rightarrow (BF_2)^+ + F + 2e^-$$

fragment molecular ion

These various positive ions produced in the ion source are accelerated by an electric voltage, V, and passed into the analyzer part of the mass spectrometer where they are separated according to their m/e values. We achieve this separation by deflecting the ions in a strong magnetic field. The equation governing their deflection is

$$\frac{m}{e} = \frac{H^2 r^2}{2\,V}$$

in which H is the strength of the magnetic field, r is the radius of the circular path followed by an ion in the field, and V is the accelerating electrical voltage. For a fixed H and V, the ions of larger mass/charge ratios are deflected less (larger r) from their line of flight than the ions of smaller mass/charge ratios; thus, only ions with a specific m/e value will strike the ion collector. However, we want to know the relative amounts of different species (different m/e values) present. So we keep H constant by using a permanent magnet and vary V (r is fixed because the collector cannot be moved). This variation in V results in the detection of different species (m/e) at the ion collector. In other words, we determine the different species present by scanning V and find (in most instances) the composition of our sample.

Problems 3–9 to 3–14 illustrate the importance of mass spectrometry in the determination of the relative abundance of the isotopes of an element and in the identification of chemical compounds. In these questions, assume that all species are singly ionized and that ionization efficiencies are equal.

PROBLEM 3-9 The mass spectrum of chlorine gas (Cl_2) showing only the parent molecular ions is given in Figure 3-5. Peaks are observed at mass numbers (that is, *m/e* values) of 70, 72, and 74. What are the naturally occurring isotopes of chlorine?

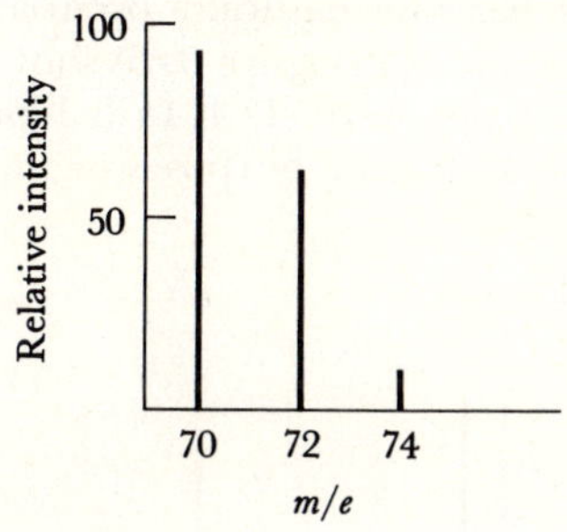

Figure 3-5

1 ^{35}Cl
2 ^{35}Cl and ^{36}Cl
3 ^{35}Cl and ^{37}Cl
4 ^{36}Cl and ^{37}Cl
5 ^{35}Cl, ^{36}Cl, and ^{37}Cl

PROBLEM 3-10 In Problem 3-9 you found that the naturally occurring isotopes of chlorine are ^{35}Cl and ^{37}Cl. If the heights of the peaks in Figure 3-5 are in the ratio 9/6/1, what is the relative abundance of ^{35}Cl and ^{37}Cl; that is, $^{35}Cl/^{37}Cl$ = ?

1 1/1
2 1/3
3 3/1
4 6/1
5 9/1

PROBLEM 3-11 Phosphorus trichloride (PCl_3) is used industrially in the manufacture of saccharin, a compound about 500 times sweeter than cane sugar but of no food value. One of the best ways of checking the purity of PCl_3 is to compare the mass spectrum of a sample with the mass spectrum of pure PCl_3. If chlorine has two naturally occurring isotopes, ^{35}Cl and ^{37}Cl, and phosphorus has one naturally occurring isotope, ^{31}P, which of the following mass spectra (Figure 3-6) showing only parent molecular ions would you expect pure PCl_3 to exhibit? (Neglect the relative intensities of the peaks for the purposes of this problem.)

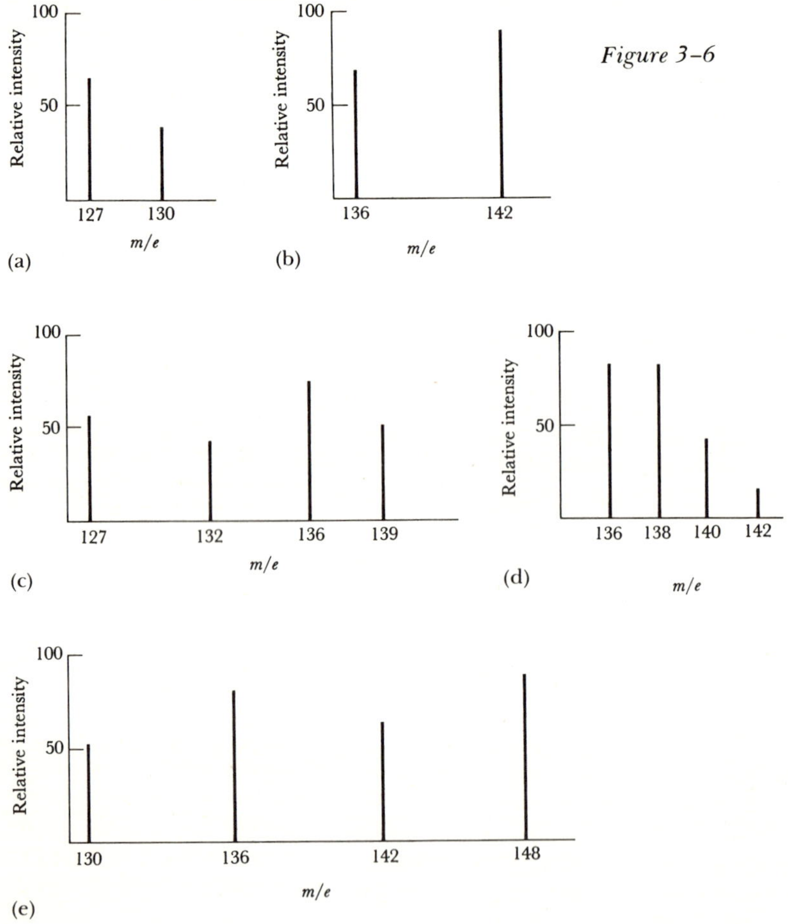

Figure 3-6

1 (a)
2 (b)
3 (c)
4 (d)
5 (e)

PROBLEM 3–12 A research chemist isolated a sample of a colorless, fuming liquid from a chemical reaction and recorded its mass spectrum. On the basis of his mass spectral data, the chemist realized that he had isolated some boron tribromide (BBr_3). If boron has two isotopes, ^{10}B and ^{11}B, and bromine has two isotopes, ^{79}Br and ^{81}Br, which of the following mass spectra (Figure 3–7) showing only parent molecular ions do you think corresponds most closely to the one obtained by the research chemist? (Neglect the relative intensities of the peaks for the purposes of this question.)

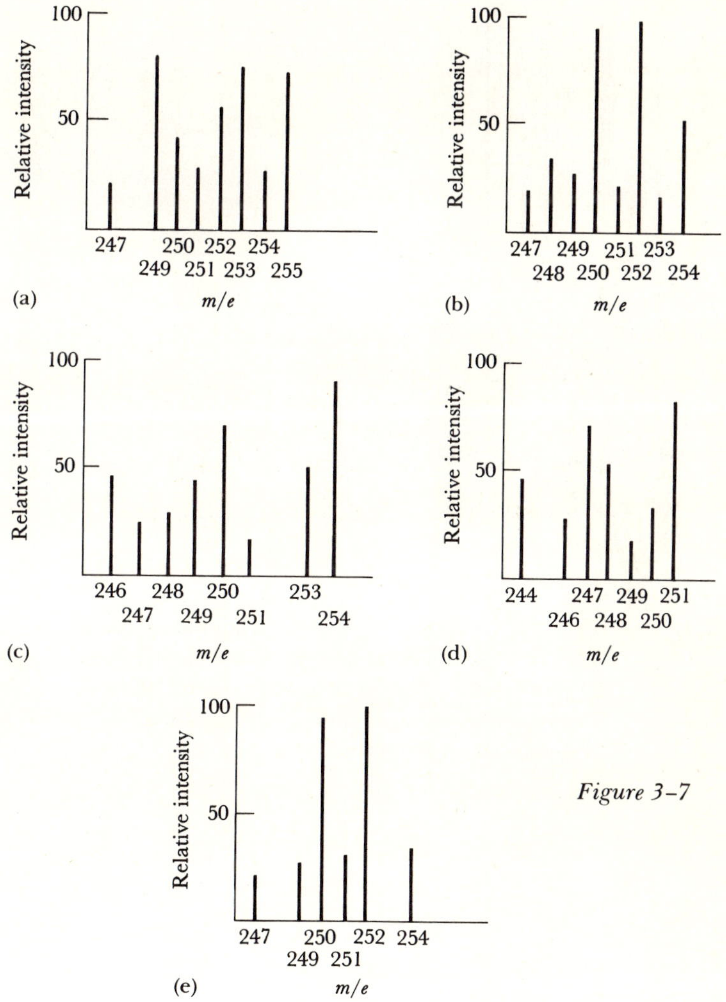

Figure 3–7

1 (a)
2 (b)
3 (c)
4 (d)
5 (e)

PROBLEM 3–13

You have been asked to identify a compound from the record of its mass spectrum between mass numbers 14 and 35 shown in Figure 3–8. Which of the five following compounds is it? (For the purposes of this question assume that the only isotopes of the elements concerned are ^{1}H, ^{12}C, ^{14}N, ^{16}O, ^{19}F, and ^{31}P and that no peaks occur at mass numbers greater than 35.)

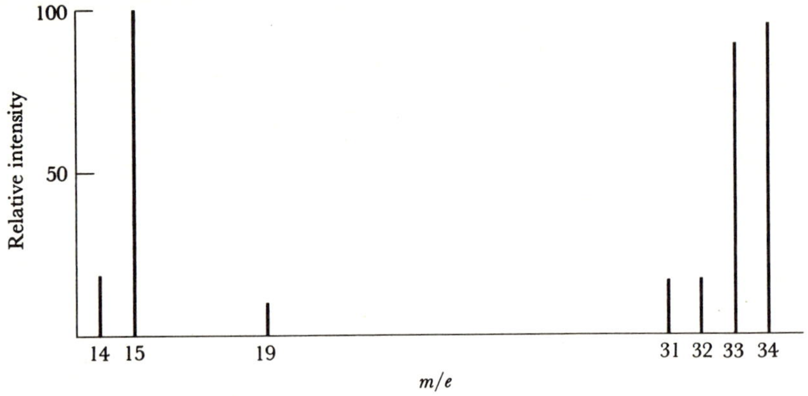

Figure 3–8

1 NH_3
2 CH_4
3 PH_3
4 H_2O_2
5 CH_3F

PROBLEM 3–14

Which of the five compounds listed below would you expect to exhibit the mass spectrum in Figure 3–9? No data were recorded below *m/e* 12 (ignore the peak at *m/e* 22 for now). The compound is commonly found in the exhaust gases of an automobile. (Assume that the only isotopes of the elements concerned are ^{1}H, ^{12}C, ^{14}N, and ^{16}O.)

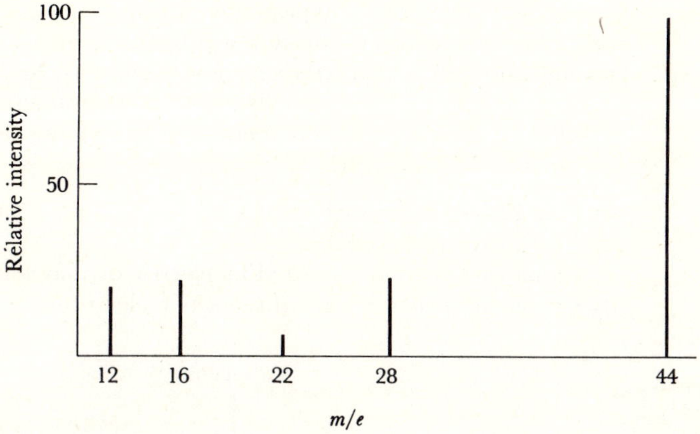

Figure 3–9

1 HCN
2 CO
3 NO
4 N_2O
5 CO_2

We will end this section on ions in the gas phase with one problem concerning Millikan's determination of the charge on an electron and one problem involving gas discharge tubes.

PROBLEM 3–15

Millikan's classical oil-drop experiment is described in Section 3–6 of Chemical Principles.

Some examples of the charges that Millikan measured in his "oil-drop experiment" are

4.32×10^{-9} esu
3.84×10^{-9} esu
2.88×10^{-9} esu
1.44×10^{-9} esu
9.60×10^{-10} esu

Using these data alone, calculate your best estimate of how many electrons were attached to an oil drop for which Millikan measured a charge of 5.76×10^{-9} esu? (Do not worry if you do not know what an esu of charge is. It is in a different system of units from the coulomb, but it is related to it by 1 coulomb $= 3.00 \times 10^9$ esu.)

1 1
2 2
3 4
4 6
5 12

PROBLEM 3–16

Figure 3–10 illustrates a typical gas discharge tube. The tube is filled with a gas (or metallic vapor) at low pressure. When a current is passed through the tube a characteristic glow is emitted. Neon signs used for advertising purposes are actually gas discharge tubes.

Which of the following statements concerning the gas discharge tube in Figure 3–10 is correct?

(a) Canal rays are positive ions.
(b) Canal rays are electrons.
(c) Canal rays are not deflected by electric or magnetic fields.
(d) Canal rays travel toward the positive electrode.

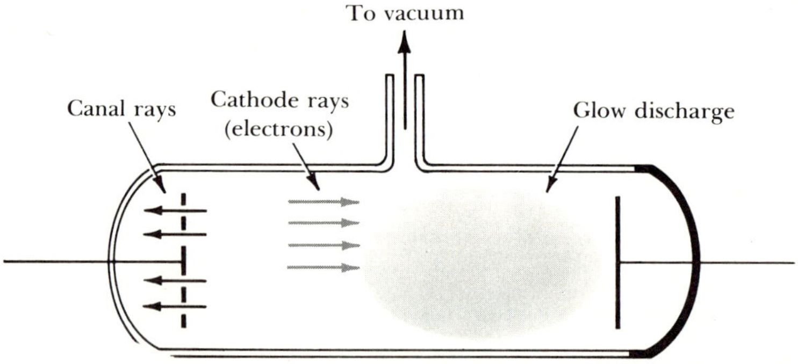

Figure 3–10

1 Only (a).
2 Only (b).
3 Only (a) and (d).
4 Only (b), (c), and (d).
5 None of them.

Thus far we have considered ions in solution and in the gas phase. Now we will turn our attention to the occurrence of ions in solids.

PROBLEM 3–17

Many metals crystallize with body-centered cubic structures like the one in Figure 3–11. Nine atoms (represented by hard spheres) are packed together. One of the atoms lies at the center of a cube, and the eight others are arranged around it so their centers are at the corners of the cube. (This cube is the smallest geometric unit that is repeated continually in all directions throughout the structure. It is referred to as the *unit cell.*) How many atoms are there in a body-centered cubic unit cell?

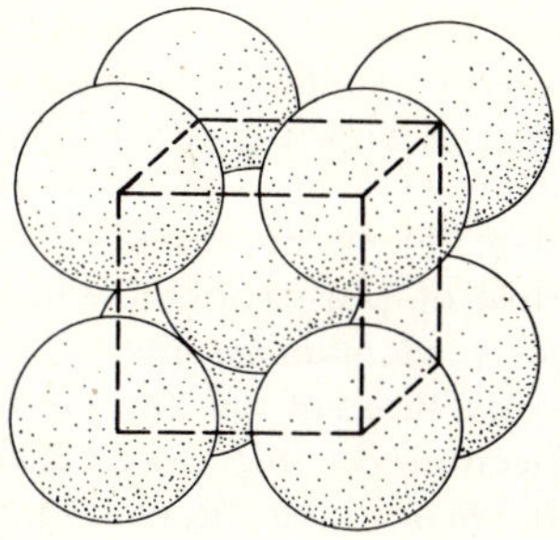

Figure 3–11

1 1
2 2
3 4
4 8
5 9

PROBLEM 3–18

Part of Surveyor 5's mission to the moon was to analyze rock samples from the Sea of Tranquility. Some of the rocks were titanium ores containing about 4.5% titanium by weight. Alloys of titanium are used principally for aircraft and missiles, in which light weight, strength, and ability to withstand extremes of temperature are vital. Pure titanium metal exists in two distinct crystalline forms, one of which adopts a body-centered cubic structure (Figure 3–11). If the distance between the centers of neighboring titanium atoms is $2\sqrt{2}$ Å, what is the density of this form of titanium metal? [*Note:* The number of Ti atoms per unit cell is 2. Take $\sqrt{\frac{2}{3}} \simeq 0.8$.]

(at. wt: Ti = 48)

1 $\simeq 0.1$ g cm^{-3}
2 $\simeq 1$ g cm^{-3}
3 $\simeq 5$ g cm^{-3}
4 $\simeq 23$ g cm^{-3}
5 $\simeq 30$ g cm^{-3}

PROBLEM 3–19

Several types of alloys are discussed in Section 3–7 of Chemical Principles.

Brass is the name given to the alloys of copper and zinc. Pure copper metal and pure zinc metal both crystallize in face-centered cubic, close packed structures such as that in Figure 3–12. Provided that the zinc content by weight is kept below 33%, brass also adopts a face-centered cubic, close packed structure. The zinc atoms replace some of the copper atoms in the structure of pure copper. The brass formed in this manner is an example of a *substitutional alloy.*

Suppose that you have such a brass of low zinc content. Its structure can be considered as a face-centered cubic, close packed lattice of copper atoms in which the eight corner atoms in every unit cell have been replaced by zinc atoms. What is the mass of the unit cell of this brass?

(at wt: Cu = 63.6, Zn = 65.4;
$N = 6.0 \times 10^{23}$ atoms g-atom^{-1})

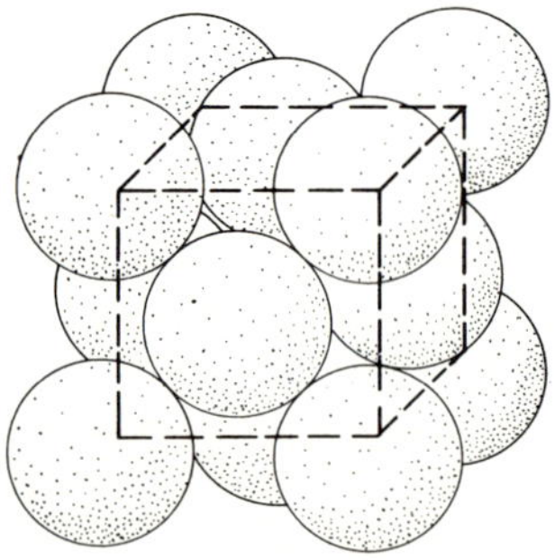

Figure 3–12

1 1.1×10^{-22} g (unit cell)$^{-1}$
2 3.2×10^{-22} g (unit cell)$^{-1}$
3 4.3×10^{-22} g (unit cell)$^{-1}$
4 6.4×10^{-22} g (unit cell)$^{-1}$
5 1.5×10^{-21} g (unit cell)$^{-1}$

PROBLEM 3–20

Using the results of Problem 3–19, what is the percent zinc content by weight of the brass?

(at. wt: Cu = 63.6, Zn = 65.4)

1 1.6%
2 2.6%
3 7.4%
4 26%
5 74%

PROBLEM 3–21

Which of the following sets of coordination numbers (C.N.) do the cation and the anion possess in the ionic crystal structure shown in Figure 3–13?

The crystal structure of ionic solids is discussed in Section 3–7 of Chemical Principles.

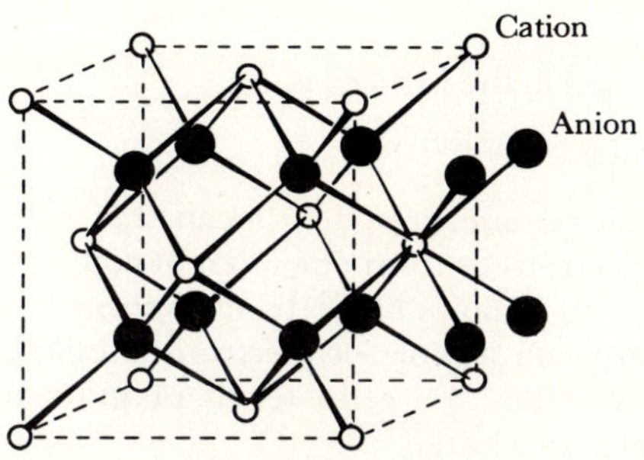

Figure 3–13

1 cation C.N. 3; anion C.N. 6
2 cation C.N. 4; anion C.N. 4
3 cation C.N. 6; anion C.N. 6
4 cation C.N. 8; anion C.N. 4
5 cation C.N. 8; anion C.N. 8

PROBLEM 3–22

Strontium oxide (SrO) crystallizes in a cubic structure and has a density of 5.00 g cm^{-3}. If the side of the unit cell is 5.20 Å, how many Sr^{2+} and O^{2-} ions are there in the unit cell?

(mol wt: SrO = 104)

1 1 Sr^{2+} and 1 O^{2-}
2 1 Sr^{2+} and 2 O^{2-}
3 4 Sr^{2+} and 4 O^{2-}
4 20 Sr^{2+} and 20 O^{2-}
5 Insufficient data; need to know if the structure is face-centered or body-centered.

SOLUTION 3-1

(5)

If you have difficulty solving problems involving electrolysis, refer to Review 3 of Programed Reviews of Chemical Principles *by Lassila* et al.

The cathode reaction in the first cell is

$$Cu^{2+} + 2e^- \rightarrow Cu(s)$$

This equation can be interpreted to mean that either one Cu^{2+} ion reacts with two electrons to form one atom of copper or that 1 mole of Cu^{2+} ions reacts with 2 moles of electrons to form 1 g-atom of copper. [*Recall:* The charge on a mole of electrons is called a faraday ($\mathscr{F}$).] So the cathode reaction above, in terms of gram-atoms of copper, involves 2 $\mathscr{F}$ of electric charge.

By definition, 1 $\mathscr{F}$ will deposit the *combining weight* of any element. Therefore, the combining weight of Cu is equal to the weight of one half of a gram-atom of Cu; that is, one half of its atomic weight, namely, $\frac{1}{2} \times 63.6 = 31.8$ g.

The cathode reaction in the second cell is

$$Ag^+ + e^- \rightarrow Ag(s)$$

From this, the combining weight of Ag is equal to its atomic weight, that is, 108 g.

Since 3.18 g of Cu were deposited, and this is one tenth of a combining weight, 0.100 $\mathscr{F}$ of electric charge was involved. This quantity of charge will deposit 0.100×108 g $= 10.8$ g of Ag. The correct answer is **5**.

1 You thought that the weights of different elements deposited by the same quantity of electricity are *inversely* proportional to their combining weights.
2 You divided the atomic weight of Ag by two instead of that of Cu.
3 You reversed the atomic weights of Cu and Ag.
4 You forgot to divide atomic weight of Cu by two.

SOLUTION 3-2

(4)

The cathode reaction in the $AgNO_3$ cell is

$$Ag^+ + e^- \rightarrow Ag(s)$$

Therefore, the combining weight of silver is its atomic weight, namely, 108 g. From Faraday's second law of electrolysis

$$\frac{\text{wt of silver deposited}}{\text{wt of gold deposited}} = \frac{\text{combining wt of silver}}{\text{combining wt of gold}}$$

The combining weight of gold in the gold salt is therefore

$$\frac{6.08 \times 108}{10.0} = 65.7 \text{ g}$$

The cathode reaction involving the gold cations of charge x^+ is

$$Au^{x^+} + xe^- \rightarrow Au(s)$$

Hence

$$x = \frac{\text{at. wt of gold}}{\text{combining wt of gold}} = \frac{197}{65.7} = 3$$

An alternative solution to this problem is

g-atoms of Ag deposited = 10.0/108 = 0.093 g-atom
g-atoms of Au deposited = 6.08/197 = 0.031 g-atom

The gram-atoms of Au deposited equals one third of g-atoms of Ag deposited, so the charge on Au equals three times the charge on Ag^+. The species is Au^{3+}.

The correct answer is **4**. Answer **1** is impossible since you were told that the solution contained gold cations.

SOLUTION 3–3

(3)

You were asked to memorize that

1 coulomb = 1 A sec

This means that the number of coulombs of electric charge passed through an electrolyte is equal to the product of the current in amperes and the time in seconds. A current of 5.36 A for 30 min is

$$5.36 \text{ A} \times 30 \text{ min} \times 60 \text{ sec min}^{-1} = 9{,}650 \text{ coulombs}$$

If we convert to faradays,

$$9{,}650 \text{ coulombs} = \frac{9{,}650 \text{ coulombs}}{96{,}500 \text{ coulombs } \mathscr{F}^{-1}} = 0.100\ \mathscr{F}$$

The combining weight of zinc is the weight deposited by 1 $\mathscr{F}$. Since 3.27 g of zinc are deposited by 0.100 $\mathscr{F}$, the combining weight of zinc is 32.7 g. [*Note:* Nowadays, most galvanizing is performed by dipping objects into molten zinc metal. This method is more expensive than electrogalvanizing, but it is preferred because it produces a thicker and more durable zinc coating.]

3 is correct.
1 You thought that the combining weight is the weight deposited by 1 A.
2 You multiplied 3.27 g by 5.36 A!
4 You forgot to convert the 30 min to seconds. You should have realized your error because a combining weight of 1960 g is impossible. Combining weights for atomic species are equal either to atomic weights or to some fraction of them. Atomic weights go up to only about 250.

SOLUTION 3-4 (4)

The battery is exhausted when all the MnO_2 in the cathode has reacted to form $ZnMn_2O_4$. The cathode reaction indicates that 2 $\mathscr{F}$ are transferred each time 2 moles of MnO_2 are consumed. The cathode contains 4.35 g MnO_2.

$$4.35 \text{ g } MnO_2 = 4.35 \text{ g}/87 \text{ g mole}^{-1} = 0.050 \text{ mole } MnO_2$$

Therefore, the battery will be exhausted when 0.050 mole MnO_2 has been used up, and 0.050 $\mathscr{F}$ will have been transferred.

$$0.050\ \mathscr{F} = 0.050\ \mathscr{F} \times 96{,}500 \text{ coulombs } \mathscr{F}^{-1}$$
$$\simeq 4.8 \times 10^3 \text{ coulombs}$$

Throughout its lifetime, the battery is supposed to deliver a current of 2.0 mA.

$$2.0 \text{ mA} = 2.0 \times 10^{-3} \text{ A} = 2.0 \times 10^{-3} \text{ coulomb sec}^{-1}$$

The time in which the battery could deliver this current is

$$\frac{4.8 \times 10^3 \text{ coulombs}}{2.0 \times 10^{-3} \text{ coulomb sec}^{-1}} = 2.4 \times 10^6 \text{ sec } (\simeq 30 \text{ days})$$

The correct answer is **4**.

1 You did not convert the number of faradays passed through the battery to coulombs and completely disregarded your units.
2 You forgot to convert the 2.0-mA current to amperes.
3 You used the molecular weight of $ZnMn_2O_4$ by mistake. The molecular weight of $ZnMn_2O_4$ was not needed for the calculation.
5 You reasoned incorrectly that 1 mole MnO_2 is consumed for every 2 $\mathscr{F}$ passed through the battery.

SOLUTION 3-5 (4)

First we have to find the number of faradays involved when our "100 ampere hour" battery delivers, for example, a current of 1.00 A for 100 h. This current is

$$1.00 \text{ A} \times 100 \text{ h} \times 60.0 \text{ min h}^{-1} \times 60.0 \text{ sec min}^{-1} = 3.60 \times 10^5 \text{ coulombs}$$

$$= \frac{3.60 \times 10^5 \text{ coulombs}}{9.65 \times 10^4 \text{ coulombs } \mathscr{F}^{-1}}$$

$$= 3.73\ \mathscr{F}$$

From the two electrode reactions given, we can see that for every 2 $\mathscr{F}$ passed through the battery, 2 moles of $PbSO_4$ will be deposited (1 mole at the anode and 1 mole at the cathode) in each of the six cells. Therefore, a total of 12 moles of $PbSO_4$ will be deposited for every 2 $\mathscr{F}$ passed through the battery. So the number of moles of $PbSO_4$ deposited by 3.73 $\mathscr{F}$ is

$$3.73\ \mathscr{F} \times \frac{12 \text{ moles}}{2\ \mathscr{F}} = 22.4 \text{ moles } PbSO_4$$

Weight of $PbSO_4$ deposited when the battery is fully discharged is

$$22.4 \text{ moles} \times 303 \text{ g mole}^{-1} = 6.79 \times 10^3 \text{ g} \simeq 6.8 \text{ kg } PbSO_4$$

The answer is **4**.

1 You converted the 100 h to minutes instead of seconds.
2 You forgot that there were six cells in the battery and did the calculation for only one cell.
3 You forgot that $PbSO_4$ was deposited at the cathodes *and* the anodes.
5 You did not realize that $2\mathscr{F}$ resulted in the deposition of 2 moles of $PbSO_4$.

SOLUTION 3–6

(3)

The current used for the silver plating is 1.0 A, but the current efficiency is only 89%. This means that only 89% of the available current is actually used for the silver plating. (The other 11% of the current is used in electrolyzing some of the water in the silver cyanide solution.) Our useful current is therefore 0.89 A. For 1.0 h this current is

$$0.89 \text{ A} \times 3600 \text{ sec} = 3.2 \times 10^3 \text{ coulombs}$$

$$= \frac{3.2 \times 10^3 \text{ coulombs}}{9.65 \times 10^4 \text{ coulombs } \mathscr{F}^{-1}}$$

$$\simeq 0.033\ \mathscr{F}$$

The cathode reaction is

$$Ag^+ + e^- \rightarrow Ag(s)$$

The number of gram-atoms of silver deposited as a result of the consumption of 0.033 $\mathscr{F}$ is therefore

$$0.033\ \mathscr{F} \times 1 \text{ g-atom } \mathscr{F}^{-1} = 0.033 \text{ g-atom Ag}$$

The weight of silver deposited ón the table fork is

$$0.033 \text{ g-atom} \times 108 \text{ g g-atom}^{-1} = 3.6 \text{ g Ag}$$

The density of silver is 10 g cm^{-3}, and so the volume of silver deposited is

$$3.6 \text{ g}/10 \text{ g cm}^{-3} = 0.36 \text{ cm}^3$$

Since the surface area of the fork is 30 cm^2, the thickness of the silver deposited is

$$0.36 \text{ cm}^3/30 \text{ cm}^2 = 0.012 \text{ cm (or 0.0047 in.)}$$

3 is correct.
1 You forgot to convert the time that the current was passed to seconds.
2 You only converted the time to minutes.
4 You forgot to allow for the current efficiency of the silver-plating process.

SOLUTION 3-7

(4)

You can find background information for Problems 3-4 to 3-7 in Sections 3-1 to 3-5 of Chemical Principles.

The amount of lead converted to $PbSO_4$ at the six anodes is 1.242 g; the amount of lead converted at one anode is 1.242/6 = 0.207 g.

$$0.207 \text{ g Pb} = 0.207 \text{ g}/207 \text{ g g-atom}^{-1} = 1.00 \times 10^{-3} \text{ g-atom Pb}$$

From the anode reaction, 1 g-atom of Pb is converted into $PbSO_4$ for every 2 $\mathcal{F}$ passed. The number of faradays required to convert 1.00×10^{-3} g-atom Pb into $PbSO_4$ is

$$1.00 \times 10^{-3} \text{ g-atom} \times 2\ \mathcal{F} \text{ g-atom}^{-1} = 2.00 \times 10^{-3}\ \mathcal{F}$$

The number of coulombs passed is

$$2.00 \times 10^{-3}\ \mathcal{F} \times 9.65 \times 10^{4} \text{ coulombs } \mathcal{F}^{-1} = 193 \text{ coulombs}$$

Since the current was passed for 1.20 sec, its value is

$$\begin{aligned} 193 \text{ coulombs}/1.20 \text{ sec} &= 161 \text{ coulombs sec}^{-1} \\ &= 161 \text{ A} \end{aligned}$$

The answer is **4**. In Answer **1**, you read 1.20 sec as 1.20 min. For Answer **2**, you thought that the 1 g-atom of Pb was converted to $PbSO_4$ by 1 $\mathcal{F}$ instead of 2 $\mathcal{F}$. Answer **3**: You used the molecular weight of $PbSO_4$ in your calculation instead of the atomic weight of Pb. For Answer **5**, you forgot that there were six cells in the battery and did the calculation for only one cell.

SOLUTION 3-8

(4)

Let us take 20.0 kW h as 20.0 kW for 1.00 h and calculate the number of coulombs required to produce 1.00 kg of Al.

$$1 \text{ W} = 1 \text{ coulomb V sec}^{-1}$$

$$\begin{aligned} 20.0 \text{ kW h} &= 20.0 \times 10^{3} \text{ coulombs V sec}^{-1} \times 1.0 \text{ h} \times 3600 \text{ sec h}^{-1} \\ &= 7.20 \times 10^{7} \text{ coulombs V} \end{aligned}$$

The electrolytic cells operate with a potential difference of 5.00 V. So the total number of coulombs required to produce 1.00 kg of Al is

$$\frac{7.20 \times 10^{7} \text{ coulombs V}}{5.00 \text{ V}} = 1.44 \times 10^{7} \text{ coulombs}$$

This material may have been new to you, but you should have realized that you had to divide by 5.00 V just to cancel units. To produce 3.40×10^{8} kg, we would need

$$3.40 \times 10^{8} \text{ kg} \times 1.44 \times 10^{7} \text{ coulombs kg}^{-1} = 4.90 \times 10^{15} \text{ coulombs}$$

This is equivalent to the electrical energy consumption of the whole of New York City for one year.

4 is correct.
1 You forgot to convert the time to seconds.
2 You forgot to include the potential difference of 5.00 V and also did not convert kilowatts to watts.
3 You only converted the time to minutes instead of seconds.
5 You used the following incorrect relationship for a watt: $1\ W = 1$ coulomb $V^{-1}\ sec^{-1}$.

SOLUTION 3–9

(3)

Mass spectrometry is discussed in Section 3–6 of Chemical Principles.

Since there are three peaks in the mass spectrum at mass numbers 70, 72, and 74, you might assume that there are three isotopes of Cl with masses 35, 36, and 37. However, it is possible to have Cl_2 molecules in which the two Cl atoms have different isotopic masses. Thus, the peak at mass number 72 could result from the species $(^{35}Cl\ ^{37}Cl)^+$ as well as $(^{36}Cl\ ^{36}Cl)^+$. But if there is a ^{36}Cl isotope, there should also be peaks at mass numbers 71 [$(^{35}Cl\ ^{36}Cl)^+$] and 73 [$(^{36}Cl\ ^{37}Cl)^+$]. These peaks did not appear in the spectrum; hence, ^{36}Cl is not a naturally occurring isotope. [*Note:* ^{36}Cl is a radioactive isotope that scientists have made artificially.]

Answer **3** is correct. In Answer **5** you fell into the trap concerning ^{36}Cl.

SOLUTION 3–10

(3)

The three peaks correspond to $(^{35}Cl\ ^{35}Cl)^+$, $(^{35}Cl\ ^{37}Cl)^+$, and $(^{37}Cl\ ^{37}Cl)^+$. Their relative heights are 9/6/1. Consider a sample containing $9 + 6 + 1 = 16$ Cl_2 molecules. In this sample there are nine $^{35}Cl\,^{35}Cl$ molecules, six $^{35}Cl\,^{37}Cl$ molecules, and one $^{37}Cl\,^{37}Cl$ molecule. The relative numbers of the two isotopes can be found from Table 3–1.

Table 3–1

Molecular species	Molecules Cl_2	Atoms ^{35}Cl	Atoms ^{37}Cl
$^{35}Cl\ ^{35}Cl$	9	18	0
$^{35}Cl\ ^{37}Cl$	6	6	6
$^{37}Cl\ ^{37}Cl$	1	0	2
Total	16	24	8

The table indicates that the isotopes ^{35}Cl and ^{37}Cl are present in the ratio 24/8 or 3/1; that is, the natural abundance of ^{35}Cl is three times that of ^{37}Cl. Answer **3** is correct.

SOLUTION 3-11 (4)

The molecular ion of PCl_3 is $(PCl_3)^+$. We must consider all the possible combinations of the two chlorine isotopes and the one phosphorus isotope in forming ions of the type $(PCl_3)^+$. The various possibilities are tabulated in Table 3-2.

Table 3-2

Molecular species	Mass number
$^{31}P\ ^{35}Cl\ ^{35}Cl\ ^{35}Cl$	136
$^{31}P\ ^{35}Cl\ ^{35}Cl\ ^{37}Cl$	138
$^{31}P\ ^{35}Cl\ ^{37}Cl\ ^{37}Cl$	140
$^{31}P\ ^{37}Cl\ ^{37}Cl\ ^{37}Cl$	142

There should, therefore, be four parent peaks in the spectrum of PCl_3 at mass numbers 136, 138, 140, and 142. The correct answer is **4**.

[*Note:* Although the appearance of peaks at the four appropriate mass numbers would be indicative that a sample is pure PCl_3, the relative abundance of the four molecular species would have to agree exactly with that for pure PCl_3, namely, 27/27/9/1, in order to say that the sample was definitely pure.]

SOLUTION 3-12 (2)

You approach this question in exactly the same way as in Problem 3-11. First you must determine all the possible molecular ions $(BBr_3)^+$ and then calculate the mass numbers of the species.

Table 3-3

Molecular species	Mass numbers
$^{10}B\ ^{79}Br\ ^{79}Br\ ^{79}Br$	247
$^{11}B\ ^{79}Br\ ^{79}Br\ ^{79}Br$	248
$^{10}B\ ^{79}Br\ ^{79}Br\ ^{81}Br$	249
$^{11}B\ ^{79}Br\ ^{79}Br\ ^{81}Br$	250
$^{10}B\ ^{79}Br\ ^{81}Br\ ^{81}Br$	251
$^{11}B\ ^{79}Br\ ^{81}Br\ ^{81}Br$	252
$^{10}B\ ^{81}Br\ ^{81}Br\ ^{81}Br$	253
$^{11}B\ ^{81}Br\ ^{81}Br\ ^{81}Br$	254

There should, therefore, be eight peaks in the spectrum at integral mass numbers from mass number 247 to mass number 254. The only spectrum that satisfies these conditions is that of Figure 3-7(b). Answer **2** is correct.

SOLUTION 3–13

(5)

First we must locate the peak caused by the parent molecular ion in order to determine the molecular weight of the unknown compound. The peak at highest mass number might be caused by the parent molecular ion, thus indicating that, in this case, the molecular weight is 34. NH_3 and CH_4 have molecular weights of 17 and 18, respectively; so Answers **1** and **2** can be quickly discarded. However, PH_3, H_2O_2, and CH_3F all have molecular weights of 34. This is when the mass numbers of the fragment ions become important in deciding which of the three compounds is correct. Consider the comparison in Table 3–4. This table lists the possible ions for each of the three compounds and their mass numbers and the mass numbers for the ions of the unknown compound. From the comparison only the fragment ions of fluoromethane (CH_3F) are identical to those of the unknown compound. Neither PH_3 nor H_2O_2 has fragment ions with mass numbers 19, 15, or 14. The correct answer is **5**.

[*Note:* Our analysis has been simple because the parent molecular ion did appear. Sometimes it is so unstable that only the fragment molecular ions are detected. In such a case the identification is more complicated.]

Table 3–4

PH_3		H_2O_2		CH_3F		Unknown
Parent ion	*m/e*	*Parent ion*	*m/e*	*Parent ion*	*m/e*	*m/e*
$(PH_3)^+$	34	$(H_2O_2)^+$	34	$(CH_3F)^+$	34	34
Fragment ions		*Fragment ions*		*Fragment ions*		
$(PH_2)^+$	33	$(HO_2)^+$	33	$(CH_2F)^+$	33	33
$(PH)^+$	32	$(O_2)^+$	32	$(CHF)^+$	32	32
P^+	31			$(CF)^+$	31	31
				F^+	19	19
		$(H_2O)^+$	18			
		$(HO)^+$	17			
		O^+	16			
				$(CH_3)^+$	15	15
				$(CH_2)^+$	14	14

SOLUTION 3-14 (5)

The peak in the spectrum at mass number 44 leads us to guess that the molecular weight of the unknown compound is 44. Answers **1**, **2**, and **3** are immediately ruled out because HCN, CO, and NO_2 have molecular weights of 27, 28, and 30, respectively. The other two compounds, N_2O and CO_2, have molecular weights of 44. The ions for these compounds that have mass numbers between 12 and 44 are tabulated in Table 3–5 together with their mass numbers and the mass numbers of the ions of the unknown compound.

Table 3–5

N_2O		CO_2		Unknown
Parent ion	*m/e*	*Parent ion*	*m/e*	*m/e*
$(N_2O)^+$	44	$(CO_2)^+$	44	44
Fragment ions		*Fragment ions*		
		$(O_2)^+$	32	
$(NO)^+$	30			
$(N_2)^+$	28	$(CO)^+$	28	28
		$(CO_2)^{2+}$	22	22
O^+	16	O^+	16	16
N^+	14			
		C^+	12	12

The critical peak is at *m/e* 12. Only CO_2 has a peak at this value. The peak expected for O_2^+ from CO_2 at *m/e* 32 was apparently too weak to be detected. The weak peak at *m/e* 22 is due to the doubly charged ion $(CO_2)^{2+}$ ($m/e = 44/2$). The unknown compound is CO_2. Answer **5** is correct.

[*Note:* In the first sentence of this solution we said that we "guessed" that molecular weight was 44. Why were we not certain? Because our mass spectrum was incomplete and did not give any information about possible peaks at high mass numbers. If we were not choosing one out of five possible compounds, but were more realistically just identifying a compound, we would have to consider a higher molecular weight compound that gives the same mass spectrum in this range as does CO_2.]

SOLUTION 3-15 (5)

First you must determine the charge associated with one electron from the data given.

Value of charge measured (esu)	Difference between adjacent values (esu)
4.32×10^{-9}	
	4.8×10^{-10}
3.84×10^{-9}	
	9.6×10^{-10}
2.88×10^{-9}	
	1.44×10^{-9}
1.44×10^{-9}	
	4.8×10^{-10}
9.60×10^{-10}	

If you look at the list of the differences on the right-hand side, you see that all of the differences are multiples of 4.8×10^{-10} esu. Since the oil drops acquire different numbers of electrons during the experiment, we would expect that the measured charges would differ from each other by multiples of the charge on one electron. This value should, therefore, be 4.8×10^{-10} esu. The number of electrons attached to an oil drop for which a charge of 5.76×10^{-9} esu was measured is given by

$$\frac{5.76 \times 10^{-9}\ \text{esu}}{4.8 \times 10^{-10}\ \text{esu electron}^{-1}} = 12\ \text{electrons}$$

The correct answer is **5**.

Because 5.76×10^{-9} esu is greater than any of the other charges given, the charge on one electron must be less than 5.76×10^{-9} esu. This immediately rules out Answer **1**. If you chose Answers **2**, **3**, or **4**, you realized that 5.76×10^{-9} esu had to be some multiple of the other charges; but you forgot that these other charges themselves had to be multiples of a specific charge.

You could have solved this problem another way if you remembered that 1 $\mathscr{F}$ represents 1 mole of electric charge and is equal to 96,500 coulombs. The charge on one electron is

$$\frac{96{,}500\ \text{coulombs}\ \mathscr{F}^{-1}}{6.0 \times 10^{23}\ \text{electrons}\ \mathscr{F}^{-1}} = 1.6 \times 10^{-19}\ \text{coulomb electron}^{-1}$$

The charge measured for the oil drop was

$$5.76 \times 10^{-9}\ \text{esu} = \frac{5.76 \times 10^{-9}\ \text{esu}}{3.00 \times 10^{9}\ \text{esu coulomb}^{-1}}$$

$$= 1.93 \times 10^{-18}\ \text{coulomb}$$

Therefore, the number of electrons attached to the oil drop

$$= \frac{1.93 \times 10^{-18}\ \text{coulomb}}{1.6 \times 10^{-19}\ \text{coulomb electron}^{-1}}$$

$$= 12\ \text{electrons}$$

SOLUTION 3-16

(1)

The key to the whole question is that the *cathode rays* are traveling from left to right in the tube. Cathode rays are electrons; therefore, the electrode on the right-hand side of the tube must be positively charged and is the *anode*. Now let us consider each statement in turn.

(a) Since cathode rays are electrons and negatively charged, canal rays must be positively charged particles because they are traveling in the opposite direction. In fact, they are positive ions of the gas molecules, for example, Ne^{+} in a neon discharge tube. This statement is *correct*.

(b) Since statement (a) is correct, this statement must be *incorrect*.

(c) This statement is *incorrect* since canal rays are positively charged particles and, as such, they are deflected by both electric and magnetic fields.

(d) This statement is *incorrect* since, from statement (a), canal rays are positive ions and must travel toward the negative electrode (cathode).

Since statement (a) is the only correct one, the required answer to the question is **1**.

SOLUTION 3–17

(2)

Close packing of spheres in metals is treated in Section 3–7 of Chemical Principles.

The central atom obviously lies inside the unit cell but what of the corner atoms? The cubic unit cell is made by joining the centers of the eight corner atoms, and so only a fraction of the volume of each corner atom is actually located in the unit cell. Each corner atom is shared equally between eight unit cells similar to the one that we are considering. Therefore, only one eighth of the volume of each of the corner atoms lies inside the unit cell. Thus, in effect, we have $8 \times \frac{1}{8} = 1$ whole atom located inside the unit cell. Since the central atom lies completely in the unit cell, there are therefore $1 + 1 = 2$ atoms occupying space in the unit cell. Answer **2** is correct.

SOLUTION 3–18

(3)

Consider the schematic diagram of a cubic unit cell of side d shown in Figure 3–14. Each black circle represents the center of a titanium atom. The distance between the center of the central titanium atom and the centers of its eight nearest neighbors is $2\sqrt{2}$ Å. This means that the length of the body diagonal BE is twice $2\sqrt{2}$ Å, that is $4\sqrt{2}$ Å. If we apply Pythagoras' theorem to triangle ABD, we obtain

$$(\mathrm{BD})^2 = (\mathrm{AB})^2 + (\mathrm{AD})^2 = d^2 + d^2 = 2d^2$$

and

$$\mathrm{BD} = \sqrt{2}\, d$$

From triangle BDE

$$(\mathrm{DE})^2 = (\mathrm{BE})^2 - (\mathrm{BD})^2$$

or

$$d^2 = (4\sqrt{2})^2 - 2d^2$$

from which $d = 4\sqrt{\frac{2}{3}} \simeq 4 \times 0.8 \simeq 3.2$ Å. The volume of the unit cell = $d^3 \simeq (3.2\ \text{Å})^3 \simeq 3.3 \times 10^{-23}\ \mathrm{cm}^3$.

There are two Ti atoms located in this volume. To calculate the density we need to know the mass of the unit cell.

$$\text{mass} = \frac{48\ \text{g g-atom}^{-1} \times 2\ \text{atoms (unit cell)}^{-1}}{6.0 \times 10^{23}\ \text{atoms}}$$

$$= 1.6 \times 10^{-22}\ \text{g (unit cell)}^{-1}$$

$$\text{density} = \frac{\text{mass}}{\text{volume}} = \frac{1.6 \times 10^{-22}\ \text{g (unit cell)}^{-1}}{3.3 \times 10^{-23}\ \text{cm}^3\ \text{(unit cell)}^{-1}}$$

$$\simeq 5\ \text{g cm}^{-3}$$

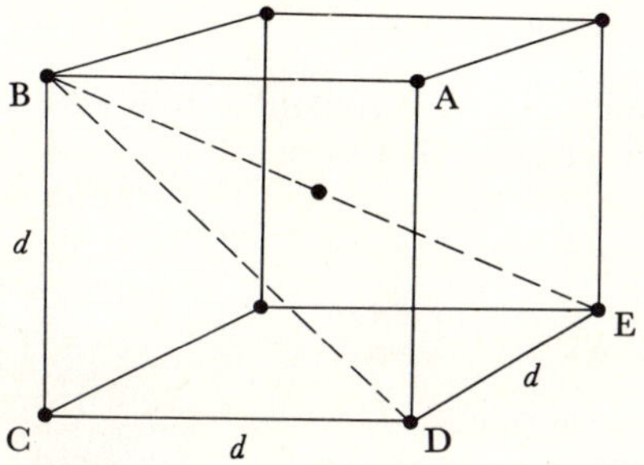

Figure 3–14

3 is correct.
1 You forgot the atomic weight of Ti.
4 You thought that there were nine Ti atoms per unit cell.
5 You forgot Avogadro's number and ignored the powers of 10.

SOLUTION 3–19

(3)

The unit cell of the brass consists of eight Zn atoms at the corners of a cube and six Cu atoms at the face centers of the cube. The corner atoms contribute one eighth of themselves to the unit cell; therefore, there is $8 \times \frac{1}{8} =$ one Zn atom effectively located inside the unit cell. The face-centered atoms contribute one half of themselves to the unit cell; therefore, there are $6 \times \frac{1}{2} = 3$ Cu atoms effectively located inside the unit cell.

The mass in the unit cell of this brass is given by the mass of one Zn atom and three Cu atoms:

$$\frac{1\ \text{atom} \times 65.4\ \text{g g-atom}^{-1}\ \text{Zn}}{6.0 \times 10^{23}\ \text{atoms g-atom}^{-1}\ \text{Zn}} + \frac{3\ \text{atoms} \times 63.6\ \text{g g-atom}^{-1}\ \text{Cu}}{6.0 \times 10^{23}\ \text{atoms g-atom}^{-1}\ \text{Cu}}$$

$$= 4.3 \times 10^{-22}\ \text{g (unit cell)}^{-1}$$

3 is correct.
1 is the mass of the Zn atom effectively located in the unit cell.
2 is the mass of the three Cu atoms effectively located in the unit cell.
4 is the mass of the six Cu atoms at the face centers of the cube.
5 is the mass of the eight Zn corner atoms and the six Cu face-centered atoms of the cube.

SOLUTION 3–20 (4)

From Problem 3–19 we know that there are three Cu atoms and one Zn atom effectively located in the unit cell of the brass.

$$\text{mass of 1 Zn atom} = \frac{1 \text{ atom} \times 65.4 \text{ g g-atom}^{-1}}{6.0 \times 10^{23} \text{ atoms g-atom}^{-1}}$$

$$= 1.1 \times 10^{-22} \text{ g}$$

The mass of the unit cell from Problem 3–19 is 4.3×10^{-22} g. The percent zinc content by weight is therefore

$$\frac{1.1 \times 10^{-22}}{4.3 \times 10^{-22}} \times 100 = 26\%$$

4 is correct.
2 You slipped a decimal point.
3 is the percent copper content except that you dropped a factor of 10.
5 You forgot that the zinc content had to be less than 33%; in any case, this is the copper content.

SOLUTION 3–21 (4)

Coordination number is the number of nearest neighbors of an atom or ion in a structure. Consider the cation on the right-hand face of the structure. This cation sits at the center of a cube of eight anions that give it a coordination number of eight. Now look at the anion at the bottom of the left-hand side of the front face of the structure. This is at the center of a tetrahedron of four cations that give the anion a coordination number of four. (Remember that the lines are only there to help you to understand the structures more easily. They are not chemical bonds!) Answer **4** is correct. The crystal structure is actually that of fluorite (CaF_2), the principal mineral of fluorine.

SOLUTION 3–22 (3)

The length of the side of the unit cell is 5.20 Å. Hence, the volume of the unit cell is $(5.20 \text{ Å})^3 = 1.40 \times 10^{-22} \text{ cm}^3$. Since the density of SrO is 5.00 g cm^{-3}, we can determine the mass of one unit cell:

$$\text{mass} = 1.40 \times 10^{-22} \text{ cm}^3 \text{ (unit cell)}^{-1} \times 5.00 \text{ g cm}^{-3}$$
$$= 7.00 \times 10^{-22} \text{ g (unit cell)}^{-1}$$

The mass of one SrO molecule is

$$\frac{104 \text{ g mole}^{-1}}{6.00 \times 10^{23} \text{ molecules mole}^{-1}} = 1.73 \times 10^{-22} \text{ g molecule}^{-1}$$

Hence, the number of SrO molecules per unit cell is

$$\frac{7.00 \times 10^{-22} \text{ g (unit cell)}^{-1}}{1.73 \times 10^{-22} \text{ g molecule}^{-1}} \simeq 4 \text{ molecules (unit cell)}^{-1}$$

This means that there are four Sr^{2+} and four O^{2-} ions in each unit cell.

Answer **3** is correct. Answer **4**: You must have used the square of density of SrO in your calculation.

CHAPTER 4 QUANTITIES IN CHEMICAL CHANGE: STOICHIOMETRY

In this chapter we will polish our stoichiometric skills, particularly for reactions involving gases and solutions, and extend them to the field of thermochemistry.

To answer the questions in this chapter, you should memorize the following relationships:

$$\text{molarity} = \frac{\text{moles of solute}}{\text{liter of } \textit{solution}}$$

$$\text{molality} = \frac{\text{moles of solute}}{\text{kg of } \textit{solvent}}$$

$$\text{normality} = N = \frac{\text{equivalents of solute}}{\text{liter of } \textit{solution}}$$

ΔH positive implies endothermic. (It is useful to memorize that the molar volume of an ideal gas at STP is 22.4 liters, but it is not necessary. Given R, you can always derive this number from $PV = nRT$, with $P = 1$ atm, $n = 1$ mole, and $T = 273°K$.) Assume that the gases are ideal and that all reactions go to completion.

Solutions to problems in Chapter 4 begin on page 100.

PROBLEMS

PROBLEM 4–1

You are told to choose a reactant for use on a lunar expedition to remove CO_2 from the cabin atmosphere. The only consideration is minimum weight. That is, which of the hydroxides below contributes the least weight to the environmental control system?

(at. wt: H = 1, Li = 7, O = 16, Mg = 24, Al = 27, K = 39, Cs = 133)

1 $2LiOH + CO_2 = Li_2CO_3 + H_2O$
2 $2KOH + CO_2 = K_2CO_3 + H_2O$
3 $2CsOH + CO_2 = Cs_2CO_3 + H_2O$
4 $Mg(OH)_2 + CO_2 = MgCO_3 + H_2O$
5 $2Al(OH)_3 + 3CO_2 = Al_2(CO_3)_3 + 3H_2O$

PROBLEM 4–2

Cyanogen is used as a welding gas for heat-resistant metals, as a rocket propellant (with ozone or fluorine), and as a fumigant. The empirical formula of cyanogen is CN. An amount of 1.73 g of this compound occupies 0.82 liter at 1 atm and 300°K. What is the molecular formula?

(at. wt: C = 12.0, N = 14.0;
$R = 0.082$ liter atm mole^{-1} deg^{-1})

1 CN
2 $(CN)_2$
3 C_3N_4
4 $(CN)_6$
5 C_2N_4

PROBLEM 4–3

An excess of NaOH is treated with 1.12 liters of dry hydrogen chloride gas measured at STP. What weight of sodium chloride is formed? (The reaction is complete.)

(at. wt: O = 16.0, Na = 23.0, Cl = 35.5;
$R = 0.082$ liter atm mole^{-1} deg^{-1})

1 0.050 g
2 1.8 g
3 2.0 g
4 2.9 g
5 22.4 g

PROBLEM 4–4

The San Francisco Police Department seized a package that they suspected contained PAX (an illegal tranquilizer, $C_{24}H_{30}N_3O$). When the compound is combusted the following reaction is expected:

$$C_{24}H_{30}N_3O + 34O_2 \rightarrow 24CO_2 + 15H_2O + 3NO_2$$

The crime laboratory combusted a 10.0-g sample. If the sample is PAX, what volume of dry CO_2 gas will be collected at STP?

(mol wt: PAX = 376;
$R = 0.082$ liter atm mole^{-1} deg^{-1})

1 1.43 liters
2 10.0 liters
3 14.3 liters
4 22.4 liters
5 22.4×24 liters $\simeq$ 540 liters

PROBLEM 4–5

The determination of chemical formulas is discussed in Section 4–1 of Chemical Principles *by Dickerson, Gray, and Haight.*

In 1962, the preparation of the first stable compound of xenon and fluorine from xenon gas (Xe) and fluorine gas (F_2) was announced. A reaction mixture was prepared at 25°C by filling a 1.0-liter nickel container with F_2 gas at 8 atm and Xe gas at 1.7 atm. The reaction mixture was maintained at 400°C for 1 hour. Then it was cooled to 25°C and the contents of the nickel container analyzed. All the Xe gas had reacted to form a solid xenon–fluorine compound, but some of the F_2 gas had not reacted. The pressure of the unreacted F_2 gas was 4.6 atm. What formula would you propose for the xenon–fluorine compound?

1 XeF_2
2 XeF_3
3 XeF_4
4 XeF_6
5 XeF_8

PROBLEM 4–6

The combustion of methane gas (CH_4) results in the formation of carbon dioxide and water. Indicate which of the chemical equations correctly describes this process.

1 $CH_4 + O_2 \rightarrow CO_2 + 2H_2O$
2 $CH_4 + 2O_2 \rightarrow CO_2 + 2H_2O$
3 $CH_4 + O_2 \rightarrow CO_2 + H_2O$
4 $\frac{1}{2}CH_4 + 2O_2 \rightarrow \frac{1}{2}CO_2 + H_2O$
5 $CH_4 + \frac{1}{2}O_2 \rightarrow CO_2 + H_2O$

PROBLEM 4-7

A mixture of *n*-hexane (C_6H_{14}) gas and O_2 gas was placed in an evacuated vessel at 297°C. The pressure inside the vessel was 340 torr. After ignition by a spark, the mixture reacted completely to form CO gas, CO_2 gas, and steam. The total pressure exerted by the gaseous products in the vessel was then 520 torr, and the temperature was unchanged. Which of the following equations describes the reaction that took place?

1 $C_6H_{14} + 8O_2 \rightarrow 3CO + 3CO_2 + 7H_2O$
2 $C_6H_{14} + 7O_2 \rightarrow 5CO + CO_2 + 7H_2O$
3 $2C_6H_{14} + 17O_2 \rightarrow 4CO + 8CO_2 + 14H_2O$
4 $2C_6H_{14} + 15O_2 \rightarrow 8CO + 4CO_2 + 14H_2O$
5 $C_6H_{14} + 9O_2 \rightarrow CO + 5CO_2 + 7H_2O$

PROBLEM 4-8

Gasoline is a complex mixture of many different compounds. Most of these compounds contain 5 to 10 carbon atoms, but they differ radically in structure. One of the chief components of gasoline is the hydrocarbon octane (C_8H_{18}). When this hydrocarbon burns completely in air, it forms carbon dioxide and water. The gas tank of an average automobile holds 60 liters [about 16 gallons (US)], and the density of octane is 0.70 g cm^{-3}. What volume of air (20% O_2 by volume) would be required at STP to combust completely the contents of an automobile gas tank? For this order-of-magnitude calculation, assume that the gas tank is filled with pure octane.

(mol wt: $C_8H_{18} = 114$;
$R = 0.082$ liter atm mole^{-1} deg^{-1})

1 2.8×10^2 liters
2 5.2×10^2 liters
3 4.2×10^4 liters
4 5.2×10^5 liters
5 1.0×10^6 liters

PROBLEM 4-9

In the United States, with regard to alcohol, the term "proof" usually is defined as twice the percentage volume of pure ethanol (C_2H_5OH) in a solution measured at 60°F. Thus, a solution that is 95% (by volume) ethanol is 190 proof. What is the molarity of ethanol in a "92 proof" alcohol water solution?

$D(C_2H_5OH, 60°F) = 0.80$ g cm^{-3}
$D(H_2O, 60°F) = 1.0$ g cm^{-3}

(mol wt: $C_2H_5OH = 46$)

1 0.46 molar
2 0.80 molar
3 0.92 molar
4 8.0 molar
5 17 molar

PROBLEM 4–10

An amount of 2.3 g of ethyl alcohol (C_2H_5OH) is added to 500 g of water. The resulting solution is ________ molal in C_2H_5OH.

(mol wt: $C_2H_5OH = 46$)

1 0.010
2 0.050
3 0.10
4 0.20
5 4.6

PROBLEM 4–11

The average concentration of sodium ions (Na^+) in human blood serum is about 3.4 g liter^{-1}. What is the molarity of Na^+?

(at. wt: Na = 23)

1 0.15
2 3.4
3 $23/3.4 \simeq 6.8$
4 23
5 $23 \times 3.4 \simeq 78$

PROBLEM 4–12

Boiling point elevation is discussed in detail in Section 15–10 of Chemical Principles.

An over-qualified short-order cook knows that a 0.25-molal aqueous solution of NaCl will boil at 100.26°C and therefore will boil eggs more quickly. If he puts 496 g of NaCl in a large pot, how many kilograms of H_2O must he add to have the 0.25-molal solution?

(mol wt: NaCl = 58.5)

1 0.030 kg
2 2.0 kg
3 4.3 kg
4 8.5 kg
5 34 kg

PROBLEM 4-13

Dilution problems are discussed in Chemical Principles, *Section 4-4.*

Over 90% of people whose blood assays show an ethanol (C_2H_5OH) concentration of 0.0030 g ml^{-1} of blood demonstrate signs of obvious intoxication. The fatal concentration is estimated as 0.0070 g ml^{-1} of blood. What is the volume of 80 proof scotch (40% C_2H_5OH by volume) that corresponds to the difference between an intoxicating and a fatal dose for a person whose blood volume is 7.0 liters? (Assume that *all* the alcohol goes directly to the blood.)

(Density: $C_2H_5OH = 0.80$ g ml^{-1})

1 0.088 ml
2 22 ml
3 35 ml
4 88 ml
5 Insufficient data; density of scotch must be given.

PROBLEM 4-14

It has been reported that the consumption of 44 g of pure ethanol taken either as 4 oz of whiskey or 5.5 oz of a martini results in an average alcohol concentration in the blood of 0.080 g/100 ml blood. If the total volume of blood in a human adult is 7.0 liters, what percentage of the ingested alcohol is in the blood?

(mol wt: $C_2H_5OH = 46$)

1 0.080%
2 1.3%
3 13%
4 100(44/46)% = 96%
5 100%

PROBLEM 4-15

To reduce water pollution, you are told to choose a base that neutralizes a waste sulfuric acid solution prior to disposal in a river. Which of the following would you choose solely in terms of the total cost of base?

Table 4-1

Base	cost, ¢/lb
$CaCO_3$	0.5
$Ca(OH)_2$	1
NH_3	3
NaOH	5

(mol wt: $CaCO_3$ = 100, $Ca(OH)_2$ = 74, NH_3 = 17, NaOH = 40, $H_2SO_4 = 98$)

1 $CaCO_3$
2 $Ca(OH)_2$
3 NH_3
4 NaOH
5 They are all equal.

PROBLEM 4-16

Acid-base titrations are discussed in Chemical Principles, *Sections 4–5 and 5–6.*

One method of peeling potatoes commercially is to soak them in a 10–20% NaOH solution at 60–88°C for 1–5 min, and to spray off the peel after the potatoes have been removed from the solution. To determine if the NaOH solution is still capable of peeling after a day's use, a peeler titrates a 10.0-ml sample to neutrality with 64.0 ml of a 0.200-molar H_2SO_4 solution. What concentration of NaOH did he find?

(mol wt: $NaOH = 40.0$, $H_2SO_4 = 98.0$)

1 0.64 molar
2 1.28 molar
3 2.56 molar
4 12.8 molar
5 25.6 molar

PROBLEM 4-17

In Problem 4–16 the potato peeler's solution was 2.56 molar in NaOH. To operate, it must be at least 10% by weight (10 g NaOH per 100 g solution). What is the weight percentage corresponding to the determined molarity?

(Density of 2.56-molar NaOH solution $\simeq 1.10$ g cm^{-3};
mol wt: $NaOH = 40.0$, $H_2O = 18.0$)

1 0.26%
2 2.56%
3 9.3%
4 10.2%
5 93%

PROBLEM 4-18

A human patient suffering from a duodenal ulcer may show an HCl concentration of 80×10^{-3} molar in his gastric juice. If his stomach receives 3 liters of gastric juice per day, how much patent medicine, containing 2.6 g of $Al(OH)_3$ per 100 ml, must he consume per day to neutralize the acid?

(mol wt: $Al(OH)_3 = 78$, $HCl = 36.5$)

1 27 ml
2 80 ml
3 240 ml
4 720 ml
5 None of the above.

PROBLEM 4–19

Water that does not easily form a lather with soap is termed "hard," whereas water that readily forms a lather with soap is termed "soft." The chief cause of hardness in water is the presence of small quantities of dissolved calcium salts such as calcium bicarbonate, $Ca(HCO_3)_2$, and calcium sulfate, $CaSO_4$. These calcium salts react with the soap before it has a chance to form a lather with the water. To soften water, the calcium compounds either must be precipitated from solution or converted to soluble sodium compounds. Calcium bicarbonate can be removed from the water by boiling because insoluble calcium carbonate, $CaCO_3$, is produced in the process. But calcium sulfate presents another problem. This compound is removed by the addition of ordinary washing soda, Na_2CO_3, to the water. The following reaction occurs:

$$CaSO_4 + Na_2CO_3 \rightarrow CaCO_3 + Na_2SO_4$$

A typical concentration of $CaSO_4$ in river water is 1.8×10^{-3} g liter^{-1}. How many kilograms of Na_2CO_3 would be required to soften 6.8×10^9 liters [about 2×10^9 gallons (US)] of water? (This is about the daily water consumption of New York City.)

(mol wt: $CaSO_4 = 136$, $Na_2CO_3 = 106$)

1 6.8 kg
2 9.5 kg
3 12 kg
4 9.5×10^3 kg
5 1.2×10^4 kg

PROBLEM 4–20

The dirigible Hindenburg had a gas volume of 1.4×10^8 liters (about 5 million cubic feet), and was filled with H_2. How much energy was released when it burned to gaseous products? (Assume that it was filled to a pressure of 1 atm at 0°C, and, for the purposes of our calculation, that the reaction proceeded at 25°C.)

Standard heat of formation (298°K)
$H_2O(g) = -57.8$ kcal mole^{-1}

1 58 kcal
2 180 million kcal
3 360 million kcal
4 720 million kcal
5 806 million kcal

PROBLEM 4–21

We are given the following data:

	ΔH^0_{298} (kcal mole^{-1})
$C(graphite) + O_2(g) \rightarrow CO_2(g)$	–94.05
$C(diamond) + O_2(g) \rightarrow CO_2(g)$	–94.50

Thus, the heat of the reaction diamond → graphite, in kcal mole^{-1} at 298°K and 1 atm pressure, is given by which one of the following?

1 $\Delta H^0 = -188.55$ kcal mole^{-1}
2 $\Delta H^0 = -0.45$ kcal mole^{-1}
3 $\Delta H^0 = 0.45$ kcal mole^{-1}
4 $\Delta H^0 = 188.55$ kcal mole^{-1}
5 $\Delta H^0 = 450$ kcal mole^{-1}

PROBLEM 4–22 Which of the following compounds has the highest caloric value in cal g^{-1}?

	Compound	Molecular weight	Heat of combustion (cal mole^{-1})
(a)	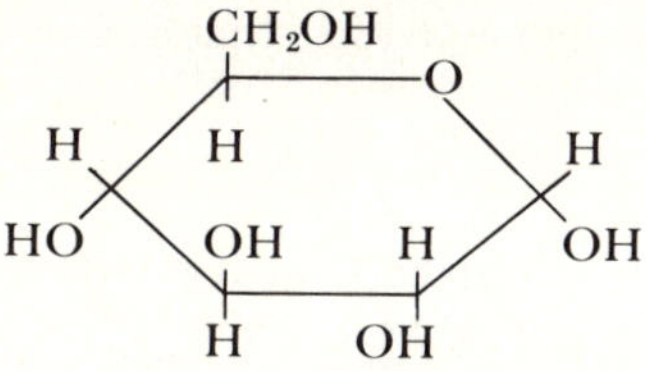 *d*-glucose (a carbohydrate)	180	−673,000
(b)	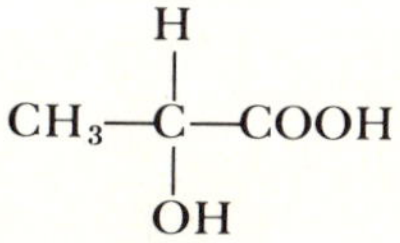 lactic acid (a carbohydrate)	90	−326,000
(c)	$CH_3(CH_2)_{14}COOH$ palmitic acid (a lipid or fat)	256	−2,380,000
(d)	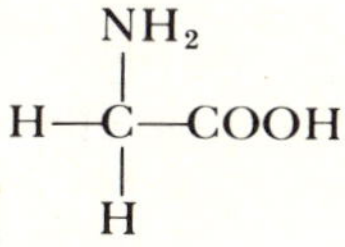 glycine (an amino acid)	75	−234,000

1 (a)
2 (b)
3 (c)
4 (d)
5 They are all equal.

PROBLEM 4-23

For the reaction $C(s) + \frac{1}{2}O_2(g) \rightarrow CO(g)$, ΔH^0_{298} is ________ kcal mole^{-1}.

Data:

$C(s) + O_2(g) \rightarrow CO_2(g)$ $\Delta H^0_{298} = -94$ kcal mole^{-1}

$CO(g) + \frac{1}{2}O_2(g) \rightarrow CO_2(g)$ $\Delta H^0_{298} = -67$ kcal mole^{-1}

1 −161
2 −27
3 0
4 +27
5 +161

PROBLEM 4-24

Thermochemical calculations are discussed in Chemical Principles, *Section 4-5.*

One of the propellant systems used in rocket engines is a mixture of hvdrazine (N_2H_4) as fuel and hydrogen peroxide (H_2O_2) as oxidizer. These reactants are called *hypergolic;* that is, they ignite on contact. The reaction that occurs is

$$N_2H_4(l) + 2H_2O_2(l) \rightarrow N_2(g) + 4H_2O(g)$$

The reactants are mixed at 25°C in the mole ratio indicated by the equation. What is the heat of reaction?

Standard heats of formation (298°K)

$N_2H_4(l) = +12$ kcal mole^{-1}
$H_2O_2(l) = -45$ kcal mole^{-1}
$H_2O(g) = -57.8$ kcal mole^{-1}

1 −309 kcal mole^{-1}
2 −153 kcal mole^{-1}
3 −24.8 kcal mole^{-1}
4 +153 kcal mole^{-1}
5 Insufficient data; the standard heat of formation (298°K) of $N_2(g)$ must be given.

PROBLEM 4-25

Many attempts are being made to develop efficient methods of using solar energy for heating houses. One such method uses crushed sodium sulfate decahydrate ($Na_2SO_4 \cdot 10H_2O$) in an enclosed bin on the roof of the house. This bin stores heat collected from the sun during the day. The heat storage involves the following endothermic reaction:

$$Na_2SO_4 \cdot 10H_2O(s) \rightarrow Na_2SO_4(s) + 10H_2O(l)$$

This reaction is reversible and is completely to the right at temperatures above 32.4°C (90.3°F). At night, when the temperature falls below 32.4°C, the reaction is reversed and is completely to the left. The heat liberated from the bin is then circulated throughout the house by a system of fans.

Assuming that the chemical reactions and the heat transfer are all 100% efficient, how much heat could a 322-kg charge (about 700 lb) of $Na_2SO_4 \cdot 10H_2O$ provide to a solar house by night?

Compound	Standard heat of formation at 298°K (kcal $mole^{-1}$)	mol wt
$Na_2SO_4 \cdot 10H_2O(s)$	−1033.5	322
$Na_2SO_4(s)$	−330.9	142
$H_2O(l)$	−68.3	18

1 19.6 kcal
2 634.3 kcal
3 1.96×10^4 kcal
4 6.34×10^5 kcal
5 2.05×10^6 kcal

SOLUTION 4–1

(1)

If you had trouble with this problem, refer to Lassila et al., Programed Reviews of Chemical Principles, *Section 4–1, Items 1–18.*

It is easy to choose among the first three answers since the stoichiometry is the same. And it is clear that LiOH is the choice here since it has the smallest molecular weight, 24 g $mole^{-1}$. But we need 2 moles of it (i.e., 48 g) to react with 1 mole of CO_2. We need only 1 mole of $Mg(OH)_2$ to do the same job; however, its molecular weight is 58 g $mole^{-1}$ and is therefore rejected. Since 2 moles of $Al(OH)_3$ consume 3 moles of CO_2, $\frac{2}{3}$ mole of $Al(OH)_3$ or $(\frac{2}{3})78 = 52$ g of $Al(OH)_3$ consume 1 mole of CO_2. The choice is still LiOH or Answer **1**. LiOH was used on Apollo 11 for this purpose.

SOLUTION 4–2

(2)

$PV = nRT$
$n = PV/RT = \text{wt/mol wt}$
$\text{mol wt} = \text{wt} \times RT/PV$

$$= \frac{1.73 \text{ g} \times 0.082 \text{ liter atm mole}^{-1} \text{ deg}^{-1} \times 300°\text{K}}{0.82 \text{ liter} \times 1 \text{ atm}}$$

$$= 1.73 \times 30 \text{ g mole}^{-1}$$
$$\simeq 52 \text{ g mole}^{-1}$$

The formula is $(CN)_x$, and the molecular weight is $x(14 + 12) = 26x$. This amount must be equal to 52. Clearly, $x = 2$ and the molecular formula is C_2N_2 or $(CN)_2$. The correct answer is **2**.

SOLUTION 4–3

(4)

If you couldn't solve this problem, try Items 19–29 in Section 4–1 of Lassila's book.

First write the equation, which is the simplest kind of acid–base reaction to balance:

$$NaOH + HCl \rightarrow NaCl + H_2O$$

Since the NaOH was in excess, the amount of NaCl produced is the same as the number of moles of HCl used:

moles NaCl = moles HCl (from the stoichiometry)

$$= \frac{1.12 \text{ liters}}{22.4 \text{ liters mole}^{-1}}$$

(Either memorize the molar volume of an ideal gas at STP or use $PV = nRT$.)

moles NaCl = 0.050 mole
wt NaCl = 0.050 mole × 58.5 g $mole^{-1}$
$\simeq 2.9$ g

4 is correct.
1 is the number of moles labeled with the wrong units.
2 is the weight of HCl used.
3 is the weight of NaOH used.

SOLUTION 4–4

(3)

The number of moles of CO_2 that would be produced from a 1-mole PAX sample is 24, and from a 10.0-g PAX sample it is

$$\frac{10.0 \text{ g PAX}}{376 \text{ g mole}^{-1} \text{ PAX}} \times \frac{24 \text{ moles } CO_2}{\text{moles PAX}} = \frac{240}{376} \text{ moles } CO_2$$

The ideal molar volume at STP is 22.4 liters mole^{-1} (refer to Problem 4–3); therefore, the volume consistent with the formula for PAX is

$$\frac{240}{376} \text{ moles} \times 22.4 \text{ liters mole}^{-1} = 14.3 \text{ liters } CO_2$$

The correct answer is **3**. Answer **1** is the volume for a 1.0-g sample. Answer **5** is the volume for a 1.0-mole sample.

(PAX is an imaginary tranquilizer—as far as we know.)

SOLUTION 4–5

(3)

All the pressures were measured at the same temperature, and the volume of the reaction vessel was unchanged. Therefore, T and V are constant in $PV = nRT$, and $P \propto n$. This means that we can take the pressures as a measure of the number of moles of each reactant present.

pressure of F_2 gas before reaction = 8 atm
pressure of F_2 gas after reaction = 4.6 atm
pressure difference = (8 − 4.6) = 3.4 atm

The pressure difference is proportional to the number of moles of F_2 consumed in the reaction.

pressure of Xe gas before reaction = 1.7 atm
pressure of Xe gas after reaction = zero
pressure difference = 1.7 atm

This pressure difference is proportional to the number of moles of Xe used in the reaction. So the ratio of moles used in forming the solid is

$$\frac{\text{moles } F_2 \text{ used}}{\text{moles Xe used}} = \frac{3.4}{1.7} = \frac{2}{1}$$

1 mole of fluorine gas contains 2 g-atoms of fluorine, whereas 1 mole of xenon gas contains 1 g-atom of xenon. The simplest formula of the solid xenon–fluorine compound formed must, therefore, be XeF_4. The answer is **3**.

SOLUTION 4–6

(2)

The number of atoms of each kind must be the same on both sides of the equation.

1 Two O atoms on left, but four on right.
3 Two O atoms on left, but three on right.
4 Four O atoms on left, but two on right.
5 One O atom on left, but three on right.
2 is the only answer that has equal numbers of O, C, and H on both sides.

SOLUTION 4–7

(4)

$PV = nRT$

$P \propto n \qquad (V, T \text{ constant})$

$$\frac{P_{\text{initial}}}{P_{\text{final}}} = \frac{n_{\text{initial}}}{n_{\text{final}}} = \frac{340}{520} = \frac{17}{26}$$

The *n*'s are the total number of moles of reactants (initial) and the total number of moles of products (final). We need only add the moles on each side of the suggested equations until the ratios match.

	$n_{\text{initial}}/n_{\text{final}}$
1	9/13
2	8/13
3	19/26
4	17/26 is correct.
5	10/13

SOLUTION 4–8

(4)

Sooner or later we must balance the equation

$C_8H_{18} + (?)O_2 \rightarrow (?)CO_2 + (?)H_2O$

We'll try to use 1 mole of octane on the left:

$C_8H_{18} + (?)O_2 \rightarrow 8CO_2 + 9H_2O$

The carbon and hydrogen are balanced; we need only balance oxygen. There are $16 + 9 = 25$ oxygen atoms on the right, so we must use $\frac{25}{2}$ moles of O_2 on the left for the number of *atoms* to balance:

$C_8H_{18} + \frac{25}{2}O_2 \rightarrow 8CO_2 + 9H_2O$

Some chemists would double the entire equation to obtain integer coefficients:

$2C_8H_{18} + 25O_2 \rightarrow 16CO_2 + 18H_2O$

That was only the first step. Now we must find the amount of C_8H_{18} that burned:

$$\text{moles } C_8H_{18} = \frac{60 \text{ liters} \times 10^3 \text{ cm}^3 \text{ liter}^{-1} \times 0.70 \text{ g cm}^{-3}}{114 \text{ g mole}^{-1}}$$

$$= \frac{42 \times 10^3}{114} \text{ moles}$$

For every 2 moles of C_8H_{18}, 25 moles of O_2 are needed:

$$\text{moles } O_2 = \frac{25 \text{ moles } O_2}{2 \text{ moles } C_8H_{18}} \times \frac{42}{114} \times 10^3 \text{ moles } C_8H_{18}$$

And with the molar volume of an ideal gas at STP = 22.4 liters,

$$\text{volume of } O_2 = \frac{25 \times 42}{2 \times 114} \times 10^3 \text{ moles} \times 22.4 \text{ liters mole}^{-1}$$

Since air is one fifth O_2, the volume of air is five times the O_2 volume:

$$\text{volume of air} = \frac{5 \times 25 \times 42}{2 \times 114} \times 22.4 \times 10^3 \text{ liters}$$

Instead of multiplying this, we'll simplify the arithmetic by making a few approximations.

$$5 \times 25 \times \frac{42}{2} \times \frac{22.4}{114} \times 10^3$$

is the term to be simplified. We take

$$\frac{42}{2} = 21 \simeq 20$$

and

$$\frac{22.4}{114} \simeq \frac{22.4}{112} = \frac{224 \times 10^{-1}}{112} = 2 \times 10^{-1}$$

The term becomes approximately equal to

$$5 \times 25 \times 20 \times 2 \times 10^{-1} \times 10^{-3}$$

which is

$$5000 \times 10^2 \simeq 5 \times 10^5 \text{ liters of air}$$

The only answer that is close is **4** (5.2×10^5 liters); this is the correct answer. Try to use such shortcuts when time is important. Answer **5**: 2 moles of C_8H_{18}. Answer **3**: Did you forget the stoichiometric ratio? Answer **2**: Conversion troubles.

SOLUTION 4–9

(4)

If you made a mistake, review this material in Lassila's book, Section 4–2, Items 1–40.

92 proof equals 46 volume % ethanol.

Consider 1 liter of *solution*:

460 cm^3 of ethanol

$460 \text{ cm}^3 \times 0.80 \text{ g cm}^{-3} = \text{g of ethanol}$

$$\frac{460 \times 0.80 \text{ g}}{46 \text{ g mole}^{-1}} = \text{moles of ethanol}$$

So there are 8.0 moles of ethanol in 1 liter of solution, and it is 8.0 molar.

The correct answer is **4**. Answer **5**: You probably thought molarity is moles of solute per liter of H_2O.

SOLUTION 4–10 (3)

Molality is moles of solute per kilogram of H_2O.

2.3 g C_2H_5OH to 500 g H_2O

4.6 g C_2H_5OH to 1000 g H_2O

$\frac{4.6\ \text{g}}{46\ \text{g mole}^{-1}}$ moles C_2H_5OH to 1000 g H_2O

0.10 mole C_2H_5OH to 1000 g H_2O

The solution is 0.10 molal. The correct answer is **3**.

SOLUTION 4–11 (1)

There are 3.4 g Na^+ in a liter of solution (serum). So there is 3.4 g/23 g $\text{mole}^{-1} \simeq 0.15$ mole Na^+ per liter, and the solution is 0.15 molar. The correct answer is **1**.

Note that the atomic weight of Na is also that of Na^+. The weight of the missing electron is negligible in these calculations.

SOLUTION 4–12 (5)

$$496\ \text{g NaCl} = \frac{496\ \text{g NaCl}}{58.5\ \text{g mole}^{-1}}$$

$$= 8.5\ \text{moles of NaCl}$$

$$0.25\text{-molal} = \frac{8.5\ \text{moles NaCl}}{x\ \text{kg}\ H_2O}$$

$$x = \frac{8.5}{0.25} = 34\ \text{kg}\ H_2O$$

Answer **5** is correct. This is about 9 gallons (US) or one *firkin*. Answer **1**: You inverted the equation for *x*. Answer **2**: You may have forgotten to convert from grams to moles of NaCl. Answer **4** is just the moles of NaCl.

SOLUTION 4–13 (4)

The difference in concentration terms is 0.0040 g ml^{-1} of blood, and the weight of alcohol in 7.0 liters of blood corresponding to this figure is 4.0×10^{-3} g $\text{ml}^{-1} \times 7.0 \times 10^3$ ml = 28 g C_2H_5OH. This is 28 g/0.80 g ml^{-1} = 35 ml of C_2H_5OH.

Let *x* equal the volume of scotch in milliliters:

$$(0.40\ \text{ml}\ C_2H_5OH/\text{ml scotch})\ (x\ \text{ml scotch}) = 35\ \text{ml}\ C_2H_5OH$$
$$x = 35/0.40 \simeq 88\ \text{ml (almost 3 oz)}$$

[*Note:* It is said that ethanol, of all commonly used drugs, has the lowest ratio of toxic to commonly used concentration levels.]

4 is correct.
1 7.0 liters = 7000 ml.
2 You multiplied by the density and forgot that scotch is not pure alcohol.
3 You made the second part of the previous error.

SOLUTION 4–14 ③

The total weight of alcohol in the blood is

$$\frac{0.080 \text{ g}}{100 \text{ ml}} \times 7000 \text{ ml} = 5.6 \text{ g } C_2H_5OH$$

The percentage that has been absorbed is

$$\frac{5.6}{44} \times 100 = 13\%$$

The correct answer is **3**. Now we can see the error in our previous calculation (Problem 4–13), since only a fraction of the ingested alcohol goes into the blood.

Macbeth, Act 2, Scene 3:
Macduff: What three things does drink especially provoke?
Porter: Marry, sir, nose-painting, sleep, and urine. Lechery, sir, it provokes, and unprovokes; it provokes the desire, but it takes away the performance . . .

SOLUTION 4–15 ①

If you did not find the correct answer, you might want to check through Section 4–4 of Lassila's book.

It is (or should be) clear that the important number is not cost on a per pound or even per kilogram basis, but on a per *equivalent* basis.

Table 4–2

Base	Reaction	Equivalent wt
$CaCO_3$	$CaCO_3 + H_2SO_4 \rightarrow CaSO_4 + H_2CO_3$	$\frac{100}{2} = 50$
$Ca(OH)_2$	$Ca(OH)_2 + H_2SO_4 \rightarrow CaSO_4 + 2H_2O$	$\frac{74}{2} = 37$
NH_3	$2NH_3 + H_2SO_4 \rightarrow (NH_4)_2SO_4$	17
NaOH	$2NaOH + H_2SO_4 \rightarrow Na_2SO_4 + 2H_2O$	40

equivalent wt	× cost/wt	= cost/equivalent
50	0.5	25
37	1	37
17	3	51
40	5	200

The above occurs because the equivalent weight is weight per equivalent and the cost per weight will have the same relative values no matter whether we consider kilograms, pounds, or tons. So **1** is the correct answer.

SOLUTION 4–16

(3)

If titration presents difficulties for you, work Items 41–60 in Section 4–2 of Lassila's book.

From the chemical equation, we see that each molecule of H_2SO_4 is equal to 2 equivalents:

$$H_2SO_4 + 2NaOH \rightarrow Na_2SO_4 + 2H_2O$$

$$N_{H_2SO_4} = 2 \times 0.200 = 0.400 \text{ normal}$$

$$N_{NaOH} \times V_{NaOH} = N_{H_2SO_4} \times V_{H_2SO_4}$$

$$N_{NaOH} = 0.400 \times \frac{64.0 \text{ ml}}{10.0 \text{ ml}}$$

$= 2.56$ *molar* since the normality and molarity are equal for NaOH

3 is correct.
1 You divided by two to find the normality instead of multiplying.
2 Review the difference between normality and molarity.
4 Mistakes **2** and **5** combined.
5 OK, if he had titrated only 1.00 ml of the NaOH solution.

SOLUTION 4–17

(3)

We must convert everything to a weight basis:

$$\frac{2.56 \text{ moles}}{1000 \text{ cm}^3 \text{ solution}} = \frac{2.56 \text{ moles} \times 40.0 \text{ g mole}^{-1}}{1000 \text{ cm}^3 \times 1.10 \text{ g cm}^{-3}}$$

$$= \frac{102.4 \text{ g NaCl}}{1100 \text{ g solution}}$$

and dividing numerator and denominator by 1.1, we obtain

$$= \frac{93 \text{ g NaCl}}{1000 \text{ g solution}}$$

$$= 9.3\% \text{ by weight}$$

The correct answer is **3**. Stop the potatoes! Answer **4**: You forgot the density.

SOLUTION 4–18

(3)

$$Al(OH)_3 + 3HCl \rightarrow AlCl_3 + 3H_2O$$

N_{HCl} = molarity of HCl since there is 1 equivalent weight per mole. But the normality of $Al(OH)_3$ is three times its molarity, since it has 3 equivalent weights per mole:

$$\text{molarity of } Al(OH)_3 = \frac{26 \text{ g}}{1000 \text{ ml}} \times \frac{1}{78 \text{ g mole}^{-1}} = 0.33 \text{ mole liter}^{-1}$$

$$N_{Al(OH)_3} = 3 \times (0.33) = 1.0 \text{ normal}$$
$$N_{HCl} \times V_{HCl} = N_{Al(OH)_3} \times V_{Al(OH)_3}$$

[*Note:* You shouldn't have to memorize this equation. It really only states what the chemical equation does: that $\frac{1}{3}$ mole (1 equivalent weight) of $Al(OH)_3$ consumes 1 mole (1 equivalent weight) of HCl.]

$$80 \times 10^{-3} \times 3 \times 10^3 \text{ ml} = 1.0 \times V_{Al(OH)_3}$$
$$V_{Al(OH)_3} = 240 \text{ ml}$$

The answer is **3**.

SOLUTION 4–19

(4)

This is really a titration problem. We don't have to worry about normalities here because the stoichiometric coefficients of the reactants are equal.

The molarity of $CaSO_4$ is

$$\frac{1.8 \times 10^{-3} \text{ g liter}^{-1}}{136 \text{ g mole}^{-1}}$$

and the number of moles of $CaSO_4$ to be softened is

$$\frac{1.8 \times 10^{-3} \text{ mole}}{136 \text{ liters}} \times 6.8 \times 10^9 \text{ liters} = 0.90 \times 10^5 \text{ moles}$$

which is the number of moles of Na_2CO_3 required.

The weight of Na_2CO_3 is

$$0.90 \times 10^5 \text{ moles} \times 106 \text{ g mole}^{-1} = 9.5 \times 10^6 \text{ g} = 9.5 \times 10^3 \text{ kg}$$

The correct answer is **4**. (Only about 10 tons.)

SOLUTION 4–20

(3)

We write the equation first·

$$H_2(g) + \tfrac{1}{2}O_2(g) \rightarrow H_2O(g)$$

For this reaction, ΔH^0_{298} is just the heat of formation of $H_2O(g)$ since the heats of formation of $H_2(g)$ and $O_2(g)$ are zero by definition:

$$\Delta H^0_{298} = -57.8 \text{ kcal mole}^{-1}$$

The number of moles of H_2 is given by $P = 1$ atm, $T = 0°C$, and the volume. But these are STP conditions; so the molar volume is 22.4 liters (remember that these are the conditions when the dirigible was first *filled*), and the number of moles of H_2 burned was

$$\frac{1.4 \times 10^8 \text{ liters}}{22.4 \text{ liters mole}^{-1}}$$

and the heat *released* (ΔH negative means exothermic) was

$$\frac{1.4 \times 10^8 \times 57.8 \text{ kcal}}{22.4} \simeq 360 \times 10^6 \text{ kcal}$$

The correct answer is **3**. Answer **1** is the heat released per mole. Answer **2**: You may have introduced the stoichiometric coefficient for O_2. We always write ΔH^0_{298} for the reaction as written. If we had written $2H_2(g) + O_2(g) \rightarrow 2H_2O(g)$, ΔH^0_{298} would have been -115.6 kcal mole^{-1}.

SOLUTION 4–21 (2)

If you did not correctly solve this problem and the preceding one, study Section 4–5 of Lassila's book; it contains programed review material of heat effects on chemical reactions.

We want a sum whose result is the equation

C(diamond) → C(graphite)

So we reverse the first equation to put C(graphite) on the right-hand side (changing the sign of ΔH^0_{298} at the same time) and keep the second equation as is.

	ΔH^0_{298} (kcal mole^{-1})
$CO_2(g) \rightarrow O_2(g) + C(\text{graphite})$	+94.05
$C(\text{diamond}) + O_2(g) \rightarrow CO_2(g)$	−94.50
Sum: C(diamond) → C(graphite)	−0.45

The correct answer is **2**.

SOLUTION 4–22 (3)

The caloric value in cal g^{-1} is simply the value in cal mole^{-1} divided by the molecular weight (g mole^{-1}). The results are

(a) *d*-glucose	3740
(b) lactic acid	3620
(c) palmitic acid	9300
(d) glycine	3120

Palmitic acid has the highest caloric value; fats are usually very high in caloric value.

SOLUTION 4–23 (2)

We reverse the second reaction to put CO(g) on the right-hand side and change the sign of ΔH^0_{298} for that reaction.

	ΔH^0_{298} (kcal mole^{-1})
$C(s) + O_2(g) \rightarrow CO_2(g)$	−94
$CO_2(g) \rightarrow CO(g) + \frac{1}{2}O_2(g)$	+67
Sum: $C(s) + \frac{1}{2}O_2(g) \rightarrow CO(g)$	−27

The correct answer is **2**. Answer **1**: You didn't reverse the sign of the reversed reaction.

SOLUTION 4–24 (2)

We'll write the standard heats of formation (298°K) below the chemical terms.

	$N_2H_4(l)$ +	$2H_2O_2(l)$ →	$N_2(g)$ +	$4H_2O(l)$
kcal mole^{-1}:	+12	−45	0	−57.8
multiply by stoichiometric coefficient:	×1	×2	×1	×4
kcal mole^{-1}:	+12	−90	0	−231

$$\begin{aligned}\Delta H^0_{298} &= \text{products} - \text{reactants}\\ &= -231 - (12 - 90)\\ &= -153 \text{ kcal mole}^{-1}\end{aligned}$$

2 is correct.
4 You must remember to subtract reactants from products. Anyway, you should know this answer is wrong because you expect a rocket fuel to liberate energy when it is ignited; hence, it is exothermic; hence, ΔH^0_{298} is negative.
1 You added reactants to products instead of subtracting.
3 You forgot the stoichiometric coefficients.
5 Heats of formation of elements in their standard states are defined to be zero.

SOLUTION 4–25 (3)

We will calculate the heat of reaction for the daytime (endothermic) reaction and determine the amount of solar energy absorbed.

	$Na_2SO_4 \cdot 10H_2O(s)$ →	$Na_2SO_4(s)$ +	$10H_2O(l)$
kcal mole^{-1}:	−1033.5	−330.9	−68.3
multiply by stoichiometric coefficients:	×1	×1	×10
kcal mole^{-1}:	−1033.5	−330.9	−683

$$\begin{aligned}\Delta H^0_{298} &= \text{products} - \text{reactants}\\ &= [-330.9 - 683] - [-1033.5]\\ &= +\ 19.6 \text{ kcal mole}^{-1}\end{aligned}$$

This is the amount of heat absorbed by 1 mole of $Na_2SO_4 \cdot 10H_2O$, but we have 322×10^3 g/322 g mole^{-1} or 1.00×10^3 moles, so the heat absorbed and therefore available for release at night is

$$19.6 \text{ kcal} \times 10^3 = 1.96 \times 10^4 \text{ kcal}$$

The correct answer is **3**. (This is equivalent to the heat available from about 6 lb of Pennsylvania anthracite.) Unfortunately, solar batteries at this time can store only a few watts per square foot of battery panel.

1 This is the heat per mole.
2 You forgot the stoichiometry and the number of moles of reactants.
4 You forgot the stoichiometry.
5 You added the heats of products and reactants.

CHAPTER 5 CHEMICAL EQUILIBRIUM

The problems in this chapter deal with systems whose equilibrium properties are the daily concern of the working chemist: gas mixtures undergoing reaction, acid–base solutions, and precipitates in equilibrium with their solutions.

To do the problems in this chapter, you will need to memorize the following relationships.

(1) For gas equilibrium,

$$aA + bB \rightleftarrows cC + dD$$

$$K_c = \frac{[C]^c[D]^d}{[A]^a[B]^b}$$ if the concentrations are measured at equilibrium

$$K_p = \frac{p_C{}^c p_D{}^d}{p_A{}^a p_B{}^b}$$ if the partial pressures are measured at equilibrium

(2) For acid–base equilibria

$K_w = [H^+][OH^-]$ for $H_2O \rightleftharpoons H^+ + OH^-$

$K_a = \dfrac{[H^+][A^-]}{[HA]}$ for $HA \rightleftharpoons H^+ + A^-$

$K_b = \dfrac{[B^+][OH^-]}{[BOH]}$ for $BOH \rightleftharpoons B^+ + OH^-$

$K_h = \dfrac{[HA][OH^-]}{[A^-]}$ for $A^- + H_2O \rightleftharpoons HA + OH^-$

or

$K_h = \dfrac{[BOH][H^+]}{[B^+]}$ for $B^+ + H_2O \rightleftharpoons H^+ + BOH$

$pH = -\log[H^+]$
$pOH = -\log[OH^-]$
$pK = -\log K$

We will use the notation H^+ and H_3O^+ interchangeably to represent the hydrated H^+. Assume that all data are at a T for which $K_w = 1.0 \times 10^{-14}$ mole² liter⁻² unless otherwise specified. Assume that the volumes are additive.

(3) For solubility product calculations,

$$A_xB_y(s) \rightleftharpoons xA^{pos}(aq) + yB^{neg}(aq)$$

$$K_{sp} = [A^{pos}]^x[B^{neg}]^y$$

when the concentrations are measured at equilibrium for a saturated solution. When units for K_{sp} are not displayed, the constant has been constructed by using mole liter^{-1} values for the concentrations.

Solutions to problems in Chapter 5 begin on page 135.

PROBLEMS

PROBLEM 5-1

For the homogeneous gas reaction

$4NH_3 + 5O_2 \rightleftharpoons 4NO + 6H_2O$

the equilibrium constant, K_c, has the units of ________.

1 $conc^{+10}$
2 $conc^{+1}$
3 It is dimensionless.
4 $conc^{-1}$
5 $conc^{-9}$

PROBLEM 5-2

The determination of equilibrium constants for gas reactions is discussed in Section 5-2 of Chemical Principles.

Two moles of NH_3 gas are introduced into a previously evacuated 1.0-liter container in which it partially dissociates at high T:

$2NH_3(g) \rightleftharpoons N_2(g) + 3H_2(g)$

At equilibrium, 1.0 mole of $NH_3(g)$ remains. What is K_c?

1 $27/64 = 0.42$ mole^2 liter^{-2}
2 $3/4 = 0.75$ mole^2 liter^{-2}
3 1.0 mole^2 liter^{-2}
4 $3/2 = 1.5$ mole^2 liter^{-2}
5 $27/16 = 1.7$ mole^2 liter^{-2}

PROBLEM 5-3

An equilibrium mixture for the reaction

$2H_2S(g) \rightleftharpoons 2H_2(g) + S_2(g)$

had 1.0 mole H_2S, 0.20 mole H_2, and 0.80 mole of S_2 in a 2.0-liter vessel. K_c is ________.

1 0.0040 mole liter^{-1}
2 0.080 mole liter^{-1}
3 0.016 mole liter^{-1}
4 0.032 mole liter^{-1}
5 0.16 mole liter^{-1}

K_p and K_c problems are solved in identical ways, but K_c uses concentrations and K_p uses partial pressures.

PROBLEM 5-4

At 1000°C and a total pressure of 1 atm, steam is approximately 0.002% dissociated according to the equation

$2H_2O(g) \rightleftharpoons 2H_2(g) + O_2(g)$

You may want to defer working problems involving partial pressures until you have studied Chapters 15 and 16 of Chemical Principles.

What is the best approximate value of K_p at this temperature?

1 2×10^{-15} atm
2 4×10^{-15} atm
3 3×10^{-14} atm
4 2×10^{-10} atm
5 2×10^{-6} atm

PROBLEM 5-5

For the gas reaction

$3H_2(g) + N_2(g) \rightleftharpoons 2NH_3(g)$

the partial pressures of H_2 and N_2 are 0.400 atm and 0.800 atm, respectively. The total pressure of the system is 2.80 atm. What is K_p when the concentrations are in atmospheres?

1 1.00
2 3.13
3 5.00
4 50.0
5 153

PROBLEM 5-6

Initially, 0.84 mole of PCl_5 and 0.18 mole of PCl_3 are mixed in a 1.0-liter vessel. It is later found that 0.72 mole PCl_5 is present at equilibrium. What is K_c for the reaction

$PCl_5(g) \rightleftharpoons PCl_3(g) + Cl_2(g)$

at this temperature?

1 0.0010 mole liter^{-1}
2 0.015 mole liter^{-1}
3 0.030 mole liter^{-1}
4 0.050 mole liter^{-1}
5 0.150 mole liter^{-1}

PROBLEM 5–7

For the reaction

$$CO(g) + H_2O(g) \rightleftharpoons CO_2(g) + H_2(g)$$

the equilibrium constant equals four at a given T. If 0.6 mole of CO and 0.6 mole of steam were put into a 2-liter vessel at this temperature, the number of moles of CO_2 present at equilibrium would be ________.

1 0.2
2 0.4
3 0.6
4 0.8
5 1.2

PROBLEM 5–8

Gaseous H_2 and I_2 are allowed to react to give gaseous HI. When equilibrium is reached, the value of the equilibrium constant K_p will depend on which one of the following?

1 The initial concentration of H_2.
2 The temperature.
3 The total pressure of the system.
4 The volume of the reaction vessel.
5 The initial concentration of HI.

PROBLEM 5–9

For the gas-phase reaction $2NO \rightleftharpoons N_2 + O_2$, $\Delta H^0 = -43.5$ kcal mole^{-1}. Therefore, for the system

$$N_2(g) + O_2(g) \rightleftharpoons 2NO(g)$$

which one of the statements below is true?

1 K is independent of T.
2 K increases as T increases.
3 K decreases as T increases.
4 K varies with addition of NO.
5 Variation in K cannot be predicted.

PROBLEM 5–10

Champagne is poured cold into chilled glasses because ________.

[*Note:* liquid + gas $\rightleftharpoons$ solution is exothermic]

1 It would change color when warm.
2 This will stop the fermentation.
3 Its molar volume is greater at the lower temperature, and it therefore provides more servings when cold.
4 It would quickly go "flat" if poured into a warm glass.
5 This is the traditional method of service.

PROBLEM 5-11 What change would be observed in the equilibrium constant, K, for the endothermic reaction $H_2(g) \rightleftharpoons 2H(g)$ when (a) the temperature is increased and (b) the volume of the system is decreased?

	(a) T increased	(b) V decreased
1	K increases	K decreases
2	K decreases	K increases
3	No change	No change
4	No change	K increases
5	K increases	No change

PROBLEM 5-12 The reaction $2NOCl(g) \rightleftharpoons 2NO(g) + Cl_2(g)$ was studied in a 1.0-liter reaction vessel. At equilibrium, the number of moles of NOCl, NO, and Cl_2 were 1.4, 0.70, and 0.40, respectively. Then 0.10 mole of NO was added and the reaction allowed to come to equilibrium again. In what way are the new equilibrium concentrations related to the old equilibrium concentrations?

	NOCl	NO	Cl_2
1	same	same	same
2	new > old	new > old	new > old
3	new < old	new > old	new < old
4	new > old	new > old	new < old
5	new > old	new < old	new < old

PROBLEM 5-13 The exothermic ammonia synthesis reaction is

$$N_2(g) + 3H_2(g) \rightleftharpoons 2NH_3(g)$$

This reaction was of great importance during World War I because NH_3 was used not only for the manufacture of fertilizer but also of high explosives. Dr. Fritz Haber found the conditions of temperature and pressure that produced the greatest yield of NH_3 from this reaction, thereby making Germany self-sufficient in this respect. What would you predict the conditions to be?

1 Low T, low P.
2 Low T, high P.
3 High T, low P.
4 High T, high P.
5 All the same.

PROBLEM 5–14

Sulfur dioxide (SO_2) has many uses: as a preservative for meats, wine, and beer; as a bleaching agent for foods; in the pulp and paper industry; and even in tanning, among others. Consider the two gaseous equilibria involving SO_2 and the corresponding equilibrium constants at 25°C:

$$SO_2(g) + \tfrac{1}{2}O_2(g) \rightleftharpoons SO_3(g) \quad K_1$$
$$2SO_3(g) \rightleftharpoons 2SO_2(g) + O_2(g) \quad K_2$$

The values of the equilibrium constants are related by which one of the following?

1 $K_2 = K_1$
2 $K_2 = K_1^2$
3 $K_2 = 1/K_1^2$
4 $K_2^2 = K_1$
5 $K_2 = 1/K_1$

PROBLEM 5–15

$COCl_2$ gas (phosgene) is put into a 2.0-liter reaction vessel and then heated so it is partially decomposed:

$$COCl_2(g) \rightleftharpoons CO(g) + Cl_2(g)$$

When equilibrium was reached, the $COCl_2$ concentration was 0.40 mole liter^{-1}. More $COCl_2$ was added to the vessel; and when equilibrium was reestablished, $[COCl_2]$ was 1.6 moles liter^{-1}. How had the CO concentration changed?

1 It was halved.
2 It was unchanged.
3 It was doubled.
4 It was quadrupled.
5 Insufficient data; K_c must be known.

PROBLEM 5–16

For the following reaction, the equilibrium concentrations in a 1.0-liter vessel are shown under the appropriate chemical formulas:

	$PCl_3(g)$ +	$Cl_2(g)$ $\rightleftharpoons$	$PCl_5(g)$	
equil:	0.20	0.10	0.40	mole liter^{-1}

Then 0.10 mole of $Cl_2(g)$ was added. What is the new equilibrium concentration of $PCl_5(g)$?

1 0.05 mole liter^{-1}
2 0.35 mole liter^{-1}
3 0.40 mole liter^{-1}
4 0.45 mole liter^{-1}
5 0.80 mole liter^{-1}

PROBLEM 5–17 The equilibrium composition for the following reaction was found to be

	$CO(g)$	+ $H_2O(g)$	⇄ $CO_2(g)$	+ $H_2(g)$	
equil:	0.10	0.10	0.40	0.10	mole liter^{-1}

Then 0.30 mole of $H_2(g)$ was added to the 1.0-liter reaction vessel. What was the new $CO_2(g)$ concentration when equilibrium was reestablished?

1 0.070 molar
2 0.33 molar
3 0.40 molar
4 0.47 molar
5 0.80 molar

Most chemical processes and all biological processes take place in aqueous solutions, and in most of these the acidity and basicity of the solution is extremely important. The following problems deal with equilibria in such systems.

PROBLEM 5–18 If the pH of an aqueous solution is less than 7, the solution ________.

($K_w = 1 \times 10^{-14}$ mole^2 liter^{-2})

1 Is acidic.
2 Is basic.
3 Is neutral.
4 Is strongly oxidizing.
5 Has a $K_w < 10^{-14}$ mole^2 liter^{-2}.

PROBLEM 5–19 Given that the dissociation constant for H_2O is $K_w = 1 \times 10^{-14}$ mole^2 liter^{-2}, what is the pH of a 0.001-molar KOH solution? [*Note:* KOH is a strong electrolyte.]

1 10^{-11}
2 10^{-3}
3 3
4 11
5 14

PROBLEM 5–20 It was found that 0.00050 mole of the strong electrolyte calcium hydroxide $Ca(OH)_2$ dissolved to form 100 ml of a saturated aqueous solution. What is the pH of this solution?

($K_w = 1.0 \times 10^{-14}$ mole^2 liter^{-2})

1 $-\log(5 \times 10^{-4}) = 3.3$
2 9.78
3 11.7
4 12
5 Cannot be calculated from these data.

PROBLEM 5-21

An amount of 40 ml of 0.10-molar NaOH is added to 10 ml of 0.45-molar HCl. What is the pH of the resulting solution? Both HCl and NaOH are strong electrolytes.

($K_w = 1 \times 10^{-14}$ mole^2 liter^{-2})

1 1
2 2
3 3
4 4
5 5

An equivalent *of an acid sometimes is defined as that amount which, if completely reacted, will produce Avogadro's number of protons; for example, 36.46 g of HCL or 49.039 g of H_2SO_4.*

PROBLEM 5-22

Urine normally has a pH of 6. If a patient eliminates 1300 ml of urine per day, how many equivalents of acid have been eliminated?

1 1.3×10^{-7}
2 1.3×10^{-6}
3 1.3×10^{-3}
4 $1300/6 \simeq 220$
5 9800

PROBLEM 5-23

The ionization constant of water, K_w, at 37°C (98.6°F) is 2.42×10^{-14} mole^2 liter^{-2}. What is the pH for a neutral solution at this normal temperature of the human body?

1 0
2 6.8
3 7.0
4 7.2
5 14

PROBLEM 5-24 In 1970, the leading cause of death for 15- to 35-year-olds in New York City was drug abuse. Acute amphetamine (uppers, dexies, bennies, meth, speed) poisoning may be treated by acidifying the urine. An orally administered compound passes into the urine where it produces its acidifying effect. Which of the following soluble electrolytes would be a suitable choice?

Data

$K_a(HCl)$ is large

$K_a(CH_3COOH$, acetic acid$) = 1.8 \times 10^{-5}$ mole liter^{-1}

$K_{a_1}(H_2CO_3) = 4.3 \times 10^{-7}$ mole liter^{-1}

$K_{a_2}(H_2CO_3) = 5.6 \times 10^{-11}$ mole liter^{-1}

$K_b(NH_4OH) = 1.8 \times 10^{-5}$ mole liter^{-1}

$K_b(NaOH)$ is large.
[*Note:* K_{a_1} represents the dissociation of the first proton: $H_2CO_3 \rightleftharpoons H^+ + HCO_3^-$ and K_{a_2} represents the dissociation of the second proton: $HCO_3^- \rightleftharpoons H^+ + CO_3^{2-}$.]

1 NH_4Cl
2 ammonium acetate, $NH_4^+CH_3COO^-$
3 NaCl
4 Na_2CO_3
5 None of the above.

PROBLEM 5-25 Which of the following statements is true?

1 An acid and its conjugate base react to form a salt and water.
2 The acid H_2O is its own conjugate base.
3 The conjugate base of a weak acid is a strong base.
4 The conjugate base of a strong acid is a strong base.
5 A base and its conjugate acid react to form a neutral solution.

PROBLEM 5-26

You have equal volumes of the following solutions:

Acid	Concentration	pH
HCl	0.1 molar	1.1
Acetic (CH_3COOH) (monoprotic)	0.1 molar	2.9
Formic (HCOOH) (monoprotic)	0.1 molar	2.3
Hydrocyanic (HCN)	0.1 molar	5.1

Which solution requires the most base to titrate?

1 HCl
2 Acetic
3 Formic
4 Hydrocyanic
5 All the same.

PROBLEM 5-27

Lemon juice normally has a pH = 2.1. If all the acid in lemon juice is citric acid (HCit $\rightleftharpoons H^+ + Cit^-$, $K_a = 8.4 \times 10^{-4}$ mole liter^{-1}) and there are no citrate salts present, what is the citric acid concentration [HCit]? (Citric acid is actually a polyprotic acid but we will assume for this problem that only the first hydrogen is important.)

1 6.3×10^{-5} molar
2 8.4×10^{-4} molar
3 7.5×10^{-2} molar
4 1.5×10^{-1} molar
5 9.5 molar

PROBLEM 5–28 The following strong electrolytes are commonly added to swimming pools:

Na_2CO_3 (Sodium carbonate)	$NaHSO_4$ (Sodium bisulfate)

What are their effects on the pH of the pool?

Data

$K_{a_1}(H_2CO_3) = 4.3 \times 10^{-7}$ mole liter^{-1}

$K_{a_2}(H_2CO_3) = 5.6 \times 10^{-11}$ mole liter^{-1}

K_{a_1} and K_{a_2} for H_2SO_4 are large.

	Na_2CO_3	$NaHSO_4$
1	pH reducer	no effect on pH
2	pH increaser	no effect on pH
3	pH increaser	pH reducer
4	pH reducer	pH increaser
5	None of the above combinations.	

PROBLEM 5–29 The bicarbonate ion (HCO_3^-) can act as a base by hydrolyzing in an aqueous solution. Given only K_{a_1} (H_2CO_3) and K_w, what is K_h for HCO_3^-?

1 $K_{a_1} \times K_w$
2 K_{a_1}/K_w
3 K_w/K_{a_1}
4 $K_w/K_{a_1}^2$
5 Insufficient data; $K_{a_2}(H_2CO_3)$ must be given.

PROBLEM 5–30 Aspirin can pass through membranes most easily in its un-ionized form. The treatment for an overdose of aspirin is oral administration of sodium bicarbonate, $NaHCO_3$. Which one of the following statements is most likely true?

Data

$K_{a_1}(H_2CO_3) = 4.3 \times 10^{-7}$ mole liter^{-1}

$K_{a_2}(H_2CO_3) = 5.6 \times 10^{-11}$ mole liter^{-1}

K_b(NaOH) is large

Sodium bicarbonate is a soluble electrolyte.

[*Hint:* Use the result of Problem 5–29.]

1 Aspirin is a strong acid.
2 Aspirin is a weak acid.
3 Aspirin is a strong base.
4 Aspirin is a weak base.
5 Aspirin does not have acid–base character.

PROBLEM 5-31

Besides carbon monoxide, barbiturates are the most popular suicide poison in the United States. A toxic dose is 0.5 g or more, and a lethal dose is usually 1.5 g. The formula for a common barbiturate, phenobarbital (Nembutal), is

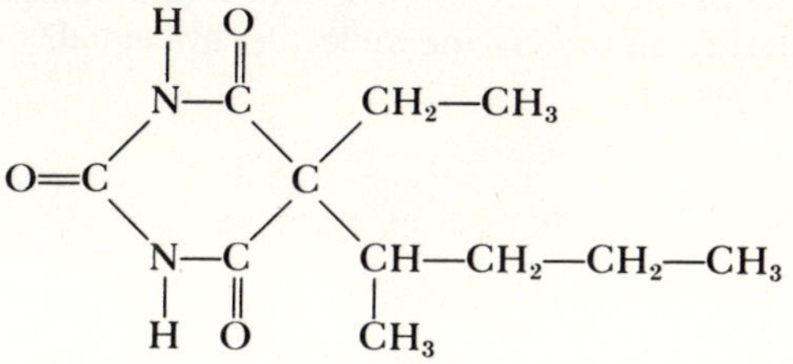

The formula for barbituric acid, which might be used in the synthesis of the barbiturate, is

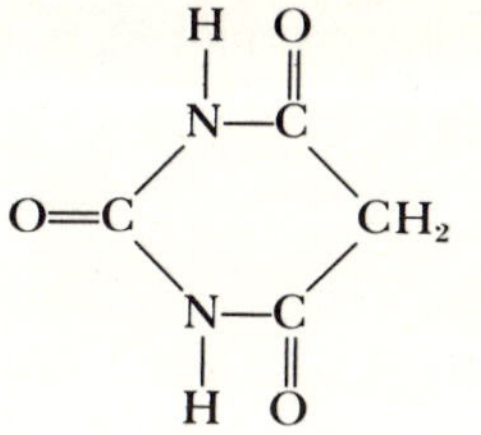

(denoted by HBar)

It is a weak acid ($pK_a = 4.01$). For 0.10-molar barbituric acid solution, approximately what percent of the acid molecules have dissociated? (Assume that the acid is monoprotic.)

1 $3 \times 10^{-3}\%$
2 3%
3 5%
4 50%
5 100%

PROBLEM 5–32 The amino acid glycine ($NH_2—CH_2—COOH$) can add and lose protons in aqueous solution to assume the forms

$NH_3^+—CH_2—COOH$ (completely protonated, a dibasic acid)

$NH_3^+—CH_2—COO^-$ (no net charge)

$NH_2—CH_2—COO^-$ (has no more protons that are easily donated)

which can participate in the reactions

$$NH_3^+—CH_2—COOH \rightleftharpoons H^+ + NH_3^+—CH_2—COO^- \quad pK_1 = 2.35$$
$$NH_3^+—CH_2—COO^- \rightleftharpoons H^+ + NH_2—CH_2—COO^- \quad pK_2 = 9.78$$

What is the pH of the solution when the total numbers of positive and negative charges on the glycine molecules are equal? (This pH is called the *isoelectric point.*)

1 7
2 $pK_1 \times pK_2$
3 $pK_1 + pK_2$
4 $2(pK_1 + pK_2)$
5 $\frac{1}{2}(pK_1 + pK_2)$

PROBLEM 5–33 Formic acid has an acid dissociation constant $K_a = 1.8 \times 10^{-4}$ mole liter^{-1} at 25°C.

$$\mathrm{H—\overset{\overset{O}{\|}}{C}—OH \rightleftharpoons H—\overset{\overset{O}{\|}}{C}—O^- + H^+}$$

You are to titrate an approximately 2.00-normal formic acid solution with a 2.00-normal NaOH solution by using a pH meter to determine the end point. What pH will correspond to the end point?

($K_w = 1.0 \times 10^{-14}$ mole^2 liter^{-2})

1 1.9
2 5.0
3 7.0
4 8.9
5 14

In the following series of problems we will build a buffer and examine the common ion effect for acids and bases.

PROBLEM 5–34

What is the pH of a 1.0-molar sodium acetate solution?

[K_a (acetic acid) = 1.8×10^{-5} mole liter^{-1};
$K_w = 1.0 \times 10^{-14}$ mole^2 liter^{-2}]

(Sodium acetate is a strong electrolyte.)

1 0.0
2 2.4
3 4.8
4 7.0
5 9.4

PROBLEM 5–35

Buffers are described in Section 5–8 of Chemical Principles.

You want to prepare a buffer solution of acetic acid (pK_a = 4.75) and sodium acetate (a strong electrolyte) such that pH = pK_a. You must accept which of the following five choices? (Refer to Problem 5–34, and assume that the effect of the ionization of water is negligible.)

[K_a(HAc) = 1.8×10^{-5} mole liter^{-1};
$K_w = 1.0 \times 10^{-14}$ mole^2 liter^{-2}]

1 Make the initial acetate concentration equal to the initial acetic acid concentration.
2 Make the initial hydronium ion (H_3O^+) concentration equal to the initial acetic acid concentration.
3 Make the initial hydronium ion (H_3O^+) concentration equal to the initial acetate ion concentration.
4 Give up since it is an impossible situation.
5 None of the above.

PROBLEM 5–36

You have prepared a buffer solution by adding 0.50 mole of acetic acid (pK_a = 4.75) and 2.5 moles of sodium acetate (a strong electrolyte) to enough pure water to make 1.0 liter of solution. What is the pH? (Refer to Problem 5–35.)

[pK_w = 14]

1 0.40
2 4.0
3 5.5
4 7.0
5 8.5

PROBLEM 5–37

If we dissolve 100 millimole (100×10^{-3} mole) of the strong acid HCl in 1 liter of pure water (pH = 7.0), pH will change to 1.0. If the same amount of HCl gas is dissolved in 1 liter of the buffer solution prepared in Problem 5–36 (pH = 5.45), what will the pH change to?

[pK_a(HAc) = 4.75]

1 4.15
2 4.74
3 5.35
4 7.00
5 10.7

Now we shall consider the more complicated situation of polyprotic acids and their buffer systems. We shall confine ourselves to the carbonic acid (H_2CO_3) system, which is primarily responsible for transport of CO_2 in the blood and for the control of blood pH. (We shall assume that all dissolved CO_2 is initially in the form of H_2CO_3 before it dissociates.)

PROBLEM 5–38

Equilibria of polyprotic acids are discussed in Appendix 3 of Chemical Principles.

We have prepared a carbonic acid (H_2CO_3) solution by dissolving 0.020 mole of CO_2 in 1.0 liter of pure water. What is the H^+ concentration?

Data

$H_2CO_3 \rightleftharpoons H^+ + HCO_3^-$ $\quad K_{a_1} = 4.3 \times 10^{-7}$ mole liter^{-1}

$HCO_3^- \rightleftharpoons H^+ + CO_3^{2-}$ $\quad K_{a_2} = 5.6 \times 10^{-11}$ mole liter^{-1}

	[H^+], molar
1	0.020
2	0.040
3	9.3×10^{-5}
4	1×10^{-7}
5	4.8×10^{-11}

PROBLEM 5–39

What is the CO_3^{2-} concentration for the solution in Problem 5–38? Use the data and results from Problem 5–38.

$[CO_3^{2-}]$, molar
1 5.6×10^{-11}
2 9.3×10^{-5}
3 0.020
4 0.040
5 4.0

PROBLEM 5-40

When $NaHCO_3$ is dissolved in pure water, the relevant equilibria are

$$HCO_3^- + H_2O \rightleftharpoons H_3O^+ + CO_3^{2-}$$
$$HCO_3^- + H_2O \rightleftharpoons H_2CO_3 + OH^-$$

or we may write one equation that is the sum of these:

$$2HCO_3^- \rightleftharpoons H_2CO_3 + CO_3^{2-}$$

where

$$K = \frac{[H_2CO_3][CO_3^{2-}]}{[HCO_3^-]^2}$$

What is the value of K if K_{a_1} and K_{a_2} for H_2CO_3 are 4.3×10^{-7} and 5.6×10^{-11}, respectively?

1 $5.6 \times 4.3 \times 10^{-18} \simeq 24 \times 10^{-18}$
2 5.6×10^{-11}
3 4.3×10^{-7}
4 $(5.6/4.3) \times 10^{-4} \simeq 1.3 \times 10^{-4}$
5 $(4.3/5.6) \times 10^{4} \simeq 0.77 \times 10^{4}$

PROBLEM 5-41

Calculate a system that is close to the one in human blood by determining the H_2CO_3 (or dissolved hydrated CO_2) concentration in human arterial blood, where the pH = 7.4 and $[HCO_3^-] = 26 \times 10^{-3}$ molar.

$$\left.\begin{aligned} K_{a_1}(H_2CO_3) &\simeq 8.0 \times 10^{-7} \text{ mole liter}^{-1} \\ K_{a_2}(HCO_3^-) &\simeq 6.0 \times 10^{-11} \text{ mole liter}^{-1} \end{aligned}\right\} \text{ at } 37°C$$

$[H_2CO_3]$, molar
1 $10^{-7.4}$
2 4.4×10^{-7}
3 1.3×10^{-3}
4 2.6×10^{-3}
5 26×10^{-3}

PROBLEM 5–42 The titration curve in Figure 5–1 shows the titrations of equal volumes of two solutions. The dashed line is the titration of 0.10-normal strong acid; the solid line is an unknown. Which one of the following statements about the unknown is correct?

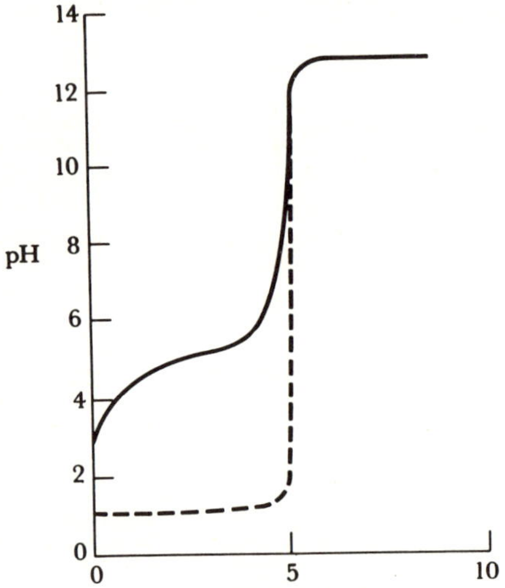

Figure 5–1

1 It is 0.10-normal strong acid.
2 It is 0.10-normal strong base.
3 It is 0.010-normal weak acid.
4 It is 0.10-normal weak base.
5 It is 0.10-normal weak acid.

PROBLEM 5–43 The titration curve in Figure 5–2 shows the pH behavior when 25 ml of an unknown were titrated with 0.1-normal NaOH. Which statement is appropriate?

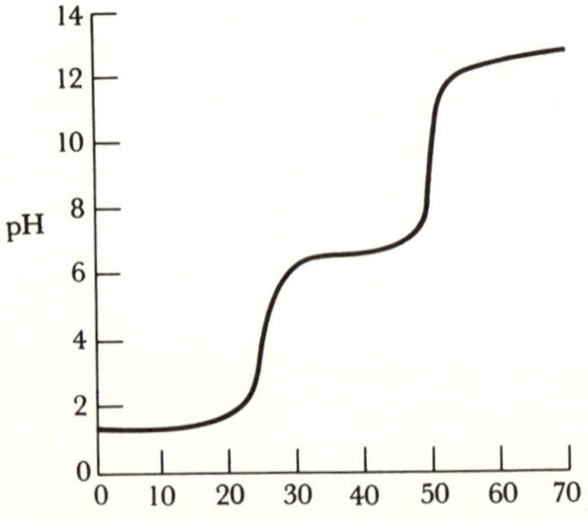

Figure 5–2

1 The unknown is a weak monoprotic acid.
2 The unknown is a strong monoprotic acid.
3 The unknown is a diprotic acid.
4 The unknown is a diprotic base.
5 The unknown is a triprotic acid.

PROBLEM 5–44

Some acid–base titration indicators are themselves weak organic acids. Thymolphthalein, for instance, changes from blue to colorless at pH = 10.5 as acid is added. At this pH the ratio of salt to acid is 0.10; that is, $[HIn]/[In^-] = 0.10$. Estimate K_a for thymolphthalein.

Notation

HIn = the acid form of thymolphthalein
In^- = the salt form of thymolphthalein

1 10^{-14}
2 $10^{-11.5}$
3 $10^{-10.5}$
4 $10^{-9.5}$
5 10^{-7}

Dissolving and precipitating used to be the sole task of a chemist. Even today these tasks may occupy a large share of his attention. The following problems treat the interaction of solids with their aqueous solutions.

PROBLEM 5–45

How many grams of $Ca(C_2O_4)$ will dissolve in pure H_2O to form 1.0 liter of saturated solution?

[$K_{sp}[Ca(C_2O_4)] = 2.5 \times 10^{-9}$ mole^2 liter^{-2};
mol wt: $Ca(C_2O_4) = 128$]

The solubility of ionic compounds is discussed in Section 5–10 of Chemical Principles *by Dickerson, Gray, and Haight.*

1 0.0064 g
2 0.016 g
3 0.032 g
4 0.11 g
5 0.26 g

PROBLEM 5-46 What are the units of K_{sp} for silver phosphate, $Ag_3(PO_4)$, if the concentrations are expressed as mole liter^{-1}?

1 mole liter^{-1}
2 mole^2 liter^{-2}
3 mole^3 liter^{-3}
4 mole^4 liter^{-4}
5 It is dimensionless; that is, it has no units.

PROBLEM 5-47 K_{sp} for lead iodate, $Pb(IO_3)_2$, is 3.2×10^{-14} at a given temperature. What is the solubility of $Pb(IO_3)_2$ in mole liter^{-1}?

1 $\sqrt{3.2} \times 10^{-7} \simeq 1.8 \times 10^{-7}$
2 $2.0 \times \sqrt{3.2} \times 10^{-7} \simeq 3.6 \times 10^{-7}$
3 2.0×10^{-5}
4 4.0×10^{-5}
5 8.0×10^{-5}

PROBLEM 5-48 Silver oxalate $Ag_2(C_2O_4)$ is dissolved in pure water. The concentration of Ag^+ ion in a saturated solution is 2.2×10^{-4} mole liter^{-1}. What is K_{sp} of $Ag_2(C_2O_4)$?

1 5.3×10^{-12}
2 11×10^{-12}
3 5.0×10^{-8}
4 2.4×10^{-8}
5 2.2×10^{-4}

PROBLEM 5-49 CaF_2 (fluorspar or fluorite) is used extensively in the metallurgical industry, particularly in the preparation of aluminum and steel. The solubility of CaF_2 in pure water is 3.5×10^{-4} mole liter^{-1}. K_{sp} is ________.

1 $(3.5 \times 10^{-4})^3 = 43 \times 10^{-12}$
2 $2(3.5 \times 10^{-4})^3 = 86 \times 10^{-12}$
3 $4(3.5 \times 10^{-4})^3 = 17 \times 10^{-11}$
4 $8(3.5 \times 10^{-4})^3 = 34 \times 10^{-11}$
5 $2(3.5 \times 10^{-4})^2 = 24 \times 10^{-8}$

PROBLEM 5–50

Silver cyanide solution is used in silver plating (Problem 3–6). The solubility product for the reaction

$$2AgCN(s) \rightleftharpoons Ag^+(aq) + Ag(CN)_2^-(aq)$$

is 4.0×10^{-12} mole^2 liter^{-2} at a given temperature. What is the equilibrium concentration of free Ag^+ in a saturated solution?

1 3.3×10^{-12} molar
2 8.0×10^{-12} molar
3 2.0×10^{-6} molar
4 4.0×10^{-6} molar
5 1.0×10^{-4} molar

PROBLEM 5–51

$BaSO_4$ is a compound that aids radiographic investigation of the gastrointestinal (GI) tract because it is opaque to radiation and only slightly soluble ($K_{sp} \simeq 1.6 \times 10^{-9}$). Suppose that you had added some solid $BaSO_4$ to pure water to obtain a saturated solution, and wished to reduce the Ba^{2+} concentration to a minimum to prepare it for ingestion by a patient uncommonly allergic to Ba^{2+}. Which procedure might be recommended?

Data

$$BaSO_4(s) \rightleftharpoons Ba^{2+}(aq) + SO_4^{2-}(aq) \text{ is endothermic}$$

1 Heat the system.
2 Add sufficient H_2SO_4 to give a $[H^+] = 1$ molar.
3 Add more solid $BaSO_4$.
4 Reduce the amount of saturated solution over the solid.
5 Add enough Na_2SO_4 to give a $[Na^+] = 10^{-2}$ molar.

PROBLEM 5–52

In preparing a barium solution for radiographic investigation, a nurse mistakenly uses a dilute $Ba(NO_3)_2$ (a soluble, strong electrolyte) solution. The error is discovered soon after the solution has been administered to a patient. The correct treatment to prevent the absorption of soluble barium and subsequent barium poisoning is given by which one of the choices below?

$[K_{sp}(BaSO_4) = 1.6 \times 10^{-9}]$

1 Injection of 15 ml of 2% Na_2SO_4 solution into the bloodstream.
2 Intestinal lavage (enema, cathartic) with 0.1-normal HNO_3.
3 Keep the patient awake and walking.
4 Intestinal lavage with 2% Na_2SO_4 solution.
5 None. No treatment is possible.

PROBLEM 5–53

An archaeologist must have salt-free water in his field laboratory, and it may be difficult to find in desert regions. One archaeological manual suggests the following recipe to test for the presence of chloride ion in water: 5.0 g of $AgNO_3$ (mol wt = 170) is dissolved in enough acidified distilled water to give 500 ml of solution. Ten drops (0.5 ml) of this silver solution are added to 10 ml of the water to be tested. What is the lowest $[Cl^-]$ that can be detected by the formation of a precipitate?

$[K_{sp}(AgCl) = 2.8 \times 10^{-10}]$

1 2.8×10^{-10} molar
2 1.4×10^{-10} molar
3 1.0×10^{-7} molar
4 $\sqrt{2.8} \times 10^{-5} = 1.7 \times 10^{-5}$ molar
5 $2 \times \sqrt{2.8} \times 10^{-5} = 3.4 \times 10^{-5}$ molar

PROBLEM 5–54

Radiometric analysis is discussed in Section 13–5 of Chemical Principles.

A 1.0×10^{-3} -molar solution of Na_2SO_4 contains a radioisotope tracer in the form of $^{35}_{16}S$, which has an 88-day half-life. A 10-ml sample of this solution gives 40,000 counts per minute (cpm) over background on a radiation detector. An equal volume of 2.0×10^{-4} molar $Pb(NO_3)_2$ is added to the Na_2SO_4 solution, and $PbSO_4$ precipitates. What is the counting rate (over background) for 10 ml of the solution over the precipitate?

$[K_{sp}(PbSO_4) = 1.3 \times 10^{-8}$;
$Pb(NO_3)_2$ and Na_2SO_4 are strong electrolytes.]

[*Note:* Assume that the experiment is performed in a time negligible compared to the 88-day half-life of the radioisotope.]

1 1,000 cpm
2 2,000 cpm
3 17,000 cpm
4 20,000 cpm
5 40,000 cpm

PROBLEM 5–55

The K_{sp} for CaF_2 is 1.7×10^{-10}. What is the solubility of CaF_2 in a solution of F^- ions having a concentration of 0.35 mole liter^{-1}?

1 2.4×10^{-10} mole liter^{-1}
2 4.9×10^{-10} mole liter^{-1}
3 1.4×10^{-9} mole liter^{-1}
4 1.6×10^{-5} mole liter^{-1}
5 2.2×10^{-5} mole liter^{-1}

PROBLEM 5-56

The K_{sp} for AgCl, silver chloride, (used in photography) is 2.8×10^{-10} at a given temperature. The solubility of AgCl in 0.010-molar HCl solution at this temperature will be ________.

1 2.8×10^{-12} mole liter^{-1}
2 2.8×10^{-8} mole liter^{-1}
3 5.6×10^{-8} mole liter^{-1}
4 2.8×10^{-4} mole liter^{-1}
5 5.6×10^{-4} mole liter^{-1}

PROBLEM 5-57

What weight of silver chromate (Ag_2CrO_4) will dissolve in 1.0 liter of a solution that is 0.010 molar in Ag^+?

[$K_{sp}(Ag_2CrO_4) = 1.9 \times 10^{-12}$;
mol wt: $Ag_2CrO_4 = 332$]

1 1.9×10^{-12} g
2 1.9×10^{-8} g
3 $\sqrt{1.9} \times 10^{-6} \simeq 1.4 \times 10^{-6}$ g
4 6.3×10^{-6} g
5 $1.4 \times 332 \times 10^{-6} = 4.7 \times 10^{-4}$ g

PROBLEM 5-58

A solution has the following concentrations:

$[Cl^-] = 1.5 \times 10^{-1}$ molar
$[Br^-] = 5.0 \times 10^{-4}$ molar
$[CrO_4^{2-}] = 1.9 \times 10^{-2}$ molar

A solution of $AgNO_3$ (100% dissociated) is added to the above solution, drop by drop. Which silver salt will precipitate first?

Data

$K_{sp}(AgCl) = 1.5 \times 10^{-10}$
$K_{sp}(AgBr) = 5.0 \times 10^{-13}$
$K_{sp}(Ag_2CrO_4) = 1.9 \times 10^{-12}$

1 AgCl
2 AgBr
3 Ag_2CrO_4
4 AgCl and AgBr together
5 None of the above.

PROBLEM 5–59

Suppose that CaF_2 is to be used as a fluoridation agent in a municipal water system. What is the $[F^-]$ if *extremely* hard water ($[Ca^{2+}] = 0.070$ molar) is saturated with CaF_2?

$[K_{sp}(CaF_2) = 1.7 \times 10^{-10}]$

1 $\sqrt{1.7} \times 10^{-5} = 1.3 \times 10^{-5}$ molar
2 2.5×10^{-5} molar
3 5.0×10^{-5} molar
4 1.0×10^{-4} molar
5 0.070 molar

PROBLEM 5–60

Hydrogen sulfide (H_2S) is a diprotic acid with

$K_{a_1} = 6.3 \times 10^{-8}$ mole liter^{-1}
$K_{a_2} = 1.0 \times 10^{-14}$ mole liter^{-1}

At 1.0 atm pressure and 25°C a solution is saturated with H_2S at $[H_2S] = 0.10$ mole liter^{-1}. If the pH = 7, what is $[S^{2-}]$ in such a solution?

[*Hint:* Review Problem 5–39.]

1 6.3×10^{-9} molar
2 6.3×10^{-2} molar
3 0.10 molar
4 $(1/6.3) \times 10^{7}$ molar $= 0.16 \times 10^{7}$ molar
5 6.3×10^{19} molar

PROBLEM 5–61

A 1.0-liter saturated solution of manganese sulfide (MnS) has 10 g of MnS solid at the bottom of the container. The pH is changed from 6.5 to 6.0 by the addition of concentrated strong acid such that the volume of the solution is essentially constant (1.0 liter). How much of the solid MnS will now dissolve into the solution? (Use the result of Problem 5–60, namely, $[H^+]^2[S^{2-}] = 6.3 \times 10^{-23}$ mole^3 liter^{-3}.)

$[K_{sp}(MnS) = 7.9 \times 10^{-13}$;
mol wt: MnS = 87]

1 $\sqrt{79} \times 10^{-14}$ g $= 8.9 \times 10^{-7}$ g
2 1.3×10^{-3} g
3 1.2×10^{-2} g
4 1.0 g
5 2.1 g

SOLUTIONS

SOLUTION 5-1

(2)

Equilibria involving homogeneous gas reactions are described in Section 5-1 of Chemical Principles.

Since this is a *homogeneous* gas phase reaction, all four substances are gases.

$$K_c = \frac{[NO]^4[H_2O]^6}{[NH_3]^4[O_2]^5} = \frac{\text{conc}^{10}}{\text{conc}^9} = \text{conc}^{+1}$$

2 is correct.
1 You forgot the reactants.
4 You inverted K_c.
3 You forgot to raise the concentrations to their stoichiometric coefficients.
5 You forgot the products.

SOLUTION 5-2

(5)

The first thing we shall do is to establish a format. First, write the equation:

$$2NH_3 \rightleftharpoons N_2 + 3H_2$$

Then write above the appropriate term the number (no.) of moles present initially:

initial	2	0	0	no. of moles
	$2NH_3 =$	$N_2 +$	$3H_2$	

and the number of moles present at equilibrium below the equation:

initial	2	0	0	no. of moles
	$2NH_3 =$	$N_2 +$	$3H_2$	
equilibrium	1	$\frac{1}{2}$	$\frac{3}{2}$	no. of moles

This result occurs because we are left with 1 mole of NH_3, so 1 mole of NH_3 must have dissociated. For every 2 moles of NH_3 that dissociate we obtain 1 mole of N_2 and 3 moles of H_2. So if 1 mole dissociates, we obtain half as much. We are asked to find K_c, which means that we should express the concentrations in units of moles per liter. Because the volume is 1 liter, however, the number of moles is numerically equal to the concentration, mole liter^{-1}. All we must do is to write the equilibrium constant,

$$K_c = \frac{[N_2][H_2]^3}{[NH_3]^2}$$

and plug in the equilibrium concentrations,

$$K_c = \frac{[\frac{1}{2}][\frac{3}{2}]^3}{[1]^2}$$

$$K_c = \frac{1}{2} \times \frac{3^3}{2^3} = \frac{27}{16} \qquad \text{(units are mole}^2\text{ liter}^{-2}\text{)}$$

The correct answer is **5**. Answer **2**: K_c is improperly constructed if this is your answer. You did not raise the concentrations to their stoichiometric coefficients. Answer **1**: You probably multiplied the $[NH_3]$ by two because you saw two as a stoichiometric coefficient.

SOLUTION 5-3

(3)

	$2H_2S \rightleftharpoons$	$2H_2$ +	S_2	
equil:	1.0	0.20	0.80	no. of moles
equil:	0.50	0.10	0.40	mole liter^{-1}

$$K_c = \frac{[H_2]^2[S_2]}{[H_2S]^2}$$

$$= \frac{(0.10)^2(0.40)}{(0.50)^2} \text{ mole liter}^{-1}$$

$$= 0.016 \text{ mole liter}^{-1}$$

Answer **3** is correct.

Answer **4**:

$$K_c \neq \frac{(0.20)^2(0.80)}{(1.0)^2} = 0.032$$

This is the error of neglecting to use concentration units.

Answer **2**:

$$K_c \neq \frac{(0.10)(0.40)}{(0.50)} = 0.080 \text{ mole liter}^{-1}$$

The stoichiometry was left out of the formulation of K_c.

Answer **5**:

$$K_c \neq \frac{(0.20)(0.80)}{1.0} = 0.16 \text{ mole liter}^{-1}$$

Both the above errors.

SOLUTION 5-4

(2)

If you have difficulty understanding calculations involving gas-phase equilibrium, try Section 5-11 of Programed Reviews of Chemical Principles *by Lassila* et al.

The quick way to do this is to realize that the total number of moles present initially does not change very much in going to equilibrium. Suppose that x moles of water are brought up to the conditions specified. Then the moles of each substance present are

	$2H_2O(g)$	$\rightleftharpoons$	$2H_2(g)$	+	$O_2(g)$	
equil:	$x - 2 \times 10^{-5}x$		$2 \times 10^{-5}x$		$1 \times 10^{-5}x$	no. of moles

So the total number of moles at equilibrium is

$$x - 2 \times 10^{-5}x + 2 \times 10^{-5}x + 1 \times 10^{-5}x = x(1 + 1 \times 10^{-5})$$

which is close to the initial value of $1x$. Then we shall say that since the total pressure at equilibrium is 1 atm, it is a good approximation that the partial pressures are

	$2H_2O(g)$	$\rightleftharpoons$	$2H_2(g)$	+	$O_2(g)$
p_{equil}	1 atm		2×10^{-5} atm		1×10^{-5} atm

$$K_p = \frac{[H_2]^2[O_2]}{[H_2O]^2} = \frac{(2 \times 10^{-5}\text{ atm})^2(1 \times 10^{-5}\text{ atm})}{(1\text{ atm})^2}$$

$$K_p = 4 \times 10^{-15}\text{ atm}$$

Answer **2** is correct.

SOLUTION 5-5

(4)

	$3H_2$	+	N_2	=	$2NH_3$	
equil:	0.400		0.800		?	atm

$$P_{total} = 2.80\text{ atm} = p_{NH_3} + p_{N_2} + p_{H_2}$$
$$= p_{NH_3} + 0.800 + 0.400$$
$$p_{NH_3} = 1.60\text{ atm}$$

	$3H_2$	+	N_2	$\rightleftharpoons$	$2NH_3$	
equil:	0.400		0.800		1.60	atm

$$K_p = \frac{p_{NH_3}^{\,2}}{p_{N_2}p_{H_2}^{\,3}} = \frac{(1.60)^2}{(0.800)(0.400)^3} = 50.0\text{ atm}^{-2}$$

4 is correct.
3 $K_p \neq p_{NH_3}/p_{H_2}p_{N_2} = 5.00$
2 $P_{total} \neq p_{N_2} + 3p_{H_2} + 2p_{NH_3}$ $\quad K_p \neq 3.13$
5 $K_p \neq P^2_{total}/p_{H_2}^{\,3}p_{N_2} = 153$

SOLUTION 5-6 (4)

init:	0.84	0.18	0	no. of moles
	$PCl_5 \rightleftharpoons$	PCl_3	$+\ Cl_2$	
equil:	0.72	(0.18 + 0.12)	0.12	no. of moles
equil:	0.72	0.30	0.12	mole liter^{-1}

$$K_c = \frac{[PCl_3][Cl_2]}{[PCl_5]}$$

$$= \frac{(0.30)(0.12)}{0.72} \text{ mole liter}^{-1}$$

$$= 0.050 \text{ mole liter}^{-1}$$

Answer **4** is correct.

SOLUTION 5-7 (2)

Let x be the number of moles of CO that react:

init:	0.6	0.6	0	0	no. of moles
	CO	$+\ H_2O$	$\rightarrow CO_2$	$+\ H_2$	
equil:	$0.6 - x$	$0.6 - x$	x	x	no. of moles
equil:	$\frac{0.6 - x}{2}$	$\frac{0.6 - x}{2}$	$\frac{x}{2}$	$\frac{x}{2}$	mole liter^{-1}

$$K = \left(\frac{x}{2}\right)\left(\frac{x}{2}\right) \Big/ \left(\frac{0.6 - x}{2}\right)\left(\frac{0.6 - x}{2}\right) = 4$$

$$= \left(\frac{x}{2}\right)^2 \Big/ \left(\frac{0.6 - x}{2}\right)^2 = 4$$

Trick: Take the square root of the entire equation. The square root of four is two, and

$$\frac{x/2}{(0.6 - x)/2} = 2$$

$$\frac{x}{0.6 - x} = 2$$

$$x = 1.2 - 2x$$
$$3x = 1.2$$
$$x = 0.4 \text{ mole of } CO_2$$

Answer **2** is correct.
[*Check:* $(0.4/2)^2/[(0.6 - 0.4)/2]^2 = 0.2^2/0.1^2 = 4$.]

[*Note:* You may have solved this problem without dividing the number of moles by 2 liters and still obtained the correct answer. This is because the equilibrium constant, K_c, for this particular reaction is *dimensionless;* it possesses no units. Usually, however, it does have units, and an omission as above would lead to an incorrect result.]

SOLUTION 5-8

(2)

The effect of temperature on equilibrium constants is discussed in Section 5-4 of Chemical Principles.

The equilibrium constant depends *only* on T (Answer **2**). But we must carefully distinguish between a change in K_p and a shift in the equilibrium concentrations that leave K_p unchanged. There are an infinite number of sets of equilibrium concentrations that will satisfy K_p at a given T, and the particular set in our system will depend on Answers **1** and **5**. But K_p will be unchanged. What about Answers **3** and **4**? Since there are equal numbers of moles of gas on each side of the equation

$$H_2(g) + I_2(g) \rightleftharpoons 2HI(g)$$

a change in total pressure or volume will not shift the equilibrium concentrations. For example, if the pressure is increased, there is no way to shift the equilibrium such that the number of moles of gas is reduced. In terms of Le Chatelier's principle, there is no way to shift the equilibrium to relieve the stress. But if the number of moles of gas on each side were not equal, then there would be a change in composition with a pressure or volume change. Be *very* careful to distinguish between a change in equilibrium composition following the addition of a reactant or product (or a P or V change) that occurs at constant temperature and fixed K_p and a change in K_p itself brought about only by a temperature change.

SOLUTION 5-9

(2)

ΔH^0 negative means exothermic:

$$2NO \rightleftharpoons N_2 + O_2 + \text{heat}$$

Increasing T (adding heat) forces equilibrium to the left. $[N_2]$ and $[O_2]$ decrease. $[NO]$ increases; therefore,

$$K = \frac{[N_2][O_2]}{[NO]^2} \text{ decreases as } T \text{ increases}$$

But the K we are interested in is for the reaction

$$N_2(g) + O_2(g) \rightleftharpoons 2NO(g)$$

$$K' = \frac{[NO]^2}{[N_2][O_2]}$$

and this K' will increase as T increases since this reaction is endothermic.

Answer **2** is correct. Answer **4:** K is a constant and changes *only* with T.

SOLUTION 5-10

(4)

Although all the answers are more or less true, the scientific reason for chilling champagne is to keep the dissolved CO_2 in solution. As the temperature increases, the exothermic reaction

$$\text{liquid} + CO_2(g) \rightleftharpoons \text{carbonated liquid} + \text{heat}$$

is driven to the left, and the champagne goes "flat."

Answer **4** is correct.

SOLUTION 5-11 $H_2 \rightleftharpoons 2H$

(5)

is a dissociation reaction. H_2 is a stable molecule, and we must supply energy (heat) to break the chemical bond. The reaction is endothermic:

$H_2 + \text{heat} \rightleftharpoons 2H$

The equilibrium *constant, K,* depends only on *T*. So a change in volume (b) gives no change in *K*. Either Answer **3** or **5** is the correct one.

When *T* is increased (heat added), the reaction is shifted to the right, more H is produced, and the equilibrium constant

$$K = \frac{[H]^2}{[H_2]}$$

is increased.

Answer **5** is correct.

SOLUTION 5-12

(4)

	$2NOCl$ $\rightleftharpoons$	$2NO$ +	Cl_2	
equil:	1.4	0.70	0.40	mole liter^{-1}
new init:	1.4	0.80	0.40	mole liter^{-1}

To find the new equilibrium concentrations, we must realize that the equilibrium will be shifted to the left:

$$K_c = \frac{[NO]^2[Cl_2]}{[NOCl]^2}$$

$$K_c = \frac{(0.70)^2(0.40)}{(1.4)^2} = 0.10 \text{ for the old equilibrium}$$

The new initial concentrations are not equilibrium concentrations because they give a *K* that is too large:

$$\frac{(0.80)^2(0.40)}{(1.4)^2} > 0.10$$

So the new equilibrium concentrations will be smaller for NO and Cl_2 as they react with each other to produce more NOCl, and it will have a concentration greater than 1.4.

We can easily establish the limits within which the new concentrations must lie:

	$2NOCl$ $\rightleftharpoons$	$2NO$ +	Cl_2	
new equil:	1.4 to 1.5	0.70 to 0.80	0.35 to 0.40	molar

How did we arrive at these numbers?

Suppose that all the added NO (0.10 mole) is consumed. Then the [NO] would be equal to its original equilibrium value of 0.70, but the [Cl_2] would decrease by 0.05 mole in producing NOCl because of the stoichiometry. However, these concentrations indicate a shift too far to the left because

$$\frac{(0.70)^2(0.35)}{(1.5)} < 0.10$$

and the true equilibrium concentrations lie between these limits.

Answer **4** is correct.

[*Note:* Nitrosyl chloride (NOCl) is sometimes used commercially as a flour bleach.]

SOLUTION 5–13

(2)

Synthesis of ammonia is discussed in Section 5–4 of Chemical Principles.

At constant T, an increase in P will force the reaction to the right because there are fewer moles of gaseous products than reactants. Since it is exothermic, the reaction can be shifted to the right by a decrease in T. So high P and low T favor high equilibrium concentrations of NH_3.

Answer **2** is correct. Actually, these are not the conditions that make ammonia production feasible because the *rate* of production must be considered.

SOLUTION 5–14

(3)

$$K_1 = \frac{[SO_3]}{[SO_2][O_2]^{1/2}}$$

$$K_2 = \frac{[SO_2]^2[O_2]}{[SO_3]^2}$$

Either we can see the answer by inspection,

$$1/K_1^2 = K_2$$

and Answer **3** is correct, or we may solve the K_1 equation for $[SO_3]$:

$$[SO_3] = K_1[SO_2][O_2]^{1/2}$$

Square it:

$$[SO_3]^2 = K_1^2[SO_2]^2[O_2]$$

and substitute for $[SO_3]^2$ in the K_2 equation:

$$K_2 = \frac{[SO_2]^2[O_2]}{K_1^2[SO_2]^2[O_2]} = \frac{1}{K_1^2}$$

Answer **3** is correct.

SOLUTION 5–15 (3)

We had originally x mole liter^{-1} of $COCl_2$ that dissociated until, at equilibrium, the equation

$$K_c = \frac{[CO][Cl_2]}{[COCl_2]}$$

was satisfied. The concentrations at this equilibrium and the new equilibrium after $COCl_2$ was added are

	$COCl_2$	$\rightleftharpoons$ CO	+ Cl_2	
init:	x	0	0	mole liter^{-1}
equil:	0.40	a	a	mole liter^{-1}
new equil:	1.6	b	b	mole liter^{-1}

in which a and b are the unknown concentrations.

$$K_c = \frac{a^2}{0.40} = \frac{b^2}{1.6}$$

since both situations are *equilibrium* conditions.

$$\frac{1.6}{0.40} = 4.0 = \frac{b^2}{a^2}$$

$$2.0 = \frac{b}{a}$$

The CO and Cl_2 concentrations have doubled.

Answer **3** is correct.

SOLUTION 5–16 (4)

	PCl_3	+ Cl_2	$\rightleftharpoons$ PCl_5	
equil:	0.20	0.10	0.40	mole liter^{-1}
new init:	0.20	0.20	0.40	mole liter^{-1}
equil:	$0.20 - x$	$0.20 - x$	$0.40 + x$	mole liter^{-1}

in which x = moles of PCl_3 that react to form PCl_5.

$$K_c = \frac{[PCl_5]}{[PCl_3][Cl_2]} = \frac{0.40}{(0.20)(0.10)} = 20 \text{ liters mole}^{-1}$$

$$K_c = \frac{0.40 + x}{(0.20 - x)(0.20 - x)} = 20$$

$$0.40 + x = 0.80 - 8.0x + 20x^2$$

$$20x^2 - 9.0x + 0.40 = 0$$

$$x = \frac{9.0 \pm \sqrt{81 - 32}}{40} = 0.40 \text{ or } 0.05 \text{ mole liter}^{-1}$$

Since the first choice makes $[PCl_3]$ negative, we choose the second.

$$[PCl_5] = 0.45 \text{ mole liter}^{-1}$$

Answer **4** is correct.

$[PCl_3] = [Cl_2] = 0.15$ mole liter^{-1}

[*Check:* $0.45/(0.15)^2 = 20$. This is correct.]

SOLUTION 5–17

(2)

	CO +	H_2O ⇄	CO_2 +	H_2	
equil:	0.10	0.10	0.40	0.10	mole liter^{-1}
new init:	0.10	0.10	0.40	0.40	mole liter^{-1}
new equil:	$0.10 + x$	$0.10 + x$	$0.40 - x$	$0.40 - x$	mole liter^{-1}

$$K_c = \frac{(0.40)(0.10)}{(0.10)(0.10)} = 4 = \frac{(0.40 - x)(0.40 - x)}{(0.10 + x)(0.10 + x)}$$

$$4 = \frac{(0.40 - x)^2}{(0.10 + x)^2}$$

Take the square root of both sides:

$$2 = \frac{0.40 - x}{0.10 + x}$$

$$0.40 - x = 0.20 + 2x$$
$$x = 0.20/3 = 0.067 \simeq 0.07 \text{ mole liter}^{-1}$$
$$[H_2] = [CO_2] = 0.40 - 0.07 = 0.33 \text{ mole liter}^{-1}$$

Answer **2** is correct.

[*Check:* Do our calculated concentrations satisfy the equilibrium constant? $(0.33)^2/(0.17)^2 \simeq 2^2 = 4$. Thus, our solution is correct.]

SOLUTION 5–18

(1)

pH, pOH, and pK are discussed in Section 5–5 of Chemical Principles *by Dickerson, Gray, and Haight.*

pH is defined to be $-\log[H^+]$. (We will use $[H^+]$ as a convenient shorthand for $[H_3O^+]$, while remembering that the physical situation in the solution does not correspond to a proton.)

When a solution is neutral,

$$[H^+] = [OH^-]$$

We know that

$$[H^+] \times [OH^-] = 1 \times 10^{-14} \text{ mole}^2 \text{ liter}^{-2}$$

so

$$[H^+] = [OH^-] = 1 \times 10^{-7} \text{ molar}$$
$$\log[H^+] = -7$$
$$-\log[H^+] = 7 = \text{pH}$$

When a solution is acidic,

$$[H^+] > [OH^-]$$
$$[H^+] > 1 \times 10^{-7}$$
$$\log[H^+] > -7$$
$$-\log[H^+] < 7$$
$$\text{pH} < 7$$

1 is correct.
2 Base: $pH > 7$.
3 Neutral: $pH = 7$.
5 K_w will be other than 1×10^{-14} mole^2 liter^{-2} only when the temperature is changed.

SOLUTION 5–19

(4)

KOH is completely ionized in solution. Thus

$$[OH^-] = 0.001 = 1 \times 10^{-3} \text{ mole liter}^{-1}$$
$$K_w = [H^+][OH^-] = 1 \times 10^{-14} \text{ mole}^2 \text{ liter}^{-2}$$

$$[H^+] = \frac{1 \times 10^{-14}}{1 \times 10^{-3}} = 1 \times 10^{-11} \text{ mole liter}^{-1}$$

$$pH = -\log[H^+]$$
$$= -(-11) = 11$$

Answer **4** is correct. Answer **3**: This is $pOH = -\log[OH^-] = -(-3) = 3$. Answer **5**: This is $pK = -\log K = -(-14) = 14$.

SOLUTION 5–20

(4)

For a review of equilibrium calculations for acids and bases, see Section 5–2 of Programed Reviews of Chemical Principles *by Lassila* et al.

$Ca(OH)_2 \rightarrow Ca^{2+} + 2OH^-$ as the strong electrolyte dissolves completely. Since there is 5.0×10^{-4} mole of $Ca(OH)_2$, it will produce $2 \times (5.0 \times 10^{-4})$ mole of $OH^- = 1.0 \times 10^{-3}$ mole OH^-.

$$[OH^-] = \frac{1.0 \times 10^{-3} \text{ mole}}{100 \text{ ml}} = 1.0 \times 10^{-2} \text{ molar}$$

$$pOH = -\log[OH^-] = 2$$

$$[H^+][OH^-] = 1.0 \times 10^{-14}$$
$$\log[H^+] + \log[OH^-] = -14$$
$$pH \quad + \quad pOH \quad = 14$$

But we know that $pOH = 2$, so

$$pH = 12$$

Answer **4** is correct. Answer **3**: You forgot about the two OH^- from each $Ca(OH)_2$.

SOLUTION 5-21

(2)

	V (liter)	$\times$	molarity (mole liter^{-1})	=	no. of moles
NaOH	0.040		0.10		4.0×10^{-3}
HCl	0.010		0.45		4.5×10^{-3}

Both substances are completely ionized in solution, and then 4.0×10^{-3} mole of each neutralizes the other. We find a solution whose total volume is $0.040 + 0.010 = 0.050$ liter and which contains 0.5×10^{-3} mole (each) of H^+ and Cl^- ions:

$$[H^+] = \frac{0.5 \times 10^{-3} \text{ mole}}{0.050 \text{ liter}}$$

$$= 1 \times 10^{-2} \text{ molar}$$

$$pH = 2$$

Answer **2** is correct.

[*Note:* We have neglected the $[H^+]$ contribution from the dissociation of water since it is so much smaller than 1×10^{-2} molar. Moreover, we have assumed that the total volume is 0.050 liter, but for some systems we could add 40 ml of one substance to 10 ml of another and obtain a total volume of only 45 ml because the forces of attraction between the different species are so strong. However, we will make this assumption of additivity here.]

SOLUTION 5-22

(2)

$$[H^+] = 1 \times 10^{-6} \text{ mole liter}^{-1}$$

$$\text{moles of } H^+ = 1 \times 10^{-6} \text{ mole liter}^{-1} \times 1.3 \text{ liters}$$

$$= 1.3 \times 10^{-6} \text{ mole}$$

$$= 1.3 \times 10^{-6} \text{ equiv.}$$

since 1 mole of H^+ is equal to one equivalent of H^+.

Answer **2** is correct. Answer **1** is correct for a solution of $pH = 7$. In Answer **3**, you erred in the milliliter to liter conversion.

SOLUTION 5-23

(2)

$$K_w = [H^+][OH^-] = 2.42 \times 10^{-14} \text{ mole}^2 \text{ liter}^{-2}$$

At neutrality

$$[H^+] = [OH^-] = x$$

$$K_w = x^2 = 2.42 \times 10^{-14}$$

$$x = 1.56 \times 10^{-7} \text{ mole liter}^{-1}$$

$$\log x = (\log 1.56) - 7$$

$$= 0.193 - 7 = -6.807$$

$$pH = 6.8$$

Answer **2** is correct.

At temperatures above 25°C, water is more highly dissociated, and the H^+ concentration is greater.

SOLUTION 5-24

(1)

Hydrolysis equilibria are discussed in Section 5-9 of Chemical Principles.

All the compounds dissolve and dissociate, but since there are no proton donors in the list, only those ions that can hydrolyze to form *weak* bases can make the solution acidic:

$$A^+ + HOH \rightarrow \underset{\text{(undissociated)}}{\underset{\text{Weak base}}{AOH}} + \underset{\text{acid character}}{\underset{\text{Gives solution}}{H^+}}$$

Of all the ions in our list, only NH_4^+ will form a weak base. Answer **1** is correct.

3: Since neither Na^+ nor Cl^- hydrolyzes, a NaCl solution is neutral.
4: CO_3^{2-} hydrolyzes to form a weak acid, HCO_3^-, and hydroxide, which gives the solution a basic character.
2: The two hydrolysis reactions

$$NH_4^+ + H_2O \rightarrow NH_4OH + H^+$$
$$CH_3COO^- + H_2O \rightarrow CH_3COOH + OH^-$$

proceed to the same extent, so the solution's pH is unchanged.

[*Note:* The reason amphetamine poisoning is treated by acidifying the urine is that this compound is involved in an acid–base equilibrium:

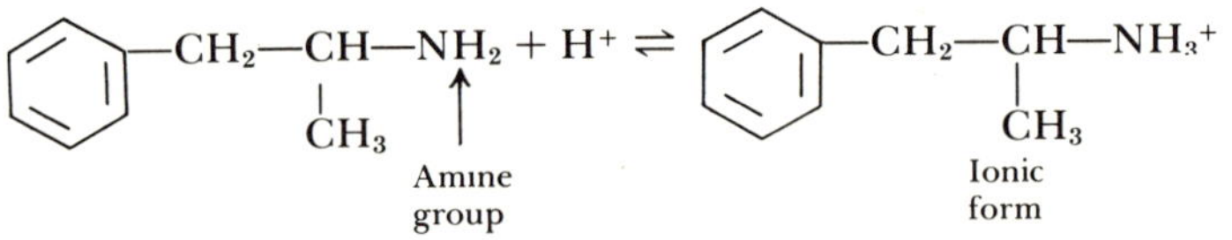

and the ionic form of the compound, which is favored when excess H^+ is added, cannot be reabsorbed by the body and is excreted. Another substance used for this treatment is ascorbic acid, vitamin C, which is not stored in the body and passes relatively quickly into the urine.]

SOLUTION 5-25

(3)

A conjugate acid–base pair differs by only a proton!

Pair (H_2O / H_3O^+)

$$HCl + H_2O \rightleftharpoons H_3O^+ + Cl^-$$

Pair (HCl / Cl^-)

Pair (H_2O / H_3O^+)

$$CH_3COOH + H_2O \rightleftharpoons H_3O^+ + CH_3COO^-$$

Pair (CH_3COOH / CH_3COO^-)

If they differ by only a proton, Answer **2** is clearly incorrect because the conjugate base of H_2O is OH^-, as, for instance, in the hydrolysis of cyanide ion:

Pair

$$CN^- + H_2O \rightleftharpoons HCN + OH^-$$

Pair

Moreover, since the pairs are on different sides of the equation, Answers **1** and **5** are not acceptable. CN^- and HCN do not react, nor do HCl and Cl^-, or CH_3COOH and CH_3COO^-. But the conjugate base of a weak acid is strong (Answer **3**) because a weak acid wants to keep its proton. Even if it is lost (to form the proton-lacking conjugate base), it will be welcomed back again. A substance that has a large affinity for protons is a strong base, and Answer **3** is correct. (Consider the second chemical equation in this solution in terms of the tendency to go in the forward and reverse directions.)

The first chemical equation in this solution is an example of the dissociation of a strong acid (Answer **4**), but the conjugate base, Cl^-, is not strong, for it does not show any tendency to push the equilibrium back to the left by reacting with the acid H_3O^+.

SOLUTION 5–26

(5)

Titration is not just a process of changing a certain pH to a pH of 7. Rather, it is the addition to the equivalents of acid (or base) present in the solution of an equal number of equivalents of base (or acid). Since we have equal volumes and identical concentration (and identical normalities because all the acids have only one dissociating hydrogen), we have equal numbers of equivalents. Thus, the solutions all require the same amount of base to titrate them.

Answer **5** is correct.

If you chose Answer **1** you may have reasoned that HCl is almost completely dissociated, whereas the other acids are not, and that it would therefore have the most free hydrogen ions to neutralize. But the weak acids are continually readjusting their equilibrium as more base is added to provide more H^+; for instance,

$$HCN \rightarrow H^+ + CN^-$$

until finally all the acid has dissociated because we have continued to add base to remove the H^+.

SOLUTION 5–27

(3)

$$K_a = \frac{[H^+][Cit^-]}{[HCit]}$$

From the reaction

$HCit \rightleftharpoons H^+ + Cit^-$

the $[H^+] = [Cit^-]$ if there is no citrate ion from any other source (as specified in the question) and if the contribution to $[H^+]$ from the dissociation of H_2O is negligible compared with that from citric acid. Since pH = 2.1, this second criterion appears to be acceptable.

$$K_a = \frac{[H^+]^2}{[HCit]} = \frac{(10^{-2.1})^2}{[HCit]} = 8.4 \times 10^{-4} \text{ mole liter}^{-1}$$

$$[HCit] = \frac{10^{-4.2}}{8.4 \times 10^{-4}}$$

$$\simeq \frac{6.3 \times 10^{-5}}{8.4 \times 10^{-4}}$$

$$= 0.75 \times 10^{-1} = 7.5 \times 10^{-2} \text{ molar}$$

Answer **3** is correct. For Answer **4**, you doubled the $[H^+]$. For Answer **5**, you did not square $[H^+]$.

SOLUTION 5–28

(3)

The species present just after solution but *before* dissociation or hydrolysis occur are

Na^+, CO_3^{2-} and Na^+, HSO_4^-

Na^+ does not hydrolyze, so it will not affect the pH. CO_3^{2-} can and will hydrolyze:

$CO_3^{2-} + H_2O \rightarrow HCO_3^- + OH^-$

because HCO_3^- is such a weak acid (we know this from the small value of K_{a_2}). The effect of this reaction is to produce OH^-, which makes the pool more alkaline; that is, raises the pH. It is a *pH increaser.* In theory, HSO_4^- can hydrolyze:

$HSO_4^- + H_2O \rightleftharpoons H_2SO_4 + OH^-$

but this reaction will not proceed to any appreciable extent because H_2SO_4 is a strong acid (large K_{a_1}) and wants to get rid of its protons, so the equilibrium lies far to the left for this reaction. The only alternative left is for HSO_4^- to dissociate: $HSO_4^- \rightarrow H^+ + SO_4^{2-}$. Since K_{a_2} is large, HSO_4^- is a strong acid and the proton-producing dissociation takes place. So HSO_4^- is a *pH reducer.*

Answer **3** is correct.

If you thought that CO_3^{2-} hydrolyzed, but that the hydrolysis products gave an acidic rather than a basic reaction, remember that an acidic or basic reaction is caused when a proton or hydroxide ion is *freed* from its parent compound.

SOLUTION 5–29

(3)

We must realize that the acid conjugate with the *base* HCO_3^- differs from it by a proton, and is H_2CO_3:

Pair

Acid — Base

$$HCO_3^- + H_2O \rightleftarrows H_2CO_3 + OH^-$$

Base — Acid

Pair

$$K_h(HCO_3^-) = \frac{[H_2CO_3][OH^-]}{[HCO_3^-]}$$

Note that $[H_2O]$ has been omitted for the usual reasons.

$$K_w = [H^+][OH^-]$$

$$K_{a_1}(H_2CO_3) = \frac{[HCO_3^-][H^+]}{[H_2CO_3]}$$

If we divide these two equations,

$$\frac{K_w}{K_{a_1}(H_2CO_3)} = \frac{[H_2CO_3][OH^-]}{[HCO_3^-]} = K_h(HCO_3^-)$$

(This result is quite general.)
Answer **3** is correct.

The numerical answer is

$$K_h(HCO_3^-) = \frac{1.0 \times 10^{-14} \text{ mole}^2 \text{ liter}^{-2}}{4.3 \times 10^{-7} \text{ mole liter}^{-1}}$$

$$= 2.3 \times 10^{-8} \text{ mole liter}^{-1}$$

SOLUTION 5–30

(2)

The first, and most difficult, thing to realize is that when sodium bicarbonate is added to water the solution will be basic. Without calculations, we can see this by comparing the values of the K_a's. H_2CO_3 is a very weak acid ($K_{a_1} = 4.3 \times 10^{-7}$ mole liter^{-1}), weaker even than acetic acid ($K_a = 1.8 \times 10^{-5}$ mole liter^{-1}). Suppose there was no second dissociation to worry about in H_2CO_3. Then we would have no trouble seeing that HCO_3^- (like acetate ion) would hydrolyze to give a weak acid, H_2CO_3, and OH^- to make the solution basic. But what about that second dissociation:

$$H_2O + HCO_3^- \rightleftharpoons H_3O^+ + CO_3^{2-}$$

K_{a_2} is so small (5.6×10^{-11} mole liter^{-1}) that HCO_3^- must be an *extremely* weak acid; thus the equilibrium lies so far to the left that it barely proceeds at all, and is negligible compared to the hydrolysis. We can look at this quantitatively, by comparing the $K_{a_2}(H_2CO_3)$ with $K_h(HCO_3^-)$:

Dissociation: $H_2O + HCO_3^- \rightleftharpoons H_3O^+ + CO_3^{2-}$
(acid reaction) $K_{a_2} = 5.6 \times 10^{-11}$ mole liter^{-1}

Hydrolysis: $H_2 + HCO_3^- \rightleftharpoons H_2CO_3 + OH^-$
(basic reaction) $K_h = 2.3 \times 10^{-8}$ mole liter^{-1}

The equilibrium constant for the hydrolysis reaction is almost three orders of magnitude larger than that for the acid dissociation, so hydrolysis predominates and the solution becomes basic.

The treatment must be to prevent aspirin from being absorbed by the body by passing through membranes. If the aspirin is ionized, its passage is restricted and it will pass harmlessly out of the intestine. The way $NaHCO_3$ solution can ionize aspirin is by affecting an acid-base equilibrium. If aspirin is a weak acid, HAsp,

$$\text{HAsp} + OH^- \text{ (from bicarbonate hydrolysis)} \rightarrow \text{Asp}^- + \text{HOH}$$

2 is correct.

3,4: If aspirin is basic, the addition of more base represses the dissociation and keeps it in its un-ionized form.

1: If aspirin is a strong acid, all of it has ionized already and the addition of base does not change the Asp^- concentration.

[*Note:* Aspirin is acetylsalicylic acid,

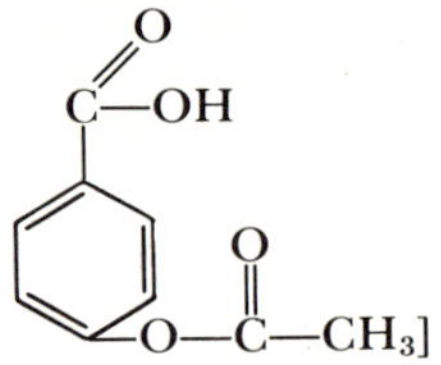

SOLUTION 5–31

(2)

Dissociation of acids and bases is discussed in Section 5–5 of Chemical Principles.

init:	0.10	0	0	mole liter^{-1}
	HBar $\rightleftharpoons$	H^+ +	Bar^-	
equil:	$0.10 - x$	x	x	mole liter^{-1}

in which x = mole liter^{-1} of HBar that dissociates.

Note that we have neglected the initial concentration of H^+ from the dissociation of water (1×10^{-7} molar) as being small compared to the H^+ contribution from barbituric acid.

$$K_a = \frac{x^2}{0.10 - x} = 10^{-4.01}$$

It is possible to solve this equation with the quadratic formula, but it is quicker to realize that x will be small because K_a is small and possibly much smaller than 0.10, so

$$0.10 - x \simeq 0.10$$

We shall try this approximation and check it at the end.

$$\frac{x^2}{0.10} \simeq 10^{-4.01}$$

$$x^2 \simeq 10^{-5.01}$$
$$x \simeq 10^{-2.50} \simeq 3.2 \times 10^{-3}$$

[*Check:* $0.10 - 3.2 \times 10^{-3} \simeq 0.10$ gives a 3% error, which is acceptable for the purposes of this calculation.] The percent dissociation is

$$100 \times \frac{3.2 \times 10^{-3}}{0.10} \simeq 3\%$$

Answer **2** is correct. Answer **1** is the number of moles per liter of HBar that have dissociated, incorrectly labeled "percent."

SOLUTION 5–32

(5)

To make the notation more wieldy, we shall make the following identifications:

NH_3^+—CH_2—COOH = H_2P^+
NH_3^+—CH_2—COO^- = HP (no net charge)
NH_2—CH_2—COO^- = P^-

Then the two equilibria become

$$H_2P^+ \rightleftharpoons HP + H^+ \qquad pK_1$$
$$HP \rightleftharpoons P^- + H^+ \qquad pK_2$$

$$K_1 = \frac{[HP][H^+]}{[H_2P^+]}$$

$$K_2 = \frac{[P^-][H^+]}{[HP]}$$

$$K_1K_2 = \frac{[H^+]^2[P^-]}{[H_2P^+]}$$

Because P^- and H_2P^+ are the only species of glycine that are charged, it follows that when $[P^-] = [H_2P^+]$, the number of positive charges on glycine molecules will equal the number of negative charges on glycine molecules.
So at the isoelectric point

$$K_1K_2 = [H^+]^2$$
$$\log[H^+] = \tfrac{1}{2}(\log K_1 + \log K_2)$$

(If this step is difficult, refer to Appendix 2.)

$$pH = \tfrac{1}{2}(pK_1 + pK_2)$$

Answer **5** is correct.

SOLUTION 5-33 (4)

Acid–base titration is treated in Section 5–6 of Chemical Principles.

At the end point, the equivalents of NaOH added will be equal to the equivalents of formic acid initially present. We will be left with H_2O, Na^+, and $HCOO^-$, and the $HCOO^-$ will hydrolyze. We might compare the solution at the end point to a solution prepared by adding sodium formate, Na^+ $(HCOO^-)$, to pure water. The correct concentration to consider for the analogy is 1.0 molar; since the initial concentrations are equal, equal volumes of formic acid solution and NaOH solution were used. So the $[Na^+]$ is halved as is the $HCOO^-$ concentration.

In this situation the formate ion immediately hydrolyzes:

init:	1.0		0	0	mole liter^{-1}
	$HCOO^-$	$+ \; H_2O \rightleftharpoons$	$HCOOH \; +$	OH^-	
equil:	$1.0 - x$		x	x	mole liter^{-1}

in which $x =$ mole liter^{-1} of formate that hydrolyzes. Note that we have neglected the contribution to the OH^- concentration from dissociation of water ($\simeq 1 \times 10^{-7}$ molar). We must check at the end to see if this assumption is justified. (See Problem 5–31 for a similar assumption.)

$$K_h = \frac{[HCOOH][OH^-]}{[HCOO^-]} = \frac{K_w}{K_a} = \frac{1.0 \times 10^{-14}}{1.8 \times 10^{-4}} = \frac{1.0}{1.8} \times 10^{-10} \text{ mole liter}^{-1}$$

(refer to Problem 5–29)

$$\frac{x^2}{1.0 - x} = \frac{1.0}{1.8} \times 10^{-10}$$

Approximation: Since K_h is so small we can assume that little of the $HCOO^-$ will hydrolyze; thus, we will further assume that x is small compared to 1.0.

$$\frac{x^2}{1.0 - x} \simeq \frac{x^2}{1.0} = \frac{1.0}{1.8} \times 10^{-10}$$

$$x \simeq 0.75 \times 10^{-5} \text{ molar} = [OH^-]$$

$$[H^+] \simeq \frac{1.0 \times 10^{-14}}{0.75 \times 10^{-5}} = 1.3 \times 10^{-9} \text{ molar}$$

$$\log[H^+] = -8.9$$
$$pH = 8.9$$

Answer **4** is correct. The correct titration end point is very much different from pH = 7.

[*Check:* Is $1.0 - x \simeq 1.0$?

$$1.0 - 0.75 \times 10^{-5} \simeq 1.0.$$

So our approximation is justified.]

SOLUTION 5-34 Let

(5)

HAc = acetic acid
Ac^- = acetate ion

The hydrolysis of acetate is

$$\begin{array}{llcccccl} \text{init:} & 1.0 & & & & 0 & 0 & \text{mole liter}^{-1} \\ & Ac^- & + & H_2O & \rightleftharpoons & HAc & + \ OH^- & \\ \text{equil:} & 1.0 - x & & & & x & x & \text{mole liter}^{-1} \end{array}$$

in which x is the number of moles per liter that hydrolyze. Again, we have neglected the contribution to $[OH^-]$ from the dissociation of H_2O.

$$K_h = \frac{[HAc][OH^-]}{[Ac^-]} = \frac{K_w}{K_a} = \frac{1.0 \times 10^{-14}}{1.8 \times 10^{-5}} \text{ mole liter}^{-1}$$

Approximation: $1.0 - x \simeq 1.0$. We shall try this because K_h is so small that we do not expect much Ac^- to hydrolyze.

$$\frac{x^2}{1.0 - x} \simeq \frac{x^2}{1.0} = \frac{1.0}{1.8} \times 10^{-9}$$

$$x \simeq 2.4 \times 10^{-5} \text{ molar} = [OH^-]$$

$$[H^+] = \frac{K_w}{[OH^-]} = \frac{1.0 \times 10^{-14} \text{ mole}^2 \text{ liter}^{-2}}{2.4 \times 10^{-5} \text{ mole liter}^{-1}}$$

$$= 4.2 \times 10^{-10} \text{ molar}$$

$$\text{pH} = 9.4$$

[*Check:* Is $1.0 - x \simeq 1.0$?

$1.0 - 2.4 \times 10^{-5} \simeq 1.0$]

Answer **5** is correct.

Answer **1** is the pH of a 1.0-normal solution of strong acid. Answer **2** is the pH of a 1.0-normal solution of acetic acid.

We can now summarize the approximate concentrations present in this solution.

Species	Concentration (mole liter^{-1})
Ac^-	1.0
OH^-	2.4×10^{-5}
HAc	2.4×10^{-5}
H^+	4.2×10^{-10}

SOLUTION 5–35 We shall use the same notation as we did in Problem 5–34.

(1)

$$HAc \rightleftharpoons H^+ + Ac^- \quad \text{(a)}$$
$$Ac^- + H_2O \rightleftharpoons HAc + OH^- \quad \text{(b)}$$

are the two processes that we must consider.

From Reaction (a), [HAc] $\simeq$ initial concentration of HAc because it is a weak acid ($K_a = 10^{-4.75}$), and its dissociation will be suppressed (shifted to the left) by the presence of Ac^-.

From Reaction (b), [Ac^-] $\simeq$ initial concentration of Ac^- since (in Problem 5–34) for 1.0-molar NaAc only 2.4×10^{-5} mole of Ac^- hydrolyzed. This equilibrium will be suppressed also by the presence of HAc.

$$K_a = \frac{[H^+][Ac^-]}{[HAc]}$$

Now we want the condition for $pK_a = pH$.

$$\log K_a = \log[H^+]$$
$$K_a = [H^+]$$

which will be satisfied if $[Ac^-] = [HAc]$. Thus, we must make $[Ac^-] = [HAc]$. By our approximations above, this equivalence can be accomplished by making the initial acetate concentration equal to the initial acetic acid concentration.

Answer **1** is correct.

[*Note:* If you solved this problem you probably have derived (or are one step from doing so) the equation

$$pH = pK_a + \log\frac{[\text{salt}]}{[\text{acid}]}$$

which is called the Henderson–Hasselbalch equation. The approximations involved in this equation are that the equilibrium [Ac^-] is equal to the original concentration of sodium acetate and that the equilibrium [HAc] is equal to the original concentration of acetic acid, as we have discussed above.]

SOLUTION 5–36 In Problem 5–35, we derived the relationship

(3)

$$pH = pK_a + \log\frac{[\text{salt}]}{[\text{acid}]}$$

and assumed that the salt and acid concentrations were the initial ones.

$$pH = 4.75 + \log\frac{2.5}{0.50}$$
$$= 4.75 + 0.70$$
$$= 5.45 \simeq 5.5$$

We can check our assumptions now.

$[H^+] \simeq 3 \times 10^{-6}$ molar

so Reaction (a) [see Solution 5-35],

$HAc \rightleftharpoons H^+ + Ac^-$

could not have proceeded very much at all. If $[H^+] = 3 \times 10^{-6}$ molar, $[OH^-]$ must be $1.0 \times 10^{-14}/3 \times 10^{-6} \simeq 3 \times 10^{-9}$ molar. Reaction (b),

$Ac^- + H_2O \rightleftharpoons H_2O + OH^-$

was also suppressed. The assumptions of our equation are valid.

Answer **3** is correct. For Answer **2**, you put the wrong sign on the log.

[*Note:* Actually there is an error in these calculations because we have neglected the H^+ and OH^- from the dissociation of water, and these are important here. But for the precision of our calculations we can continue to neglect this contribution.]

SOLUTION 5-37

(3)

Consider Equation (a):

$$HAc \rightleftharpoons H^+ + Ac^- \qquad \text{(a)}$$

When acid (H^+) is added, Equation (a) is shifted to the left, thereby removing the H^+:

$H^+ + Ac^- \rightarrow HAc$

And in our buffer we have a relatively large concentration of Ac^- that can continue to combine with and remove added H^+.

We can approximate the acetate and acetic acid concentrations after the HCl has been added by assuming that *all* the H^+ combined with Ac^- to produce HAc.

$[HAc] = 0.50 + 0.10 = 0.60$ mole liter^{-1}

The $[Ac^-]$ has been lowered by the same amount.

$[Ac^-] = 2.5 - 0.10 = 2.4$ mole liter^{-1}

$$\frac{[Ac^-]}{[HAc]} = \frac{2.4}{0.60} = 4.0$$

$$pH = pK_a + \log\frac{[Ac^-]}{[HAc]} = 4.75 + \log 4.0 = 5.35$$

[*Note:* See Problem 5-35 for the derivation of this equation.]
Answer **3** is correct. Answer **1**: Wrong sign on the log.

SOLUTION 5–38

(3)

Since K_{a_2} is much smaller than K_{a_1}, we can assume that the major process will be

$$H_2CO_3 \rightleftharpoons H^+ + HCO_3^-$$

and ignore the second dissociation for the moment.

init:	0.020	0	0	mole liter^{-1}
	H_2CO_3	$\rightleftharpoons$ H^+	+ HCO_3^-	
equil:	$0.020 - x$	x	x	mole liter^{-1}

in which x is the number of moles per liter of bicarbonate (HCO_3^-) formed by this dissociation and in which we have neglected the contribution to $[H^+]$ from the dissociation of water.

$$K_{a_1} = \frac{[H^+][HCO_3^-]}{[H_2CO_3]} = \frac{x^2}{0.020 - x} = 4.3 \times 10^{-7} \text{ mole liter}^{-1}$$

Approximation: Since K_{a_1} is small, x must be small. We shall assume that $x \ll 0.020$ (x negligible compared to 0.020), so

$$\frac{x^2}{0.020 - x} \simeq \frac{x^2}{0.020} \simeq 4.3 \times 10^{-7}$$

$$x^2 \simeq 8.6 \times 10^{-9}$$
$$x \simeq 9.3 \times 10^{-5} \text{ molar}$$

[*Check:* Is $0.020 - x \simeq 0.020$?

$$0.020 - 9.3 \times 10^{-5} \simeq 0.020$$

and our approximation is accurate to better than 1%.]

Answer 3 is correct. Answer 2 is the $[H^+]$ if H_2CO_3 were a strong, completely dissociated acid.

Now recall that we have ignored the second dissociation. We shall consider it in Problem 5–39 to see if we were justified in this procedure.

SOLUTION 5–39

(1)

In Problem 5–38, we found that the concentrations for the first dissociation of H_2CO_3 are approximately

	$H_2CO_3 \rightleftharpoons$	H^+	+ HCO_3^-	
equil:	0.020	9.3×10^{-5}	9.3×10^{-5}	mole liter^{-1}

For the second dissociation, we shall assume that initially the above concentrations are present, and then we suddenly allow the dissociation to proceed until equilibrium is established.

init:	9.3×10^{-5}	9.3×10^{-5}	0	mole liter^{-1}
	HCO_3^- $\rightleftharpoons$	H^+	+ CO_3^{2-}	
equil:	$9.3 \times 10^{-5} - x$	$9.3 \times 10^{-5} + x$	x	mole liter^{-1}

in which x is the number of moles per liter of CO_3^{2-} formed.

Approximation: Since K_{a_2} is so small, we can assume that $x \ll 9.3 \times 10^{-5}$ and $9.3 \times 10^{-5} + x \simeq 9.3 \times 10^{-5}$.

$$K_{a_2} = \frac{[H^+][CO_3^{2-}]}{[HCO_3^-]} = \frac{(9.3 \times 10^{-5})(x)}{9.3 \times 10^{-5}} = 5.6 \times 10^{-11}$$

$$x = 5.6 \times 10^{-11} \text{ molar} = [CO_3^{2-}]$$

[*Check:*

$$9.3 \times 10^{-5} - x = 9.3 \times 10^{-5} - 5.6 \times 10^{-11} \simeq 9.3 \times 10^{-5}$$

and the approximation is correct.]

We can check this result again by using the overall equilibrium constant for the sum of the two dissociation reactions.

$$H_2CO_3 \rightleftharpoons 2H^+ + CO_3^{2-}$$

$$K = \frac{[H^+]^2[CO_3^{2-}]}{[H_2CO_3]} = K_{a_1}K_{a_2}$$

$$[CO_3^{2-}] = K_{a_1}K_{a_2}\frac{[H_2CO_3]}{[H^+]^2}$$

$$= \frac{4.3 \times 10^{-7} \times 5.6 \times 10^{-11} \times 0.020}{(9.3 \times 10^{-5})^2}$$

$$\simeq 5.6 \times 10^{-11} \text{ molar}$$

Answer **1** is correct.

[*Note:* This K, for the sum of the two dissociation reactions, must be satisfied, but it is not very useful. For instance, in solving Problem 5-38 it would not have given the pH directly. Caution must be exercised in the use of these equations, which are sums.]

So the solution has the following composition:

$[H_2CO_3] \simeq 0.020$ mole liter^{-1}
$[HCO_3^-] \simeq [H^+] \simeq 9.3 \times 10^{-5}$ mole liter^{-1}
$[CO_3^{2-}] \simeq [OH^-] \simeq 5.6 \times 10^{-11}$ mole liter^{-1}

Note that this problem justifies our neglect of the second dissociation in Problem 5-38.

For Answer **2**, this is either the $[HCO_3^-]$ or $[OH^-]$. For Answer **3**, this is $[H_2CO_3]$. For Answer **5**, this is the pH.

SOLUTION 5-40

(4)

$$K_{a_1} = \frac{[HCO_3^-][H_3O^+]}{[H_2CO_3]}$$

$$K_{a_2} = \frac{[CO_3^{2-}][H_3O^+]}{[HCO_3^-]}$$

$$K = \frac{[H_2CO_3][CO_3^{2-}]}{[HCO_3^-]^2}$$

$$K = \frac{[H_2CO_3]}{[HCO_3^-]} \times \frac{[CO_3^{2-}]}{[HCO_3^-]}$$

$$\frac{K_{a_1}}{[H_3O^+]} = \frac{[HCO_3^-]}{[H_2CO_3]}$$

$$\frac{K_{a_2}}{[H_3O^+]} = \frac{[CO_3^{2-}]}{[HCO_3^-]}$$

$$K = \frac{[H_3O^+]}{K_{a_1}} \times \frac{K_{a_2}}{[H_3O^+]}$$

$$K = \frac{K_{a_2}}{K_{a_1}} = \frac{5.6 \times 10^{-11}}{4.3 \times 10^{-7}} \simeq 1.3 \times 10^{-4}$$

Answer **4** is correct. For Answer **1**, you inverted K_{a_1}. For Answer **5**, you inverted K.

SOLUTION 5-41

(3)

For each individual dissociation we can still employ the Henderson–Hasselbalch equation, subject to the usual assumptions (see Problem 5–35) and the additional conditions that the K's (K_{a_1} and K_{a_2}) are sufficiently different that the two equilibria can be considered individually and separately.

$$K_{a_1} = \frac{[H^+][HCO_3^-]}{[H_2CO_3]}$$

The Henderson–Hasselbalch equation for this system is

$$\log K_{a_1} = \log[H^+] + \log \frac{[HCO_3^-]}{[H_2CO_3]}$$

$$pH = pK_{a_1} + \log \frac{[HCO_3^-]}{[H_2CO_3]}$$

Inserting the values

$pH = 7.4$
$pK_{a_1} = 6.1$
$[HCO_3^-] = 26 \times 10^{-3}$

in our equation gives

$$\log[H_2CO_3] = 6.1 - 7.4 + \log(26 \times 10^{-3})$$
$$= -2.9$$
$$[H_2CO_3] = 1.3 \times 10^{-3} \text{ molar}$$

Answer **3** is correct.

SOLUTION 5-42

(5)

The sharp rise in pH at 5 ml is the end point, so equal volumes of base were used to neutralize the two solutions, which are both acidic. Therefore, the unknown has the same concentration and is 0.1-normal acid. The question is whether it is strong or weak. The pH of the strong acid will be two when $[H^+] = 1 \times 10^{-2}$ or when most of the acid has been neutralized. For a weak acid, however,

$$pH = pK_a + \log \frac{[\text{salt}]}{[\text{acid}]}$$

The ratio of [salt]/[acid] is increased as base is added. The equilibrium

$$HA \rightleftharpoons H^+ + A^-$$

is forced to the right as H^+ is removed from the solution by neutralization. Hence, we expect the pH to change markedly in the titration of a weak acid even before the end point is reached, as shown in Figure 5-1.

Answer **5** is correct.

SOLUTION 5-43

(3)

There are two sudden changes of pH (when about 25 and 50 ml of NaOH are added). The first sudden rise corresponds to the depletion of the first H^+ that dissociates. The second rise corresponds to the second H^+.

$$H_2A \rightleftharpoons H^+ + HA^-$$
$$HA^- \rightleftharpoons H^+ + A^{2-}$$

Finally, the pH tends toward 13, which corresponds to a 0.1-normal NaOH solution as more and more base is added.

Answer **3** is correct.

SOLUTION 5-44

(4)

$$HIn \rightleftharpoons H^+ + In^-$$

$$K_a = \frac{[H^+][In^-]}{[HIn]}$$

$$K_a = [H^+] \times \frac{[In^-]}{[HIn]}$$

$$K_a = \frac{[H^+]}{0.10}$$

At this condition, pH = 10.5, and $[H^+] = 10^{-10.5}$.

$$K_a = \frac{10^{-10.5}}{0.10} = 10^{-9.5}$$

Answer **4** is correct. Answer **2**: You inverted the [salt]/[acid] condition.

SOLUTION 5-45

(1)

init:	0	0	mole liter^{-1}
$Ca(C_2O_4)(s) \rightleftharpoons$	$Ca^{2+}(aq)$ +	$C_2O_4^{2-}(aq)$	
equil:	x	x	mole liter^{-1}

in which x is the concentration of the saturated solution.

$K_{sp} = [Ca^{2+}][C_2O_4^{2-}] = 2.5 \times 10^{-9}$ mole^2 liter^{-2}
$x^2 = 2.5 \times 10^{-9} = 25 \times 10^{-10}$ mole^2 liter^{-2}
$x = 5.0 \times 10^{-5}$ mole liter^{-1} = solubility of $Ca(C_2O_4)$
wt of $Ca(C_2O_4)$ per liter
of saturated solution $= 5.0 \times 10^{-5}$ mole liter$^{-1} \times 128$ g mole^{-1}
$= 0.0064$ g liter^{-1}

Answer **1** is correct. Answer **4**: If you thought that the dissociation was

$Ca(C_2O_4) \rightarrow Ca^{2+} + 2CO_2^-$

then K_{sp} would have been $[Ca^{2+}][CO_2^-]^2$ and the units of K_{sp} would be mole^3 liter^{-3}, which is contrary to the units given in the question.

SOLUTION 5-46

(4)

$Ag_3(PO_4)(s) \rightleftharpoons 3Ag^+(aq) + PO_4^{3-}(aq)$
$K_{sp} = [Ag^+]^3[PO_4^{3-}]$
K_{sp} = (mole liter^{-1})^3(mole liter^{-1})
K_{sp} = mole^4 liter^{-4}

Answer **4** is correct.

Activity is defined in Section 16-1 of Chemical Principles.

Remember that we are using concentrations here, not dimensionless activities.

SOLUTION 5-47

(3)

Since we obtain two ions of IO_3^- for each Pb^{2+} ion, the concentration of IO_3^- at equilibrium will be twice that of Pb^{2+}. If x is the number of moles of $Pb(IO_3)_2$ that dissolve to form a liter of solution, then

init:	0	0	mole liter^{-1}
$Pb(IO_3)_2(s) \rightleftharpoons$	$Pb^{2+}(aq)$ +	$2IO_3^-(aq)$	
equil:	x	$2x$	mole liter^{-1}

$K_{sp} = [Pb^{2+}][IO_3^-]^2 = (x)(2x)^2 = 3.2 \times 10^{-14}$ mole^3 liter^{-3}
$4x^3 = 3.2 \times 10^{-14} = 32 \times 10^{-15}$
$x^3 = 8.0 \times 10^{-15}$
$x = 2.0 \times 10^{-5}$ molar = solubility of $Pb(IO_3)_2$

since it is the number of moles of the salt that dissolve to form 1.0 liter of solution.

Answer **3** is correct. Answer **4** is the IO_3^- concentration.

SOLUTION 5–48

(1)

K_{sp} calculations for slightly soluble salts are performed in Section 5–10 of Chemical Principles.

init:		0	0	mole liter^{-1}
	$Ag_2(C_2O_4)(s) \rightleftharpoons$	$2Ag^+(aq)$ +	$C_2O_4^{2-}(aq)$	
equil:		2.2×10^{-4}	1.1×10^{-4}	mole liter^{-1}

Since for every two Ag^+ ions we obtain only one oxalate ion, $[C_2O_4^{2-}] = \frac{1}{2}[Ag^+] = 1.1 \times 10^{-4}$ mole liter^{-1}.

$$K_{sp} = [Ag^+]^2[C_2O_4^{2-}]$$
$$= (2.2 \times 10^{-4})^2(1.1 \times 10^{-4})$$
$$= 5.3 \times 10^{-12} \text{ mole}^3 \text{ liter}^{-3}$$

1 is correct.
4 $K_{sp} \neq [Ag^+][C_2O_4^{2-}] = 2.4 \times 10^{-8}$
2 $[C_2O_4^{2-}] \neq 2.2 \times 10^{-4}$; $K_{sp} \neq (2.2 \times 10^{-4})^2(2.2 \times 10^{-4}) \simeq 11 \times 10^{-12}$
3 Combines the two above mistakes.

SOLUTION 5–49

(3)

init:		0	0	mole liter^{-1}
	$CaF_2(s) \rightleftharpoons$	$Ca^{2+}(aq)$ +	$2F^-(aq)$	
equil:		3.5×10^{-4}	7.0×10^{-4}	mole liter^{-1}

The concentration of F^- must be twice that of Ca^{2+} because we obtain two F^- ions for each Ca^{2+} when one CaF_2 molecule dissolves, and there are no other sources of either species.

$$K_{sp} = [Ca^{2+}][F^-]^2$$
$$= (3.5 \times 10^{-4})(7.0 \times 10^{-4})^2$$
$$= 2^2 \times (3.5 \times 10^{-4})^3$$
$$= 4 \times (3.5 \times 10^{-4})^3 = 17 \times 10^{-11} \text{ mole}^3 \text{ liter}^{-3}$$

3 is correct.
1 $[F^-] \neq [Ca^{2+}] = 3.5 \times 10^{-4}$; therefore, $K_{sp} \neq (3.5 \times 10^{-4})^3$
2 $K_{sp} \neq \frac{1}{2}[Ca^{2+}][F^-]^2 = 2 \times (3.5 \times 10^{-4})^3$
5 $K_{sp} \neq [Ca^{2+}][F^-] = 2 \times (3.5 \times 10^{-4})^2$

SOLUTION 5–50

(3)

init:		0	0	mole liter^{-1}
	$2AgCN(s) \rightleftharpoons$	$Ag^+(aq)$ +	$Ag(CN)_2^-(aq)$	
equil:		x	x	mole liter^{-1}

$$K_{sp} = [Ag^+][Ag(CN)_2^-] = 4.0 \times 10^{-12} \text{ mole}^2 \text{ liter}^{-2}$$
$$= x^2 = 4.0 \times 10^{-12}$$
$$x = 2.0 \times 10^{-6} \text{ molar} = [Ag^+] = [Ag(CN)_2^-]$$

Answer **3** is correct. Answer **4** is the concentration of Ag^+ plus that of the complexed silver, $Ag(CN)_2^+$. Answer **5**: $K_{sp} \neq (x)(2x)^2$.

SOLUTION 5-51

(5)

Answer **1**: Since the reaction is endothermic, heating the system will drive the equilibrium to the right, and more $BaSO_4$ will dissolve.

Answers **3** and **4**: The concentration of a saturated solution does not depend on the relative amounts of solid and solution present. Since the solution is *saturated,* no more solid can dissolve, and any amount of saturated solution above the solid has the same properties.

Answers **2** and **5**: By increasing the SO_4^{2-} concentration the equilibrium is shifted to the left and $[Ba^{2+}]$ is reduced. Certainly, the sulfuric acid will suppress the solution of $BaSO_4$ more than the Na_2SO_4 because the latter is at a lower concentration. But since the solution is being prepared for a patient whose GI tract is presumably not in the best condition, it is not recommended procedure to fill him full of 1-molar H_2SO_4. So the best solution is Answer **5**.

[*Note:* This is a "trick" question. We can allow ourselves this luxury because you are not being marked and, more importantly, because the invalid is imaginary.]

SOLUTION 5-52

(4)

If you have difficulty with solubility product calculations, try Section 5-3 of Programed Reviews of Chemical Principles *by Lassila* et al.

The solution of $Ba(NO_3)_2$ is represented by

$$Ba(NO_3)_2(s) \rightarrow Ba^{2+}(aq) + 2NO_3^-(aq)$$

and the treatment is to prevent the Ba^{2+} from being absorbed by the body. One way is to precipitate the Ba^{2+} as $BaSO_4$, which is very insoluble. However, doing this in the bloodstream (Answer **1**) would have disastrous clogging effects, and is therefore incorrect. Intestinal lavage is the technique, but the HNO_3 (Answer **2**) will have no effect because the $Ba(NO_3)_2$ solution is dilute. [Think of the analogy to a solution of common salt (NaCl) in water. Acidifying the salt water with HCl will not precipitate NaCl unless the solution is saturated.] However, a sodium sulfate solution will precipitate $BaSO_4$ in the intestine where it is easily voided.

Answer **4** is correct.

SOLUTION 5-53

(3)

First we calculate the concentration of the silver solution:

$$[Ag^+] = \frac{5.0 \text{ g}/170 \text{ g mole}^{-1}}{0.500 \text{ liter}}$$

$$= \frac{1}{17} \text{ molar}$$

Now we ask: What is the $[Ag^+]$ in the mixture of the 10-ml water to be tested and the 0.5-ml silver solution just *before* precipitation starts? We call it $[Ag^+]^*$.

$$[Ag^+]^* = \frac{1}{17} \text{ molar} \times \frac{0.5 \text{ ml}}{10.5 \text{ ml}}$$

$$= \frac{0.1}{17(2.1)} \text{ molar} \simeq \frac{1}{360} \text{ molar}$$

$$K_{sp} = [Ag^+][Cl^-] = 2.8 \times 10^{-10}$$

$$[Cl^-] \simeq 2.8 \times 10^{-10} \Big/ \frac{1}{360} \simeq 1 \times 10^{-7} \text{ molar}$$

as precipitation starts. At a lower $[Cl^-]$ concentration, no precipitate will form.

Answer **3** is correct.

[*Note:* The solution is acidified to prevent formation of $AgOH(s)$, which might obscure the test.]

SOLUTION 5–54

(3)

We can calculate the concentrations of SO_4^{2-} and Pb^{2+} just after the two solutions have been mixed, but *just before* $PbSO_4$ starts precipitating.

$$[SO_4^{2-}] = (1.0 \times 10^{-3} \text{ molar}) \frac{10 \text{ ml}}{20 \text{ ml}} = 0.50 \times 10^{-3} \text{ molar}$$

$$[Pb^{2+}] = (2.0 \times 10^{-4} \text{ molar}) \frac{10 \text{ ml}}{20 \text{ ml}} = 1.0 \times 10^{-4} \text{ molar}$$

Now we allow the precipitate to form until equilibrium is reached.

init:	1.0×10^{-4}		0.50×10^{-3}	mole liter^{-1}
	$Pb^{2+}(aq)$	+	$SO_4^{2-}(aq) \rightleftharpoons PbSO_4(s)$	
equil:	$1.0 \times 10^{-4} - x$		$1.0 \times 10^{-4} - x$	mole liter^{-1}

in which x = mole liter^{-1} of Pb^{2+} and SO_4^{2-} that precipitate.

$$K_{sp} = [Pb^{2+}][SO_4^{2-}] = 1.3 \times 10^{-8} \text{ mole}^2 \text{ liter}^{-2}$$
$$= (1.0 \times 10^{-4} - x)(0.50 \times 10^{-3} - x) = 1.3 \times 10^{-8}$$
$$x^2 - 6.0 \times 10^{-4}x + 5.0 \times 10^{-8} = 1.3 \times 10^{-8}$$
$$x^2 - 6.0 \times 10^{-4}x + 3.7 \times 10^{-8} = 0$$

If we use the quadratic formula

$$x = \frac{+6.0 \times 10^{-4} \pm \sqrt{36 \times 10^{-8} - 15 \times 10^{-8}}}{2}$$

$$= \frac{+6.0 \times 10^{-4} \pm 4.6 \times 10^{-4}}{2}$$

$$= 5.3 \times 10^{-4} \quad \text{or} \quad 0.7 \times 10^{-4} \text{ mole liter}^{-1}$$

we must choose the second root because the first root is physically meaningless: It means that more lead sulfate precipitates than we have originally. So $x = 0.7 \times 10^{-4}$ mole liter^{-1}.

$$[Pb^{2+}] = 1.0 \times 10^{-4} - 0.7 \times 10^{-4} = 0.3 \times 10^{-4} \text{ molar}$$
$$[SO_4^{2-}] = 5.0 \times 10^{-4} - 0.7 \times 10^{-4} = 4.3 \times 10^{-4} \text{ molar}$$

[*Check:* Do these concentrations give the correct K_{sp}?

$$K_{sp} = 1.3 \times 10^{-8} \stackrel{?}{=} (0.3 \times 10^{-4})(4.3 \times 10^{-4}) = 1.3 \times 10^{-8}.$$ This is right.]

A 10-ml sample of 1.0×10^{-3} molar sulfate gave 40,000 cpm; so a 10-ml sample of 4.3×10^{-4} molar sulfate will give a counting rate of

$$\frac{4.3 \times 10^{-4}}{1.0 \times 10^{-3}} \times 40{,}000 \simeq 17{,}000 \text{ cpm}$$

Answer **3** is correct.

SOLUTION 5–55 (3)

init:		0	0.35	mole liter^{-1}
	$CaF_2(s) \rightleftharpoons$	$Ca^{2+}(aq)$ +	$2F^-(aq)$	
equil:		x	$0.35 + 2x$	mole liter^{-1}

in which x is the number of moles of CaF_2 that dissolves per liter and is the solubility of CaF_2.

$K_{sp} \simeq [Ca^{2+}][F^-]^2 = 1.7 \times 10^{-10}$ mole^3 liter^{-3}
$(x)(0.35 + 2x)^2 = 1.7 \times 10^{-10}$

Now is the time to try an approximation. If we don't, we will be able to solve the problem (one equation, one unknown) but will have to solve a cubic equation:

$$4x^3 + 1.4x^2 + 0.35^2x - 1.7 \times 10^{-10} = 0$$

This could be tiresome.

An evident approximation is that the amount of F^- that dissolves from CaF_2 ($2x$) will be small compared with the amount that is there initially (0.35), or

$$0.35 + 2x \simeq 0.35$$

Our equation becomes

$$x(0.35 + 2x)^2 \simeq x(0.35)^2 \simeq 1.7 \times 10^{-10}$$

$$x \simeq \frac{1.7}{(0.35)^2} \times 10^{-10} \simeq 1.4 \times 10^{-9}$$

[*Check:* Is our assumption correct? Is $0.35 + 2x \simeq 0.35$?

$0.35 + 2x = 0.35 + 2.8 \times 10^{-9} = 0.35$. This is right.]

3 is correct.
1 $K_{sp} \neq 2[Ca^{2+}][F^-] = 2x(0.35)$
2 $K_{sp} \neq [Ca^{2+}][F^-] = x(0.35)$
5 $K_{sp} \neq [Ca^{2+}]^2[F^-] \simeq x^2(0.35)$

SOLUTION 5–56 (2)

init:		0	0.010	mole liter^{-1}
	$AgCl(s) \rightleftharpoons$	$Ag^+(aq)$ +	$Cl^-(aq)$	
equil:		x	$0.010 + x$	mole liter^{-1}

in which x is the amount of AgCl that dissolves per liter of solution, and is therefore the solubility.

$K_{sp} = [Ag^+][Cl^-] = x(0.010 + x) = 2.8 \times 10^{-10}$ mole^2 liter^{-2}

If we try to solve this problem exactly with the quadratic formula, we shall get into trouble:

$x^2 + 0.010x - 2.8 \times 10^{-10} = 0$
$x = [-0.010 \pm (1.0 \times 10^{-4} + 11.2 \times 10^{-10})^{1/2}]/2$

We know that x equals something very close to zero; and if our square-root-taking apparatus is not good, we cannot tell how close to zero. So let us try an approximation. If x is small, $0.010 + x \simeq 0.010$; hence, our K_{sp} equation becomes

$$x(0.010 + x) \simeq x(0.010) = 2.8 \times 10^{-10}$$
$$x \simeq 2.8 \times 10^{-8} \text{ molar}$$

[*Check:* Our approximation is that $0.010 + x$, which is $0.010 + 2.8 \times 10^{-8}$, is $\simeq 0.010$. So it is correct.]

Answer **2** is correct.

SOLUTION 5–57

(4)

init:		0.010	0	mole liter^{-1}
	$Ag_2CrO_4(s) \rightleftharpoons$	$2Ag^+(aq)$ +	$CrO_4^{2-}(aq)$	
equil:		$2x + 0.010$	x	mole liter^{-1}

in which x = moles of Ag_2CrO_4 that dissolve per liter of solution.

$$K_{sp} = [Ag^+]^2[CrO_4^{2-}]$$
$$K_{sp} = (2x + 0.010)^2(x) = 1.9 \times 10^{-12}$$

We shall assume that

$$(2x + 0.010)^2x \simeq (0.010)^2x = 1.9 \times 10^{-12}$$
$$x \simeq 1.9 \times 10^{-8} \text{ molar}$$
$$\text{wt of } Ag_2CrO_4 = 1.9 \times 10^{-8} \text{ mole liter}^{-1} \times 332 \text{ g mole}^{-1} \times 1.0 \text{ liter}$$
$$= 6.3 \times 10^{-6} \text{ g}$$

Answer **4** is correct. Answer **2** is the molarity, incorrectly labeled grams.

[*Check:* Our approximation is $2x + 0.010 \simeq 0.010$. $2x + 0.010$ is $2 \times 1.9 \times 10^{-8} + 0.010$, and our approximation is valid.]

SOLUTION 5–58

(4)

We can calculate the $[Ag^+]$ for which each salt will just begin to precipitate. This is the concentration that makes the ion product equal to K_{sp}.

AgCl: $$[Ag^+][Cl^-] = K_{sp}(AgCl) = 1.5 \times 10^{-10}$$
$$[Ag^+] = 1.5 \times 10^{-10}/1.5 \times 10^{-1}$$
$$= 1.0 \times 10^{-9} \text{ molar}$$

AgBr: $$[Ag^+][Br^-] = K_{sp}(AgBr) = 5.0 \times 10^{-13}$$
$$[Ag^+] = 5.0 \times 10^{-13}/5.0 \times 10^{-4}$$
$$= 1.0 \times 10^{-9} \text{ molar}$$

Ag_2CrO_4: $$[Ag^+]^2[CrO_4^{2-}] = K_{sp}(Ag_2CrO_4) = 1.9 \times 10^{-12}$$
$$[Ag^+]^2 = 1.9 \times 10^{-12}/1.9 \times 10^{-2} = 1.0 \times 10^{-10}$$
$$[Ag^+] = 1.0 \times 10^{-5} \text{ molar}$$

So the first salts to precipitate—the ones that require the lowest $[Ag^+]$ to do so—will be AgCl and AgBr, which will precipitate together at the same $[Ag^+]$.

Answer **4** is correct.

SOLUTION 5–59

(3)

$$\begin{array}{lcccl} \text{init:} & & 0.070 & 0 & \text{mole liter}^{-1} \\ & CaF_2(s) \rightleftharpoons & Ca^{2+}(aq) & + \; 2F^-(aq) & \\ \text{equil:} & & 0.070 + x & 2x & \text{mole liter}^{-1} \end{array}$$

$$K_{sp} = [Ca^{2+}][F^-]^2 = (0.070 + x)(2x)^2 = 1.7 \times 10^{-10}$$

Approximation: $0.070 + x \simeq 0.070$ (because K_{sp} is so small, we guess that little of the salt will dissolve).

$$x^2 \simeq \frac{1.7 \times 10^{-10}}{4 \times 0.070} = 6.0 \times 10^{-10}$$

$$x \simeq 2.5 \times 10^{-5} \text{ molar}$$
$$[F^-] = 2x = 5.0 \times 10^{-5} \text{ molar}$$

Answer **3** is correct.

[*Check:* Is our approximation valid?

$$0.070 + x = 0.070 + 2.5 \times 10^{-5}$$

Thus, the error is small.]

For Answer **2**, you forgot that twice as many F^- dissolve as Ca^{2+}.

[*Note:* 5.0×10^{-5} molar in fluoride is about 1 part per million (ppm) by weight of fluoride ion in solution, an amount which is about one tenth that required for fluoridation for drinking water supplies. Moreover, hard water usually has a Ca^{2+} concentration less than one twentieth that specified in this problem.]

SOLUTION 5–60

(1)

From Problem 5–39, we have

$$K = \frac{[H^+]^2[S^{2-}]}{[H_2S]} = K_{a_1} \times K_{a_2} = 6.3 \times 10^{-8} \times 1.0 \times 10^{-14}$$
$$= 6.3 \times 10^{-22} \text{ mole}^2 \text{ liter}^{-2}$$

$$[H^+]^2[S^{2-}] = 6.3 \times 10^{-22}\,[H_2S] = 6.3 \times 10^{-22} \times 0.10$$
$$[S^{2-}] = 6.3 \times 10^{-23}/10^{-14} \text{ because } [H^+] = 10^{-7}$$
$$[S^{2-}] = 6.3 \times 10^{-9} \text{ molar}$$

1 is correct.
2 $[S^{2-}] \neq [HS^-] = K_{a_1}\,[H_2S]/[H^+]$
4 $K \neq K_{a_2}/K_{a_1}$
5 $K \neq K_{a_1}/K_{a_2}$

SOLUTION 5-61

(4)

$[H^+]^2[S^{2-}] = 6.3 \times 10^{-23}$

$$[S^{2-}] = 6.3 \times 10^{-23}/[H^+]^2 \text{ mole liter}^{-1} \qquad \text{(a)}$$

and

$$K_{sp} = [Mn^{2+}][S^{2-}] = 7.9 \times 10^{-13} \text{ mole}^2 \text{ liter}^{-2} \qquad \text{(b)}$$

If we combine equations (a) and (b),

$$[Mn^{2+}] = 7.9 \times 10^{-13}\ [H^+]^2/6.3 \times 10^{-23}$$
$$\simeq 1.3 \times 10^{10}\ [H^+]^2$$

At pH = 6.5

$$[Mn^{2+}] = 1.3 \times 10^{10} \times 10^{-13}$$
$$= 1.3 \times 10^{-3} \text{ molar}$$

At pH = 6.0

$$[Mn^{2+}] = 1.3 \times 10^{10} \times 10^{-12}$$
$$= 1.3 \times 10^{-2} \text{ molar}$$

The change in the concentration of the solution is an increase from 1.3×10^{-3} molar to 1.3×10^{-2} molar, or about 1.2×10^{-2} molar net increase. The weight of solid MnS that must dissolve to effect this change is 1.2×10^{-2} mole liter^{-1} $\times$ 87 g mole^{-1} = 105×10^{-2} g liter^{-1} $\simeq$ 1.0 g for the 1.0-liter solution.

4 is correct.
1 You neglected the H_2S equilibrium.
2 is just the molarity at pH = 6.5.
3 is the molarity change, not the weight change.

CHAPTER 6 CLASSIFICATION OF THE ELEMENTS AND PERIODIC PROPERTIES

The problems in this chapter deal with the periodic table and some of the properties of the elements such as ionization energy and electron affinity. To answer the problems, you will find it necessary to refer to the list of names of the elements (inside back cover) and the periodic table (inside front cover). You need not memorize the names of the elements or their positions in the periodic table, although with frequent use you will find that you have done so. However, it is necessary for you to memorize the common valences of the important families (or groups) of elements and the names associated with some of these families.

Group	Family name	Common valence	
IA	alkali metals	1	same as the group number
IIA	alkaline earth metals	2	
IIIA		3	
IVA		4	
VA		3	eight minus the group number
VIA	chalcogens	2	
VIIA	halogens	1	
VIIIA (or 0)	noble gases	?	

The valence of hydrogen (H) is always 1. The "B" group elements are known as the *transition* metals, whereas the elements beneath the main body of the periodic table are the *lanthanides* (rare earth metals) [cerium (Ce) to lutecium (Lu)] and the *actinides* [thorium (Th) to lawrencium (Lr)].

For the purpose of this chapter, we will neglect the difference between ΔH^0 and ΔE^0.

Solutions to problems in Chapter 6 begin on page 180.

PROBLEMS

PROBLEM 6–1

See Sections 6–1 and 6–2 of Chemical Principles *by Dickerson, Gray, and Haight for a discussion of early classification schemes for elements.*

In 1829, Johann Döbereiner observed that several groups of three elements (*triads*) had similar chemical properties. In addition, in every case the atomic weight of one element in the triad was nearly the average of the atomic weights of the other two. Which one of the following groups constitutes a Döbereiner triad?

(at. wt: Li = 7, Na = 23, S = 32, K = 39, Ca = 40, Se = 79, Sr = 88, Te = 127, Ba = 137)

1 Ca, Na, Te
2 S, Sr, K
3 Te, S, Se
4 Li, Se, Ba
5 Ca, Sr, Na

PROBLEM 6–2

From a knowledge of the physical properties of silicon (Si) and tin (Sn), Dmitri Mendeleev predicted the properties of "ekasilicon" (germanium, Ge) in 1869, before the element had been isolated. He placed Ge below Si and above Sn in his periodic table and assumed that the properties of Ge and its compounds were approximately the average of those observed for Si and Sn and their compounds. With the data given in Table 6–1, which of the physical properties indicated for Si and $SiCl_4$ would you expect to be correct on the basis of Mendeleev's assumption?

Table 6–1

	Ge	Sn
Melting point of metal (°C)	938	232
Density of metal (g cm^{-3})	5.36	7.31
Boiling point of the chloride MCl_4(°C)	84	114

1 Melting point of Si is 750°C.
2 Density of Si is 1.00 g cm^{-3}.
3 Density of Si is 7.10 g cm^{-3}.
4 Boiling point of $SiCl_4$ is 52°C.
5 Boiling point of $SiCl_4$ is 100°C.

PROBLEM 6–3

In 1913, Moseley discovered an approximately linear relationship between the square roots of the frequencies, ν, of the x rays emitted by a number of different elements and the atomic numbers, Z, of the elements. His plot for the elements titanium through zinc is shown in Figure 6–1. The general equation for the line in the diagram can be expressed as

$$\nu^{1/2} = A^{1/2}(Z - 1)$$

or

$$\nu = A(Z - 1)^2$$

in which A is a constant. The line on the graph is referred to as the K_α line. Moseley found a break in this line for $Z = 43$ because there was no element known for that particular value of Z. On the basis of his linear relationship he proposed that an element with $Z = 43$ should exist. If the frequency of the K_α x ray emitted by copper (Cu, $Z = 29$) is approximately 2.0×10^{18} sec^{-1}, at what frequency would you look for the K_α x radiation emitted by the missing element?

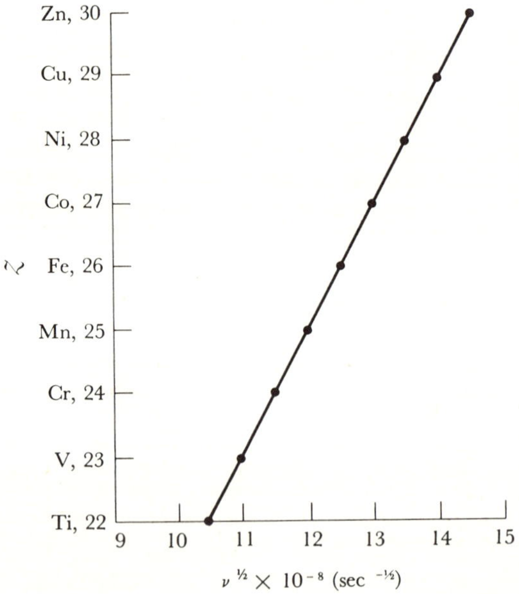

Figure 6–1

1 2.2×10^{17} sec^{-1}
2 2.0×10^{18} sec^{-1}
3 3.0×10^{18} sec^{-1}
4 4.5×10^{18} sec^{-1}
5 2.0×10^{19} sec^{-1}

PROBLEM 6–4

Which one of the following elements does *not* belong to the family of elements indicated?

1 Rubidium (Rb)—alkali metal.
2 Strontium (Sr)—alkaline earth metal.
3 Xenon (Xe)—noble gas.
4 Barium (Ba)—alkaline earth metal.
5 Iridium (Ir)—noble gas.

PROBLEM 6–5

Which one of the following elements does *not* belong to the family indicated?

1 Fluorine (F)—halogen
2 Cerium (Ce)—lanthanide or rare earth metal
3 Ruthenium (Ru)—transition metal
4 Astatine (At)—chalcogen
5 Uranium (U)—actinide

PROBLEM 6–6

Which of the following statements concerning the element iodine (I, $Z = 53$) is *incorrect?*

(a) Iodine has a greater atomic weight than tellurium (Te, $Z = 52$).
(b) Iodine is in the fifth period of the periodic table.
(c) Iodine has chemical properties similar to chlorine (Cl, $Z = 17$).
(d) Iodine is a halogen.
(e) One mole of iodine contains 2 g-atoms.

1 Only (a).
2 (b) and (e).
3 (c) and (d).
4 (a), (c), and (e).
5 None of them.

PROBLEM 6–7

Which one of the following statements concerning the elements and the periodic table is *incorrect?*

1 The modern periodic law may be stated as: Properties of elements are periodic functions of their atomic numbers.
2 There are more nonmetallic elements than metallic elements.
3 There are a few elements that have properties intermediate between those of metals and nonmetals.
4 Hydrogen can be considered to be a member of more than one group.
5 Metallic behavior decreases as you go from left to right across a period.

PROBLEM 6–8

The first ionization energy for a mole of gaseous neutral atoms M is $\Delta H°$ for the process ________.

Ionization energies are discussed in Section 6–4 of Chemical Principles.

1 $M(s) \rightarrow M(g)$
2 $M^+(g) \rightarrow M^{2+}(g) + e^-$
3 $M(g) \rightarrow M^+(g) + e^-$
4 $M(g) \rightarrow M^{2+}(g) + 2e^-$
5 $M(l) \rightarrow M(g)$

PROBLEM 6–9

The second ionization energy of a gaseous, neutral atom M is defined by ΔH^0 for the process

$M^+(g) \rightarrow M^{2+}(g) + e^-$

Which statement is correct?

1 The second ionization energy is greater than the first ionization energy.
2 The second ionization energy is zero.
3 The fourth ionization energy would be defined by ΔH^0 for the process $M^{2+}(g) \rightarrow M^{3+}(g) + e^-$.
4 The third ionization energy of helium (He, $Z = 2$) would be greater than its second ionization energy.
5 The mass of $M^{2+}(g)$ would be greater than that of $M^+(g)$.

PROBLEM 6–10

Which of the following reactions requires the greatest amount of energy before it would take place?

(a) $Si(g) \rightarrow Si^+(g) + e^-$
(b) $Si^+(g) \rightarrow Si^{2+}(g) + e^-$
(c) $Si^{2+}(g) \rightarrow Si^{3+}(g) + e^-$
(d) $Si^{3+}(g) \rightarrow Si^{4+}(g) + e^-$

1 (a)
2 (b)
3 (c)
4 (d)
5 All of these reactions take place spontaneously; no energy would have to be supplied to any of them.

PROBLEM 6–11

Three units of energy commonly used by chemists are the calorie, the erg, and the electron volt. The electron volt (eV) is the kinetic energy that *one* electron gains when accelerated through a potential difference of 1 V in an electric field. With the data given, calculate which one of the five answers shown below is correct.

Data

$1 \text{ eV} = 1.6 \times 10^{-12}$ erg
$1 \text{ cal} = 4.2 \times 10^{7}$ ergs
$1 \text{ kcal} = 10^{3}$ cal

1 $1 \text{ eV} = 3.8 \times 10^{-23}$ kcal mole^{-1}
2 $1 \text{ eV} = 1.6$ kcal mole^{-1}
3 $1 \text{ eV} = 23$ kcal mole^{-1}
4 $1 \text{ eV} = 42$ kcal mole^{-1}
5 $1 \text{ eV} = 2.3 \times 10^{4}$ kcal mole^{-1}

PROBLEM 6–12

Figure 6–2 is a plot of the first ionization energies (in electron volts) of the elements against atomic number, Z. Which one of the following elements has the largest first ionization energy?

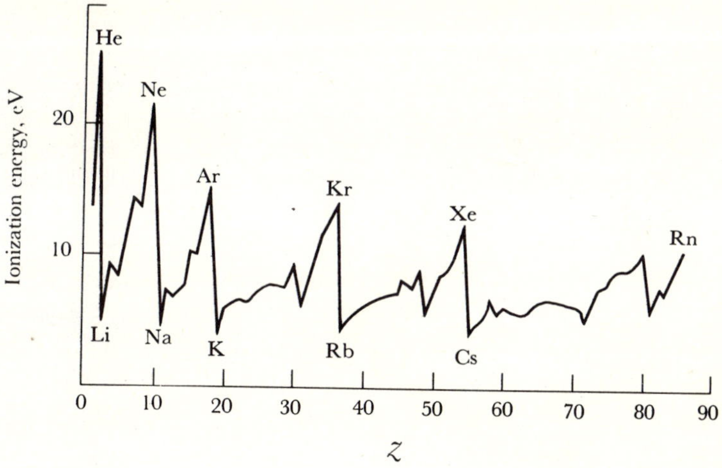

Figure 6–2

1 Lithium
2 Carbon
3 Nitrogen
4 Oxygen
5 Fluorine

PROBLEM 6–13 Figure 6–3 illustrates schematically the operation of an ionization gauge, which is a device for measuring low pressure in the range 10^{-4} to 10^{-10} torr. Its design is similar to that of a triode vacuum tube. Electrons travel from a heated filament through a grid (whose voltage determines the electron energy) into a chamber connected to the high vacuum system whose pressure you wish to measure. These electrons ionize some of the neutral species there and form positive ions that are attracted to a negative collector plate. The positive ions cause a small current that can be measured and related to the pressure in the high vacuum system. The current varies proportionally to the number of positive ions arriving at the collector plate. The lower the pressure in the vacuum system, the fewer neutral molecules there are in the chamber and fewer positive ions will be formed. The fewer positive ions formed, the smaller is the current measured.

You have calibrated an ionization gauge whose electron energy is 15 eV by measuring the pressure of a system containing sodium vapor (Na). How would your gauge reading be affected if you replaced the sodium vapor in the system by neon gas (Ne) at the same pressure? [*Hint:* Refer to Figure 6–2.]

(Ne, $Z = 10$; Na, $Z = 11$)

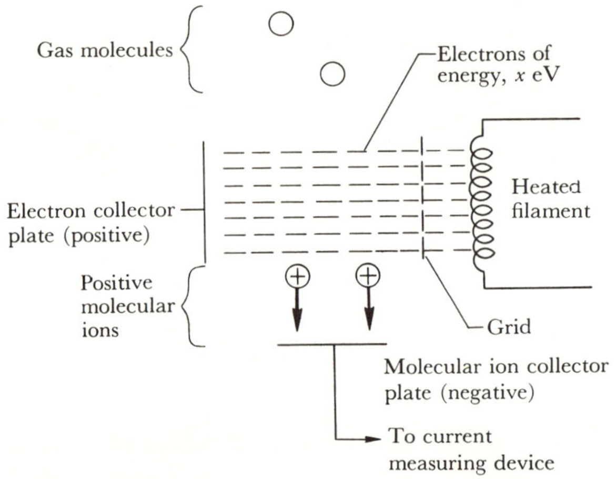

Figure 6–3

1 Unaffected; same gauge reading.
2 Gauge reading would be higher.
3 Gauge reading would be lower but not zero.
4 Gauge reading would be zero.
5 None of the above is correct.

PROBLEM 6–14

You want to design an ion propulsion engine for rocketry. If you were to choose the fuel solely on the basis that it required the least energy to ionize, which one of the elements below would you select? [*Hint:* Refer to Figure 6–2.]

1 He
2 Li
3 Ne
4 Cs
5 Rn

PROBLEM 6–15

Consider the process

$X(g) + e^- \rightarrow X^-(g)$

in which X is any atomic gas. $\Delta H°$ for this process is called ________.

1 The vaporization energy.
2 The solvation energy.
3 The hydration energy.
4 The electron affinity.
5 The sublimation energy.

PROBLEM 6–16

Which one of the following elements would you expect to have the largest electron affinity?

Electron affinity is discussed in Section 6–4 of Chemical Principles.

1 Aluminum (Al).
2 Chlorine (Cl).
3 Phosphorus (P).
4 Silicon (Si).
5 Sulfur (S).

PROBLEM 6–17

An *isoelectronic series* is a series in which all members have the same number of electrons but differ in nuclear charge (i.e., the total proton charge resulting from the protons in the nucleus). Consider an isoelectronic series in which the members have *nuclear* charges in units of proton charge 9, 10, 11, 12, 13, and 14. If the member of the series with nuclear charge 14 is Si^{5+}, what are the other members of the series in order of *increasing* nuclear charge?

1 Al^{4+}, Mg^{3+}, Na^{2+}, Ne^{+}, F
2 F, Ne^{+}, Na^{2+}, Mg^{3+}, Al^{4+}
3 P^{5+}, S^{6+}, Cl^{7+}, Ar^{8+}, K^{9+}
4 Be^{5-}, B^{4-}, C^{3-}, N^{2-}, O^{-}
5 O^{-}, N^{2-}, C^{3-}, B^{4-}, Be^{5-}

PROBLEM 6–18

If the electron affinities of O and O^- are −34 kcal $mole^{-1}$ and +202 kcal $mole^{-1}$, respectively, what is $\Delta H°$ for the following reaction?

$$O(g) + 2e^- \rightarrow O^{2-}(g)$$

1 −34 kcal $mole^{-1}$
2 −168 kcal $mole^{-1}$
3 +168 kcal $mole^{-1}$
4 +202 kcal $mole^{-1}$
5 +236 kcal $mole^{-1}$

PROBLEM 6–19

Which one of the following isoelectronic ions would you expect to have the largest radius?

Sizes of atoms and ions are discussed in Section 6–4 of Chemical Principles.

1 Mn^{7+}
2 P^{3-}
3 S^{2-}
4 Sc^{3+}
5 Ti^{4+}

PROBLEM 6–20

On the basis of group valences, predict the formula of the compound formed between tin (Sn) and fluorine (F).

1 SnF
2 SnF_2
3 SnF_4
4 SnF_5
5 SnF_6

PROBLEM 6–21

Recently, the two organomercury compounds, $(CH_3)_2Hg$ and CH_3Hg^+, have achieved notoriety because they have been identified as the principal sources of mercury contamination in fish. Many nations have instituted controls on the level of inorganic mercury that industrial companies may discharge into the rivers and lakes, because it has been established that certain bacteria convert inorganic to organic mercury that is concentrated in fish. People eating these fish in large quantities may suffer nerve damage or ultimately death. Which one of the following elements would you expect to be *most like mercury* in its chemical behavior?

1 Gold (Au)
2 Silicon (Si)
3 Iodine (I)
4 Cadmium (Cd)
5 Oxygen (O)

PROBLEM 6–22

The effect of hydrogen ion on the solubility of compounds is discussed in Section 5–10 of Chemical Principles.

An aqueous solution known to contain only an aluminum salt in acidic solution (pH = 4) is neutralized with NaOH solution. At pH ≃ 5 a solid appears, and, with continued addition of base, the solid dissolves at pH ≃ 12. Which statement provides the most likely explanation? (Assume that the temperature remains constant.)

1 The aluminum compound is the salt of a weak acid.
2 Aluminum hydroxide is a weak base.
3 The sodium salt of the anion of the aluminum salt is sparingly soluble.
4 Aluminum hydroxide is amphoteric.
5 Insufficient information; normality of the NaOH solution must be known.

SOLUTION 6–1

(3)

First, we have to pick out from each set of three elements those with the lowest and highest atomic weights and average these atomic weights. Then, we have to compare this average with the atomic weight of the remaining member of each set.

1 [at. wt Na + at. wt Te]/2 = (23 + 127)/2 = 75
Since the atomic weight of calcium (Ca) is 40, Ca, Na, and Te do not form a Döbereiner triad. Similarly, Answers **2**, **4**, and **5** are incorrect.

3 [at. wt S + at. wt Te]/2 = (32 + 127)/2 = 79.5
The atomic weight of selenium (Se) is 79, so this answer is correct. S, Se, and Te have similar chemical properties, and for this reason they are placed in the same group (Group VIA) of the periodic table (the so-called "chalcogens"). For example, all three elements form the same type of hydride: H_2S, H_2Se, and H_2Te. The odor of the first compound is well known to you; it is the odor of rotten eggs. The other two hydrides have even worse odors, so much so that few chemists have chosen to work with them.

SOLUTION 6–2

(4)

Since we assume the physical properties of Ge, P_{Ge}, are supposed to be approximately the average of the physical properties of Si and Sn, P_{Si} and P_{Sn}, respectively, it follows that

$$P_{Ge} = \frac{P_{Si} + P_{Sn}}{2}$$

or

$$2P_{Ge} - P_{Sn} = P_{Si}$$

Table 6–2

	P_{Ge}	P_{Sn}	$2P_{Ge} - P_{Sn} = P_{Si}$
Melting point of metal (°C)	938	232	2(938) − 232 = 1644
Density (g cm^{-3})	5.36	7.31	2(5.36) − 7.31 = 3.41
Boiling point of the chloride MCl_4 (°C)	84	114	2(84) − 114 = 54

Table 6–2 shows that Answer **4** is correct. The observed boiling point of $SiCl_4$ is 57.6°C, in good agreement with the predicted 54°C. However, note that the relationship between the physical properties of the three elements is only approximate because the predicted boiling point of Si is 1644°C, whereas the observed value is 1410°C. Just think of the relationship as a good "rule of thumb."

SOLUTION 6–3 (4)

You can read more about Moseley's experiments in Section 6–2 of Chemical Principles.

The general equation is

$$\nu = A(Z - 1)^2$$

For Cu ($Z = 29$):

$$2.0 \times 10^{18} \text{ sec}^{-1} = A(29 - 1)^2$$

For $Z = 43$:

$$\nu_{43} = A(43 - 1)^2$$

Dividing the two equations yields

$$\frac{2.0 \times 10^{18} \text{ sec}^{-1}}{\nu_{43}} = \left(\frac{28}{42}\right)^2$$

or

$$\nu_{43} = 4.5 \times 10^{18} \text{ sec}^{-1}$$

[The missing element was discovered in 1937 and named technetium (Tc). It is radioactive and was the first element produced artificially. The frequency of its K_α x ray is 4.452×10^{18} sec^{-1}, in good agreement with our predicted value, showing the validity of Moseley's equation.]

4 is correct.
1 is the reciprocal of Answer **4** except that you fiddled the powers of 10.
2 is the frequency for Cu.
3 You forgot to square 28/42.

SOLUTION 6–4 (5)

If you were unable to solve this problem, refer to Section 6–1, Items 1–54, of Programed Reviews of Chemical Principles *by Lassila* et al.

Rb is in Group IA and is therefore an alkali metal. Answer **1** is correct. (The word "alkali" is derived from an Arabic word meaning "ashes"; compounds of the Group IA metals were obtained from wood ashes.)

Sr and Ba are both in Group IIA and are therefore alkaline earth metals. Answers **2** and **4** are both correct. (Alchemists used to give the name "earth" to many nonmetallic substances. The Group IIA oxides, particularly MgO and CaO, produced an alkaline reaction to moist litmus paper; hence, these oxides were called the *alkaline earths*. The name "alkaline earths" soon became associated with pure Mg and Ca metal and eventually with all the Group IIA metals.)

Xe is in Group 0 and is therefore a noble gas. Answer **3** is correct. (The noble gases used to be referred to as the *inert gases* because of their complete chemical inactivity. However, since 1962, with the discovery of fluorine compounds of xenon, they are more correctly referred to as the noble gases.)

Ir is *not* in Group 0 and is therefore not a noble gas. It is actually in the third transition metal series. Answer **5** is incorrect and is therefore the required answer.

SOLUTION 6–5

(4)

Answer **1** is correct; fluorine is at the top of Group VIIA and is the most reactive of the halogens.

Cerium is the first member of the rare earth metals (or lanthanides). These elements first were called the rare earth metals because of their apparent scarcity. However, we now know that they occur in the earth's crust to a greater extent than some of the more familiar elements such as lead (Pb). Answer **2** is correct.

Answer **3** is correct; ruthenium is a member of the second transition metal series with chemical properties similar to those of iron (Fe) and osmium (Os).

Answer **5** is also correct; uranium is one of the better known actinides. The actinides were so called because of the similarity of their chemical properties to those of actinium (Ac).

The incorrect answer is **4.** Astatine is *not* a chalcogen, but is the heaviest (and only naturally radioactive) halogen.

SOLUTION 6–6

(1)

The elements in the periodic table are arranged in order of increasing atomic number. With a few exceptions, the same order results when the elements are arranged in order of increasing atomic weight. The exceptions are argon (Ar, $Z = 18$) and potassium (K, $Z = 19$), cobalt (Co, $Z=27$) and nickel (Ni, $Z=28$), and tellurium (Te, $Z=52$) and iodine (I, $Z = 53$). The reversal in these cases results from the masses and relative abundances of the isotopes of the different elements. For instance, I has one naturally occurring isotope, $^{127}_{53}I$, whereas Te has eight, the two most abundant of which, $^{128}_{52}Te$ and $^{130}_{52}Te$, have masses significantly above that of $^{127}_{53}I$. The high relative abundances of $^{128}_{52}Te$ and $^{130}_{52}Te$ (31.8% and 34.5%, respectively) result in the atomic weight of Te being higher than that of I. Returning to the problem, then, statement (a) is *incorrect.*

Statement (b) is *correct;* check the periodic table. Remember the difference between a group and a period.

Since I and Cl are both in the same group, they have similar chemical properties. Furthermore, since the group in question is Group VIIA, they are halogens. Statements (c) and (d) are both, therefore, *correct.* According to some etymologists, the word "halogen" is derived from the Greek words "*hals,*" "salt," and "*genes,*" "producing."

I and Cl have similar chemical properties and therefore, in the gas phase, I exists as a diatomic molecule, I_2. One mole of I_2 contains 2 g-atoms, so statement (e) is *correct.*

Only statement (a) is incorrect, and the required answer is **1**.

SOLUTION 6–7

(2)

Answer **1** is correct. This is the whole basis of the periodic table.

Elements having the properties of high electrical and thermal conductivity, luster, malleability (ability to be rolled into sheets or hammered flat), and ductility (ability to be drawn into wire) are referred to as *metals.* Such elements comprise most of the periodic table. The rest

The modern periodic table is discussed in Section 6–3 of Chemical Principles.

of the elements are called *nonmetals* with the exception of a few (such as B, Si, Ge, As, Te, and Po), which are intermediate in their properties and are known as *semimetals* or *metalloids.* This means that Answer **2** is incorrect and Answer **3** is correct.

Answer **4** is correct because H is unique in its properties. It sometimes behaves as an alkali metal and sometimes as a halogen.

Answer **5** is also correct. On the left-hand side of a period the elements exhibit metallic properties. But as you go across the period toward the noble gas at the extreme right-hand side, metallic properties start to decrease and nonmetallic behavior becomes evident.

The required answer is **2**.

SOLUTION 6–8

③

The first ionization energy of a gaseous, neutral atom M is defined as the energy *required* to remove an electron completely from the atom; so, per mole, it is ΔH^0 for the endothermic (ΔH^0 positive) process

$$M(g) \rightarrow M^+(g) + e^-$$

Answer **3** is correct.

The force holding an electron in an atom is electrostatic; the electron is negatively charged and the protons in the nucleus are positively charged. The ease with which the electron can be removed completely from the atom must depend on the number of protons in the nucleus and their resultant charge; that is, it must depend on the *nuclear charge* of the atom. The greater the nuclear charge, the harder it should be to remove the electron from the atom. However, this is not the whole story because every atom except H has more than one electron, and the electrons closest to the nucleus shield the outermost, or *valence,* electrons from the nucleus. (The valence electrons also shield each other from the nucleus.) The electron that we are trying to remove from the neutral atom is a valence electron. As a result of the electronic shielding, the nuclear charge of the atom attracting the electron is reduced somewhat and we speak of an *effective nuclear charge.* The ease with which the electron is removed from the atom depends on the effective nuclear charge of the atom and, as we shall see later in Problem 9–16, the distance of the electron from the nucleus.

1 is the sublimation energy of M.

2 is energy required for the removal of a second electron from M (the *second* ionization energy). Obviously, it is possible to define third, fourth, and higher ionization energies. These different ionization energies may often be obtained from the atomic spectrum of M (if there are enough electrons).

4 is the sum of the first and second ionization energies.

5 is the vaporization energy of M.

SOLUTION 6-9

(1)

As soon as you remove one electron from an atom, the effective nuclear charge is increased because the electronic shielding is reduced. So it becomes more difficult to remove another electron. This means that the second ionization energy is greater than the first, the third ionization energy is greater than the second, and so on.

Answer **1** is therefore correct and Answer **2** is incorrect.

Answer **3** is incorrect because the process shown would involve the *third* ionization energy and not the fourth. Answer **4** is incorrect because He only has two electrons and therefore there is no third ionization energy because there is no electron to remove. Answer **5** is incorrect because the electron has a definite mass; hence, the mass of $M^+(g)$ must be greater than that of $M^{2+}(g)$ or else the law of conservation of mass would be violated.

SOLUTION 6-10

(4)

Reactions (a), (b), (c), and (d) describe the first, second, third, and fourth ionization, respectively, of a gaseous, neutral Si atom. The greatest amount of energy would be required to complete the fourth ionization because the effective nuclear charge on Si^{3+} would be much greater than on Si^{2+}, Si^+, or Si because of the greatly reduced shielding of the one remaining valence electron from the nucleus after the removal of the first three. Answer **4** is correct.

[The actual energies required for the four reactions are (a) 187.9 kcal mole^{-1}, (b) 376.8 kcal mole^{-1}, (c) 771.6 kcal mole^{-1}, and (d) 1040.7 kcal mole^{-1}.]

If we tried to remove a fifth electron as in

$$Si^{4+}(g) \rightarrow Si^{5+}(g) + e^-$$

then 3844 kcal mole^{-1} would be needed! This is because we would no longer be removing a valence electron but an electron from an inner shell of the atom. You will soon learn about the idea of shells of electrons around the nucleus.

SOLUTION 6-11

(3)

$$1 \text{ eV} = \frac{1.6 \times 10^{-12} \text{ erg}}{4.2 \times 10^7 \text{ ergs cal}^{-1} \times 10^3 \text{ cal kcal}^{-1}}$$

$$= 3.8 \times 10^{-23} \text{ kcal}$$

This is the kinetic energy that one electron gains when accelerated through a potential difference of 1 V in an electric field. We are interested in the energy associated with a *mole* of electrons; therefore, we have to multiply 3.8×10^{-23} kcal by Avogadro's number.

$$1 \text{ eV} = 3.8 \times 10^{-23} \times 6.0 \times 10^{23} = 23 \text{ kcal mole}^{-1}$$

Answer **3** is correct. For Answer **1**, you forgot to multiply by N. For Answer **5**, you did not convert to kilocalories.

SOLUTION 6–12

(5)

Li, C, N, O, and F are all in the second period, and, from Figure 6–2, F ($Z = 9$) has the largest first ionization energy of the five elements. Therefore, Answer **5** is correct. As you go across a period the nuclear charge on the atoms increases, but so does the amount of electronic shielding owing to the presence of the additional electrons. These two factors oppose each other. The net result, however, is that the effective nuclear charge increases across the period. Hence, the first ionization energies increase across a period because the valence electrons are more tightly held as a result of the increasing effective nuclear charge. (Do not worry for now about the fine structure on the line joining the first ionization energies of Li and Ne. This kind of fine structure will be important in Problem 9–15.)

[*Note:* The experimental first ionization energies are Li, 124.3 kcal $mole^{-1}$; C, 258.0 kcal $mole^{-1}$; N, 335.3 kcal $mole^{-1}$; O, 313.8 kcal $mole^{-1}$; and F, 401.7 kcal $mole^{-1}$.]

SOLUTION 6–13

(4)

Sodium is an alkali metal and, from Figure 6–2, has a much lower ionization energy than neon, which is the nearest noble gas to it in the periodic table. Since the electrons in the gauge have an energy of 15 eV, they have sufficient energy to ionize Na (ionization energy $\simeq$ 5 eV, from Figure 6–2) but not enough to ionize Ne (ionization energy > 20 eV). Therefore, the gauge reading when Ne gas is in the system is zero because no Ne^+ ions are formed. Answer **4** is correct.

[*Note:* The electron energy in most ionization gauges is about 100 eV. This energy is high enough to ensure that all species present in a system are ionized; thus, the gauges can be used for any type of system. They must still be calibrated for different gases, however.]

SOLUTION 6–14

(4)

The element requiring the least energy to be ionized will be the one with the lowest first ionization energy. And from Figure 6–2, Cs has the lowest ionization energy of the five choices given.

Answer **4** is correct.

[*Note:* There is a decrease in ionization energy downward in *any* group, owing to the increase in the number of electrons around the nuclei of the atoms. The more electrons there are around a nucleus, the larger the atom is and the less attraction there is between the nucleus and the outermost electrons of the atom. The increase in atomic size and the concomitant increase in the shielding of the outermost electrons from the nuclei by the inner electrons more than compensate for the increasing nuclear charge in a group. As a result the ionization energies decrease down a group.]

SOLUTION 6–15

(4)

Answer **1** is the energy change associated with processes like

$H_2O(l) \rightarrow H_2O(g)$

Answer **2** is the energy change associated with the processes of the type

$M^{n+}(g) + x(\text{solvent}) \rightarrow M^{n+}(\text{solvent})_x$

in which $n+$ is the charge on any gaseous M^{n+} cation, and x is the number of solvent molecules attached to the cation and is not usually known. Answer **3** is a special case of Answer **2** when the solvent involved is water.

Answer **5** is the energy change associated with processes like

$Li(s) \rightarrow Li(g)$

So Answer **4** must be the correct answer. Electron affinity is defined as the energy change when a gaseous neutral atom acquires an additional electron to form an anion in the process

$X(g) + e^- \rightarrow X^-(g)$

Electron affinities (in ΔH terms) are generally negative; that is, energy is released in the acquisition of an additional electron by an atom. Some examples of electron affinities are fluorine (F, $Z = 9$), −83.5 kcal mole^{-1}; hydrogen (H, $Z = 1$), −16.1 kcal mole^{-1}; lithium (Li, $Z = 3$), −18.4 kcal mole^{-1}; and zinc (Zn, $Z = 30$), +20.7 kcal mole^{-1}.

SOLUTION 6–16

(2)

Al, Cl, P, Si, and S are all in the third period. Cl lies farthest to the right. As a result of the increasing effective nuclear charge in going from left to right across a period, it should have the greatest tendency to attract an additional electron; that is, it should have the largest electron affinity. Answer **2** is correct.

(The experimental values for Cl and S are −85.1 kcal mole^{-1} and −47.7 kcal mole^{-1}, respectively; the other elements do not form anions as readily as Cl and S; consequently, their electron affinities have not been measured.)

SOLUTION 6–17

(2)

Since nuclear charge is the number of units of proton charge resulting from the protons in the nucleus, nuclear charges and atomic numbers are equal. The nuclear charges of the elements involved in the isoelectronic series are 9, 10, 11, 12, 13, and 14, respectively; so it follows that the elements themselves must be F, Ne, Na, Mg, Al, and Si, respectively. The atomic number of a neutral atom is equal to the number of electrons around the nucleus of the atom. This means that F, Ne, Na, Mg, Al, and Si have 9, 10, 11, 12, 13, and 14 electrons, respectively, around their nuclei. The last member of the isoelectronic series is

Si^{5+}, in which there are 14 − 5 = 9 electrons. Therefore, the other members of the series must also have nine electrons. Aluminum has 13 electrons and thus needs to lose four electrons. In doing this it forms the Al^{4+} cation. Similarly, the other members of the series are F, Ne^{+}, Na^{2+}, and Mg^{3+}. In order of increasing nuclear charge the series is

F, Ne^{+}, Na^{2+}, Mg^{3+}, Al^{4+}, Si^{5+}

Answer **2** is correct. Answer **1** is the reverse of Answer **2**; it is the order of *decreasing* nuclear charge. The species in Answers **3**, **4**, and **5** are all isoelectronic with those in Answer **2**, but they involve elements with the wrong nuclear charges.

SOLUTION 6–18

(3)

We need to know the energy change associated with the process

$$O(g) + 2e^- \rightarrow O^{2-}(g) \qquad \Delta H^0 = ?$$

This equation can be derived by the addition of the two following equations:

$$\begin{array}{ll} O(g) + e^- \rightarrow O^-(g) & \Delta H^0 = -34 \text{ kcal mole}^{-1} \\ O^-(g) + e^- \rightarrow O^{2-}(g) & \Delta H^0 = +202 \text{ kcal mole}^{-1} \\ \hline O(g) + 2e^- \rightarrow O^{2-}(g) & \Delta H^0 = +168 \text{ kcal mole}^{-1} \end{array}$$

The correct answer is **3**.

We know from our data that ΔH^0 for the addition of one electron to $O(g)$ is negative. Therefore, why should ΔH^0 for the addition of two electrons be positive?

Imagine that we have already added one electron to $O(g)$ to form $O^-(g)$. We now wish to add a second electron to form $O^{2-}(g)$. In this case, energy is required to overcome the repulsion between like charges: the electron we are adding and the extra electron already on the O^- ion. The energy needed for the addition of the second electron is so much greater than that released in the addition of the first electron that ΔH^0 for the formation of $O^{2-}(g)$ from $O(g)$ is positive. Similarly, ΔH^0 is positive in the formation of other multicharged anions such as $S^{2-}(g)$ from $S(g)$, $N^{3-}(g)$ from $N(g)$, and $P^{3-}(g)$ from $P(g)$.

1 is the electron affinity of O.
2 is **3** with the wrong sign.
4 is the electron affinity of O^-.
5 is **4** minus **1**.

SOLUTION 6–19

(2)

Since the five ions are isoelectronic, the one with the smallest nuclear charge should be the largest because the electrons will not be as strongly attracted to the nucleus in this case as in the others. From the periodic table the atomic numbers (or nuclear charges) of Mn, P, S, Sc, and Ti are 25, 15, 16, 21, and 22. Each of the ions has 18 electrons (e.g., Mn^{7+}, 25 minus 7 for the net charge), but P^{3-} has the smallest nuclear charge; hence, it should be the largest. The correct answer is **2**.

The observed ionic radii from x-ray measurements (assuming spherical ions) are P^{3-}, 2.12 Å; S^{2-}, 1.84 Å; Sc^{3+}, 0.73 Å; Ti^{4+}, 0.70 Å; and Mn^{7+}, 0.46 Å.

[*Note:* In any isoelectronic series, the largest species will be the anion with the greatest negative charge; the smallest will be the cation with the greatest positive charge.]

SOLUTION 6–20

(3)

You were asked to memorize the common valences of the Group IA to Group VIIA elements. Tin (Sn) is in Group IVA; thus its valence is 4; fluorine (F) is in Group VIIA and its valence is $8-7=1$. Therefore, to satisfy the valences of both elements the formula of the compound must be SnF_4. Answer **3** is correct. The name of the compound is stannic fluoride.

2 This is the formula for stannous fluoride, the compound present in fluoride toothpaste to reduce cavities. Tin is one of the elements that exhibits more than one valence when it forms compounds. The endings "-ic" and "-ous" in the names "stannic" and "stannous" indicate that tin is in the 4 and 2 valence states, respectively.

SOLUTION 6–21

(4)

It should be obvious that the solution to this problem lies in the use of the periodic table. Cadmium is the only element noted that belongs to the same group (IIB) as mercury. This means that cadmium should form compounds similar to the mercury compounds mentioned in the problem, for example, $(CH_3)_2Cd$ and CH_3Cd^+. These compounds should exhibit similar chemical properties. Indeed, they do. In fact, it already is apparent that cadmium is a real environmental hazard because it is believed that its compounds may be responsible for certain cardiovascular diseases and hypertension. Cadmium has been detected in polluted water, the wastes from coal and zinc mining, and even tobacco smoke!

The correct answer is **4**. You should have been able to eliminate Answers **2**, **3**, and **5** (silicon, iodine, and oxygen, respectively) because all of these elements are typical nonmetals. Moreover, gold (Answer **1**) is well known to you as a relatively inert metal because of its use in precious ornaments and jewelry since antiquity; consequently, it is unlikely that organogold compounds such as $(CH_3)_2Au$ and CH_3Au^+ would be produced under the relatively mild conditions of the ecosystem that favor the formation of organomercury compounds.

SOLUTION 6–22

(4)

Amphoteric oxides are discussed in Section 6–7 of Chemical Principles.

First we precipitate the base $Al(OH)_3$:

$$Al^{3+}(aq) + 3OH^-(aq) \rightleftharpoons Al(OH)_3(s)$$

and then we redissolve it with excess base to form $Al(OH)_4^-$:

$$Al(OH)_3(s) + OH^-(aq) \rightleftharpoons Al(OH)_4^-(aq)$$

Hence, $Al(OH)_3$ will react with either a base, as in the second equation, or with an acid in the usual way:

$$Al(OH)_3(s) + 3H^+(aq) \rightleftharpoons Al^{3+}(aq) + 3H_2O$$

Such a compound is called *amphoteric.*

Answer **4** is correct. Answer **1**: If we have such a compound, for example, aluminum acetate, there is no mechanism whereby the acetate–acetic acid equilibrium will redissolve the solid $Al(OH)_3$. Answer **3**: Even if we thought that we were precipitating a sodium compound, there is no explanation in this answer for its dissolving again.

[*Note:* One method of preparing alumina (Al_2O_3) is to heat aluminum hydroxide. Alumina is used for furnace insulators and electrical insulators because of its high melting point and resistance.]

CHAPTER 7 REDOX REACTIONS

In Chapter 3 we considered the transfer of electrons in the reactions occurring at electrodes during electrolysis. In this chapter we are concerned with the direct exchange of electrons between molecules in a chemical reaction. The problems will introduce the ideas of oxidation number and redox (or oxidation–reduction) reactions. Many important chemical reactions, especially in the field of metallurgy, involve reactions of this type.

To answer some of the problems you will need to memorize the five rules for oxidation numbers immediately following Problem 7–3.

Solutions to problems in Chapter 7 begin on page 201.

PROBLEMS

PROBLEM 7–1

Section 7–2 of Chemical Principles *by Dickerson, Gray, and Haight deals with the calculation of oxidation numbers.*

Chemists introduced the idea of combining number for elements to help them to write chemical formulas for compounds. The *combining number* of an element in a particular compound is defined as the ratio of the atomic weight of the element to its combining weight in the compound in question. The combining number of a metal in its compound with oxygen is called the *oxidation number.* Sodium forms a compound with oxygen named sodium monoxide, Na_2O. What is the oxidation number of sodium in Na_2O?

(at. wt: Na = 23, O = 16)

1 $\frac{1}{2}$
2 1
3 2
4 3
5 23

PROBLEM 7–2

What are the oxidation numbers of Ca and F in fluorite, $Ca^{2+}(F^-)_2$?

	Ca	F
1	−2	+1
2	+1	+2
3	+2	−1
4	+2	−2
5	+8	+4

PROBLEM 7–3

Among the representative elements, metals often have positive oxidation numbers equal to their group numbers in the periodic table, whereas nonmetals often exhibit negative oxidation numbers equal to their group numbers minus eight. On this basis, which one of the following representative elements has been assigned an incorrect oxidation number?

1 Lithium (Li, $Z = 3$); +1
2 Nitrogen (N, $Z = 7$); +4
3 Sulfur (S, $Z = 16$); −2
4 Chlorine (Cl, $Z = 17$); −1
5 Strontium (Sr, $Z = 38$); +2

Although most of the representative elements have oxidation numbers in accordance with Problem 7–3, some of them have several oxidation numbers depending on which compounds they are in. Before chemists could write the correct formulas for both ions *and* neutral *compounds, several rules had to be stated. These rules are as follow.*

(1) The oxidation number for any free element is zero; thus, H_2, O_2, *Na, Fe, and* Cl_2 *have zero oxidation numbers.*

(2) The oxidation number for any simple one-atom ion is equal to its charge; thus, Na^+ *has an oxidation number of +1,* Ca^{2+} *of +2, and* Cl^- *of −1.*

(3) The oxidation number of hydrogen is +1, but in metal hydrides such as NaH the oxidation number is −1.

(4) The oxidation number of oxygen is −2 in all compounds except fluorine–oxygen compounds such as F_2O*, where the fluorine assignment of −1 takes precedence; the peroxides such as* BaO_2*, where oxygen is given an oxidation number of −1; and the superoxides such as* KO_2*, where it has an oxidation number of* $-\frac{1}{2}$.

(5) In combinations of nonmetals not involving hydrogen or oxygen, the nonmetal that is either above or to the right of the other in the periodic table is considered negative. Its oxidation number is given the same value as the charge on its most commonly encountered negative ion. For instance, in BrF_3*, the oxidation number for fluorine is −1 and for bromine it is +3; whereas in* $AsBr_3$*, bromine is −1 and arsenic is +3.*

PROBLEM 7–4

In writing the chemical formulas of *neutral* compounds, the following principle must be adhered to: *The sum of the oxidation numbers of all the atoms in a properly written formula must be zero.* Keeping this in mind, give the formula of a neutral compound that contains one hydrogen atom, one chlorine atom, and an unknown number of oxygen atoms, if the oxidation number of the chlorine atom in the compound is +7.

1 HClO
2 $HClO_2$
3 $HClO_3$
4 $HClO_4$
5 $HClO_8$

PROBLEM 7–5 Phosphorus forms three well-known compounds containing chlorine: phosphorus trichloride (PCl_3), phosphorus pentachloride (PCl_5), and phosphorus oxychloride ($POCl_3$). What are the oxidation numbers of phosphorus in these compounds?

	PCl_3	PCl_5	$POCl_3$
1	−3	−5	−1
2	+1	+1	+3
3	+3	+5	+4
4	+3	+5	+5
5	+4	+6	+5

PROBLEM 7–6 Manganese (Mn, $Z = 25$) exhibits a larger number of oxidation numbers in its compounds than any of the other first-row transition metals. What are the most important oxidation numbers of manganese if they are illustrated by the following oxides: (a) manganous oxide, MnO; (b) manganese dioxide, MnO_2; (c) manganese heptoxide, Mn_2O_7?

1 −2, −4, and −7
2 +2 and +7
3 +2, +8, and +9
4 +2, +4, and +7
5 +2, +4, and +14

PROBLEM 7–7 Sodium thiosulfate, $Na_2S_2O_3 \cdot 5H_2O$ (commonly known as "hypo" from an old name, sodium hyposulfite) is used extensively in photography and volumetric analysis. Iodine (I_2, $Z = 53$) reacts quantitatively with sodium thiosulfate, and sodium tetrathionate is formed:

$$2Na_2S_2O_3 + I_2 \rightarrow Na_2S_4O_6 + 2NaI$$

What are the oxidation numbers of sulfur in the thiosulfate ion, $S_2O_3^{2-}$, and the tetrathionate ion, $S_4O_6^{2-}$?

[*Note:* For an ion, the sum of the oxidation numbers of all the atoms in a properly written formula must equal the net charge on the ion.]

	$S_2O_3^{2-}$	$S_4O_6^{2-}$
1	−2	$-2\frac{1}{2}$
2	$-\frac{1}{2}$	−1
3	$+\frac{1}{2}$	+1
4	+2	$+2\frac{1}{2}$
5	+3	+3

PROBLEM 7-8

One of the chief ores of uranium (U, $Z = 92$) is pitchblende (U_3O_8). When pitchblende is treated with nitric acid (HNO_3), uranyl nitrate [$UO_2(NO_3)_2$] is formed. What is the change in the oxidation number per atom of uranium when pitchblende is converted into uranyl nitrate?

1 -10
2 $-1\frac{1}{3}$
3 $+\frac{2}{3}$
4 $+6$
5 No change.

Chemical reactions in which the oxidation numbers of some atoms change are called redox (or oxidation–reduction) reactions. Let's see how our knowledge of oxidation numbers will help us to balance redox equations.

PROBLEM 7-9

Electron and atom transfer reactions—types of redox reactions—are discussed in Section 7-3 of Chemical Principles.

If the oxidation number of an atom is increased in a chemical reaction, the atom has been oxidized. Which of the following reactions are *not* oxidation reactions?

(a) $Zn \rightarrow Zn^{2+} + 2e^-$
(b) $Cl_2 + 2e^- \rightarrow 2Cl^-$
(c) $S^{2-} \rightarrow S + 2e^-$
(d) $Fe^{2+} \rightarrow Fe^{3+} + e^-$
(e) $Ce^{4+} + e^- \rightarrow Ce^{3+}$

1 (a)
2 (b) and (e)
3 (a), (c), and (d)
4 (b), (c), and (e)
5 All of them.

PROBLEM 7-10

Consider the following equation that describes the rusting of iron:

$4Fe + 3O_2 \rightarrow 4Fe^{3+} + 6O^{2-}$ (Fe^{3+} is in the form of Fe_2O_3)

Which one of the following statements is *incorrect?*

1 This is an example of a redox reaction.
2 Metallic iron is a reducing agent.
3 Fe^{3+} is an oxidizing agent.
4 Metallic iron is reduced to Fe^{3+}.
5 O^{2-} is a reducing agent.

PROBLEM 7–11

Iodine pentoxide, I_2O_5, is a very important reagent because it can oxidize the extremely toxic gas, carbon monoxide, CO, to carbon dioxide, CO_2, in the absence of water. Consequently, it is used for the detection and determination of CO, a major pollutant of the atmosphere, in small quantities when mixed with other gases. The reaction involved is

$$(?)I_2O_5 + (?)CO \rightarrow (?)I_2 + (?)CO_2$$

What are the stoichiometric coefficients in this equation when it is properly balanced?

	I_2O_5	CO	I_2	CO_2
1	1	5	1	5
2	2	10	2	10
3	5	1	5	1
4	1	1	1	1

5 The equation cannot be balanced.

We will sometimes equate change in oxidation number with gain or loss of electrons. This is more an equation-balancing device than a physical reality.

PROBLEM 7–12

One of the steps in the manufacture of nitric acid from ammonia (NH_3) involves the disproportionation (self-oxidation and self-reduction) of nitrous acid (HNO_2), according to the following equation:

$$(?)HNO_2 \rightarrow (?)NO + (?)HNO_3 + (?)H_2O$$

What is the stoichiometric coefficient of HNO_3 when the equation is properly balanced?

1 1
2 2
3 3
4 6
5 7

PROBLEM 7-13

You can read more on how to balance redox equations in Section 7-4 of Chemical Principles.

Hydrogen peroxide, H_2O_2, is used extensively by many different industries. For instance, it is used for the bleaching of cotton fabrics, for the propulsion of rockets (both as a fuel and as an oxidizer), for certain washing powders and denture-cleaning compositions, and for bleaching hair. A convenient method for determining the concentration of H_2O_2 in some of these processes is to titrate a known volume of the peroxide solution with an aqueous solution of potassium permanganate ($KMnO_4$) of known concentration. The titration is carried out under acid conditions (dilute H_2SO_4), and the equation describing the reaction is

$$(?)H_2O_2 + (?)KMnO_4 + (?)H_2SO_4 \rightarrow (?)K_2SO_4 + (?)MnSO_4 + (?)H_2O + (?)O_2$$

Which set of stoichiometric coefficients correctly balances the equation?

	H_2O_2	$KMnO_4$	H_2SO_4	K_2SO_4	$MnSO_4$	H_2O	O_2
1	1	1	1	1	1	1	1
2	1	2	3	1	2	4	3
3	2	5	3	2	1	8	5
4	5	2	3	1	2	8	5
5	10	8	12	4	8	22	15

PROBLEM 7-14

The percentage by weight of chromium (Cr) in a compound is determined quantitatively by converting a sample of known weight into sodium dichromate, $Na_2Cr_2O_7$, and titrating an acidified solution of the sodium dichromate with a standard solution of ferrous sulfate. The reaction involved in the titration is

$$(?)Na_2Cr_2O_7 + (?)FeSO_4 + (?)H_2SO_4 \rightarrow (?)Fe_2(SO_4)_3 + (?)Cr_2(SO_4)_3 + (?)Na_2SO_4 + (?)H_2O$$

Which of the following sets of stoichiometric coefficients correctly balances the equation?

	$Na_2Cr_2O_7$	$FeSO_4$	H_2SO_4	$Fe_2(SO_4)_3$	$Cr_2(SO_4)_3$	Na_2SO_4	H_2O
1	1	1	1	1	1	1	1
2	1	1	7	1	1	1	7
3	1	6	7	3	1	1	7
4	7	1	6	1	3	7	1
5	21	3	18	3	9	21	3

PROBLEM 7-15

A convenient method for the laboratory preparation of nitric oxide, NO, involves the oxidation of metallic copper, Cu, by nitrate ion, NO_3^-, in acid solution. Copper is oxidized to Cu^{2+}, and NO_3^- is reduced to NO. Which one of the following equations correctly describes the preparation of NO?

1 $Cu + NO_3^- + H^+ \rightarrow Cu^{2+} + NO + H_2O$
2 $Cu + 4NO_3^- + 4H^+ \rightarrow Cu^{2+} + 2NO_2 + 2H_2O$
3 $Cu + NO_3^- + 4H^+ \rightarrow Cu^{2+} + NO + 2H_2O$
4 $3Cu + 2NO_3^- + 8H^+ \rightarrow 3Cu^{2+} + 2NO + 4H_2O$
5 $3Cu + 12NO_3^- + 12H^+ \rightarrow 3Cu^{2+} + 6NO_2 + 6H_2O$

PROBLEM 7-16

The Marsh test indicates the presence of arsenic (As) in compounds such as arsenic acid (H_3AsO_4) by reaction with metallic zinc (Zn) in dilute H_2SO_4 solution to give arsine (AsH_3), zinc sulfate ($ZnSO_4$), and water. What is the ratio of the stoichiometric coefficients of H_3AsO_4 and Zn in the balanced equation for the Marsh test (i.e., coefficient of H_3AsO_4/coefficient of Zn)?

1 1/1
2 1/4
3 4/1
4 8/1
5 8/7

PROBLEM 7-17

One of the steps in the extraction of cobalt metal from its ores involves treating a solution of a cobaltous chloride with an alkaline hypochlorite solution to precipitate cobaltic hydroxide. The "skeleton equation" for the reaction is

$$CoCl_2 + OCl^- \rightarrow Co(OH)_3 + Cl^-$$

Which of the following is the complete, balanced equation?

1 $CoCl_2 + OH^- + OCl^- + H_2O \rightarrow Co(OH)_3 + 3Cl^-$
2 $CoCl_2 + 3OCl^- + 3H^+ \rightarrow Co(OH)_3 + 5Cl^-$
3 $2CoCl_2 + 4OH^- + OCl^- + H_2O \rightarrow 2Co(OH)_3 + 5Cl^-$
4 $2CoCl_2 + 3OH^- + OCl^- + 2H_2O \rightarrow 2Co(OH)_3 + 5Cl^- + H^+$
5 $3CoCl_2 + 3OH^- + 3OCl^- + 3H_2O \rightarrow 3Co(OH)_3 + 9Cl^-$

PROBLEM 7–18

The gram equivalent weight of an oxidizing or a reducing agent in a redox reaction is defined as the weight, in grams, of compound that will produce 1 mole of oxidation number change (i.e., accept or donate 1 mole of electrons). What are the gram equivalent weights of $Na_2Cr_2O_7$ and $FeSO_4$ under the conditions of the following reaction:

$$Na_2Cr_2O_7 + 6FeSO_4 + 7H_2SO_4 \rightarrow 3Fe_2(SO_4)_3 + Cr_2(SO_4)_3 + Na_2SO_4 + 7H_2O$$

[*Note:* This is the equation that you balanced in Problem 7–14.]

(mol wt: $FeSO_4 = 152$, $Na_2Cr_2O_7 = 262$.)

	$Na_2Cr_2O_7$	$FeSO_4$
1	43.7 g	152 g
2	152 g	262 g
3	262 g	25.3 g
4	262 g	152 g
5	1570 g	152 g

PROBLEM 7–19

You have been asked to determine the percentage of chromium in a known weight of a chromium-containing mineral by the method described in Problem 7–14. You wish to use a 0.400-normal solution of ferrous sulfate ($FeSO_4$) for the titration. Your source of $FeSO_4$ is crystalline $FeSO_4{\cdot}7H_2O$. How many grams of $FeSO_4{\cdot}7H_2O$ will you need to make a liter of 0.400-normal solution?

(mol wt: $FeSO_4 = 152$, $FeSO_4{\cdot}7H_2O = 278$)

1 0.400 g
2 60.8 g
3 111 g
4 695 g
5 1000 g

PROBLEM 7–20

What is the molarity of a liter of a 0.400-normal solution of $Na_2Cr_2O_7$ when it reacts under the conditions of the reaction given in Problem 7–18?

1 0.067 molar
2 0.400 molar
3 1.00 molar
4 2.40 molar
5 24.0 molar

PROBLEM 7-21

The weight of the mineral sample in Problem 7–19 was exactly 1.70 g. If 50.0 ml of a 0.400-normal ferrous sulfate solution were required to titrate the resultant sodium dichromate solution to an end point, what would be the approximate percentage of chromium in the mineral sample?

(at. wt: Cr = 52.0;
mol wt: $Na_2Cr_2O_7 = 262$;
g-equiv wt: $Na_2Cr_2O_7 = 43.7$ g equiv^{-1})

1 1.7%
2 10%
3 20%
4 34%
5 44%

PROBLEM 7-22

Redox titrations are discussed in Section 7–5 of Chemical Principles.

Ten milliliters of blood from an adult male patient were diluted to 100.0 ml. From a 10.0-ml portion of this diluted sample, the Ca^{2+} was precipitated as CaC_2O_4. After this precipitate had been dissolved in H_2SO_4, it required 1.0 ml of 0.00500-normal $KMnO_4$ to titrate to an end point. What is the Ca^{2+} concentration expressed as milligrams of Ca^{2+} per 10.0 ml of blood? The equation for the titration reaction is

$$5CaC_2O_4 + 2KMnO_4 + 8H_2SO_4 \rightarrow K_2SO_4 + 2MnSO_4 + 5CaSO_4 + 10CO_2 + 8H_2O$$

(at. wt: Ca = 40.00;
mol wt: $CaC_2O_4 = 128.00$)

1 0.10
2 1.00
3 2.00
4 3.20
5 10.0

SOLUTIONS

SOLUTION 7–1

(2)

If you need background drill in balancing redox equations, study Review 7 in Programed Reviews of Chemical Principles *by Lassila* et al.

In Na_2O, 2×23, or 46 g of sodium are combined with 16 g of oxygen. The combining weight of sodium is that weight which combines with 8.0 g of oxygen, that is, 46/2, or 23 g Na.

$$\text{Oxidation number of Na in } Na_2O = \frac{\text{at. wt of Na}}{\text{combining wt of Na in } Na_2O}$$

$$= \frac{23}{23}$$

$$= 1$$

The correct answer is **2**.

When two sodium atoms combine with one oxygen atom to form Na_2O, each sodium atom is regarded as losing one electron and being *oxidized* to Na^+, whereas the oxygen atom gains two electrons and is *reduced* to O^{2-}. From our calculation, the oxidation number of Na in Na_2O is 1. This is equal to the positive charge on the Na^+ ions, and by convention we assign a positive sign to the oxidation number. To write the proper formula for sodium monoxide, oxygen must be assigned an oxidation number of −2 to balance the +2 from the two sodium atoms. Note that the oxidation number of −2 for oxygen is equal to the charge on the O^{2-} ion.

Answer **1**: You thought that the combining weight of oxygen is the same as its atomic weight.

SOLUTION 7–2

(3)

In $Ca^{2+}(F^-)_2$, the oxidation number of Ca is equal to the charge on the Ca^{2+} ion, that is, +2. This +2 must be balanced by −2 from the two F^- ions to produce neutral CaF_2. This means that the oxidation number of each F is −1 (i.e., equal to the charge on each F^- ion).

Answer **3** is correct. Answer **4**: You neglected the stoichiometry of the F's. Answer **5** gives the coordination numbers of Ca^{2+} and F^- in the fluorite structure (see Problem 3–21).

SOLUTION 7–3

(2)

Lithium and strontium are metals in Groups IA and IIA, respectively, and should exhibit oxidation numbers of +1 and +2, respectively. Answers **1** and **5** are correct.

Sulfur and chlorine are nonmetals in Groups VIA and VIIA, respectively, and should therefore have oxidation numbers of $6 - 8 = -2$ and $7 - 8 = -1$, respectively. Answers **3** and **4** are correct.

Nitrogen is a nonmetal in Group VA and should have an oxidation number of $5 - 8 = -3$. Answer **2** is therefore incorrect, and the required answer.

SOLUTION 7–4

(4)

Let the formula of the compound be $HClO_n$, in which n is the unknown number of oxygen atoms. The compound does not contain a metal; thus, it is not an ionic metal hydride. The oxidation number of H is therefore +1. The oxidation number of Cl in the compound is +7. Hence, we can set up the following equation:

$$(+1) + (+7) + n(-2) = 0$$

from which $n = 4$. The formula of the compound is $HClO_4$. The compound is perchloric acid. The correct answer is **4**.

The oxidation number of Cl in $HClO_4$ is +7. Does this mean that one chlorine atom loses seven electrons to oxygen? No! Oxidation numbers cannot be taken as indicators of actual electron transfers in bonding. They are simply a means of keeping track of chemical formulas and, as you will see shortly, chemical reactions.

You would have chosen Answer **3** if you assigned H an oxidation number of −1; you would have picked Answer **5** if you assigned O an oxidation number of −1. Answer **1** is hypochlorous acid, Answer **2** is chlorous acid, Answer **3** is chloric acid, and Answer **5** is unknown.

$HClO$, $HClO_2$, $HClO_3$, and $HClO_4$ are referred to as the *oxyacids* of chlorine. The salts of $HClO$ and $HClO_3$ have many important uses. For example, sodium hypochlorite ($NaClO$) is a bleaching agent and a germicide, and potassium chlorate ($KClO_3$) is used medicinally in throat lozenges and in the manufacture of matches, fireworks, and explosives.

SOLUTION 7–5

(4)

Let x, y, and z be the oxidation numbers of P in PCl_3, PCl_5, and $POCl_3$, respectively. The oxidation number of Cl is −1, and of O is −2.

PCl_3	PCl_5	$POCl_3$
$x + 3(-1) = 0$	$y + 5(-1) = 0$	$z + (-2) + 3(-1) = 0$
$x = +3$	$y = +5$	$z = +5$

4 is correct.
1 You assigned Cl an oxidation number of +1.
2 You forgot the stoichiometry of Cl.
3 You assigned O an oxidation number of −1.
5 is simply the number of atoms in each compound.

SOLUTION 7–6

(4)

Let x, y, and z be the oxidation numbers of Mn in MnO, MnO_2, and Mn_2O_7, respectively.

MnO	MnO_2	Mn_2O_7
$x + (-2) = 0$	$y + 2(-2) = 0$	$2z + 7(-2) = 0$
$x = +2$	$y = +4$	$z = +7$

Therefore, the three important oxidation numbers of Mn are +2 (referred to as the manganous state), +4, and +7. The other oxidation numbers exhibited by Mn are 0, +1, +3, +5, and +6.

4 is correct.
1 is the negative of **4**.
2 You thought MnO_2 contained a peroxide ion, O_2^{2-}.
3 is just the number of atoms in each compound.
5 You neglected the stoichiometry of Mn in Mn_2O_7.

SOLUTION 7–7

(4)

Let x and y be the oxidation numbers of S in the $S_2O_3^{2-}$ and $S_4O_6^{2-}$ ions, respectively.

$S_2O_3^{2-}$	$S_4O_6^{2-}$
$2x + 3(-2) = -2$	$4y + 6(-2) = -2$
$x = +2$	$y = +2\frac{1}{2}$

The correct answer is **4**. Notice that an oxidation number need not be an integer. Answer **3**: You assigned O an oxidation number of −1. Answer **5**: You neglected the charges on the ions.

SOLUTION 7–8

(3)

Both nitric acid, HNO_3, and uranyl nitrate, $UO_2(NO_3)_2$, contain the nitrate ion, NO_3^-. There are two NO_3^- ions in uranyl nitrate; hence, for neutrality, the uranyl cation must be UO_2^{2+}. If x and y are the oxidation numbers of U in U_3O_8 and UO_2^{2+}, respectively, we have

U_3O_8	UO_2^{2+}
$3x + 8(-2) = 0$	$y + 2(-2) = +2$
$x = +5\frac{1}{3}$	$y = +6$

Therefore, the change in oxidation number of U when U_3O_8 is converted to $UO_2(NO_3)_2$ is $+6 - (+5\frac{1}{3}) = +\frac{2}{3}$.

The correct answer is **3**. Answer **1**: You forgot the stoichiometry of U in U_3O_8. Answer **2**: You thought that the UO_2 part of $UO_2(NO_3)_2$ is a peroxide.

SOLUTION 7–9

(2)

Table 7–1

Reaction	Atom concerned	Oxidation number of atom before reaction	Oxidation number of atom after reaction	Difference in oxidation number before and after reaction
(a)	Zn	0	+2	+2
(b)	Cl	0	−1	−1
(c)	S	−2	0	+2
(d)	Fe	+2	+3	+1
(e)	Ce	+4	+3	−1

Only in Reactions (a), (c), and (d) does the oxidation number of the atom concerned increase. Therefore, Zn, S^{2-}, and Fe^{2+} are oxidized. In Reactions (b) and (e), the oxidation numbers of Cl and Ce decrease. Therefore, Cl_2 and Ce^{4+} are *reduced*, not oxidized. Answer **2** is correct.

Zn, S^{2-}, and Fe^{2+} are examples of *reducing agents:* They donate electrons to electron acceptors (or *oxidizing agents*). Since Cl_2 and Ce^{4+} accept electrons, they are both oxidizing agents. A substance that undergoes oxidation is a reducing agent. One that undergoes reduction is an oxidizing agent. [*Note:* In an oxidation reaction the electrons are usually put on the right-hand side of the equation; in a reduction reaction, they are on the left-hand side.]

SOLUTION 7–10

(4)

What happens to Fe and O_2 in the reaction?

$$Fe \rightarrow Fe^{3+} + 3e^-$$
$$O_2 + 4e^- \rightarrow 2O^{2-}$$

If you had difficulty with this problem, refer to Review 7, Items 1–59, of Lassila's book.

Fe is oxidized to Fe^{3+} (increase of +3 in oxidation number) while O_2 is reduced to $2O^{2-}$ (decrease of −2 in oxidation number per O atom); so the rusting of iron is an example of a redox reaction. Answer **4** is incorrect and Answer **1** is correct. Fe donates $3e^-$ in forming Fe^{3+}; so Fe is a reducing agent, and Answer **2** is correct. Fe^{3+} is an oxidizing agent because it can accept three electrons to form Fe:

$$Fe^{3+} + 3e^- \rightarrow Fe$$

Answer **3** is therefore correct. O_2 accepts $4e^-$ in forming $2O^{2-}$; hence, O_2 is an oxidizing agent. O^{2-} is a reducing agent because it can donate electrons:

$$2O^{2-} \rightarrow O_2 + 4e^-$$

Thus, **5** is correct. The required answer is **4**.

SOLUTION 7–11

(1)

The equation is a fairly simple one, and you should be able to balance it merely by inspection. However, as it involves a redox reaction, let's employ it to illustrate the use of oxidation numbers in balancing redox equations. There are four steps involved.

Step 1. Write down the reactants and products and determine which atoms change in oxidation number. Write the oxidation numbers below the atoms in question. These atoms are in boldface type.

$$(?)\mathbf{I}_2O_5 + (?)\mathbf{C}O \rightarrow (?)\mathbf{I}_2 + (?)\mathbf{C}O_2$$
$$2(+5) \qquad +2 \qquad 2(0) \qquad +4$$

Step 2. Determine the change in oxidation number per mole.

The oxidation numbers of the I atoms change from +5 in I_2O_5 to zero in I_2. (I_2O_5 is the oxidizing agent in the reaction.) Each I atom has to lose 5 in the reaction, and the loss in oxidation number per mole of I_2O_5 is $2 \times 5 = 10$. The C atom, in going from +2 in CO to +4 in CO_2, has to gain 2 for each molecule of CO. (CO is the reducing agent in the reaction.)

(10 per I_2O_5)
(reduction)

$$(?)\mathbf{I}_2O_5 \quad + \quad (?)\mathbf{C}O \quad \rightarrow \quad (?)\mathbf{I}_2 \quad + \quad (?)\mathbf{C}O_2$$
$$2(+5) \qquad +2 \qquad 2(0) \qquad +4$$

(2 per CO)
(oxidation)

Step 3. Make the gain in oxidation number equal the loss by adjusting the stoichiometric coefficients of the formulas for the reducing agent and the oxidation agent in the equation.

(10 per I_2O_5)
(reduction)

$$\mathbf{I}_2O_5 \quad + \quad 5\mathbf{C}O \quad \rightarrow \quad (?)\mathbf{I}_2 \quad + \quad (?)\mathbf{C}O_2$$
$$2(+5) \qquad +2 \qquad 2(0) \qquad +4$$

(2 per CO) $\times$ 5 = 10
(oxidation)

Step 4. Our equation at present is

$$I_2O_5 + 5CO \rightarrow (?)I_2 + (?)CO_2$$

One I_2O_5 on the left requires one I_2 on the right, so we put a 1 (understood) before I_2. Ten O atoms on the left require ten O atoms on the right. These are provided by $5CO_2$. Five C atoms on the left require five C atoms on the right. These are provided by the $5CO_2$ molecules. The final balanced equation is

$$I_2O_5 + 5CO \rightarrow I_2 + 5CO_2$$

Answer **1** is correct. Answer **2** is twice **1**: You must always write the simplest possible equation.

SOLUTION 7–12

(1)

Since there are at the very least two N atoms on the right, we must have a minimum of two HNO_2 molecules on the left. Let's write our equation as

$$(?)HNO_2 + (?)HNO_2 \rightarrow (?)NO + (?)HNO_3 + (?)H_2O$$

One HNO_2 molecule is reduced while the other is oxidized. Let's say that the HNO_2 molecule on the extreme left is reduced. The oxidation number of the N atom in this molecule changes from +3 to +2 when the molecule is reduced to NO, a gain of $1e^-$ for each molecule of HNO_2. The oxidation number of the N atom in the other HNO_2 molecule changes from +3 to +5 when the molecule is oxidized to HNO_3, a loss of $2e^-$ for each molecule of HNO_2. To balance the total loss and total gain of electrons, two HNO_2 molecules are needed for the reduction process and one for the oxidation process:

(gain $1e^-$ per HNO_2) $\times 2 = 2e^-$

(reduction)

$$2HNO_2 + HNO_2 \rightarrow (?)NO + (?)HNO_3 + (?)H_2O$$

$$\quad +3 \qquad\quad +3 \qquad\qquad +2 \qquad\quad +5$$

(lose $2e^-$ per HNO_2)

(oxidation)

The $2HNO_2$ on the left must be reduced to 2NO on the right. There are three N atoms on the left; thus, we need a "1" before HNO_3 on the right to go with the 2NO to balance the N atoms. Three H atoms on the left require three H atoms on the right, one of which is already accounted for in the $1HNO_3$; so, we need only to put a "1" before H_2O on the right and our equation is balanced:

$$3HNO_2 \rightarrow 2NO + HNO_3 + H_2O$$

The stoichiometric coefficient of HNO_3 is 1.

Answer **1** is correct. If you gave Answer **3**, you assigned H an oxidation number of −1 in HNO_2 and HNO_3. You should have realized your mistake because the resulting equation doesn't balance.

SOLUTION 7–13

(4)

In this problem you should remember that the oxidation number of an oxygen atom in a peroxide is −1. The oxidation number of the O atoms changes from −1 in H_2O_2 to zero in elemental O_2, a loss of $2e^-$ for each molecule of H_2O_2. Manganese changes in oxidation number from +7 in $KMnO_4$ to +2 in $MnSO_4$, a gain of $5e^-$ for each molecule of $KMnO_4$. Five H_2O_2 molecules lose $10e^-$ and $2KMnO_4$ gain $10e^-$.

(lose $2e^-$ per H_2O_2) $\times 5 = 10e^-$

(oxidation)

$$5H_2\mathbf{O}_2 + 2K\mathbf{Mn}O_4 + (?)H_2SO_4 \rightarrow (?)K_2SO_4 + (?)\mathbf{Mn}SO_4 + (?)H_2O + (?)\mathbf{O}_2$$

2(−1)   +7   +2   2(0)

(gain $5e^-$ per $KMnO_4$) $\times 2 = 10e^-$

(reduction)

Since there are two K atoms on the left, we need to put a "1" before K_2SO_4 on the right. There are two Mn atoms on the left, and so we need to put a "2" before $MnSO_4$ on the right. Since there are to be three SO_4 units on the right, there must be three SO_4 units on the left; thus, we need to put a "3" before H_2SO_4 on the left.

There are now 16 H atoms on the left; hence, we need to put an "8" before H_2O on the right. Now we lack 10 O atoms on the right, so we put a "5" before the O_2. Our completely balanced equation is

$$5H_2O_2 + 2KMnO_4 + 3H_2SO_4 \rightarrow K_2SO_4 + 2MnSO_4 + 8H_2O + 5O_2$$

4 is correct.

2 You assigned to the O's in $KMnO_4$ an oxidation number of −1 instead of −2.

3 You interchanged some of the correct stoichiometric coefficients.

5 You forgot that the oxidation number of an O atom in a peroxide is −1 and used the normal oxidation number of −2.

SOLUTION 7–14

(3)

The oxidation number of the Cr atoms changes from +6 in $Na_2Cr_2O_7$ to +3 in $Cr_2(SO_4)_3$. (SO_4^{2-} is the sulfate ion.) Each Cr atom must gain $3e^-$; hence, the electron gain for each molecule of $Na_2Cr_2O_7$ is $2 \times 3e^- = 6e^-$. The Fe atom, in going from +2 to +3, must lose $1e^-$ for each molecule of $FeSO_4$. Hence, to balance the gain and loss of electrons we need a ratio of $6FeSO_4$ to $1Na_2Cr_2O_7$. There are seven O atoms on the left that are not bound in SO_4^{2-} groups. We add $7H_2O$ on the right to balance them. Now there are 14H atoms to be added to the left as $7H_2SO_4$ molecules. Finally, check that the SO_4^{2-} groups balance.

(gain $6e^-$ per $Na_2Cr_2O_7$)

(reduction)

$$Na_2\mathbf{Cr}_2O_7 + 6\mathbf{Fe}SO_4 + 7H_2SO_4 \rightarrow 3\mathbf{Fe}_2(SO_4)_3 + \mathbf{Cr}_2(SO_4)_3 + Na_2SO_4 + 7H_2O$$

2(+6) +2 2(+3) 2(+3)

(lose $1e^-$ per $FeSO_4$)
$\times 6 = 6e^-$

(oxidation)

Answer **3** is correct. Answer **4** happens if you muddled the correct stoichiometric coefficients.

SOLUTION 7–15

(4)

If you do not understand the ion–electron method of balancing redox equations, refer to Review 7, Items 60–75 of Lassila's book.

We will solve this question by the ion–electron half-reaction method.

First, let's write down the oxidation half-reaction and the reduction half-reaction in terms of the ionic, atomic, and molecular species involved:

$Cu \rightarrow Cu^{2+}$ (oxidation)
$NO_3^- \rightarrow NO$ (reduction)

Next, we balance the various atoms present. The oxidation half-reaction is already balanced; there is one Cu atom on each side of the equation. In the reduction half-reaction there are two more O atoms on the left than on the right. In acid solution, H^+ ions will be present, and in the reaction the two extra O atoms will become H_2O:

$$NO_3^- + 4H^+ \rightarrow NO + 2H_2O$$

Our two half-reactions are now balanced in terms of atoms; next we must balance the *charges* involved by adding electrons. The reactions become

$$Cu \rightarrow Cu^{2+} + 2e^- \quad \text{(oxidation)}$$
$$NO_3^- + 4H^+ + 3e^- \rightarrow NO + 2H_2O \quad \text{(reduction)}$$

Now we multiply the half-reactions by coefficients that make the number of electrons in these equations the same, and then we add the equations so the electrons cancel:

$$\begin{array}{r} 3Cu \rightarrow 3Cu^{2+} + 6e^- \\ 2NO_3^- + 8H^+ + 6e^- \rightarrow 2NO + 4H_2O \\ \hline 3Cu + 2NO_3^- + 8H^+ \rightarrow 3Cu^{2+} + 2NO + 4H_2O \end{array}$$

This is the required balanced equation.

4 is correct.
1 and **3** are not properly balanced.
2 NO_2, not NO, is formed in this reaction.
5 is just three times Answer **2**.

SOLUTION 7–16

(2)

Let's solve this problem by the ion–electron half-reaction method.

First, we must set up the oxidation and reduction half-reactions in terms of atoms only and balance the atoms involved. For an acid solution there will be H^+ on the left and H_2O on the right of the reduction half-reaction:

$$Zn \rightarrow Zn^{2+} \quad \text{(oxidation)}$$
$$H_3AsO_4 + 8H^+ \rightarrow AsH_3 + 4H_2O \quad \text{(reduction)}$$

Now we must complete the balancing of the half-reactions by adding electrons:

$$Zn \rightarrow Zn^{2+} + 2e^- \quad \text{(oxidation)}$$
$$H_3AsO_4 + 8H^+ + 8e^- \rightarrow AsH_3 + 4H_2O \quad \text{(reduction)}$$

Multiplying the oxidation half-reaction by four and adding the product to the reduction half-reaction, we have

$$\begin{array}{r} 4Zn \rightarrow 4Zn^{2+} + 8e^- \\ H_3AsO_4 + 8H^+ + 8e^- \rightarrow AsH_3 + 4H_2O \\ \hline H_3AsO_4 + 4Zn + 8H^+ \rightarrow AsH_3 + 4Zn^{2+} + 4H_2O \end{array}$$

To complete the balancing of the equation we must add the uninvolved SO_4^{2-} ions to give us H_2SO_4's on the left and $ZnSO_4$'s on the right. The final equation is

$$H_3AsO_4 + 4Zn + 4H_2SO_4 \rightarrow AsH_3 + 4ZnSO_4 + 4H_2O$$

Thus the ratio of the stoichiometric coefficients of H_3AsO_4 and Zn is 1/4.

2 is correct.
1 Didn't multiply the oxidation half-reaction by four.
3 Reversed Answer **2**.
4 Ratio of number of atoms in H_3AsO_4 and Zn.
5 Ratio of number of atoms in H_3AsO_4 and H_2SO_4.

SOLUTION 7-17 (3)

Our skeleton equation is

$$CoCl_2 + OCl^- \rightarrow Co(OH)_3 + Cl^-$$

In one half-reaction the change is

$$CoCl_2 \rightarrow Co(OH)_3$$

The cobalt atoms are balanced. We need $2Cl^-$ on the right to balance the chlorine atoms and $3OH^-$ on the left to balance the oxygen and hydrogen atoms. We obtain

$$CoCl_2 + 3OH^- \rightarrow Co(OH)_3 + 2Cl^-$$

To balance charge, we need $1e^-$ on the right. The result is

$$CoCl_2 + 3OH^- \rightarrow Co(OH)_3 + 2Cl^- + e^-$$

which is valid for basic (i.e., alkaline) solution. The change in the other half-reaction is

$$OCl^- \rightarrow Cl^-$$

The chlorine atoms are balanced. We need $1OH^-$ on the right to balance the oxygen atoms:

$$OCl^- \rightarrow Cl^- + OH^-$$

Now we must balance hydrogen. We do this by putting H^+ on the left. We find that

$$H^+ + OCl^- \rightarrow Cl^- + OH^-$$

This is no longer valid for basic solution; we have to eliminate H^+. We can do this by adding $1OH^-$ to both sides and by combining $1H^+$ and $1OH^-$ into $1H_2O$ on the left:

$$OCl^- + H_2O \rightarrow Cl^- + 2OH^-$$

To balance charges, we must add $2e^-$ to the left side. The final, balanced half-reaction for basic solution is

$$2e^- + OCl^- + H_2O \rightarrow Cl^- + 2OH^-$$

To combine the two balanced half-reactions we must multiply the first by two and add the product to the second. This gives

$$\begin{array}{rl} 2CoCl_2 + 6OH^- & \rightarrow 2Co(OH)_3 + 4Cl^- + 2e^- \\ 2e^- + OCl^- + H_2O & \rightarrow Cl^- + 2OH^- \\ \hline 2CoCl_2 + 4OH^- + OCl^- + H_2O & \rightarrow 2Co(OH)_3 + 5Cl^- \end{array}$$

3 is correct.
1 is balanced in terms of stoichiometry but not in terms of charge.
2 In acid solution and charge unbalanced.
4 Balanced but in acid solution.
5 is three times Answer **1**.

SOLUTION 7–18

(1)

First, we write the half-reactions:

$Cr_2O_7^{2-} + 6e^- + 14H^+ \rightarrow 2Cr^{3+} + 7H_2O$

and

$Fe^{2+} \rightarrow Fe^{3+} + e^-$

The half-reactions indicate that 1 mole of $Na_2Cr_2O_7$ accepts 6 moles of electrons, and 1 mole of $FeSO_4$ donates 1 mole of electrons. Therefore, the gram equivalent weight of $Na_2Cr_2O_7$ is 262/6 = 43.7 g, and the gram equivalent weight of $FeSO_4$ is 152/1 = 152 g. [*Note:* The gram equivalent weight of an oxidizing or reducing agent depends on the change in oxidation number that the agent undergoes in the specific reaction in which it is involved.]

1 is correct.
2 is the reverse of the gram molecular weights.
3 Divided the gram molecular weight of $FeSO_4$ by six.
4 are the gram molecular weights.
5 Multiplied the gram molecular weight of $Na_2Cr_2O_7$ by six instead of dividing.

SOLUTION 7–19

(3)

From Problem 7–18 the gram equivalent weight of $FeSO_4 \cdot 7H_2O$ in this reaction is equal to its gram molecular weight, that is, 278 g. A 1.00-normal solution of $FeSO_4 \cdot 7H_2O$ contains 278 g liter^{-1}. A 0.400-normal solution would, therefore, contain 278 × 0.400 = 111 g liter^{-1}. The correct answer is **3**. Answer **2** is the wrong gram equivalent weight: $FeSO_4$ instead of $FeSO_4 \cdot 7H_2O$. Answer **4**: You divided by 0.400 instead of multiplying.

SOLUTION 7–20

(1)

The half-reaction involving dichromate ion is

$Cr_2O_7^{2-} + 6e^- + 14H^+ \rightarrow 2Cr^{3+} + 7H_2O$

The equivalent weight of $Na_2Cr_2O_7$ is one sixth of its molecular weight. So the molarity of a 0.400-normal solution is 0.400/6 = 0.067 molar.

1 is correct.
4 Multiplied 0.400 by six instead of dividing.
2 is the normality.

SOLUTION 7–21

(3)

Number of equivalents of $FeSO_4$ required

$= 0.400$ equiv liter^{-1} × 0.0500 liter
$= 0.0200$ equiv

The number of equivalents of $Na_2Cr_2O_7$ titrated is equal to the number of equivalents of $FeSO_4$ required, namely, 0.0200. From Problem 7–20 the gram equivalent weight of $Na_2Cr_2O_7$ under the conditions of the experiment is equal to one sixth of its gram molecular weight. So the number of moles of $Na_2Cr_2O_7$ titrated is

$$\frac{0.0200}{6} = 0.0033 \text{ mole}$$

One mole of $Na_2Cr_2O_7$ contains 2 g-atoms of Cr. The number of gram-atoms of Cr in 0.0033 mole is therefore $2 \times 0.0033 = 0.0066$ g-atom. Hence, the weight of Cr in the mineral sample

$$= 0.0066 \text{ g-atom} \times 52 \text{ g g-atom}^{-1}$$
$$= 0.34 \text{ g}$$

The percentage of Cr in the mineral sample is $(0.34/1.7) \times 100 = 20\%$.

The correct answer is **3**. Answer **2** happens if you forgot that there are two Cr atoms in $Na_2Cr_2O_7$. Answer **4**: Neglected to divide by the weight of the mineral sample.

SOLUTION 7–22

(2)

For our titration we are effectively using 1.00 ml of the original blood specimen.

$$N_{KMnO_4} \times V_{KMnO_4} = \text{equivalents of } C_2O_4^{2-}$$

So

$$0.00500 \text{ normal} \times 1.00 \times 10^{-3} \text{ liter} = 5.00 \times 10^{-6} \text{ equiv}$$

The half-reaction for the oxidation of $C_2O_4^{2-}$ to CO_2 is

$$C_2O_4^{2-} \rightarrow 2CO_2 + 2e^-$$

So our titration involved half as many moles, or 2.50×10^{-6} mole $C_2O_4^{2-}$ per 1.00-ml blood sample. A 1-molar solution of CaC_2O_4 contains 1 g-atom of Ca^{2+} ions. Therefore, the weight of Ca^{2+} in the blood sample is

$$2.50 \times 10^{-6} \text{ g-atom ml}^{-1} \times 40.00 \text{ g g-atom}^{-1} \times 10.0 \text{ ml}$$
$$= 1.00 \times 10^{-3} \text{ g}$$
$$= 1.00 \text{ mg}$$

This is the same weight that was in the original 10.0-ml blood sample taken from the patient. The Ca^{2+} content in the blood of an average adult is 0.90–1.15 mg per 10.0 ml blood; so the patient has a normal Ca^{2+} content.

2 is correct.
1 is the weight of Ca^{2+} in 1.00 ml of blood.
4 You multiplied by the molecular weight of CaC_2O_4 instead of the atomic weight of Ca^{2+}.
5 Decimal point trouble.

CHAPTER 8 QUANTUM THEORY AND ATOMIC STRUCTURE

The modern concept of the atom is provided by quantum theory and wave mechanics, which were developed in response to analyses of electromagnetic radiation and the photoelectric effect. In this chapter we will study these topics as a prelude to chemical bonding and structure. To answer some of the problems you will need to memorize the following relationships.

$1 \text{ cm} = 10^8$ angstroms (Å)

$\lambda\nu = c$

in which

λ = wavelength of any electromagnetic radiation
ν = frequency of any electromagnetic radiation
c = velocity of light *in vacuo*

$E = h\nu$ = energy of one quantum of any electromagnetic radiation

in which

h = Planck's constant

Solutions to problems in Chapter 8 begin on page 228.

PROBLEMS

PROBLEM 8–1

The properties of electromagnetic waves are described in Section 8–2 of Chemical Principles *by Dickerson, Gray, and Haight.*

Consider the schematic drawings of two electromagnetic waves A and B in the infrared region shown in Figure 8–1. Which one of the following statements is *incorrect?*

($c = 3.0 \times 10^{10}$ cm sec^{-1})

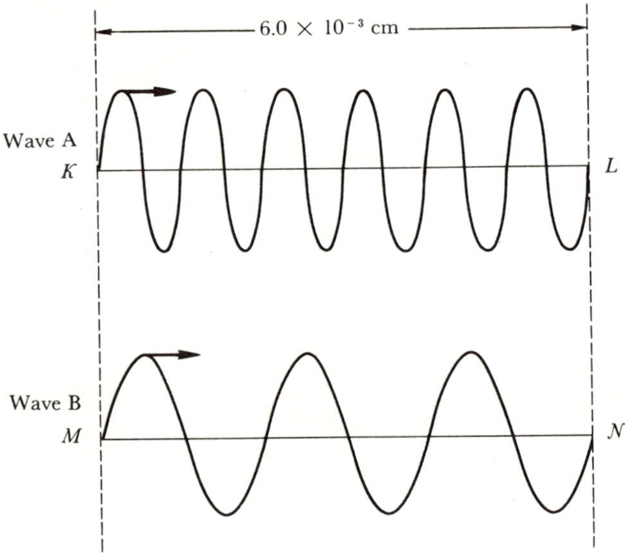

Figure 8–1

1 The wavelength of wave A is 1.0×10^5 Å.
2 The wavelength of wave B is 2.0×10^5 Å.
3 The frequency of wave A is 3.0×10^{13} sec^{-1}.
4 The frequency of wave B is 1.5×10^{13} sec^{-1}.
5 Wave A reaches point *L* before wave B reaches point *N*.

PROBLEM 8–2

What is the approximate wavelength of the radiation emitted from a mercury arc sunlamp if the frequency of the radiation is about 1.2×10^{15} sec^{-1}?

($c = 3.0 \times 10^{10}$ cm sec^{-1})

1 2.5×10^3 Å
2 4.0×10^4 Å
3 4.0×10^{12} Å
4 1.2×10^{15} Å
5 3.6×10^{25} Å

PROBLEM 8-3

Cobalt-60 ($^{60}_{27}Co$) is an artificial radioisotope that is produced in a nuclear reactor. It is an important gamma-ray source, and is used as a radio-therapeutic agent in the treatment of cancer and in the manufacture of some plastics. What is the frequency of the gamma radiation from a radioactive cobalt-60 source if its wavelength is 1.0×10^{-8} Å?

($c = 3.0 \times 10^{10}$ cm sec^{-1})

1 3.3×10^{-25} sec^{-1}
2 1.0×10^{-8} sec^{-1}
3 3.0×10^{10} sec^{-1}
4 3.0×10^{18} sec^{-1}
5 3.0×10^{26} sec^{-1}

PROBLEM 8-4

How long would it take a radiowave of frequency of 6.0×10^5 sec^{-1} to travel from the planet Mars to earth, a distance of approximately 8.0×10^7 km ($\simeq$ 50 million miles)? [*Note:* Radiowaves are electromagnetic waves.]

($c = 3.0 \times 10^{10}$ cm sec^{-1};
1 km $= 10^5$ cm)

1 0.45 min
2 2.7 min
3 4.5 min
4 270 min
5 Insufficient data; need to know the wavelength of the radiowave.

PROBLEM 8-5

When the soprano hits high C in the "mad" scene of "Lucia di Lammermoor," there is an appreciative sigh throughout the audience. High C on the musical scale has a frequency of 1024 sec^{-1}. What is the wavelength of the high C sound wave?

(Assume a velocity of sound in air $= 3.317 \times 10^4$ cm sec^{-1})

1 $\simeq$ 0.034 cm
2 $\simeq$ 32 cm
3 $\simeq$ 100 cm
4 $\simeq$ 320 cm
5 $\simeq$ 1024 cm

PROBLEM 8–6 The wavelength ranges for the different colors in the visible spectrum are shown in Figure 8–2. The characteristic colors of light emitted by alkali metal- and alkaline earth metal-containing materials when they are heated in a bunsen burner flame are used as qualitative tests for the presence of these elements. If the frequency of light emitted by potassium-containing materials is 7.41×10^{14} sec^{-1}, what color flame would you expect to see when you heat a simple potassium compound such as potassium chloride in a bunsen burner flame?

($c = 3.00 \times 10^{10}$ cm sec^{-1})

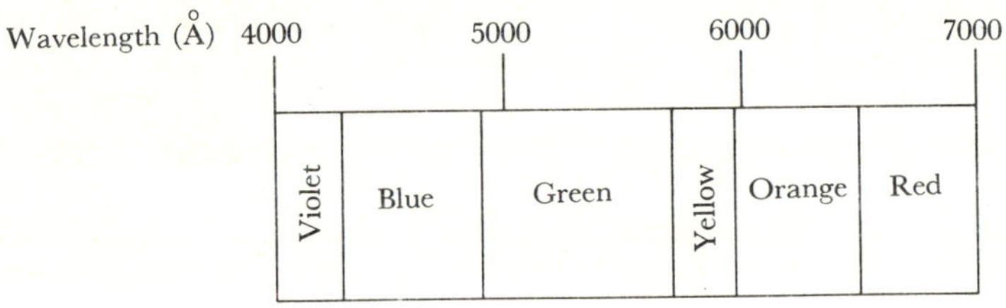

Figure 8–2

1 Red
2 Yellow
3 Green
4 Blue
5 Violet

PROBLEM 8–7

Blackbody radiation is discussed in Section 8–2 of Chemical Principles.

Figure 8–3 illustrates graphically the energy radiated (E_λ) at five different temperatures from a source that is approximately a blackbody as a function of the wavelength (λ) of the electromagnetic radiation emitted. Which one of the following statements concerning blackbody radiation is *incorrect?*

	Approximate wavelength ranges
visible	4×10^{-5} to 7×10^{-5} cm (or 4000–7000 Å)
infrared	7×10^{-5} to 1×10^{-2} cm (or 7000–1×10^6 Å)

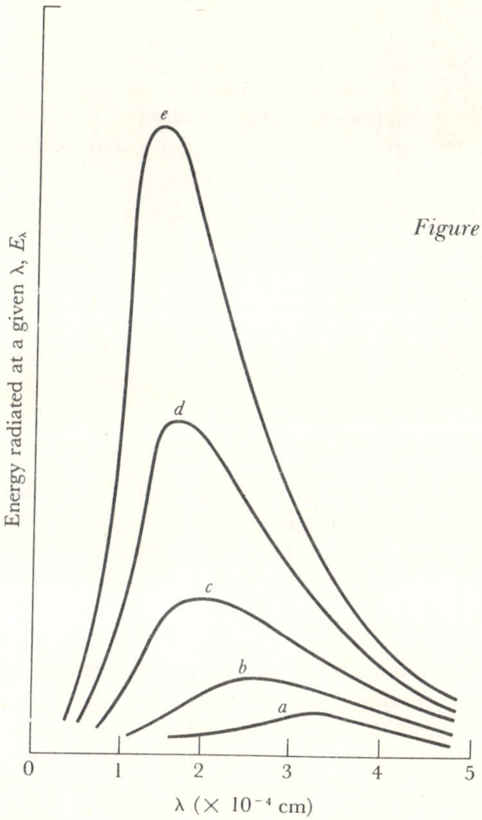

Figure 8–3. Curve a, 1000°K; b, 1200°K; c, 1500°K; d, 1800°K; e, 2000°K.

1 An *ideal* blackbody radiator is a hypothetical one that will emit all wavelengths.
2 The total energy emitted by a blackbody at all wavelengths (the total area under an experimental curve) increases with increasing absolute temperature of the blackbody.
3 The peaks of the experimental curves move toward shorter wavelengths as the temperature is increased.
4 The lower the temperature of a blackbody the more likelihood there is of high frequency rather than low frequency radiation being emitted.
5 A blackbody at 1000°K emits more infrared than visible radiation.

PROBLEM 8–8

The solar radiation falling on the outer atmosphere of the Earth is essentially that of a blackbody radiator heated to 6000°K. Using Figure 8–3 and Wien's displacement law ($\lambda_{max}T = k =$ constant), estimate λ_{max} for the solar radiation.

1 $\simeq 5 \times 10^{-5}$ cm
2 $\simeq 2 \times 10^{-4}$ cm
3 $\simeq 6 \times 10^{3}$ cm
4 $\simeq 2 \times 10^{4}$ cm
5 Insufficient information; need to know the value of the Wien displacement-law constant, k.

PROBLEM 8–9

A normal star radiates an energy of about 1.0×10^{52} ergs during its lifetime of 10^{10} years. This energy is equivalent to how many kilowatt hours (kWh)?

(1 kW $= 1 \times 10^{3}$ W; 1 W $= 1 \times 10^{7}$ ergs sec^{-1})

1 1.0×10^{7} kWh
2 7.8×10^{38} kWh
3 1.7×10^{40} kWh
4 7.8×10^{41} kWh
5 1.0×10^{42} kWh

PROBLEM 8–10

Figure 8–4 graphically illustrates plots of the kinetic energies of the electrons photoelectrically ejected from cesium, Cs, and tungsten, W, versus the frequency, ν, of the incident light. The general equation for the lines is

$$KE = \tfrac{1}{2}mv^2 = h[\nu - (\nu_0)_M] = h\nu - \phi_M$$

in which m is the mass of an electron, v is its velocity, h is Planck's constant, and $(\nu_0)_M$ is the critical frequency for metal M. The critical frequency, $(\nu_0)_M$, and the associated critical wavelength, $(\lambda_0)_M$, for metal M refer to the *minimum* frequency and the *maximum* wavelength, respectively, of the incident light required to eject an electron with zero kinetic energy from the surface of the metal. ϕ_M is equal to $h(\nu_0)_M$ and is known as the *work function* for metal M. It is the *minimum* energy required to eject an electron from the surface of the metal and leave the electron at rest. Which one of the following statements is *incorrect?*

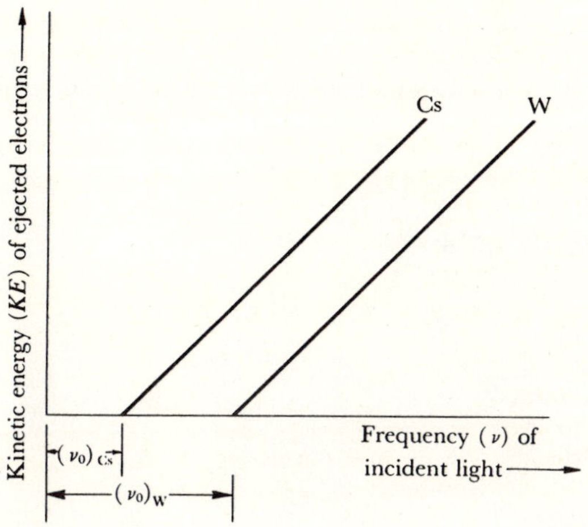

Figure 8–4

1 The critical frequency for tungsten, $(\nu_0)_W$, is higher than that for cesium, $(\nu_0)_{Cs}$.
2 The critical wavelength for cesium, $(\lambda_0)_{Cs}$, is shorter than that for tungsten, $(\lambda_0)_W$.
3 It takes less energy to eject an electron photoelectrically from cesium than from tungsten.
4 The higher the frequency of the incident light, the greater is the kinetic energy of the ejected electrons.
5 An electron ejected from cesium by a certain frequency of light has a higher velocity than that ejected from tungsten by the same frequency of light.

PROBLEM 8–11

The photoelectric effect is applied in some burglar alarms. A beam of light shines on a metal electrode in a vacuum tube (photoelectric cell) and causes electrons to be photoelectrically ejected from the surface of the metal. This electrode thus becomes the cathode. The ejected electrons are attracted toward the anode in the vacuum tube, and the electrical circuit is completed by means of a battery. If the light beam is blocked by a burglar's arm, the electrical circuit is broken. This sets off the alarm system.

What is the maximum wavelength of light that you could use for a burglar alarm if the cathode of the photoelectric cell is made of tungsten, and electrons are ejected from tungsten with a kinetic energy of

8.0×10^{-12} erg when the wavelength of the incident light is exactly 1.25×10^3 Å?

[*Note:* Kinetic energy of electrons ejected from a metal M $= h[\nu - (\nu_0)_M]$, in which $(\nu_0)_M$ is the critical frequency for metal M (refer to Problem 8–10 for the definitions of critical frequency and critical wavelength).]

($h = 6.6 \times 10^{-27}$ erg sec;
$c = 3.0 \times 10^{10}$ cm sec^{-1})

1 2.5×10^{-5} Å
2 8.3×10^{2} Å
3 1.25×10^{3} Å
4 2.5×10^{3} Å
5 6.6×10^{3} Å

PROBLEM 8–12

Cesium metal is used extensively in photocells and in television cameras because it has the lowest ionization energy of all the elements and is readily ionized by light. What is the kinetic energy of a photoelectric electron ejected from cesium by 5000 Å light? The critical *wavelength* for the photoelectric effect in cesium is 6600 Å.

[*Note:* Kinetic energy of a photoelectric electron ejected from cesium $= h[\nu - (\nu_0)_{Cs}]$, in which $(\nu_0)_{Cs}$ is the critical frequency for cesium.]

($c = 3.0 \times 10^{10}$ cm sec^{-1};
$h = 6.6 \times 10^{-27}$ erg sec)

1 6.6×10^{-27} erg
2 9.9×10^{-13} erg
3 6.9×10^{-12} erg
4 6.6×10^{-5} erg
5 1.5×10^{14} ergs

PROBLEM 8–13

A spectroscope is an instrument for measuring the wavelengths of lines in emission and absorption spectra. Two of the lines in the yellow emission from heated sodium metal, the so-called sodium-D lines, are used to calibrate spectroscopes. The wavelength of one of these lines is 5890 Å. What is the energy of the electronic transition associated with this line?

($h = 6.62 \times 10^{-27}$ erg sec;
$c = 3.00 \times 10^{10}$ cm sec^{-1})

1 6.62×10^{-27} erg
2 3.37×10^{-20} erg
3 1.12×10^{-12} erg
4 3.37×10^{-12} erg
5 5.89×10^{3} ergs

PROBLEM 8–14

In a bunsen burner flame, the barium atoms arising from the decomposition of barium-containing materials can undergo an electronic transition of energy 3.62×10^{-12} erg. Refer to Figure 8–2 in Problem 8–6 and decide what color flame would indicate to you the presence of barium in a sample of unknown composition.

1 Red
2 Orange
3 Yellow
4 Green
5 Blue

PROBLEM 8–15

The hydrogen spectrum is described in Section 8–2 of Chemical Principles.

Wave numbers are the reciprocals of wavelengths and are given by the expression $\bar{\nu} = 1/\lambda$. The equation from the Bohr theory for the wave numbers of the spectral lines observed in the emission spectrum of the hydrogen atom is

$$\bar{\nu} = R_H\left(\frac{1}{n_1^2} - \frac{1}{n_2^2}\right)$$

in which R_H is the Rydberg constant, and n_1 and n_2 are integers associated with the lower and the upper electronic levels, respectively, involved in the electron transitions responsible for the spectral lines.

In what region of the electromagnetic spectrum (see Figure 8–5) would you look for the spectral line resulting from the electronic transition from the fifth to the tenth electronic level in the hydrogen atom?

$(R_H = 1.10 \times 10^5\ \text{cm}^{-1})$

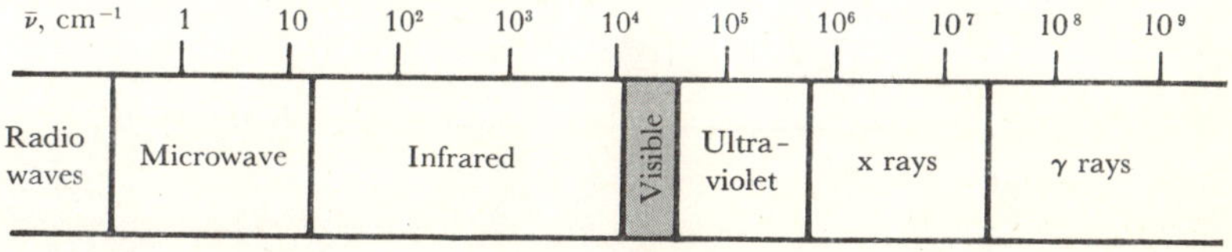

Figure 8–5

1 Microwave
2 Infrared
3 Visible
4 Ultraviolet
5 X ray

PROBLEM 8–16 The general equation for the lines in the emission spectra of one-electron systems such as H, He^+, Li^{2+}, and Be^{3+} is

$$\bar{\nu} = Z^2 R_H \left(\frac{1}{n_1^2} - \frac{1}{n_2^2} \right)$$

in which $\bar{\nu}$ are the wave numbers of emission lines and Z is the nuclear charge of the atom concerned.

If the ionization energy of a gaseous neutral atom or a gaseous ion is the energy needed to remove completely an electron from the atom or the ion in its ground state and the ionization energy of the hydrogen atom is 13.527 eV, what would you expect the second ionization energy of a helium atom to be?

1 3.382 eV
2 6.764 eV
3 13.527 eV
4 27.054 eV
5 54.108 eV

PROBLEM 8–17 What would be the maximum number of emission lines for atomic hydrogen that you would expect to see in a spectroscope if the only electronic energy levels involved are those shown in Figure 8–6?

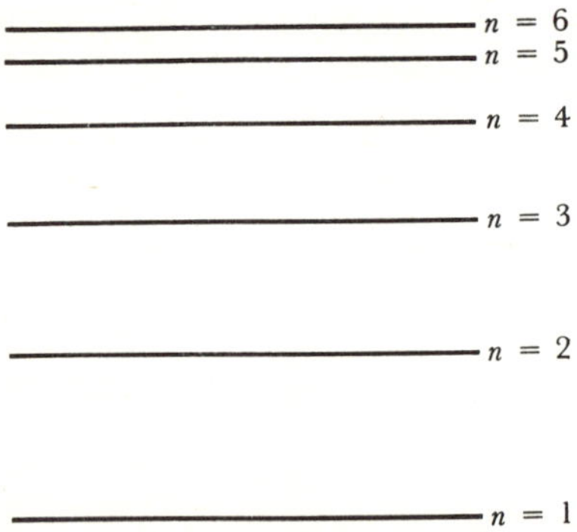

Figure 8–6

1 6
2 12
3 15
4 21
5 None of the above.

PROBLEM 8–18

In 1924, de Broglie proposed that every particle possesses wave properties with a wavelength λ given by

$$\lambda = \frac{h}{mv}$$

in which m is the mass of the particle, v is its velocity, and h is Planck's constant. What is the de Broglie wavelength of an 88-kg (≃200 lb) man skiing down Mt. Tremblant in Québec at 5.0×10^5 cm sec^{-1} (≃35 miles h^{-1})?

($h = 6.6 \times 10^{-27}$ erg sec;
1 erg = 1 g cm^2 sec^{-2})

1 1.5×10^{-37} cm
2 1.5×10^{-34} cm
3 6.6×10^{-27} cm
4 3.3×10^{-21} cm
5 5.0×10^{5} cm

PROBLEM 8–19

Radium is a radioactive element that emits alpha particles (helium nuclei, He^{2+}), beta particles (electrons), and gamma rays. One of its uses is in self-luminous paints. The alpha particles emitted from radium have an energy of 4.8 MeV (million electron volts). What is the de Broglie wavelength of one of these alpha particles?

(mass of alpha particle $= 6.6 \times 10^{-24}$ g;
$h = 6.6 \times 10^{-27}$ erg sec;
1.0 MeV $= 1.6 \times 10^{-6}$ erg;
1 erg = 1 g cm^2 sec^{-2};
$\lambda = h/mv$)

1 6.6×10^{-27} cm
2 6.6×10^{-24} cm
3 8.3×10^{-16} cm
4 6.6×10^{-13} cm
5 9.4×10^{-13} cm

PROBLEM 8–20

The Schrödinger wave equation and quantum numbers are discussed in Sections 8–6 and 8–7 of Chemical Principles.

n and *l* are two of the four quantum numbers that characterize the solutions of the Schrödinger wave equation for the hydrogen atom. *n* is the *principal quantum number* and can be any positive integer: $n = 1, 2, 3, 4, \ldots, \infty$, whereas *l* is the *angular momentum quantum number* and can have an integral value from 0 to $n - 1$. The solutions of the Schrödinger wave equation may be illustrated by electron probability diagrams known as orbitals. Six of these orbitals are shown in Figure 8–7. Which of the following statements concerning the function of the two quantum numbers *n* and *l* with respect to an orbital is/are *correct?* [*Hint:* Each quantum number has only one function.]

(a) *n* determines the shape of an orbital.
(b) *l* determines the size of an orbital.
(c) *n* determines the size of an orbital.
(d) *l* determines the shape of an orbital.
(e) *n* determines the number of lobes of electron density in an orbital.

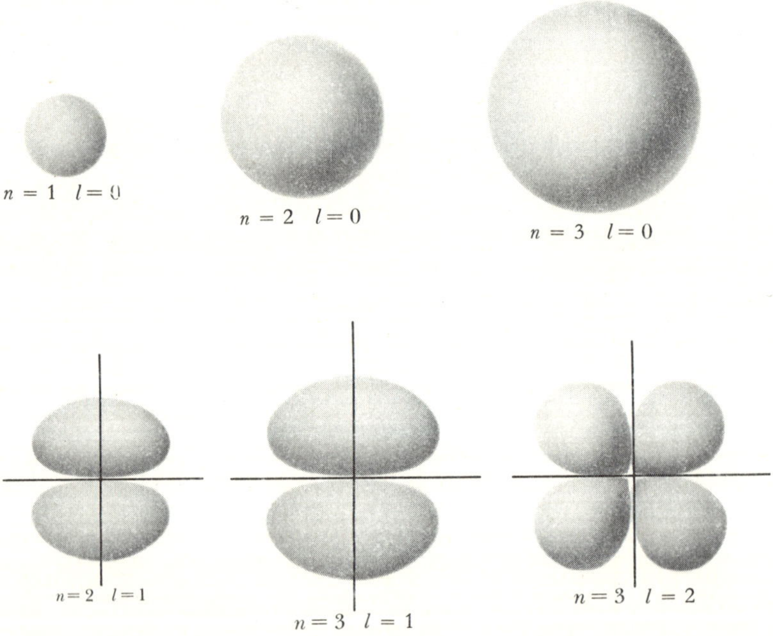

Figure 7–9

1 (a)
2 (b) and (c)
3 (e)
4 (a) and (b)
5 (c) and (d)

PROBLEM 8–21

There are two other quantum numbers in addition to n and l that we need to characterize the solutions of the Schrödinger equation for the hydrogen atom: m and s. Quantum number m describes the orientation of an orbital in space and is known as the *magnetic* or *orbital orientation quantum number.* It can have any integral value from $-l$ to $+l$ including 0. Quantum number s is the *spin quantum number.* Each electron can rotate clockwise or counterclockwise and has a choice of two spin states, namely, $s = +\frac{1}{2}$ or $-\frac{1}{2}$ in any given orbital.

It has been shown that no two electrons can have the same four quantum numbers (Pauli exclusion principle). Thus, for a complete description of an electron in an orbital of the hydrogen atom, all four quantum numbers must be specified.

Which of the following combinations of quantum numbers *do not* represent permissible solutions of the Schrödinger wave equation for the hydrogen atom? [*Note:* Refer to Problem 8–20 for the allowed values of n and l.]

	n	l	m	s
(a)	3	0	1	$-\frac{1}{2}$
(b)	2	2	0	$+\frac{1}{2}$
(c)	4	3	-4	$-\frac{1}{2}$
(d)	5	2	2	$+\frac{1}{2}$
(e)	3	2	-2	$-\frac{3}{2}$

1 (c)
2 (d) and (e)
3 (b), (d), and (e)
4 (a), (b), (c), and (e)
5 None of them is permissible.

PROBLEM 8–22

The only permissible solution of the Schrödinger wave equation for the hydrogen atom in Problem 8–21 had $n = 5$ and $l = 2$. What type of orbital do these quantum numbers describe?

1 $3s$
2 $5p$
3 $4f$
4 $5d$
5 $5s$

PROBLEM 8–23 Which one of the drawings in Figure 8–8 represents the shape of the electron density distribution in a p_z orbital? [*Hint:* Refer to Problems 8–20 and 8–22.]

Probability of electron distribution and the shapes of atomic orbitals are described in Section 8–7 of Chemical Principles.

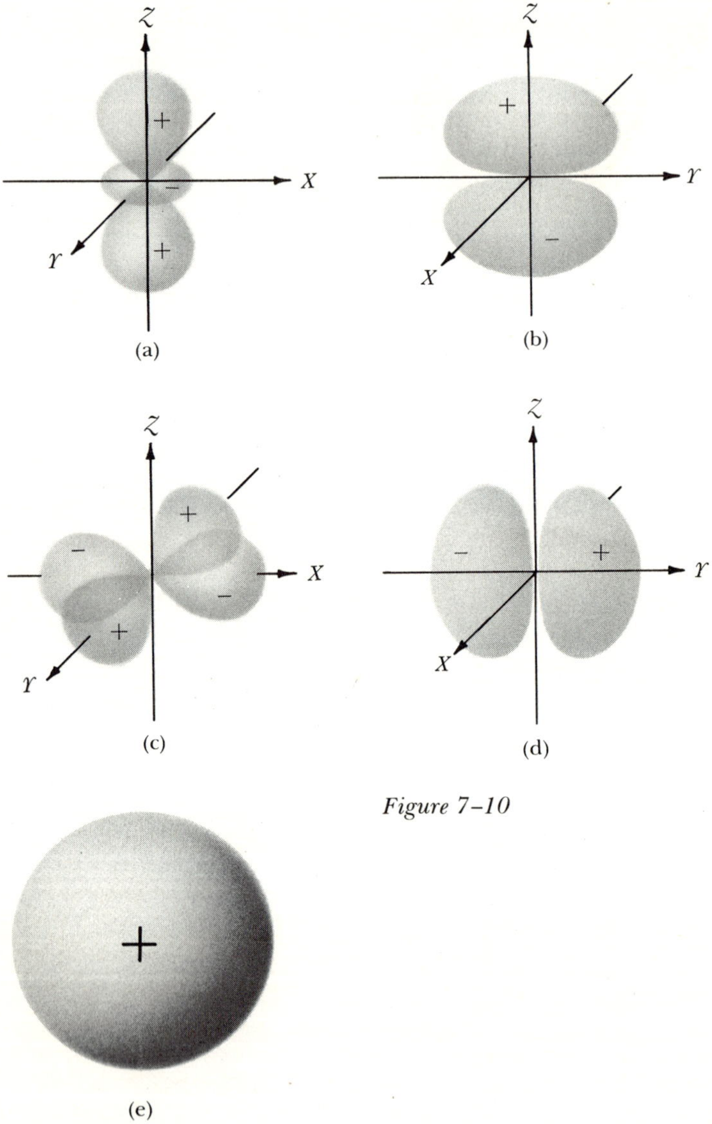

Figure 7–10

1 (a)
2 (b)
3 (c)
4 (d)
5 (e)

PROBLEM 8–24 Which one of the drawings in Figure 8–9 represents the shape of the electron density distribution in a d_{xz} orbital? [*Hint:* Refer to Problem 8–23.]

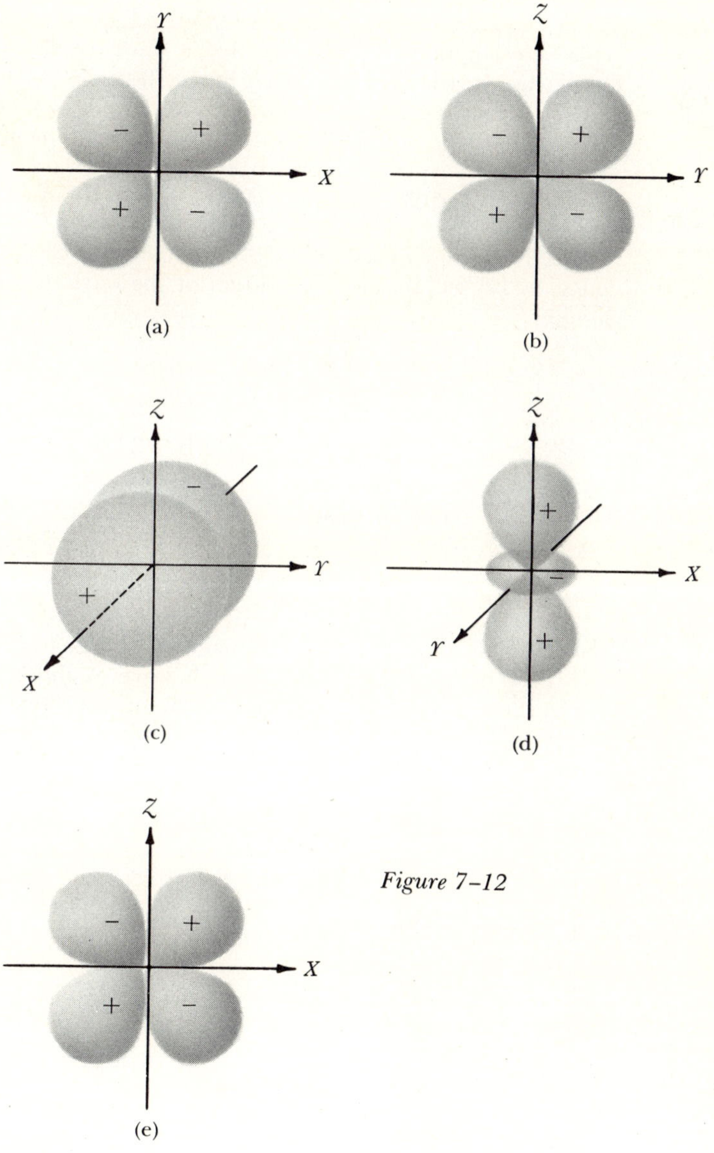

Figure 7–12

1 (a)
2 (b)
3 (c)
4 (d)
5 (e)

SOLUTIONS

SOLUTION 8–1

(5)

One wavelength is the distance that a wave travels in one complete cycle, that is, the distance *XY* in Figure 8–10. Wave A completes six cycles in traveling 6.0×10^{-3} cm. The wavelength of wave A is therefore 6.0×10^{-3} cm/6 = 1.0×10^{-3} cm = 1.0×10^{5} Å. (Remember that 1 cm = 1×10^{8} Å.) Answer **1** is correct. Similarly, the wavelength of wave B is 6.0×10^{-3} cm/3 = 2.0×10^{-3} cm = 2.0×10^{5} Å. Answer **2** is also correct.

The frequency of wave A is the number of cycles that pass a given point in 1.0 sec, that is, the velocity of the wave divided by its wavelength:

$$\nu_A = \frac{c}{\lambda_A} = \frac{3.0 \times 10^{10}\ \text{cm sec}^{-1}}{1.0 \times 10^{-3}\ \text{cm}} = 3.0 \times 10^{13}\ \text{sec}^{-1}$$

Answer **3** is correct. Similarly, the frequency of wave B is

$$\nu_B = \frac{c}{\lambda_B} = \frac{3.0 \times 10^{10}\ \text{cm sec}^{-1}}{2.0 \times 10^{-3}\ \text{cm}} = 1.5 \times 10^{13}\ \text{sec}^{-1}$$

Answer **4** is correct.

Remembering that all electromagnetic waves travel in vacuum at the speed of light (3.0×10^{10} cm sec^{-1}), waves A and B travel 6.0×10^{-3} cm in the same time. Wave A reaches point *L* at the same time as wave B reaches point *N*. So Answer **5** is incorrect and is the required answer.

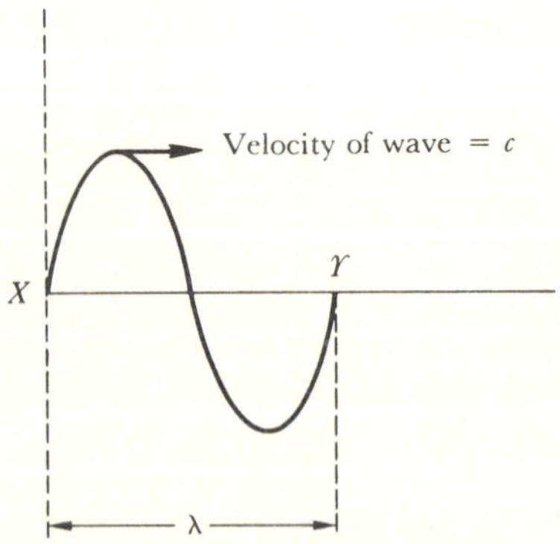

Figure 8–10

SOLUTION 8–2

(1)

The approximate wavelength of the radiation emitted by the mercury lamp is given by

$$\lambda = \frac{c}{\nu}$$

$$= \frac{3.0 \times 10^{10}\ \text{cm sec}^{-1}}{1.2 \times 10^{15}\ \text{sec}^{-1}}$$

$$= 2.5 \times 10^{-5}\ \text{cm} \times 10^{8}\ \text{Å cm}^{-1}$$
$$= 2.5 \times 10^{3}\ \text{Å (or 2500 Å)}$$

The wavelength observed experimentally is 2537 Å.

Answer **1** is correct. Answer **2**: You used $\lambda c = \nu$ instead of $\lambda\nu = c$ and did not bother about units.

SOLUTION 8–3

(5)

Gamma rays are electromagnetic radiation; hence, the frequency of the gamma radiation from the radioactive cobalt-60 source is given by

$$\nu = \frac{c}{\lambda} = \frac{3.0 \times 10^{10}\ \text{cm sec}^{-1}}{1.0 \times 10^{-8}\ \text{Å} \times 10^{-8}\ \text{cm Å}^{-1}}$$

$$= 3.0 \times 10^{26}\ \text{sec}^{-1}$$

The correct answer is **5**. Answer **1**: You used $\nu = \lambda/c$. For answer **4**, you forgot to convert angstroms to centimeters.

SOLUTION 8–4

(3)

Since all electromagnetic radiation travels in vacuum at the same speed (3.0×10^{10} cm sec^{-1} or 186,000 miles sec^{-1}), it does not matter what the frequency of our radiowave is.

Time for radiowave to travel from Mars to earth

$$= \frac{8.0 \times 10^{7}\ \text{km} \times 10^{5}\ \text{cm km}^{-1}}{3.0 \times 10^{10}\ \text{cm sec}^{-1}}$$

$$= 2.7 \times 10^{2}\ \text{sec}$$
$$= 4.5\ \text{min}$$

(Conversation will certainly be laconic. Compare 4.5 min with the 1.3 sec for a radiowave to reach earth from the moon.) The correct answer is **3**.

SOLUTION 8–5

(2)

Sound waves do not travel at the same speed as electromagnetic radiation, but there is a relationship between the wavelength (λ), frequency (ν), and velocity (v) of a sound wave similar to that for an electromagnetic wave, namely,

$$\lambda\nu = v$$

$$\text{wavelength of high C} = \frac{v}{\nu}$$

$$= \frac{3.317 \times 10^{4} \text{ cm sec}^{-1}}{1024 \text{ sec}^{-1}}$$

$$\simeq 32 \text{ cm}$$

[*Note:* The velocity of sound in air depends on a number of variables such as temperature, humidity, atmospheric pressure, and CO_2 content.]

2 is the correct answer.
1 used $\lambda v = \nu$ instead of $\lambda\nu = v$.
4 Decimal point error.
5 is the frequency of high C with the wrong units.

SOLUTION 8–6

(5)

$$\lambda = \frac{c}{\nu} = \frac{3.00 \times 10^{10} \text{ cm sec}^{-1}}{7.41 \times 10^{14} \text{ sec}^{-1}}$$

$$= 4.05 \times 10^{-5} \text{ cm}$$

$$= 4.05 \times 10^{-5} \text{ cm} \times 10^{8} \text{ Å cm}^{-1}$$

$$\simeq 4050 \text{ Å}$$

So potassium-containing materials impart a violet (actually more of a lilac) color to the bunsen burner flame. The correct answer is **5**.

[*Note:* This principle can be applied to pyrotechnic displays (for July 4 or Guy Fawkes' Day, November 5) or just coloring home fireplace flames.]

SOLUTION 8–7

(4)

Answer **1** is *correct.* Solids are not perfect radiators because they do not emit all wavelengths. A very good approximation that we can find for an ideal blackbody radiator is a small pinhole in a furnace constructed of graphite.

Answer **2** is *correct.* For example, compare the experimental curves at 1000°K (curve *a*) and 2000°K (curve *e*). [*Note:* The mathematical relationship between the total energy radiated by a blackbody, E, and the absolute temperature, T, is $E = \sigma T^4$ (Stefan's law), in which σ is the Stefan–Boltzmann constant.]

Answer **3** is *correct.* For instance, for curve *a* the maximum wavelength (λ_{max}) is about 3×10^{-4} cm; for curve *e* it is about 1.3×10^{-4} cm.

[*Note:* λ_{max} is inversely proportional to the absolute temperature, T (Wien's displacement law).]

The most likely wavelength to be emitted at a given temperature is λ_{max}, which increases with decreasing temperature. In other words, the lower the temperature the more there is a likelihood of low frequency rather than high frequency radiation being emitted. [*Recall:* $\lambda_{max}\nu_{max} = c$.] Answer **4** is *incorrect.*

At 1000°K, λ_{max} for the blackbody radiator is approximately 3×10^{-4} cm (curve *a*). This maximum lies in the infrared region (7×10^{-5} to 1×10^{-2} cm), not the visible region (4×10^{-5} to 7×10^{-5} cm). Thus, Answer **5** is *correct.*

The required answer is **4**.

SOLUTION 8–8

(1)

At 1500°K (from **Figure** 8–3), $\lambda_{max} \simeq 2 \times 10^{-4}$ cm. Thus, substituting for λ_{max} and T in Wien's displacement law, $\lambda_{max}T = k$, we have

$$2 \times 10^{-4} \text{ cm} \times 1500 \text{ deg} = k \qquad \text{(A)}$$

Similarly, at 6000°K:

$$\lambda_{max} \times 6000 \text{ deg} = k \qquad \text{(B)}$$

Equating Equations A and B, we obtain

$$\lambda_{max} \simeq 5 \times 10^{-5} \text{ cm (at 6000°K)}$$

Thus, on quadrupling the temperature of a blackbody radiator from 1500°K to 6000°K, the maximum wavelength emitted is reduced to approximately one-quarter of its original value. The correct answer is **1**. If you got Answer **4**, you probably inverted Answer **1**.

At both 1500°K and 6000°K, most of the energy radiated is in the infrared rather than the visible region. Actually, solar radiation is not characteristic of a perfect blackbody at 6000°K because part of the radiation is absorbed by the various gases in the earth's atmosphere. Some of the infrared radiation is absorbed by water vapor and CO_2. X rays and ultraviolet light with $\lambda < 2 \times 10^{-5}$ cm are absorbed by O_2 and N_2, while the majority of the radiation with $\lambda = 2 \times 10^{-5}$ cm to 3×10^{-5} cm is absorbed by ozone (O_3) present in the outer atmosphere. The ozone layer is vital to our existence on this planet; it is ultraviolet light that gives us sunburns and probably contributes to the incidence of skin cancer.

SOLUTION 8-9 (2)

There are 10^3 W per kilowatt, and using our conversion between watts and erg sec^{-1},

$$\begin{aligned}1.0\ \text{kW} &= 1.0 \times 10^3\ \text{W} \times 10^7\ \text{ergs sec}^{-1}\ \text{W}^{-1}\\ &= 1.0 \times 10^{10}\ \text{ergs sec}^{-1} \times 3600\ \text{sec h}^{-1}\\ &= 3.6 \times 10^{13}\ \text{ergs h}^{-1}\end{aligned}$$

Therefore,

$$1.0\ \text{kWh} = 3.6 \times 10^{13}\ \text{ergs}$$

is the relationship between the two units of energy. The star radiates an energy of 1.0×10^{52} ergs, that is,

$$\frac{1.0 \times 10^{52}\ \text{ergs}}{3.6 \times 10^{13}\ \text{ergs (kWh)}^{-1}}$$

$$= 7.8 \times 10^{38}\ \text{kWh}$$

Answer **2** is correct.

[*Note:* You can get some feel for the enormous amounts of energy involved when you realize that the estimated electrical energy requirements of the world in the year 2000 is only 3×10^{13} kWh.]

3 You only converted seconds to minutes instead of hours.
4 You forgot to convert kilowatts to watts.
5 You did not convert seconds to hours.

SOLUTION 8-10 (2)

For more on the photoelectric effect, see Section 8-2 of Chemical Principles.

From Figure 8–4, $(\nu_0)_W > (\nu_0)_{Cs}$. So Answer **1** is *correct.* Since $\lambda\nu = c$,

$$\frac{c}{(\lambda_0)_W} > \frac{c}{(\lambda_0)_{Cs}}$$

That is,

$$(\lambda_0)_{Cs} > (\lambda_0)_W$$

The critical wavelength for cesium is *longer* (not shorter) than that for tungsten. Answer **2** is therefore *incorrect.*

From Answer **1**, $(\nu_0)_W > (\nu_0)_{Cs}$. Thus, $h(\nu_0)_W > h(\nu_0)_{Cs}$; that is, $\phi_W > \phi_{Cs}$. This means that less energy is required to eject an electron from cesium than from tungsten; consequently, Answer **3** is *correct.*

From Figure 8–4, as ν increases so does the kinetic energy of the ejected electrons. Therefore, Answer **4** is *correct.*

For a given frequency of light to cause the ejection of an electron from both cesium and tungsten, the frequency must be greater than the critical frequency for tungsten, $(\nu_0)_W$. From Figure 8–4, the kinetic energy of an electron ejected from cesium for any frequency $\nu > (\nu_0)_W$

is always greater than the kinetic energy of an electron ejected from tungsten by light of the same frequency, ν. Since the kinetic energy of an electron is $\frac{1}{2}mv^2$, it follows that the velocity of an electron ejected from cesium by light of a given frequency travels faster than does an electron ejected from tungsten by light of the same frequency. Therefore, Answer **5** is *correct.*

The required answer is **2**.

SOLUTION 8-11

(4)

The kinetic energy of electrons ejected from tungsten, W, is given by

$$KE = h[\nu - (\nu_0)_W] = h\left(\frac{c}{\lambda} - \frac{c}{(\lambda_0)_W}\right)$$

After rearranging we obtain

$$(\lambda_0)_W = \frac{\lambda hc}{hc - (\lambda \times KE)}$$

Hence the maximum wavelength of light $(\lambda_0)_W$ that can be used to operate the burglar alarm is

$$\frac{1.25 \times 10^3\ \text{Å} \times 10^{-8}\ \text{cm Å}^{-1} \times 6.6 \times 10^{-27}\ \text{erg sec} \times 3.0 \times 10^{10}\ \text{cm sec}^{-1}}{(6.6 \times 10^{-27}\ \text{erg sec} \times 3.0 \times 10^{10}\ \text{cm sec}^{-1}) - (1.25 \times 10^3\ \text{Å} \times 10^{-8}\ \text{cm Å}^{-1} \times 8.0 \times 10^{-12}\ \text{erg})}$$

$$= \frac{2.5 \times 10^{-21}\ \text{erg cm}^2}{1.0 \times 10^{-16}\ \text{erg cm}}$$

$$= 2.5 \times 10^{-5}\ \text{cm} = 2.5 \times 10^3\ \text{Å (or 2500 Å)}$$

Thus, provided that you use light of wavelength shorter than 2500 Å, your burglar alarm will operate with tungsten metal as the cathode in the photocell. Since the visible range is 4000–7000 Å, you could not use visible light to operate the photocell. You would need to use ultraviolet light (1000–4000 Å).

The correct answer is **4**.

1 Did not convert centimeters to angstroms and dropped some powers of 10.
2 $\lambda = c/\nu$, not ν/c.
3 is impossible. You know that the kinetic energy for this wavelength is greater than zero; the kinetic energy should be exactly equal to zero for the maximum wavelength.
5 is h with the wrong powers of 10 and wrong units.

SOLUTION 8-12 (2)

The critical frequency $(\nu_0)_{Cs}$ is given by

$$(\nu_0)_{Cs} = \frac{c}{(\lambda_0)_{Cs}} = \frac{3.0 \times 10^{10} \text{ cm sec}^{-1}}{6.6 \times 10^3 \text{ Å} \times 10^{-8} \text{ cm Å}^{-1}}$$

$$= 4.5 \times 10^{14} \text{ sec}^{-1}$$

For light of wavelength 5000 Å, the frequency is

$$\nu = \frac{c}{\lambda} = \frac{3.0 \times 10^{10} \text{ cm sec}^{-1}}{5.0 \times 10^3 \text{ Å} \times 10^{-8} \text{ cm Å}^{-1}}$$

$$= 6.0 \times 10^{14} \text{ sec}^{-1}$$

Kinetic energy of a photoelectric electron

$= h[\nu - (\nu_0)_{Cs}]$
$= (6.6 \times 10^{-27} \text{ erg sec}) [(6.0 - 4.5) \times 10^{14} \text{ sec}^{-1}]$
$= 9.9 \times 10^{-13}$ erg

2 is correct.
1 is h with the wrong units.
3 You added ν and $(\nu_0)_{Cs}$ instead of subtracting them.
4 is 6600 Å with the wrong units.
5 You did not multiply by h.

SOLUTION 8-13 (4)

$$E = h\nu = \frac{hc}{\lambda}$$

The energy associated with the 5890-Å line of sodium is

$$E = \frac{6.62 \times 10^{-27} \text{ erg sec} \times 3.00 \times 10^{10} \text{ cm sec}^{-1}}{5.89 \times 10^3 \text{ Å} \times 10^{-8} \text{ cm Å}^{-1}}$$

$$= 3.37 \times 10^{-12} \text{ erg}$$

4 is the correct answer.
1 is h with the wrong units.
3 Did not multiply h/λ by c and ignored the powers of 10.
5 is the wavelength with the wrong units.

SOLUTION 8-14 (4)

$$E = h\nu = \frac{hc}{\lambda}$$

Therefore,

$$\lambda = \frac{hc}{E} = \frac{6.62 \times 10^{-27} \text{ erg sec} \times 3.00 \times 10^{10} \text{ cm sec}^{-1} \times 10^8 \text{ Å cm}^{-1}}{3.62 \times 10^{-12} \text{ erg}}$$

$$= 5490 \text{ Å}$$

Hence, you would expect to see a green flame (actually more of an apple-green) if your sample contained a compound of barium. Thus, the correct answer is **4**.

SOLUTION 8–15

(2)

$$\bar{\nu} = R_H\left(\frac{1}{n_1^2} - \frac{1}{n_2^2}\right)$$

Substituting $R_H = 1.10 \times 10^5\ \text{cm}^{-1}$, $n_1 = 5$, and $n_2 = 10$, we obtain

$$\bar{\nu} = 1.10 \times 10^5\ \text{cm}^{-1}\left(\frac{1}{5^2} - \frac{1}{10^2}\right)$$

$$= 3.3 \times 10^3\ \text{cm}^{-1}$$

This line would appear in the infrared region of the electromagnetic spectrum. It is a member of the Pfund series, for which $n_1 = 5$ and $n_2 = 6, 7, 8, 9, \ldots$. The other spectral line series observed for the hydrogen atom are

Lyman	$n_1 = 1$	$n_2 = 2, 3, 4, 5, \ldots$
Balmer	$n_1 = 2$	$n_2 = 3, 4, 5, 6, \ldots$
Paschen	$n_1 = 3$	$n_2 = 4, 5, 6, 7, \ldots$
Brackett	$n_1 = 4$	$n_2 = 5, 6, 7, 8, \ldots$

The five series are named after their discoverers. When the electron is in the first electronic level ($n_1 = 1$), the hydrogen atom is in its *ground state.* All other electronic levels are referred to as *excited states.* Only the Lyman series involves transitions from the ground state.

The correct answer is **2**.

SOLUTION 8–16

(5)

The helium atom is composed of a nucleus of charge +2 ($Z = 2$) and two electrons. The second ionization energy of helium (I_{He^+}) is the energy required to remove the lone electron from $He^+(g)$:

$$He^+(g) \rightarrow He^{2+}(g) + e^-$$

$$E = h\nu = \frac{hc}{\lambda} = hc\bar{\nu} \qquad \left(\textit{Remember: } \bar{\nu} = \frac{1}{\lambda}\right)$$

The energies associated with the electronic transitions in any one-electron system are therefore given by

$$E = hcZ^2R_H\left(\frac{1}{n_1^2} - \frac{1}{n_2^2}\right)$$

The ionization energy of the hydrogen atom is the energy needed to remove completely the electron from the atom in its ground state. For the ground state, $n_1 = 1$; for the state in which the electron is completely removed from the atom, $n_2 = \infty$. Thus, the ionization energy of the hydrogen atom, I_H, is

$$I_H = hcR_H \qquad (Z = 1)$$

Similarly, the second ionization energy of helium, I_{He^+}, is

$$\begin{aligned} I_{He^+} &= 4hcR_H \qquad (Z = 2) \\ &= 4I_H \\ &= 4 \times 13.527 \text{ eV} \\ &= 54.108 \text{ eV} \end{aligned}$$

(The experimental value is 54.416 eV.)

5 is the correct answer.
1 Divided I_H by four instead of multiplying.
3 is just I_H.
4 Multiplied I_H by two instead of four.

SOLUTION 8–17

(3)

Emission lines arise by electronic transitions from high electronic energy levels to lower ones. The transitions resulting in emission lines for atomic hydrogen when $n = 1, 2, 3, 4, 5,$ or 6 are shown in Figure 8–11. There are 15 transitions; consequently, 15 emission lines would be the maximum number expected.

The correct answer is **3**. [In general, the maximum number of emission lines expected is given by

$$\tfrac{1}{2}n(n - 1)$$

in which n is the quantum number of the highest energy level concerned. In our case, $n = 6$. Therefore, the maximum number of emission lines is $\frac{1}{2} \times 6 \times (6 - 1) = 15$.]

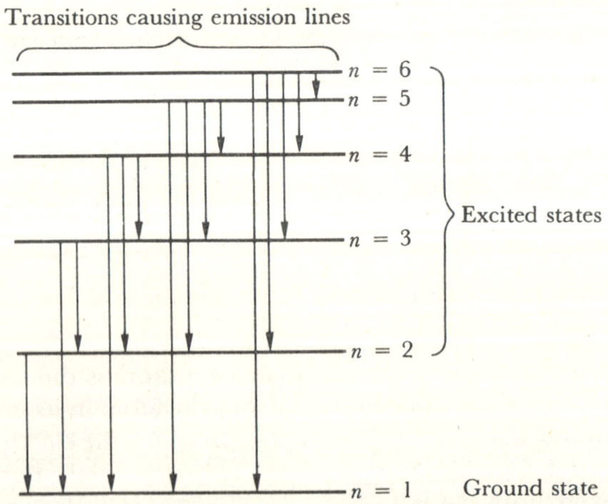

Figure 8–11

SOLUTION 8-18

(1)

$$\lambda = \frac{h}{mv}$$

$$h = 6.6 \times 10^{-27} \text{ erg sec} = 6.6 \times 10^{-27} \text{ g cm}^2 \text{ sec}^{-1}$$

Therefore

$$\lambda = \frac{6.6 \times 10^{-27} \text{ g cm}^2 \text{ sec}^{-1}}{8.8 \times 10^4 \text{ g} \times 5.0 \times 10^5 \text{ cm sec}^{-1}}$$

$$= 1.5 \times 10^{-37} \text{ cm}$$

1 is correct.
2 Didn't convert kilograms to grams.
3 is h with the wrong units.
4 is the product hv with the wrong units.
5 is v with the wrong units.

SOLUTION 8-19

(4)

The alpha particles are emitted with a kinetic energy of 4.8 MeV. So

$$\tfrac{1}{2}mv^2 = 4.8 \text{ MeV}$$
$$= 4.8 \times 1.6 \times 10^{-6} \text{ erg}$$
$$= 4.8 \times 1.6 \times 10^{-6} \text{ g cm}^2 \text{ sec}^{-2}$$
$$(mv)^2 = 2(\tfrac{1}{2}mv^2)m$$
$$= 2 \times 4.8 \times 1.6 \times 10^{-6} \text{ g cm}^2 \text{ sec}^{-2} \times 6.6 \times 10^{-24} \text{ g}$$
$$= 1.0 \times 10^{-28} \text{ g}^2 \text{ cm}^2 \text{ sec}^{-2}$$

Now

$$\lambda = \frac{h}{mv}$$

so

$$\lambda = \frac{6.6 \times 10^{-27} \text{ g cm}^2 \text{ sec}^{-1}}{(1.0 \times 10^{-28} \text{ g}^2 \text{ cm}^2 \text{ sec}^{-2})^{1/2}}$$

$$= 6.6 \times 10^{-13} \text{ cm}$$

This wavelength is much larger than that of the skier in Problem 8-18. The smaller the momentum (mv) of a "particle," the larger is its wavelength. We have to worry about quantum effects in our measurements when the wavelength of the particles involved approaches some characteristic length for the system under investigation: for instance, atomic spacing in a lattice or mean free path in a dense cold gas. But the wavelengths of a skier or even a highly energetic alpha particle are still too small to be used to study quantum effects such as electron diffraction or the strange behavior of liquid helium. For these studies the particles must be of small mass *and* velocity so their de Broglie wavelengths are long enough.

4 is the correct answer.
1 is h with the wrong units.
2 is m with the wrong units.
3 Did not convert million electron volts to ergs and ignored units.
5 Dropped the factor of 2 from $\frac{1}{2}mv^2$.

SOLUTION 8–20

(5)

If you need additional help to understand quantum numbers, refer to Review 8 of Programed Reviews of Chemical Principles *by Lassila* et al.

As n increases while l is kept constant, the size of an orbital type increases, but there is no change in either the shape or in the number of lobes of electron density involved. The principal quantum number n controls the size of all orbitals, whether they are spherical, dumbbell (two lobes of electron density), or cloverleaf (four lobes of electron density) in shape. (There are other orbital shapes but they need not concern us here.) Statement (c) is *correct* and (a) and (e) are *incorrect.*

As l increases while n is kept constant, the shapes of the orbitals change. Statement (d) is *correct* and (b) is *incorrect.* When $l = 0$, the orbital is known as an s orbital; $l = 1$, a p orbital; $l = 2$, a d orbital; $l = 3$, an f orbital. To keep track of the size as well as the shape of an orbital, we give it a number and a letter (e.g., a $3d$ orbital means $n = 3$ and $l = 2$). So only statements (c) and (d) are *correct.* The required answer is **5**.

[*Note:* Other names for the l quantum number are *azimuthal* and *orbital-shape*.]

SOLUTION 8–21

(4)

Permissible values of the quantum numbers:

n	$1, 2, 3, \ldots$
l	$0, 1, 2, \ldots, n-1$
m	$-l, \ldots, 0, \ldots, +l$
s	$-\frac{1}{2}$ or $+\frac{1}{2}$

(a) When $l = 0$, m cannot be greater than zero. Invalid solution.
(b) When $n = 2$, l cannot be greater than 1. Invalid solution.
(c) When $l = 3$, m cannot be less than -3. Invalid solution.
(d) is permissible.
(e) s can never be anything other than $\pm\frac{1}{2}$. Invalid solution.

Hence, only (d) is allowed; all the rest are invalid. The correct answer is **4**.

SOLUTION 8–22

(4)

The n value is retained and the l value is assigned a letter as follows:

$l = 0$	s
$l = 1$	p
$l = 2$	d
$l = 3$	f

Therefore, the type of orbital described by the quantum numbers $n = 5$ and $l = 2$ is a $5d$ orbital. (The shapes of the d orbitals will be discussed and shown in the subsequent problems.) The correct answer is **4**.

SOLUTION 8–23

(2)

For p orbitals, $l = 1$, and so $m = -1, 0,$ or $+1$. This means that there are three types of p orbital possible which differ in their orientations in space: p_x, p_y, and p_z. Each type is symmetric with respect to rotation about the Cartesian coordinate axis indicated by the subscript; that is, the signs [plus (+) or minus (–)] of the wave function on the electron density lobes of the orbital are unchanged upon any rotation about that axis. Every p orbital has two lobes of electron density [one where the wave function sign is plus (+) and the other where it is minus (–)] separated by a *nodal plane* of zero electron density. A p_z orbital including the nodal plane is shown in Figure 8–12. Drawing (b) in Figure 8–8 is therefore correct.

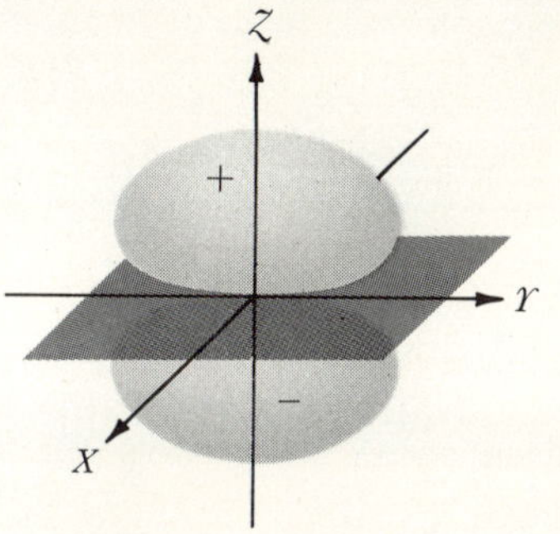

Figure 8–12

(a) is a d_{z^2} orbital with two lobes of electron density [both with a plus (+) wave function sign] along the z axis and a "doughnut" of electron density [minus (–) wave function sign] about the Z axis in the XY plane.

(c) is a $d_{x^2-y^2}$ orbital with four lobes of electron density [two with plus (+) wave function signs and two with minus (–)] located along the X and Y axes.

(d) is a p_y orbital with lobes of electron density along the Y axis.

(e) is an s orbital with a spherically symmetric electron density distribution [plus (+) wave function sign].

The correct answer is **2**.

SOLUTION 8–24

(5)

There are five types of d orbitals: d_{xy}, d_{yz}, d_{xz}, $d_{x^2-y^2}$, and d_{z^2}. The latter two were discussed in Problem 8–23. The remaining three are identical in shape but differ in orientation. Each has four lobes of electron density bisecting the angles between the Cartesian coordinate axes indicated by the subscripts. The d_{xz} orbital is shown in Figure 8–9(e).

(a) is a d_{xy} orbital. (b) is a d_{yz} orbital. (c) is a p_x orbital. (d) is a d_{z^2} orbital.

The correct answer is **5**.

CHAPTER 9 ELECTRONIC STRUCTURE AND CHEMICAL PROPERTIES

There are many experimental data such as electronic spectra and ionization energies which indicate that the behavior of electrons in larger atoms is governed by the quantum numbers n, l, m, and s, and that there are orbitals present in the larger many-electron atoms like the s, p, d, and f orbitals of the one-electron atom. The problems in this chapter are concerned with the electronic structures of many-electron atoms and how knowledge of these electronic structures enables us to explain the periodic table and some of the properties of the elements. To answer many of the problems you will need to refer to the periodic table.

Memorize:

Dipole moment $= \mu = q \times r$
in which

$q =$ some fraction or multiple of the charge on an electron, e^-.
$r =$ separation between the positive and negative poles of an electrical dipole.

Solutions to problems in Chapter 9 begin on page 252.

PROBLEMS

PROBLEM 9-1

Buildup of many-electron atoms is discussed in Section 9–1 of Chemical Principles *by Dickerson, Gray, and Haight.*

How many electrons can fit into the orbitals that comprise the third quantum shell ($n = 3$)? [*Hint:* Refer to Problems 8–21 and 8–22.]

1 -
2 8
3 18
4 32
5 None of the above is correct.

PROBLEM 9-2

The process in which electrons are assigned to atomic orbitals in a many-electron atom to give the ground state (lowest energy) electronic configuration of the atom is known as the *Aufbau* process. Which of the following rules must be adhered to in the *Aufbau* process?

(a) An orbital can accommodate a maximum of two electrons. To do so, the electrons must be of opposite spin.
(b) Electrons have to be fed into the orbitals in order of increasing energy.
(c) Each set of orbitals of the same energy has to be completely filled before proceeding to the next orbital or set of orbitals.
(d) Whenever electrons are added to orbitals of the same energy, if possible, each orbital is allocated one electron before any pairing occurs.
(e) Whenever electrons are allocated singly to separate orbitals of the same energy, the electrons must all have the same spin.

1 Only (d).
2 Only (b) and (c).
3 All but (a).
4 All of them.
5 None of them applies to the *Aufbau* process.

Before we try our hand at applying the Aufbau *process to determine the ground state electronic configurations of a few atoms and ions, let us look at the relative energies of some of the orbitals in hydrogen and many-electron atoms.*

PROBLEM 9–3

When an element is bombarded with a beam of high-energy electrons, x rays are emitted from the element. We know that sometimes the bombarding electrons cause the ejection of a $1s$ electron from an atom and then a $2s$ or $2p$ electron jumps down to fill the vacancy in the $1s$ orbital. The latter processes result in the emission of electromagnetic radiation as x rays, and the energy difference between the orbitals concerned determines the wavelength of the x rays emitted. For copper, x radiation of 1.54 Å wavelength is emitted when an electron changes from the $2p$ to the $1s$ orbital. What is the energy difference between these orbitals in copper?

($1.00\ \text{eV} = 1.60 \times 10^{-12}$ erg;
$h = 6.62 \times 10^{-27}$ erg sec;
$c = 3.00 \times 10^{10}$ cm sec^{-1})

1 $\simeq 8 \times 10^{-5}$ eV
2 $\simeq 5 \times 10^{3}$ eV
3 $\simeq 8 \times 10^{3}$ eV
4 $\simeq 13 \times 10^{3}$ eV
5 $\simeq 3 \times 10^{10}$ eV

PROBLEM 9–4

Electronic distribution of nitrogen and other representative elements is discussed in Section 9–6 of Chemical Principles.

Nitrogen gas exists as a diatomic gas (N_2) under ordinary conditions of temperature and pressure. The two nitrogen atoms are so strongly bonded that it is necessary to employ extreme conditions before N_2 will react with other species. The inertness of N_2 is a consequence of the electronic configuration of the nitrogen atoms. Which one of the following electronic distributions represents the ground state electronic configuration of a nitrogen atom? [*Hint:* Refer to Problem 9–2 and use Figure 9–4 in Solution 9–2.]

(N, $Z = 7$)

1
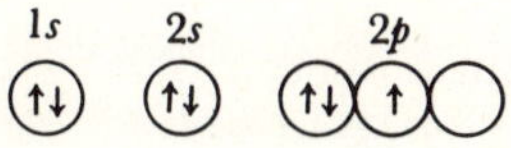

2
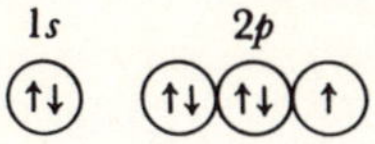

3
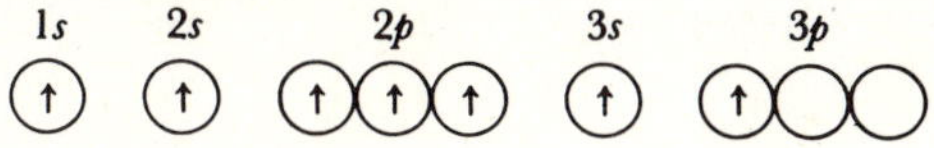

4
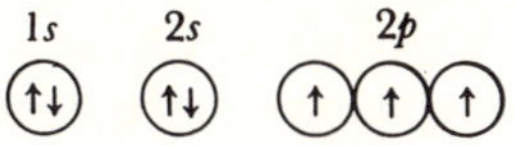

5
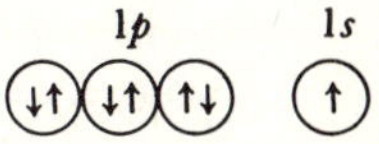

PROBLEM 9–5

What is the second most abundant element in the earth's crust (rocks only; atmosphere and oceans excluded) if the element has the ground state electronic configuration of $1s^2 2s^2 2p^6 3s^2 3p^2$?

1 Oxygen (O, Z = 8).
2 Magnesium (Mg, Z = 12).
3 Aluminum (Al, Z = 13).
4 Silicon (Si, Z = 14).
5 Iron (Fe, Z = 26).

PROBLEM 9–6

Germanium (Ge, Z = 32) is an important semiconductor. Doped with arsenic (As, Z = 33) or gallium (Ga, Z = 31), it is used as a transistor element, which is used extensively in the electronics industry. The ground state electronic configuration of germanium is important to its semiconductor properties. What is this ground state electronic configuration? [*Hint:* Refer to Problem 9–5 and to Figure 9–4.]

[*Note:* [Ar] is a shorthand way of writing the inner core of electrons, $1s^2 2s^2 2p^6 3s^2 3p^6$, corresponding to the ground state electronic configuration of the noble gas, argon (Ar, Z = 18).]

1 $[Ar]3d^{10}4p^4$
2 $[Ar]4s^2 3d^{10}4p^1$
3 $[Ar]4s^2 3d^{10}4p^2$
4 $[Ar]4s^2 3d^{10}4p^3$
5 $[Ar]3d^6 4s^2 4p^6$

PROBLEM 9–7

Uranium-235 ($^{235}_{92}U$) is a fissionable material used as a source of energy in nuclear reactors and atomic bombs. What is the ground state electronic configuration of uranium-235? [*Hint:* Use Figure 9–4 and the periodic table.]

(Kr, Z = 36; Xe. Z = 54; Rn, Z = 86)

1 $[Kr]4d^{10}4f^{14}5s^2 5p^6 5d^{10}5f^{14}$
2 $[Kr]4d^{10}4f^{14}5d^{10}6p^6 7s^2 5f^{14}$
3 $[Kr]5s^2 4d^{10}4f^{14}5p^6 5d^{10}5f^{14}$
4 $[Rn]7s^2 5f^4$
5 $[Xe]4f^{14}5d^{10}6s^2 6p^6 5f^6$

PROBLEM 9–8

All the silicon and most of the oxygen in the earth's crust is in the form of silicate minerals, such as granite. The basic chemical unit of all silicates is the SiO_4^{4-} ion, which consists of an Si^{4+} ion and four O^{2-} ions. What are the ground state electronic configurations of the Si^{4+} and O^{2-} ions in silicate minerals? [*Hint:* Use Figure 9–4 and the periodic table.]

(O, Z = 8; Si, Z = 14)

	Si^{4+}	O^{2-}
1	$[Ne]3s^23p^2$	$[He]2s^22p^4$
2	[Ar]	$[He]2s^22p^2$
3	[Ne]	[Ne]
4	$[Ar]3s^23p^2$	$[Ne]3s^23p^4$
5	[Ne]	$[He]2s^22p^4$

PROBLEM 9–9

Copper (Cu, $Z = 29$) is used extensively by the electrical industry because of its high electrical conductivity. The high electrical conductivity of copper is related to its ground state electronic configuration. What is the ground state electronic configuration of copper? [*Hint:* Use Figure 9–4 and remember that the ground state electronic configuration is the most stable (lowest energy) one.]

For a review of the electronic configurations of the transition elements, see Review 9 of Programed Reviews of Chemical Principles *by Lassila* et al.

1 $[Ar]3d^{10}4p^1$
2 $[Ar]4s^13d^94p^1$
3 $[Ar]4s^13d^{10}$
4 $[Ar]4s^23d^{10}$
5 $[Ar]4s^23d^8$

PROBLEM 9–10

The copper cation, Cu^{2+}, is of importance in some biological systems. For instance, hemocyanin, a copper–protein complex, is used by many invertebrates for the transport of oxygen. What is the ground state electronic configuration of Cu^{2+} in hemocyanin? [*Hint:* Use Figure 9–4 in Solution 9–2. For first-row transition metal cations the 3*d* orbitals are of lower energy than the 4*s* orbital.]

(Cu, $Z = 29$)

1 $[Ar]4s^13d^8$
2 $[Ar]4s^23d^7$
3 $[Ar]4s^23d^{10}4p^1$
4 $[Ar]4s^13d^{10}$
5 $[Ar]3d^9$

PROBLEM 9–11

Titanium(III)(Ti^{3+}) is the active species in the Ziegler–Natta catalysts used in the polymerization of ethylene to form polyethylene. How many unpaired electrons are there in the Ti^{3+} ion? [*Hint:* Refer to Problem 9–10.]

(Ti, $Z = 22$)

1 0
2 1
3 2
4 3
5 4

PROBLEM 9–12

Some properties of the transition metals are discussed in Section 9–4 of Chemical Principles.

There are many stable transition metal complexes for which the transition metal cation has the ground state electronic configuration $[Ar]3d^6$. Consider the cations Mn^+, Fe^{2+}, Co^{3+}, Ni^{4+}, and Zn^{2+}. Which one of them does *not* have a ground state electronic configuration of $[Ar]3d^6$?

(Mn, $Z = 25$; Fe, $Z = 26$; Co, $Z = 27$; Ni, $Z = 28$; Zn, $Z = 30$)

1 Mn^+
2 Fe^{2+}
3 Co^{3+}
4 Ni^{4+}
5 Zn^{2+}

You should now be able to appreciate the reasoning behind the periodic classification of the elements. Let us see how the ground state electronic configuration of the elements tie in with some of their physical and chemical properties.

PROBLEM 9–13

The first ionization energies of the representative elements with atomic numbers from $Z = 3$ to $Z = 10$ are plotted against atomic number in Figure 9–1. The plot is typical of the variation in first ionization energy

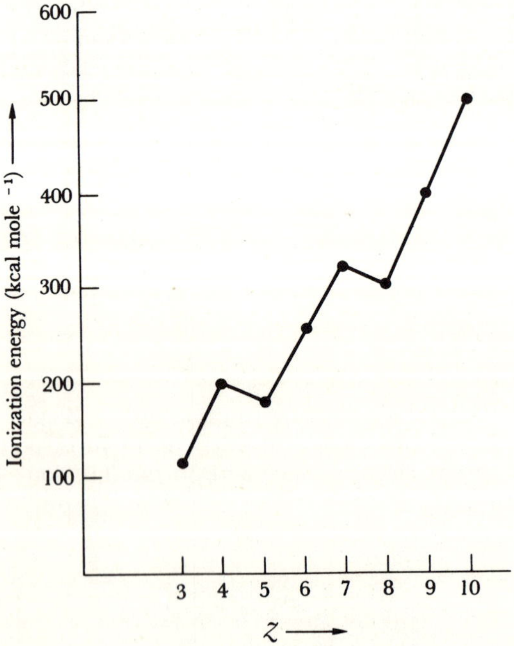

Figure 9–1

of the representative elements across any period. Which arrangement of valence electrons do you associate with the representative element exhibiting the lowest first ionization energy in any period? [*Note: n* is the principal quantum number.]

1 ns^1
2 ns^2
3 ns^2np^1
4 ns^2np^3
5 ns^2np^6

PROBLEM 9–14

Refer to Figure 9–1. Which arrangement of valence electrons do you associate with the representative element exhibiting the highest first ionization energy in any period for which $n \geqslant 2$?

1 ns^2
2 ns^2np^1
3 ns^2np^4
4 ns^2np^5
5 ns^2np^6

PROBLEM 9–15

The first ionization energy of aluminum (Al, $Z = 13$) is less than that of magnesium (Mg, $Z = 12$) because ________.

1 Aluminum has a higher atomic weight than magnesium.
2 Aluminum has a higher specific heat than magnesium.
3 Aluminum has a lower nuclear charge than magnesium.
4 The electron that is removed from aluminum is an unpaired $3p$ electron, whereas the one that is removed from magnesium is a paired $3s$ electron.
5 Aluminum has a higher melting point than magnesium.

PROBLEM 9–16 Figure 9–2 is a plot of the first ionization energies of the elements against atomic number. The first ionization energies of the noble gases *decrease* in the order ________.

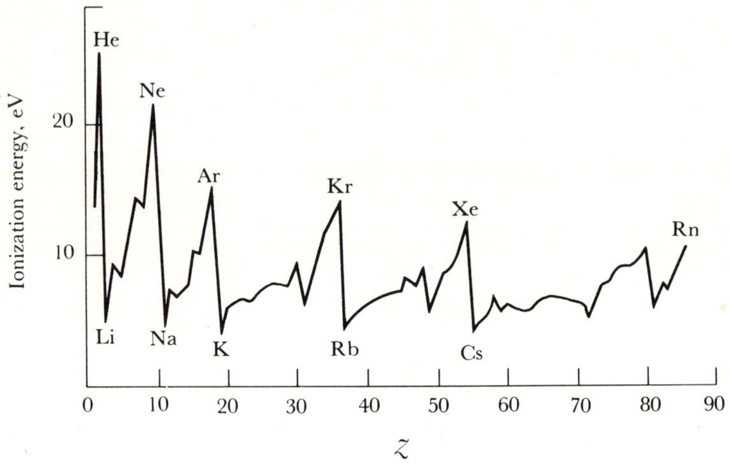

Figure 9–2

1 Li > Na > K > Rb > Cs > Fr
2 Rn > Xe > Kr > Ar > Ne > He
3 Fr > Cs > Rb > K > Na > Li
4 Ne > Ar > He > Xe > Rn > Kr
5 He > Ne > Ar > Kr > Xe > Rn

PROBLEM 9–17 The melting points of the elements with atomic numbers $Z = 57$ and $Z = 72$ to $Z = 80$ are given in Table 9–1. Which one of the following statements is *incorrect?* [*Hint:* Refer to Figure 9–4 in Solution 9–2 and the periodic table.]

Table 9–1

Z	Melting point (°C)
57	960
72	2150
73	3000
74	3410
75	3180
76	3000
77	2410
78	1770
79	1060
80	−39

1 The elements comprise the third-row transition series.
2 The elements are all metals.
3 The element melting at 2150°C has two unpaired electrons in its ground state electronic configuration.
4 The elements melting at 3000°C or above have three or more unpaired electrons in their ground state electronic configurations.
5 The element melting below 0°C has a ground state electronic configuration with one unpaired electron.

PROBLEM 9–18

Properties of the coinage metals are given in Section 9–4 of Chemical Principles.

The electrical conductivity of the solid elements with atomic numbers $Z = 55$ to $Z = 57$ and $Z = 72$ to $Z = 85$ are plotted against atomic number in Figure 9–3. If the plot is typical of the variation in electrical conductivity of the solid elements across any related period, which one of the following arrangements of valence electrons do you particularly associate with high electrical conductivity? [*Hint:* Refer to Figure 9–4 and the periodic table.]

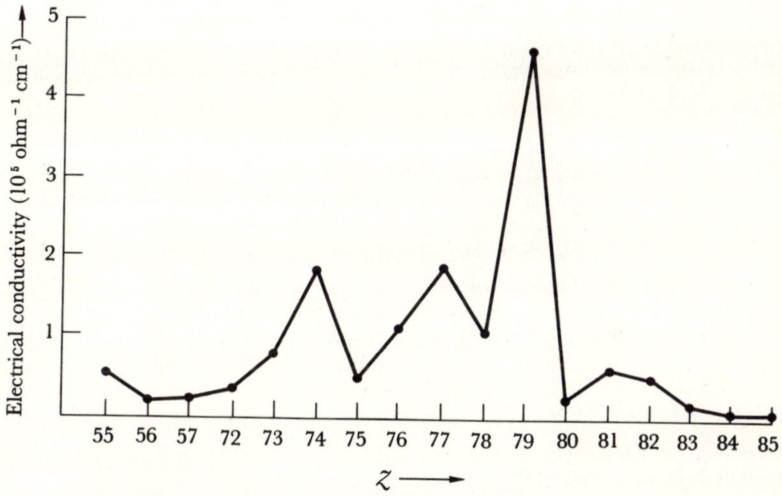

Figure 9–3

1 ns^1
2 ns^2
3 $ns^1(n-1)d^5$
4 $ns^1(n-1)d^{10}$
5 ns^2np^5

PROBLEM 9–19

The *electronegativities* (Pauling scale) of oxygen (O, $Z = 8$) and potassium (K, $Z = 19$) are 3.44 and 0.82, respectively. Which one of the following statements concerning the electronegativities of the elements is *incorrect?*

Electronegativity is discussed in Section 9–2 of Chemical Principles.

1 The electronegativity of an element is a measure of the charge on its most commonly found ion.
2 The most electronegative element is fluorine (F).
3 The least electronegative elements are cesium (Cs) and francium (Fr).
4 Compounds of elements with large differences in electronegativity will be ionic.
5 Compounds of elements of roughly equal electronegativities will be covalent.

PROBLEM 9–20

Consider the following Pauling electronegativity data.

Atom	Electronegativity value
H	2.2
B	2.0
C	2.5
N	3.0
F	4.0
S	2.6

Which of the orders listed below would you expect to be the correct one for *decreasing* ionic character of the bonds indicated?

1 H—F > H—N > H—B > H—C > H—S
2 H—N > H—F > H—B > H—C > H—F
3 H—S > H—N > H—F > H—C > H—B
4 H—N > H—S > H—B > H—F > H—C
5 H—F > H—N > H—S > H—C > H—B

PROBLEM 9–21

Because of its acidic properties, hydrogen chloride (HCl) is used industrially in the manufacture of fabric dyes. Its acidic properties arise because the H—Cl bond is polar as a result of the electronegativity difference between H and Cl. The observed dipole moment, μ_{obs}, of HCl is 1.07 D and the H—Cl bond distance is 1.27 Å. By calculating μ_{ionic}, the hypothetical dipole moment for HCl if there were complete transfer of one electron from H to Cl, estimate the percent ionic character of the HCl bond.

(electronic charge, $e = 4.80 \times 10^{-10}$ esu;
1 debye (D) $= 1 \times 10^{-18}$ esu cm)

[*Hint:* Percent ionic character $\simeq \dfrac{\mu_{obs}}{\mu_{ionic}} \times 100$.]

1 1.3%
2 17%
3 21%
4 48%
5 78%

PROBLEM 9–22

Which one of the following properties would you *not* expect silver (Ag, $Z = 47$) to exhibit? [*Hint:* Refer to the solution of Problem 9–17.]

1 High first ionization energy.
2 High electrical conductivity.
3 Luster.
4 Ductility.
5 Malleability.

PROBLEM 9–23

The most probable arrangement of the *valence* electrons in the ground state electronic configurations of several inner transition elements are shown below. Which two of these elements would you expect to have the most stable +2 oxidation states? [*Hint:* For the cations of the inner transition elements the $4f$ orbitals are of lower energy than the $6s$ orbital.]

La: $6s^2 5d^1$
Ce: $6s^2 4f^1 5d^1$
Pr: $6s^2 4f^3$
Eu: $6s^2 4f^7$
Gd: $6s^2 4f^7 5d^1$
Dy: $6s^2 4f^{10}$
Er: $6s^2 4f^{12}$
Tm: $6s^2 4f^{13}$
Yb: $6s^2 4f^{14}$
Lu: $6s^2 4f^{14} 5d^1$

1 La and Lu
2 Ce and Tm
3 Pr and Er
4 Eu and Yb
5 Gd and Dy

SOLUTIONS

SOLUTION 9-1

(3)

The permissible combinations of the four quantum numbers n, l, m, and s for the third quantum shell ($n = 3$) are shown in Table 9–2.

Table 9–2

n	Allowed values of l	Allowed values of m	Allowed values of s	Orbital labels	Total number of electrons for each set of orbitals
3	0	0	$+\frac{1}{2}$	$3s$	2
			$-\frac{1}{2}$		
	1	+1	$+\frac{1}{2}$	$3p$	6
			$-\frac{1}{2}$		
		0	$+\frac{1}{2}$		
			$-\frac{1}{2}$		
		−1	$+\frac{1}{2}$		
			$-\frac{1}{2}$		
	2	+2	$+\frac{1}{2}$	$3d$	10
			$-\frac{1}{2}$		
		+1	$+\frac{1}{2}$		
			$-\frac{1}{2}$		
		0	$+\frac{1}{2}$		
			$-\frac{1}{2}$		
		−1	$+\frac{1}{2}$		
			$-\frac{1}{2}$		
		−2	$+\frac{1}{2}$		
			$-\frac{1}{2}$		

From Table 9–2 you can see that the total number of electrons that can occupy the third quantum shell is 18, obtained by adding two for the $3s$ orbital, six for the three $3p$ orbitals, and ten for the five $3d$ orbitals. Each orbital can accommodate only two electrons; thus, regardless of the value of n, an ns orbital can hold two electrons, three np orbitals can hold six electrons, five nd orbitals can hold ten electrons, and seven nf orbitals can hold fourteen electrons. (n must be large enough for the p, d, and f, orbitals to exist.) In general, the number of orbitals in a subshell of given l is $2l+1$, which are capable of holding a total number of $4l+2$ electrons.

The correct answer is **3**.

SOLUTION 9–2

4

From Problem 9–1 you know that rule (a) must be obeyed because it is a consequence of the Pauli exclusion principle. However, Answer **4** is the only one that allows for rule (a) being correct. Therefore, even if you did not know that all five rules must be strictly obeyed in the *Aufbau* process, you should have realized that they must be for Answer **4** to be correct.

Rules (b) and (c): The sequence of orbitals in the many-electron atom is different from the one for the hydrogen atom. The energy sequence usually followed is shown in Figure 9–4. The orbitals of lowest energy are filled first, and each degenerate set of orbitals (i.e., orbitals of the same energy) must be completely filled before proceeding to the next orbital or degenerate set of orbitals.

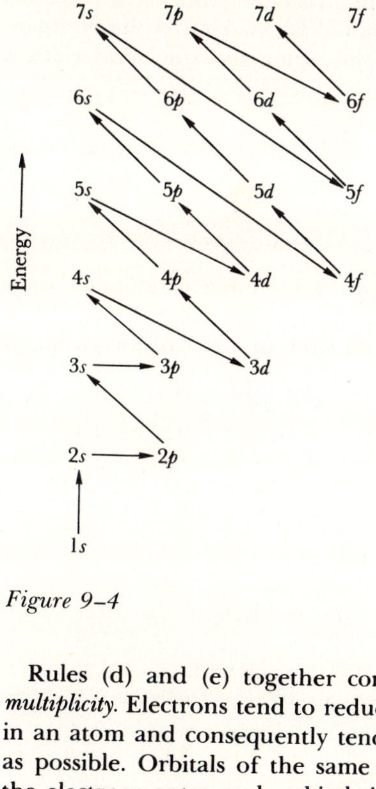

Figure 9–4

Rules (d) and (e) together comprise Hund's *principle of maximum multiplicity*. Electrons tend to reduce the repulsion between themselves in an atom and consequently tend to be as far away from each other as possible. Orbitals of the same energy differ in orientation; hence, the electrons enter each orbital singly, if possible. Furthermore, electrons in separate orbitals of the same energy interact with each other to a lesser extent if they have the same spin (parallel spin) than they do if they have opposite spin (antiparallel spin).

SOLUTION 9–3

(3)

The energy difference, E, between the $1s$ and $2p$ orbitals of copper is given by

$$E = h\nu$$

$$= \frac{hc}{\lambda}$$

$$= \frac{6.62 \times 10^{-27} \text{ erg sec} \times 3.00 \times 10^{10} \text{ cm sec}^{-1}}{1.54 \text{ Å} \times 10^{-8} \text{ cm Å}^{-1} \times 1.60 \times 10^{-12} \text{ erg eV}^{-1}}$$

$$\simeq 8 \times 10^{3} \text{ eV} \quad (\text{actually } 8.06 \times 10^{3} \text{ eV})$$

This is almost a thousand times more than the energy difference between the $1s$ and $2p$ orbitals of the hydrogen atom (10.1 eV). Even though the *orbitals* of the hydrogen atom can be used to describe many-electron atoms, you should be aware that the energies of the various orbitals in the hydrogen atom and the many-electron atoms differ greatly.

The correct answer is **3**. For Answer **1**, you did not convert angstroms to centimeters.

SOLUTION 9–4

(4)

Since nitrogen is a neutral atom and its atomic number is 7, we have seven electrons to place in atomic orbitals. From Figure 9–4, the orbitals of the many-electron atom in order of increasing energy are

$$1s < 2s < 2p < 3s < 3p < 4s < 3d < 4p \ldots$$

Remember the capacities of these various orbitals. The seven electrons are distributed as

$$1s^2 2s^2 2p^3$$

in which the superscripts represent the number of electrons in each set of orbitals.

We have three electrons to locate in the three $2p$ orbitals: $2p_x$, $2p_y$, and $2p_z$. Hund's principle (see Problem 9–2) dictates that the electrons enter the three orbitals singly and with parallel spin. No pairing occurs; hence, the ground state electronic distribution for nitrogen is

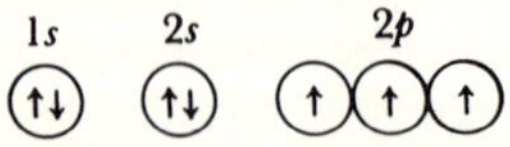

The correct answer is **4**.

SOLUTION 9–5

(4)

The number of electrons associated with the neutral atom of the unknown element is the sum of the orbital indices; $2+2+6+2+2=14$. Thus, the atomic number of the unknown element must also be 14; that is, the element is silicon (Si). [Silicon constitutes 27.7% by weight of the earth's crust; the most abundant element is oxygen (46.6% by weight).]

The correct answer is **4**.

SOLUTION 9–6

(3)

The atomic number of germanium is 32; thus, we have 32 electrons to distribute in the orbitals of the neutral atom. These are allocated according to the orbital sequence in Figure 9–4 and the *Aufbau* procedure. The ground state electronic configuration of Ge is therefore

$1s^2 2s^2 2p^6 3s^2 3p^6 4s^2 3d^{10} 4p^2$

or

$[Ar]4s^2 3d^{10} 4p^2$

3 is correct.
1 has the right number of electrons, but the $4s$ orbital has been left out.
2 is the ground state electronic configuration of gallium.
4 is the ground state electronic configuration of arsenic.
5 has the right number of electrons but contravenes the *Aufbau* rules.

SOLUTION 9–7

(4)

The atomic number of uranium-235 is given by the subscript on $^{235}_{92}U$, that is, 92. Since uranium-235 is a neutral atom, there are 92 electrons around the nucleus. Feeding 92 electrons into the orbitals of Figure 9–4 by following the *Aufbau* procedure, we eventually arrive at the ground state electronic configuration of

$1s^2 2s^2 2p^6 3s^2 3p^6 4s^2 3d^{10} 4p^6 5s^2 4d^{10} 5p^6 6s^2 4f^{14} 5d^{10} 6p^6 7s^2 5f^4$

or in the more convenient shorthand form

$[Rn]7s^2 5f^4$

Answer **4** is correct.

[*Note:* Although we have predicted that the six *valence* electrons of uranium should be in the arrangement $7s^2 5f^4$, experiments have proved that the more probable arrangement is $7s^2 5f^3 6d^1$. You could not have known this from the *Aufbau* process. However, in spite of this apparent setback, the *Aufbau* process is much more often right than wrong.]

SOLUTION 9–8

(3)

The ground state electronic configurations of Si and O are

Si: $[Ne]3s^23p^2$
O: $[He]2s^22p^4$

For a cation, electrons are removed; for an anion, electrons are added. In Si^{4+}, the Si atom has lost all four of its valence electrons. In O^{2-}, the O atom has gained two electrons. These go into the two vacancies in the $2p$ orbitals, thereby completing the $n = 2$ shell. The ground state electronic configurations of the ions are therefore

Si^{4+}: [Ne]
O^{2-}: $[He]2s^22p^6 = [Ne]$

The important point to notice here is that both Si^{4+} and O^{2-} have the same ground state electronic configuration as the noble gas, neon. When atoms form ions they tend to achieve the same electronic configuration as a noble gas because of the great stability associated with such electronic arrangements. Notice also that Si^{4+}, O^{2-}, and Ne are isoelectronic; each has ten electrons.

The correct answer is **3**. Answer **1** is the ground state electronic configuration of the parent atoms. Answer **5**: Si^{4+} is correct but you forgot to add two electrons to the O atom for O^{2-}.

SOLUTION 9–9

(3)

There are 29 electrons in a neutral copper atom, and from the *Aufbau* process the distribution of these electrons in the ground state is

$[Ar]4s^23d^9$

However, this configuration is not one of the five choices given. This must mean that for some reason the configuration $[Ar]4s^23d^9$ is *not* the most stable. We can discard configurations **4** and **5** because they do not have 29 electrons. Configurations **1** and **2** both have one valence electron in the $4p$ orbitals; consequently, these configurations will not be as stable (i.e., will not have as low an energy) as configuration **3**, in which the valence electrons are located in the lower energy $4s$ and $3d$ orbitals.

The correct answer is **3**.

Although the $4s$ orbital penetrates closer to the nucleus than the $3d$ and therefore has lower energy, the majority of the electron density in the $4s$ is farther from the nucleus. An electron in a $4s$ orbital is simultaneously farther from the nucleus on the average than a $3d$ electron, and more stable because of the small but appreciable probability that it will be very close to the nucleus. In copper, an electron can remain paired in a $4s$ orbital or complete the $3d$ shell. The electron does the latter because of the lower electron–electron repulsions in the $3d$ orbitals. Copper adopts the ground state electronic configuration $[Ar]4s^13d^{10}$

[*Note:* There are other examples throughout the periodic table of elements having apparently anomalous ground state electronic configurations because of the added stability of a half-filled or filled subshell. For example, chromium (Cr, $Z=24$) has a ground state electronic configuration of $[Ar]4s^13d^5$ rather than the *Aufbau* prediction of $[Ar]4s^23d^4$. The extra stability arises because the half-filled $3d$ subshell contains the maximum number of unpaired electrons with parallel spin (Hund's principle).]

SOLUTION 9–10

(5)

The copper atom has 29 electrons, and therefore the Cu^{2+} ion has $29-2=27$ electrons: an argon core of 18 and 9 valence electrons. This immediately rules out configurations **3** and **4**, which have 31 and 29 electrons, respectively. Since the $3d$ orbitals in first-row transition metal cations are of lower energy (i.e., more stable) than the $4s$ orbital, the nine valence electrons of the Cu^{2+} ion have a configuration of $3d^9$ instead of $4s^13d^8$ or $4s^23d^7$. Thus, configuration **5** is the correct answer.

[*Note:* In the Cu atom the $4s$ level is slightly below the $3d$ level owing to the more effective penetration of the inner electron screen around the nucleus by the $4s$ orbital compared with the $3d$. However, this energy sequence is reversed for the Cu^{2+} ion and, in fact, for most first-row transition metal cations. The $3d$ orbitals are apparently drawn closer to the nucleus in the Cu^{2+} ion than they are in the Cu atom, and the extra penetration of the $4s$ orbital is insufficient to outweigh the increased stability of the $3d$ orbitals.]

SOLUTION 9–11

(2)

There are 22 electrons in the titanium atom, and from the *Aufbau* process the ground state electronic configuration is $[Ar]4s^23d^2$. To form Ti^{3+}, three electrons must be removed from Ti. From Problem 9–10 the relative stability of the $4s$ and $3d$ orbitals in a first-row transition metal cation is reversed from what is in the neutral atom. Hence, the ground state electronic configuration of the Ti^{3+} ion is $[Ar]3d^1$. Since the 18 electrons in the Ar core are all paired, only the electron in the $3d$ is unpaired. So there is *one* unpaired electron in the Ti^{3+} ion.

The correct answer is **2**.

SOLUTION 9–12

(5)

The ground state electronic configurations of the five parent atoms predicted by the *Aufbau* process are

Mn: $[Ar]4s^2 3d^5$
Fe: $[Ar]4s^2 3d^6$
Co: $[Ar]4s^2 3d^7$
Ni: $[Ar]4s^2 3d^8$
Zn: $[Ar]4s^2 3d^{10}$

For the ions, the $3d$ orbitals are more stable than the $4s$ orbital. Consequently, the ground state electronic configurations of the five ions are

Mn^+: $[Ar]3d^6$
Fe^{2+}: $[Ar]3d^6$
Co^{3+}: $[Ar]3d^6$
Ni^{4+}: $[Ar]3d^6$
Zn^{2+}: $[Ar]3d^{10}$

The only one without a ground state electronic configuration of $[Ar]3d^6$ is Zn^{2+}. The correct answer is **5**.

(An example of a stable transition metal complex in which the metal cation has the $[Ar]3d^6$ ground state electronic configuration is hemoglobin; the metal cation is Fe^{2+}.)

[*Note:* Mn^+, Fe^{2+}, Co^{3+}, and Ni^{4+} have 24 electrons each, whereas Zn^{2+} has 28. Thus, this problem could have also been solved by simply counting the number of electrons in each cation.]

SOLUTION 9–13

(1)

The element with the lowest first ionization energy in Figure 9–1 has an atomic number of 3; that is, the element is lithium, Li, which has a ground state electronic configuration of $[He]2s^1$. The helium core so effectively screens the solitary $2s$ valence electron from the nucleus that it is not very strongly attracted to the nucleus and is easily removed. Similarly, all the other alkali metals are readily ionized. The general ground state electronic configuration of the valence electrons for the alkali metals is ns^1.

The correct answer is **1**.

SOLUTION 9–14

(5)

The element with the highest first ionization energy in Figure 9–1 has an atomic number of 10. The element is neon (Ne), which has a ground state electronic configuration of $[He]2s^22p^6$. The effective nuclear charge for the second period is largest for neon, and the exceptional stability associated with its filled shells makes neon very difficult to ionize. Similarly, all the other noble gases are difficult to ionize. The general ground state electronic configuration for the noble gases is ns^2np^6, where $n \geqslant 2$ ($n = 1$ for helium, which has a ground state electronic configuration of $1s^2$).

The correct answer is **5**.

SOLUTION 9–15

(4)

All five statements are correct in themselves, but only statement **4** accounts for the lower first ionization energy of aluminum compared with that of magnesium. The ground state electronic configurations of aluminum and magnesium are

Al: $[Ne]3s^23p^1$
Mg: $[Ne]3s^2$

The nuclear charge increases from +12 to +13 in going from Mg to Al; consequently, it should be more difficult to remove an electron from Al than from Mg. However, the first electron to be removed from Al will be the single $3p$ electron, whereas from Mg it will be one of the *paired* $3s$ electrons. The $3p$ electron in Al does not approach the nucleus as closely as the $3s$ electrons in Mg do. Furthermore, the $3p$ electron in Al experiences greater shielding from the nucleus by the other electrons in the atom than the $3s$ electrons in Mg. Thus, the *effective* nuclear charge in Al is less than that in Mg; consequently, it is easier to remove the $3p$ electron from Al than it is to remove one of the paired $3s$ electrons from Mg.

The correct answer is **4**.

SOLUTION 9–16

(5)

From Figure 9–2, Answer **5** is correct. The first ionization energies of the noble gases decrease with increasing atomic number. In fact, there is a decrease ın ionization energy downward in *any* group owing to the increase in the number of electrons around the nuclei of the atoms. The more electrons there are around a nucleus, the larger the atom is and the less attraction there is between the nucleus and the outermost electrons of the atom. The increase in atomic size and the concomitant increase in the shielding of the outermost electrons from the nuclei by the inner electrons more than compensate for the increasing nuclear charge in a group. As a result the ionization energies decrease down a group.

Answer **1** is the order of decreasing ionization energy of the *alkali metals.*

SOLUTION 9–17

(5)

Answers **1**, **2**, and **3** are all *correct.* The elements with $Z = 57$ (lanthanum, La) and $Z = 72$ (hafnium, Hf) to $Z = 80$ (mercury, Hg) comprise the third row of transition metals in the periodic table. The ground state electronic configuration of hafnium (melting point 2150°C) is $[Xe]6s^2 4f^{14} 5d^2$. All the orbitals except the $5d$ are filled, and from Hund's rule the two electrons in the $5d$ orbitals will be located in different orbitals. Thus, there are two unpaired electrons in the ground state electronic configuration of hafnium.

The elements melting at 3000°C or above are tantalum (Ta, $Z = 73$), tungsten (W, $Z = 74$), rhenium (Re, $Z = 75$), and osmium (Os, $Z = 76$). The ground state electronic configurations of these elements are

Ta: $[Xe]6s^2 4f^{14} 5d^3$
W: $[Xe]6s^2 4f^{14} 5d^4$
Re: $[Xe]6s^2 4f^{14} 5d^5$
Os: $[Xe]6s^2 4f^{14} 5d^6$

From Hund's rule, tantalum has three, tungsten four, rhenium five, and osmium four unpaired $5d$ electrons. The rest of the elements in the series all have less than three unpaired $5d$ electrons in their ground states. Answer **4** is *correct.*

[*Note:* Tungsten, unlike the other Group VIB elements, chromium (Cr, $Z = 24$) and molybdenum (Mo, $Z = 42$), obeys the *Aufbau* rules and does not adopt a ground state electronic configuration with half-filled s and d subshells (i.e., $6s^2 5d^4$ and not $6s^1 5d^5$). In general, the more unpaired electrons a transition metal has, the higher will be its melting point because of the increased metal–metal bonding. However, there are other factors involved such as the size of the atoms and the packing structure adopted by the metal.]

Answer **5** is *incorrect.* The ground state electronic configuration of mercury (melting point, −39°C) is $[Xe]6s^2 4f^{14} 5d^{10}$. There are *no* unpaired electrons, and hence the metal–metal bonding is very weak; so weak in fact that mercury is a liquid at room temperature, whereas all other metals are solids.

The required answer is **5**.

SOLUTION 9–18

(4)

The element with the highest electrical conductivity in Figure 9–3 has an atomic number of 79. The element is gold (Au), which has a ground state electronic configuration of $[Xe]6s^1 4f^{14} 5d^{10}$. The $4f$ electrons are not involved in the chemistry of gold; thus, the valence electrons are $6s^1 5d^{10}$. The unpaired $6s$ electrons of the gold atoms in a crystal are free to move throughout the crystal and are responsible for the high electrical conductivity of gold. Similarly, the other members of the *coinage group*, copper (Cu, $Z = 29$) and silver (Ag, $Z = 47$), are exceptionally good conductors. These elements also have a valence electron distribution of one unpaired s electron and a completely filled d shell, that is, $ns^1(n-1)d^{10}$.

The correct answer is **4**.

[*Note:* Because of its high electrical conductivity, gold is used as the contacts in high performance connectors. Gold occurs in sea water to the extent of 0.1 to 2 mg per ton. Unfortunately, no method has yet been devised for recovering gold from sea water profitably.]

SOLUTION 9–19

(1)

Answer **1** is *incorrect.* The most commonly found ion of oxygen is O^{2-} (charge −2), whereas that of potassium is K^+ (charge +1).

Answer **2** is *correct.* Fluorine has the greatest tendency of all the elements to acquire an extra electron and is the most electronegative element. The Pauling electronegativity value is 4.0.

Answer **3** is *correct.* The least electronegative elements (sometimes referred to as the most electropositive) are cesium and francium; they readily lose an electron. Both elements have an electronegativity value of 0.7.

Answer **4** is *correct.* If an element such as sodium, with low ionization energy and low electron affinity, combines with an element such as chlorine, with high ionization energy and high electron affinity, the compound formed, sodium chloride, is ionic (Na^+Cl^-). Sodium has a low electronegativity (0.9, Pauling scale), whereas chlorine has a high electronegativity (3.2). Therefore, compounds formed from elements with a large difference in their electronegativities will be ionic.

Answer **5** is *correct.* Elements with similar ionization energies and electron affinities form covalent bonds. For example, carbon and hydrogen combine to form the covalent compound methane, CH_4. The Pauling electronegativities for carbon and hydrogen are 2.5 and 2.2, respectively; that is, they are almost equal.

In general, compounds will have varying degrees of ionic or covalent character depending on the relative electronegativities of the elements forming the compounds.

The required answer is **1**.

SOLUTION 9–20

(5)

The differences in the electronegativities of the atoms forming the bonds are tabulated below.

Bond	Electronegativity differences
H—B	−0.2
H—C	0.3
H—N	0.8
H—F	1.8
H—S	0.4

The greater the difference, the more ionic the bond will be. Therefore, in order of decreasing ionic character, the bonds are

H—F > H—N > H—S > H—C > H—B

The correct answer is **5**.

[*Note:* The order of decreasing ionic character of the bonds is also the order of *decreasing acidity* of the H atom in the bonds, that is, the order of decreasing tendency to form H^+ ions.]

SOLUTION 9–21

(2)

The electronegativity of Cl is greater than that of H. Therefore, if the bond in HCl were completely ionic, one electron would be transferred from H to Cl to give the ionic structure H^+Cl^-. The separation between the two atoms is 1.27 Å. So the dipole moment for the purely ionic species H^+Cl^- would be given by

$$\begin{aligned}\mu_{\text{ionic}} &= 4.80 \times 10^{-10}\ \text{esu} \times 1.27\ \text{Å} \times 10^{-8}\ \text{cm Å}^{-1}\\ &= 6.10 \times 10^{-18}\ \text{esu cm} \times 10^{18}\ \text{D (esu cm)}^{-1}\\ &= 6.10\ \text{D}\end{aligned}$$

The observed dipole moment, μ_{obs}, is 1.07 D. If the molecule did not exhibit a dipole moment at all, it would be completely covalent. However, since it does, it follows that the ratio of the observed and purely ionic dipole moments can be taken as a measure of the ionic character in the H–Cl bond.

$$\text{percent ionic character} = \frac{\mu_{\text{obs}}}{\mu_{\text{ionic}}} \times 100 = \frac{1.07\ \text{D}}{6.10\ \text{D}} \times 100 \simeq 17\%$$

Answer **2** is correct.

1 is the H–Cl bond distance with the wrong units.

3 You forgot to multiply the electronic charge (e) by the H–Cl bond distance when you calculated μ_{ionic}.

SOLUTION 9–22

(1)

Silver is a typical metal and therefore exhibits the properties of luster, malleability, and ductility. Answers **3**, **4**, and **5** are *correct.* Its ground state electronic configuration, by the *Aufbau* process is [Kr]$5s^2 4d^9$. However, the extra stability associated with a filled *d* shell leads to the configuration of [Kr]$5s^1 4d^{10}$. The presence of the single *s* electron results in high electrical conductivity (see Problem 9–18). Answer **2** is *correct.* The *s* electron is extremely well screened from the nucleus by the filled *d* shell; consequently, it is easy to remove, and silver has a low, not a high, first ionization energy.

Thus, the correct answer is **1**.

SOLUTION 9–23

(4)

Although the 6*s* orbital is more stable than the 4*f* orbitals in atoms of the inner transition elements, the stability is reversed for the cations. This is similar to the outer *d* and *s* orbitals of the transition metals (Problem 9–9). The most stable ions will be formed when the seven 4*f* orbitals are either half-filled or completely filled. Since we are interested in finding the atoms with the most stable +2 states (M^{2+} ions), we need to look for the atoms with $7+2=9$ valence electrons and $14+2=16$ valence electrons. The atoms in question are europium (Eu) and ytterbium (Yb), respectively.

The correct answer is **4**.

CHAPTER 10 COVALENT BONDING

Until now we have considered only the electronic structure of atoms and their ions. But what of the electronic structure of molecules? How do atoms bond together to form molecules? The problems in this chapter will help you to answer these questions. You will be introduced to the ideas of Lewis structure, covalent bonding, and molecular orbital theory. You will be able to predict the shapes of simple molecules and ions purely from a knowledge of the number of valence electrons on each of the atoms in the molecule or ion.

To answer some of the problems you will need to memorize the following orbital hybridizations assigned to atoms and the geometrical arrangement of hybrid orbitals produced.

Type of orbital hybridization on an atom		Geometrical arrangement of the hybrid orbitals
sp		Linear
sp^2	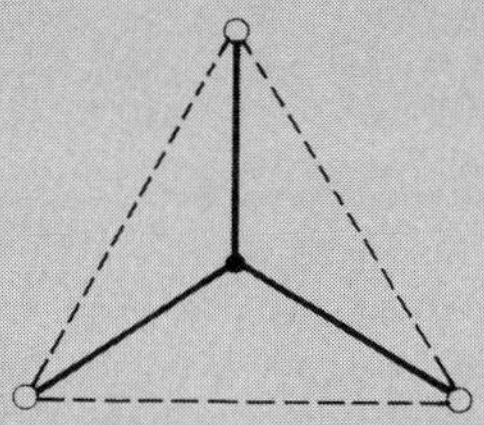	Triangle

sp^3 Tetrahedron

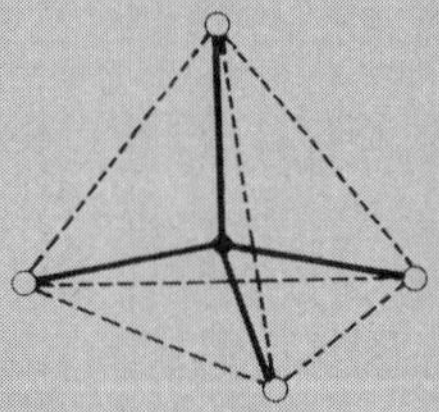

sp^3d Trigonal bipyramid

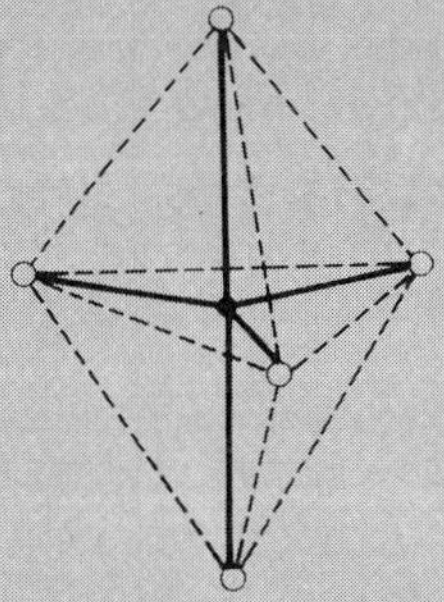

sp^3d^2 Octahedron

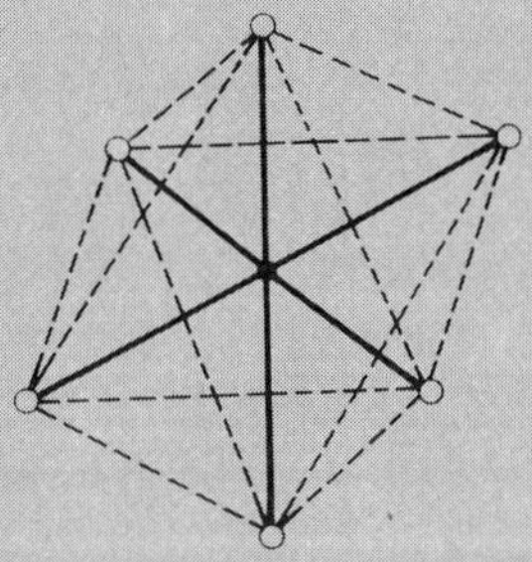

sp^3d^3 Pentagonal bipyramid

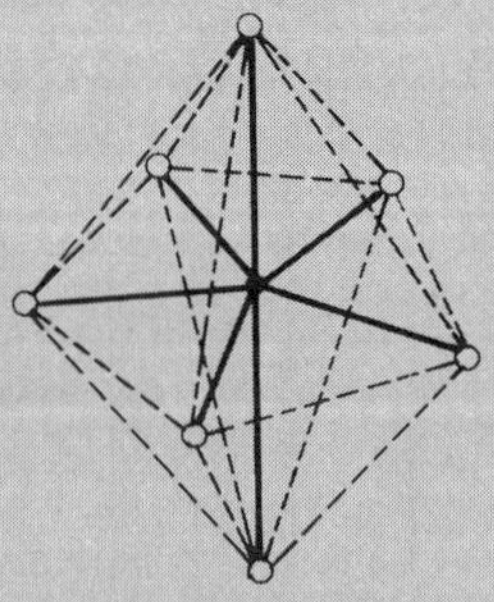

For the purposes of this chapter, assume that ΔH^0 and ΔE^0 are the same. [*Note:* You may refer to Figure 10-3 for the subsequent problems if necessary.]

Solutions to problems in Chapter 10 begin on page 281.

PROBLEMS

PROBLEM 10–1

Hydrogen trioxide (H_2O_3), a close relative of hydrogen peroxide (H_2O_2), a rocket fuel and hair bleach, has recently been synthesized. It is a highly unstable compound and can be isolated only in very small quantities. Which one of the following Lewis electron dot structures do you think represents the structure of hydrogen trioxide?

(H, $Z = 1$; O, $Z = 8$)

1 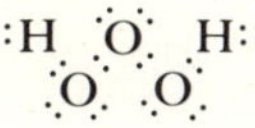

2 

3 

4 

5 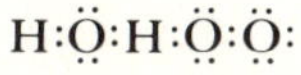

PROBLEM 10–2

"Wood alcohol" (methyl alcohol, CH_4O) is one of the *denaturants* that are added to make industrial ethyl alcohol (C_2H_6O) toxic. (Drinking methyl alcohol or breathing it for prolonged periods of time can lead to blindness or death).

Lewis electron dot formulas for compounds are described in Section 10–1 of Chemical Principles *by Dickerson, Gray, and Haight.*

Lewis structures portray some of the structural similarities and differences between molecules. If the Lewis structure for ethyl alcohol is

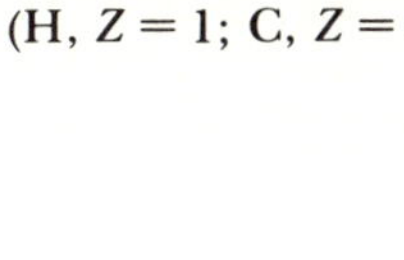

which one of the following diagrams do you think is the Lewis structure for toxic methyl alcohol?

(H, $Z = 1$; C, $Z = 6$; O, $Z = 8$)

1 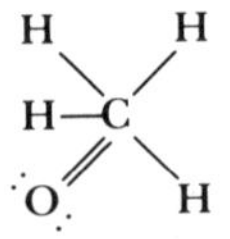

2 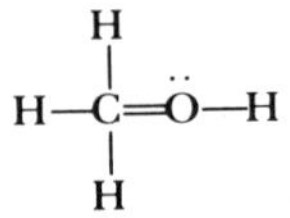

3 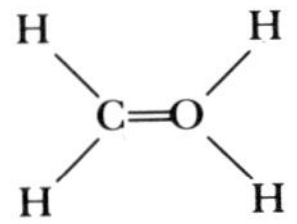

4 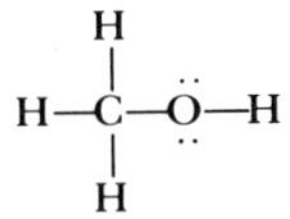

5 None of the above structures is correct; no Lewis structure can be drawn for methyl alcohol.

PROBLEM 10–3

Iodine (I_2) and iodine-containing compounds are widely used in medicine because of their antiseptic properties. Iodine was used to treat battle wounds during the U.S. Civil War and is still popular. Although it is a poisonous compound when ingested, it is not true that it is highly toxic: Fatalities occur only if large amounts (more than 30 ml of commercial tincture of iodine) are taken internally. One of the iodine-containing compounds used as an antiseptic is iodole (HC_4NI_4). Identify the Lewis structure of iodole.

(H, $Z = 1$; C, $Z = 6$; N, $Z = 7$; I, $Z = 53$)

1 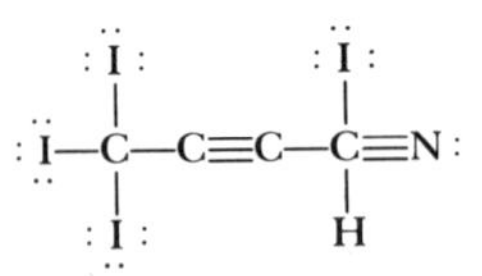

2 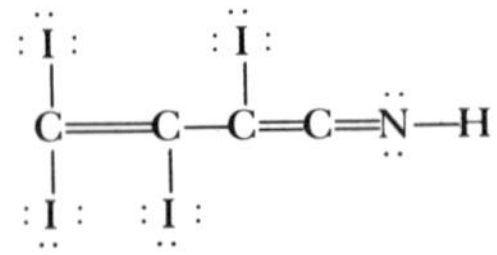

3 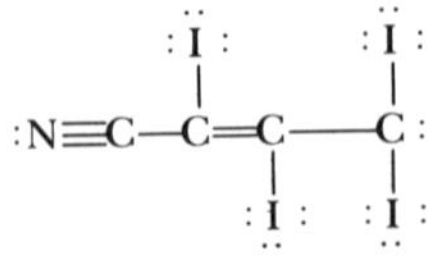

4 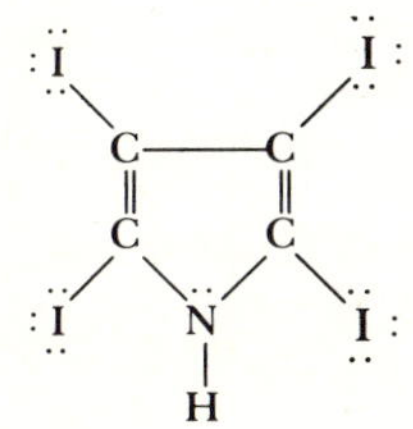

5 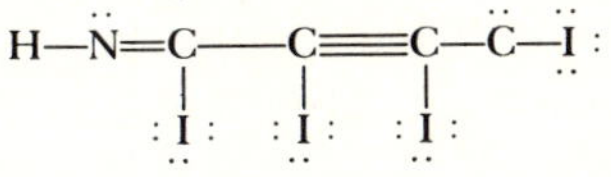

PROBLEM 10–4 Which of the following Lewis diagrams are *incorrect?*

(a) 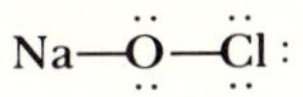 for NaOCl (sodium hypochlorite, used as bleaching agent)

(b) 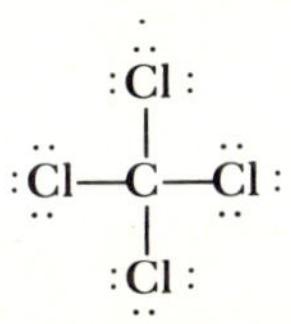 for CCl_4 (carbon tetrachloride, used widely in dry cleaning industry to remove grease from fabrics)

(c) 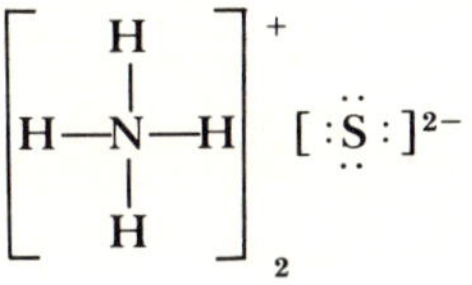 for $(NH_4)_2S$ (ammonium sulfide, used in qualitative analysis of the elements)

(d) 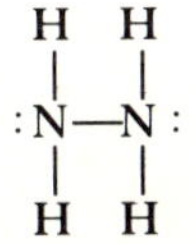 for N_2H_4 (hydrazine, used as rocket fuel)

(H, $Z=1$; C, $Z=6$; N, $Z=7$; O, $Z=8$; Na, $Z=11$; S, $Z=16$; Cl, $Z=17$)

1 (a)
2 (c)
3 (d)
4 (a) and (c)
5 (b) and (d)

PROBLEM 10–5 The starting material in the preparation of one important type of silicone rubber is dimethyldichlorosilane, $(CH_3)_2SiCl_2$. Which one of the following structures is the Lewis structure for dimethyldichlorosilane? [*Hint:* Silane is SiH_4 (analogous to methane, CH_4).]

(H, $Z = 1$; C, $Z = 6$; Si, $Z = 14$; Cl, $Z = 17$)

1 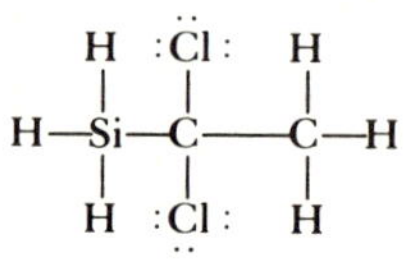

2 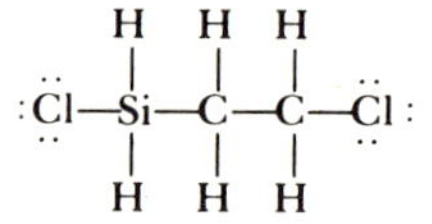

3 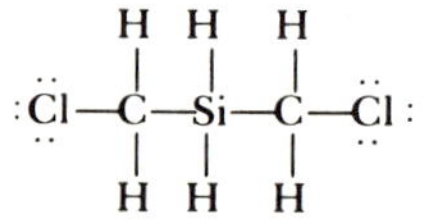

4 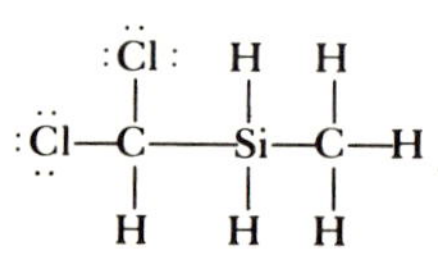

5 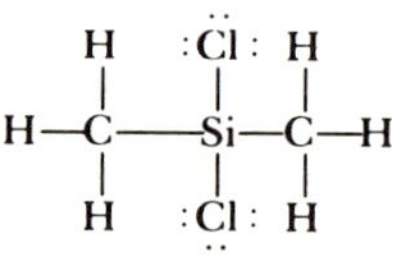

PROBLEM 10–6 Chlorine dioxide (ClO_2) is a yellow-red gas used in the manufacture of sodium chlorite ($NaClO_2$) for use as a bleaching agent for flour. Which one of the following diagrams is the Lewis structure of chlorine dioxide?

(O, $Z = 8$; Cl, $Z = 17$)

1 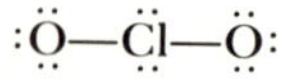

2 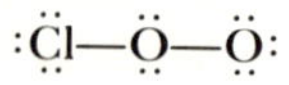

3 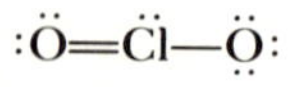

4 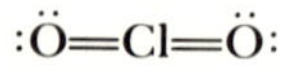

5 No Lewis structure can be drawn for chlorine dioxide.

PROBLEM 10–7

For a discussion of isoelectronic structures, see Section 10–1 of Chemical Principles.

Which one of the following species is *not* isoelectronic with the rest?

(F, Z = 9; S, Z = 16; Cl, Z = 17; Ge, Z = 32;
Br, Z = 35; I, Z = 53; Os, Z = 76)

1 Br_2
2 ICl
3 Os^{6+}
4 SF_6
5 GeF_4

PROBLEM 10–8

Resonance is discussed in Section 10–10 of Chemical Principles.

Prior to 1842, the two most common anaesthetics used by dentists were whiskey and a blow on the head. In the early 1840's nitrous oxide (laughing gas), N_2O, was used for the first time.

The chemistry of anaesthesia is not well understood even today. However, in any study of anaesthesia it is vitally important to know the electronic structures of the molecules that cause it.

N_2O has a linear, unsymmetrical structure that may be thought of as a hybrid of *two resonance forms.* If a resonance form must have a satisfactory Lewis structure, which two of the five structures shown below are the two resonance forms of N_2O?

(N, Z = 7; O, Z = 8)

(a) $:\!N\equiv N=\ddot{O}\!:$

(b) $:\!\ddot{N}=N=\ddot{O}\!:$

(c) $:\!\ddot{N}-N\equiv O\!:$

(d) $:\!\ddot{N}=\ddot{N}-\ddot{O}\!:$

(e) $:\!N\equiv N-\underset{\cdot\cdot}{\ddot{O}}\!:$

1 (a) and (b)
2 (c) and (d)
3 (b) and (e)
4 (a) and (c)
5 (d) and (e)

PROBLEM 10–9

Explosive charges containing TNT (trinitrotoluene) do not explode when subjected to mechanical shock, but they do explode violently when detonated properly. The most common kind of detonator is a percussion cap containing mercury fulminate, $Hg(CNO)_2$, that explodes when struck.

The fulminate ion $(CNO)^-$ is isoelectronic and isostructural (three atoms bonded together in a linear fashion) with the azide ion, N_3^-. Both anions are best formulated as resonance hybrids. The exceptional similarity between the fulminate ion and the azide ion may have something to do with the explosive nature of mercury fulminate and the azides of heavy metals, which are also shock sensitive explosives [lead azide, $Pb(N_3)_2$, is also used as a detonator]. Which of the following resonance structures could contribute to the structure of the fulminate ion?

(C, $Z = 6$; N, $Z = 7$; O, $Z = 8$)

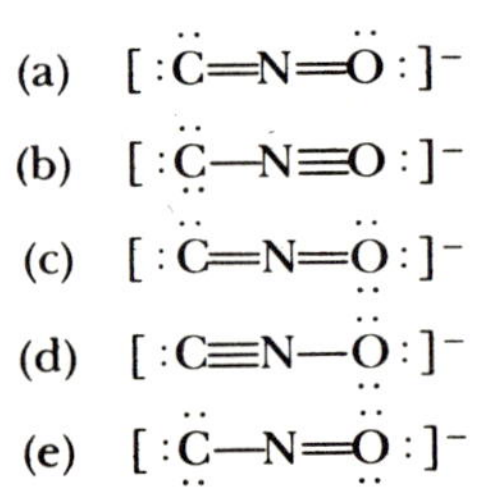

1 (a) and (d)
2 (c) and (e)
3 (b) and (c)
4 (a), (b), and (d)
5 (c), (d), and (e)

PROBLEM 10–10

The unshared pairs of valence electrons on atoms in molecules are called *lone pairs.* These lone pairs are sometimes shared with other atoms, thereby resulting in the formation of covalent bonds in which both electrons come from the same atom. (Remember the other kind of covalent bond in which one electron comes from each of the two atoms participating in the bond.) Carbon monoxide (CO) bonds to a variety of metals in this manner to form metal carbonyls. For instance, in carbon monoxide poisoning, CO bonds to the iron atoms in hemoglobin to form carboxyhemoglobin. Molecular oxygen (O_2) is then no longer able to bond to the iron atoms, and asphyxiation occurs rapidly. In how many different ways could carbon monoxide be bonded to the iron atoms in carboxyhemoglobin?

(C, $Z = 6$; O, $Z = 8$)

1 1
2 2
3 3
4 4
5 5

PROBLEM 10–11

Lewis acids and bases are discussed in Section 10–1 of Chemical Principles.

An electron pair donor is called a *Lewis base.* Which one of the following species would be the *least* likely to act as a Lewis base?

(C, $Z = 6$; N, $Z = 7$; P, $Z = 15$;
S, $Z = 16$; Cl, $Z = 17$; I, $Z = 53$)

1 PCl_3
2 CN^-
3 I^-
4 I^+
5 SCl_2

PROBLEM 10–12

An electron pair acceptor is called a *Lewis acid.* Which one of the following would *not* be regarded as a Lewis acid? [*Hint:* Many elements for which $Z > 10$ have empty *d* orbitals that can be utilized in bonding; consequently, it is possible to have more than eight valence electrons around an atom of such an element. Some stable molecules exist, in which one or more of the atoms in the molecule have less than eight electrons around them.]

(H, $Z = 1$; B, $Z = 5$; O, $Z = 8$; Al, $Z = 13$;
Si, $Z = 14$; S, $Z = 16$; Br, $Z = 35$; Ag, $Z = 47$; Cl, $Z = 17$)

1 $AlBr_3$
2 Ag^+
3 SO_3
4 BH_4^-
5 $SiCl_4$

Now that you have some idea of the relationship between Lewis structures and covalent bonding, let us look at the more modern concept of covalent bonding, the molecular orbital theory.

PROBLEM 10–13

The use of molecular orbital theory to explain bonding is described in Section 10–3 of Chemical Principles.

Which one of the following combinations of orbitals on atoms $A_{(a)}$ and $A_{(b)}$ do you think would result in the molecular orbital in Figure 10–1?

[*Hint:* For the formation of a *bonding* molecular orbital the algebraic signs of the wave function in the two overlapping atomic orbital lobes must be the *same*. If the signs are different, an antibonding molecular orbital is formed.]

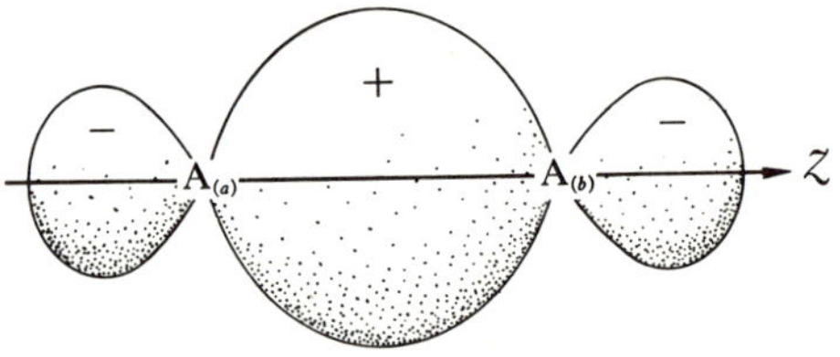

Figure 10–1

1

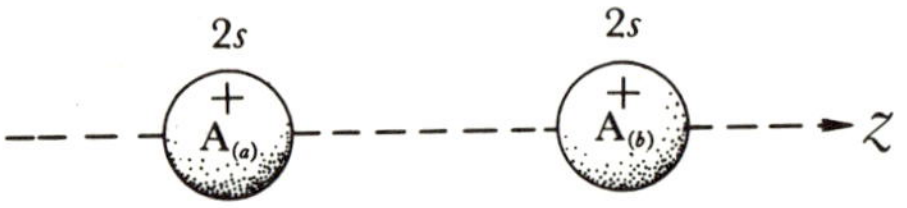

2

3

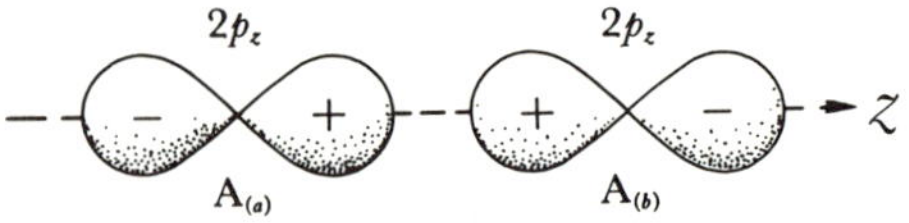

4

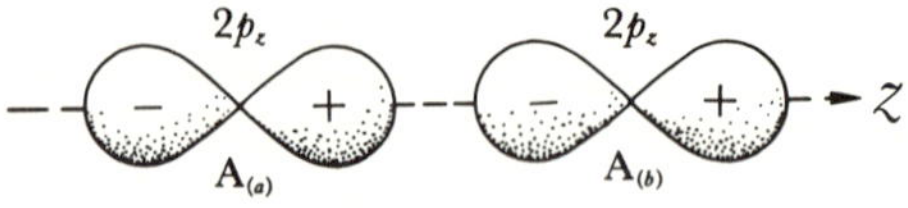

5

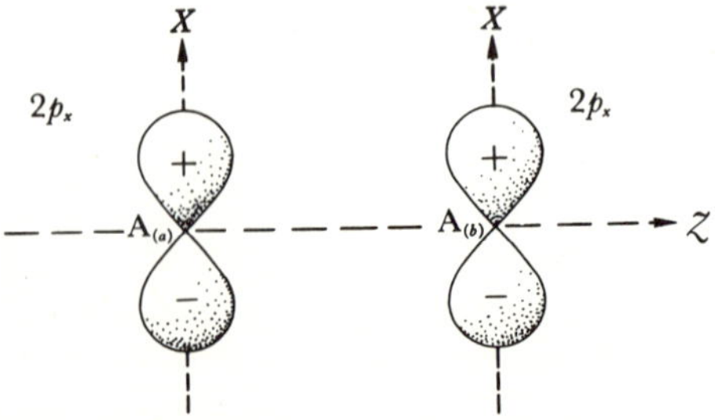

PROBLEM 10–14 The four-lobed π_x* antibonding molecular orbital of a molecule A_2 (Figure 10–2) is formed by the subtraction of the wave function of a certain orbital for which $n = 2$ on atom $A_{(b)}$ from the wave function of a similar orbital on atom $A_{(a)}$. What are these orbitals? [*Hint:* Refer to Problem 10–13.]

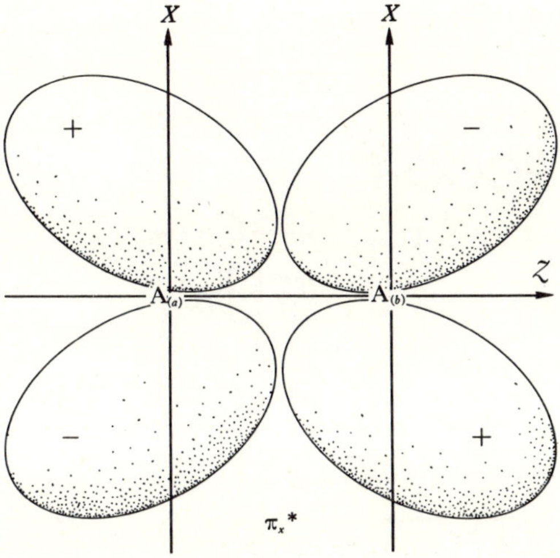

Figure 10–2

1 2*s* on $A_{(a)}$ and 2*s* on $A_{(b)}$
2 $2p_x$ on $A_{(a)}$ and $2p_x$ on $A_{(b)}$
3 $2p_y$ on $A_{(a)}$ and $2p_y$ on $A_{(b)}$
4 $2p_z$ on $A_{(a)}$ and $2p_z$ on $A_{(b)}$
5 None of the above.

PROBLEM 10–15

If you have difficulty understanding the ideas of bonding and antibonding molecular orbitals and their relative energies, refer to Section 10–2 of Programed Reviews of Chemical Principles.

Figure 10–3 is a skeleton of the energy-level diagram for the molecular orbitals of the A_2 molecule discussed in Problems 10–13 and 10–14. If an antibonding molecular orbital is always at a higher energy than its corresponding bonding molecular orbital, the molecular orbitals of the A_2 molecule in order of *increasing energy* are ________.

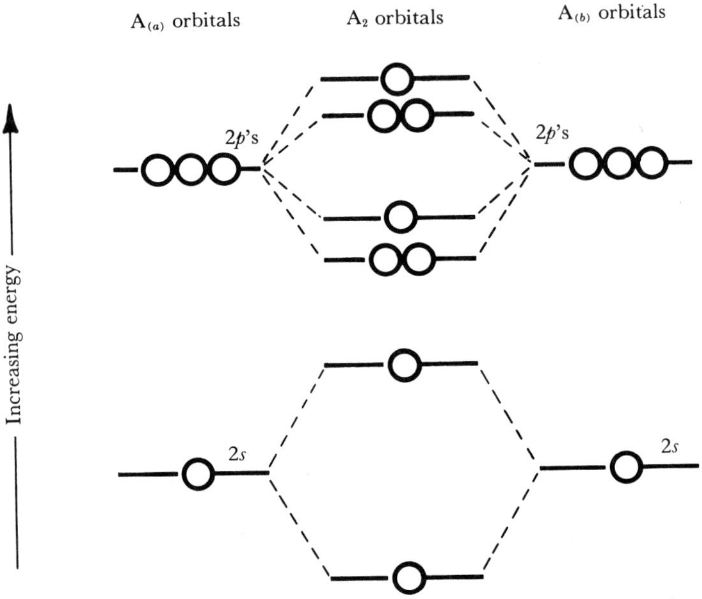

Figure 10–3

1 $\sigma_z^* < \pi_x^* = \pi_y^* < \sigma_z^b < \pi_x^b = \pi_y^b < \sigma_s^* < \sigma_s^b$
2 $\sigma_s^b < \sigma_s^* < \pi_x^b = \pi_y^b < \sigma_z^b < \pi_x^* = \pi_y^* < \sigma_z^*$
3 $\sigma_s^b < \sigma_s^* < \sigma_z^* < \pi_x^b = \pi_y^b$
4 $\pi_x^b < \pi_y^b < \pi_x^* < \pi_y^*$
5 $\sigma_s^b < \sigma_s^* < \sigma_z^b = \pi_y^b$

PROBLEM 10–16

When potassium metal (K) is burned in dry air or dry oxygen, bright yellow potassium superoxide (KO_2) is formed. What is the ground state electronic configuration of the superoxide ion (O_2^-) in potassium superoxide on the basis of molecular orbital theory?

[*Note:* Use the orbital labels and the energy sequence given in the preceding problem for the A_2 molecule. $(\pi_{x,y}^b)^4$ is an abbreviation for $(\pi_x^b)^2\ (\pi_y^b)^2$. Note also that the symbol *KK* represents the four inner electrons (two from each O atom) that are not involved in bonding.]

(O, $Z = 8$)

1 $KK\ (\sigma_s^b)^2(\sigma_s^*)^2(\sigma_z^b)^2(\pi_x^b)^1(\pi_y^b)^1$
2 $KK\ (\sigma_s^b)^2(\sigma_s^*)^2(\pi_{x,y}^b)^4(\sigma_z^*)^2(\pi_x^*)^1(\pi_y^*)^1$
3 $KK\ (\sigma_s^b)^2(\sigma_s^*)^2(\pi_{x,y}^b)^4(\sigma_z^*)^2(\pi_x^*)^2(\pi_y^*)^1$
4 $KK\ (\sigma_s^b)^2(\sigma_s^*)^2(\pi_{x,y}^b)^4(\sigma_z^b)^2(\pi_x^*)^2(\pi_y^*)^1$
5 $KK\ (\sigma_s^b)^2(\sigma_s^*)^2(\sigma_z^b)^2(\pi_{x,y}^b)^4(\sigma_z^*)^2(\pi_x^*)^1$

PROBLEM 10–17

The use of molecular orbital theory to explain paramagnetism and diamagnetism in molecules such as O_2 is described in Section 10–4 of Chemical Principles.

How many unpaired electrons are there in molecular oxygen? [*Hint:* Refer to Problem 10–16.]

(O, $Z = 8$)

1 1
2 2
3 3
4 4
5 None; the oxygen molecule is *diamagnetic.*

PROBLEM 10–18

For which one of the following endothermic processes would ΔH^0 be largest? [*Hint:* Refer to Problems 10–16 and 10–17.]

(a) $O_2^-(g) \rightarrow O^-(g) + O(g)$
(b) $O_2(g) \rightarrow O(g) + O(g)$
(c) $O_2^+(g) \rightarrow O^+(g) + O(g)$

1 (a)
2 (b)
3 (c)
4 ΔH^0 is the same for all three reactions.
5 Insufficient data; need to know the first ionization energies of O_2^-, O_2, and O_2^+.

PROBLEM 10–19

Molecular orbital theory can be used to describe bonding in polyatomic molecules; see Section 10–6 of Chemical Principles.

Which one of the following molecules would not be classed as *heteronuclear?*

1 Te_2
2 BeH
3 CO
4 BF
5 NH

PROBLEM 10–20

The E region of the ionosphere is 50–90 miles above the surface of the earth. This region of the earth's atmosphere has been probed by means of satellites. A variety of positive ions has been detected and identified. If you assume that all five of the following positive ions occur in the E region of the ionosphere, which one do you think will be the most stable?

(Li, $Z=3$; Be, $Z=4$; N, $Z=7$; O, $Z=8$)

1 N_2^+
2 NO^+
3 O_2^+
4 Li_2^+
5 Be_2^+

PROBLEM 10–21

Which one of the following species would you expect to have the longest bond?

(C, $Z=6$; N, $Z=7$; O, $Z=8$)

1 CN^+
2 CN
3 CN^-
4 NO^+
5 All four of the above species have the same bond length.

PROBLEM 10–22

How many σ and π bonds are there in N_2?

(N, $Z=7$)

1 No σ and three π.
2 One σ and one π.
3 One σ and two π.
4 One σ and three π.
5 Two σ and one π.

PROBLEM 10–23

How many σ and π bonds are there in the tetracyanoethylene molecule, $C_2(CN)_4$? [*Hint:* Refer to Problem 10–22.]

(C, $Z=6$; N, $Z=7$)

The molecular orbital treatment is given to diatomic molecules of different atoms in Section 10–5 of Chemical Principles.

1 Five σ and nine π.
2 Six σ and eight π.
3 Nine σ and seven π.
4 Nine σ and nine π.
5 Five σ and eight π.

The last four problems in this chapter will show you how you can predict the shapes of some simple molecules and ions.

PROBLEM 10–24

What type of hybridization does the sulfur atom adopt in sulfur hexafluoride, SF_6?

(F, $Z = 9$; S, $Z = 16$)

Hybrid orbitals are described in Section 10–7 of Chemical Principles.

1 *sp*
2 sp^2
3 sp^3
4 sp^3d
5 sp^3d^2

PROBLEM 10–25

If you have not grasped the ideas of hybrid orbitals, study Section 10–3 of Programed Reviews of Chemical Principles *before you try the next three problems.*

What type of hybridization does the sulfur atom have in sulfur tetrafluoride, SF_4?

(F, $Z = 9$; S, $Z = 16$)

1 *sp*
2 sp^2
3 sp^3
4 sp^3d
5 sp^3d^2

PROBLEM 10–26

How electron repulsion affects molecular structure is discussed in Section 10–8 of Chemical Principles.

What is the actual shape of the SF_4 molecule in Problem 10–25?

[*Hint:* Consider all the possible repulsions between the five valence shell electron pairs (four bond pairs and one lone pair) and use the fact that the repulsions vary in strength as lone pair–lone pair repulsions > lone pair–bond pair repulsions > bond pair–bond pair repulsions. Ignore the repulsions between pairs of electrons that lie at angles greater than 90° to each other. The most stable arrangement of the electron pairs is the one with the least amount of electron pair repulsion. Then the shape of the molecule is determined by the spatial arrangement of the bond pairs around the central atom in the molecule.]

1 Square plane.
2 Trigonal bipyramid.
3 Triangular pyramid.
4 Irregular tetrahedron.
5 Tetrahedron.

PROBLEM 10-27 What is the shape of the ICl_4^- ion? [*Note:* I is the central atom in this ion.]

(Cl, $Z = 17$; I, $Z = 53$)

1 Tetrahedron.
2 Triangle.
3 Square plane.
4 Octahedron.
5 Pentagonal bipyramid.

SOLUTIONS

SOLUTION 10–1

(3)

The ground state electronic configurations of H and O are

H: $1s^1$
O: $[He]2s^2 2p^4$

The chemical formula for hydrogen trioxide is H_2O_3. This means that we have to account for $2(1) + 3(6) = 20$ *valence* electrons (electron dots in the diagrams) in its Lewis structure.

Structures **1**, **2**, and **4** can be eliminated immediately. Structures **1** and **2** have 24 valence electrons and the H atoms are surrounded by four, not two, electrons. [Remember that with the exception of hydrogen every atom in a Lewis structure must be surrounded by eight electrons (an octet). Hydrogen needs only two electrons to give it the stable electronic configuration of the noble gas helium (He).] Structure **4** has 32 valence electrons and the H's are surrounded by eight instead of two electrons.

Both structures **3** and **5** have 20 valence electrons. However, the second H atom in structure **5** is surrounded by four electrons, so this structure is also incorrect.

The correct structure is **3** since every atom obeys the rules for Lewis structures. Note the similarity between this structure and that of hydrogen peroxide (H_2O_2)
The great reactivity of H_2O_2 and the instability of H_2O_3 result from the weak O—O bonds in these molecules.

SOLUTION 10–2

(4)

If you had difficulty with this problem, refer to Section 10–1 of Programed Reviews of Chemical Principles, *by Lassila* et al., *for practice in writing Lewis diagrams.*

The ground state electronic configurations of C, H, and O are

C: $[He]2s^2 2p^2$
H: $1s^1$
O: $[He]2s^2 2p^4$

The formula for methyl alcohol is CH_4O. Thus, we have $4 + 4(1) + 6 = 14$ valence electrons to account for in our Lewis structure.

In structure **1**, there are two pairs of dots (two unshared electron pairs) and six dashes (six shared electron pairs) giving $2(2) + 6(2) = 16$ valence electrons. That is, there are two too many electrons. The carbon atom is surrounded by ten instead of eight electrons. Structure **1** is incorrect.

In structure **2**, there are one unshared electron pair and six shared electron pairs; that is, $2 + 6(2) = 14$ valence electrons. This is the required number, but the carbon atom is surrounded by ten electrons, thereby violating the octet rule. This structure is incorrect.

In structure **3**, there are no unshared pairs but six shared electron pairs; that is, $6(2) = 12$ valence electrons. Each atom appears to obey the octet rule, but we are two electrons short of the required number. This structure is also incorrect.

In structure **4**, there are two unshared and five shared electron pairs; that is, $2(2) + 5(2) = 14$ valence electrons. Each atom obeys the octet rule; hence, structure **4** is the correct one for methyl alcohol. The important structural similarity between ethyl alcohol and methyl alcohol is in the hydroxyl group (OH), which is common to all alcohols.

The correct answer is **4**.

[*Note:* Just because we can write a Lewis diagram for a molecule based on its molecular formula does not necessarily mean that we have the correct structure. To determine molecular structures we need further information such as chemical and physical properties. For instance, there are two common compounds with the molecular formula C_2H_6O:

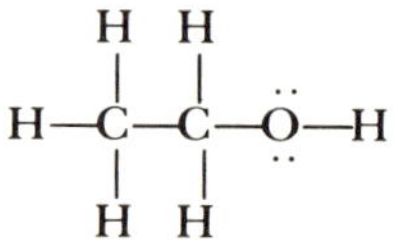

ethyl alcohol

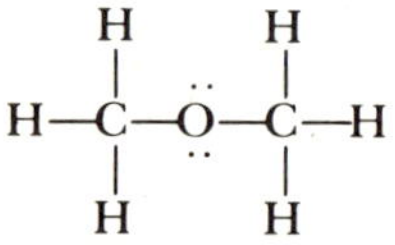

dimethyl ether

SOLUTION 10–3

(4)

The ground state electronic configurations of H, C, N, and I are

H: $1s^1$
C: $[He]2s^22p^2$
N: $[He]2s^22p^3$
I: $[Kr]5s^24d^{10}5p^5$

The $4d$ electrons in I are not valence electrons and consequently are not involved in bonding. Since the formula of iodole is HC_4NI_4, there are $1 + 4(4) + 5 + 4(7) = 50$ valence electrons to be accounted for.

Structure **1** is incorrect: There are 12 electrons around the fourth carbon atom and 52 valence electrons in all. Structure **2** is incorrect: There are 10 electrons around the nitrogen atom, and there are 52 valence electrons in the structure. Structure **3** is incorrect: There is no hydrogen atom! Structure **4** is correct: There are 50 valence electrons, and each atom satisfies the octet rule. Structure **5** is incorrect: The second and third carbon atoms are surrounded by ten electrons, and there are six electrons around the fourth carbon atom.

The required answer is **4**.

SOLUTION 10-4

1

(a) is incorrect. Sodium is a typical metal that loses its one valence electron to form ionic, not covalent compounds. The correct Lewis structure for sodium hypochlorite is

$$Na^+[:\ddot{\underset{..}{O}}—\ddot{\underset{..}{Cl}}:]^-$$

(b) is correct: The carbon atom and the chlorines all have properly completed octets. (c) is correct: All the atoms obey the octet rule. The ammonium ion (NH_4^+) forms a variety of ionic compounds. (d) is correct; the octet rule is obeyed.

Only (a) is incorrect. The required answer is **1**.

SOLUTION 10-5

5

The ground state electronic configurations of H, C, Si, and Cl are

H: $1s^1$
C: $[He]2s^22p^2$
Si: $[Ne]3s^23p^2$
Cl: $[Ne]3s^23p^5$

The molecular formula of dimethyldichlorosilane is $C_2H_6SiCl_2$; thus, there are $2(4) + 6(1) + 4 + 2(7) = 32$ valence electrons.

All five structures have the correct molecular formula and 32 valence electrons. Furthermore, in each case every atom satisfies the octet rule. How, then, do we identify the correct structure? The name of the compound is "dimethyldichlorosilane," and its formula is $(CH_3)_2SiCl_2$. To satisfy both the name and the formula, the four H atoms attached to the silicon atom in a silane molecule (SiH_4) must have been replaced by two methyl (CH_3) groups and two chlorine atoms. Only structure **5** has two CH_3 groups and two Cl atoms attached to a Si atom.

The correct answer is **5**.

[*Note:* The five different molecules are referred to as *structural isomers.* They have the same molecular formula but different arrangements of the various atoms involved. The name of a compound and its formula give an important insight into its structure. This applies particularly to organic compounds; often there are many different organic structural isomers.]

SOLUTION 10–6

(5)

The ground state electronic configurations of O and Cl are

O: $[He]2s^22p^4$
Cl: $[Ne]3s^23p^5$

There are 2(6) + 7 = 19 valence electrons in ClO_2. Every atom in each of the four structures proposed has a complete octet but none of the structures has 19 valence electrons! Structures **1** and **2** both have 20 valence electrons (two shared and eight unshared electron pairs); structure **3** has 18 valence electrons (three shared and six unshared electron pairs); and structure **4** has 16 valence electrons (four shared and four unshared electron pairs). Because ClO_2 has an odd number of valence electrons it is impossible to draw a Lewis structure for it. We cannot satisfy the eight electron (octet) rule with an odd number of electrons. A theory of bonding based on the sharing of electron pairs cannot account for such molecules. It is necessary to resort to a more complex bonding theory known as the molecular orbital theory to do so. Problems 10–13 through 10–17 illustrate some of the principles involved in the molecular orbital theory.

The required answer is **5**.

SOLUTION 10–7

(5)

All we have to do in this problem is to determine the total number of electrons in each species and them compare them. We must not forget the charge on the Os^{6+} ion.

Species	Total number of electrons
Br_2	2(35) = 70
ICl	53 + 17 = 70
Os^{6+}	76 − 6 = 70
SF_6	16 + 6(9) = 70
GeF_4	32 + 4(9) = 68

Only GeF_4 does not have 70 electrons. So the correct answer is **5**. If you chose Answer **3**, you realized that there were 70 electrons in the isoelectronic species, but you forgot to subtract six electrons from the atomic number of osmium and then probably did not bother to check GeF_4.

SOLUTION 10–8

(3)

The total number of available valence electrons in N_2O is 16 (five in each of the two N atoms and six in the O atom).

Structure (a) is incorrect. There are 16 valence electrons, but the central N atom has 10 electrons surrounding it. Structure (b) is correct. There are 16 electrons, and for each atom to satisfy the octet rule, one electron is transferred from the central N atom to the other N atom. If we indicate the loss and gain of the one electron by plus and minus signs, the structure can be written as

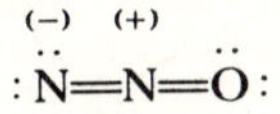

Structure (c) is incorrect: 14 electrons, with only six around the N atom. Structure (d) is incorrect: 16 electrons but only six around the O atom. Structure (e) is correct. There are 16 electrons, and for all three atoms to obey the octet rule, one electron is transferred from the central N atom to the O atom. Therefore, the charge distribution is

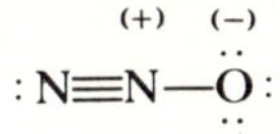

The actual structure of N_2O is a hybrid of the two resonance forms (b) and (e), and it is lower in energy (i.e., more stable) than either of them.

The correct answer is **3**.

[*Note:* You should have realized that there is a third possible resonance form of N_2O that was not given, namely

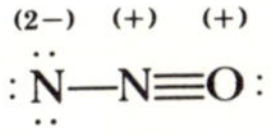

This form does not make as important a contribution to the actual structure of N_2O as the other two forms do, because of the unfavorable charge separation present in it.]

SOLUTION 10–9

(4)

There are 16 valence electrons in the fulminate ion, CNO^- (four from C, five from N, six from O, and one from the net ionic charge).

Structure (c) can be ruled out immediately because it has 18 electrons (10 electrons around the O atom). Structure (e) is also incorrect; there are six electrons around the N atom and ten around the O atom. The three remaining structures, (a), (b), and (d), are all correct.

The required answer is **4**.

[*Note:* The resonance forms of the isoelectronic azide ion, N_3^-, are exactly the same as those for the fulminate ion with the C and O atoms being replaced by N's. Notice that the fulminate and azide ions are isoelectronic with the N_2O molecule in Problem 10–8.]

SOLUTION 10–10 (2)

There are 10 valence electrons in CO (four from C and six from O). The Lewis structure is

$:C\equiv O:$

Since there are two lone pairs (one on C and one on O), there are two possible ways in which CO could bond to the iron atoms in hemoglobin.

[*Note:* Recent spectroscopic evidence suggests that the CO molecules in carboxyhemoglobin are attached only by way of the oxygen atom. This situation is strange because all x-ray data to date have shown that the CO groups in metal carbonyls are always attached by way of the carbon atom. The reason for the difference in carboxyhemoglobin is not understood.]

The correct answer is **2**.

SOLUTION 10–11 (4)

The trick here is to write down the electronic structures to see if there are any available lone pairs.

PCl_3 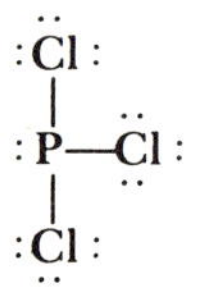

CN^- $[:C\equiv N:]^-$

I^- $[:\ddot{\underset{..}{I}}:]^-$

I^+ $[:\dot{\underset{..}{I}}]^+$

SCl_2 $:\ddot{\underset{..}{Cl}}—\ddot{\underset{..}{S}}—\ddot{\underset{..}{Cl}}:$

There are lone pairs in all five species. However, I^+ does not have a complete octet and will certainly rather gain two more electrons than lose them. So **4** is the required answer.

[*Note:* The two chief types of species that exhibit Lewis base activity are anions and molecules with one or two lone pairs of electrons. Even though there are lone pairs on both P and Cl in PCl_3 and on both S and Cl in SCl_2, the atoms that act as Lewis bases in these compounds are P and S, respectively. In CN^-, both the C and the N can act as Lewis bases.]

SOLUTION 10–12 First let us write down the electronic structures for the various species.

(4)

$AlBr_3$

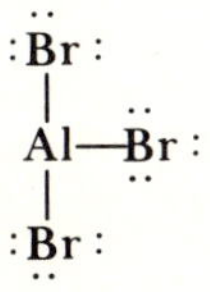

Ag^+ $[Ag]^+$

SO_3

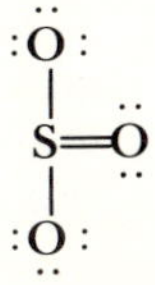

BH_4^-

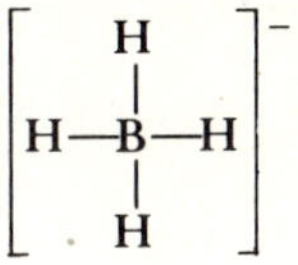

$SiCl_4$

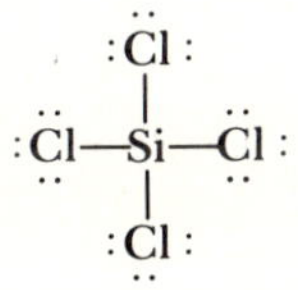

$AlBr_3$ is electron deficient (only six electrons around the Al atom) and is therefore quite likely to accept a lone pair. Ag^+ has no valence electrons (filled inner orbitals only) and is equally likely to accept lone pairs in its empty valence orbitals. So Answers **1** and **2** are Lewis acids; but what about the rest, since they all show completely filled octets? Elements for which $Z > 10$ often have vacant d orbitals that they can use in bonding. Sulfur and silicon are both in the third period ($n = 3$) and have vacant $3d$ orbitals that are low enough in energy to accept electron pairs. In doing so, sulfur and silicon will have more than eight electrons around them; that is, they will expand their octets. Boron, however, is in the second period ($n = 2$); and because there are no $2d$ orbitals, boron is unable to expand its octet. Thus, SO_3 and $SiCl_4$ are Lewis acids but BH_4^- is not.

The required answer is **4**.

[*Note:* Three of the main types of Lewis acids are positive ions, molecules with incomplete octets, and molecules in which an atom can expand its octet.]

SOLUTION 10–13 (3)

The molecular orbital in Figure 10–1 is a σ_z^b bonding molecular orbital for two reasons. First, it is rotationally symmetrical about the bonding Z axis (the signs of the wave function in the three lobes of electron density are unchanged upon rotation by any angle about the Z axis). Such molecular orbitals are referred to as sigma (σ) molecular orbitals. Second, there is a large buildup of electron density between the two atoms $A_{(a)}$ and $A_{(b)}$ that bonds them together. At first sight, there are two combinations that could possibly result in the formation of the σ_z^b molecular orbital: combinations **3** and **4**. However, combination **4** would not give a bonding molecular orbital because the signs of the wave function in the two lobes to be overlapped are not the same. It results in a σ_z^* antibonding molecular orbital that has very little electron density located between the atoms; thus, it occurs at higher energy than the σ_z^b bonding molecular orbital.

Combination **3** is the correct one.

[*Note:* Combination **1** results in a σ_s^b orbital, combination **2** in a σ_s^* antibonding orbital and combination **5** in a π_x^b bonding orbital. Any π orbital (bonding or antibonding) is rotationally asymmetrical; that is, 180° rotation about the bonding Z axis produces an orbital that looks exactly the same as the original one except that the algebraic signs of the wave function in the various lobes have been changed (plus has become minus and vice versa).]

SOLUTION 10–14 (2)

From Problem 10–13 we know that the π_x^* antibonding molecular orbital does not arise from the subtraction of either two $2s$ or two $2p_z$ wave functions. Furthermore, the corresponding π_x^b bonding molecular orbital is formed by the addition of the wave functions for two $2p_x$ orbitals. So the π_x^* antibonding molecular orbital must be formed by the subtraction of the wave function for the $2p_x$ orbital on atom $A_{(b)}$ from the wave function for the $2p_x$ orbital on atom $A_{(a)}$. This is shown diagrammatically in Figure 10–4.

[*Note:* The shapes of the π_y^b bonding and π_y^* antibonding molecular orbitals are exactly the same as those for the π_x^b bonding and π_x^* antibonding molecular orbitals, respectively. These orbitals are formed from the wave functions of two $2p_y$ orbitals and lie in the YZ rather than the XZ plane. The π_x^b and the π_y^b bonding molecular orbitals are *degenerate;* that is, they occur at the same energy. Similarly, the π_x^* and the π_y^* antibonding molecular orbitals are degenerate.]

The correct answer is **2**.

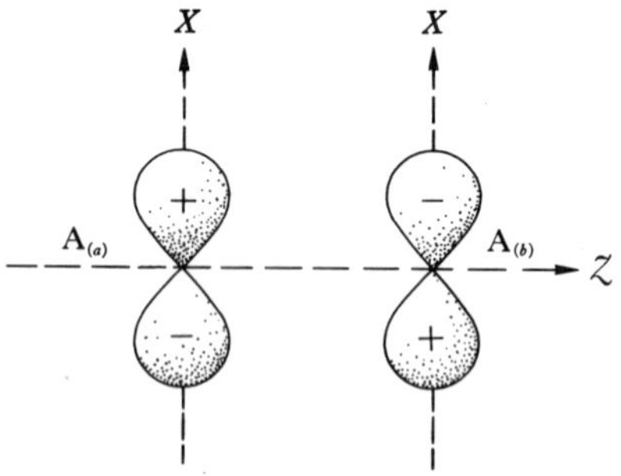

Figure 10–4

SOLUTION 10–15 (2)

From Problems 10–13 and 10–14 we know the origin of the various orbitals for the A_2 molecule. We also know that an antibonding molecular orbital lies at higher energy than the corresponding bonding molecular orbital, and, furthermore, that the π_x^b and π_y^b bonding molecular orbitals are degenerate, as are the π_x^* and π_y^* antibonding molecular orbitals. Thus, it is easy to assign the molecular orbitals to their appropriate place in Figure 10–3. In order of increasing energy, the molecular orbitals are

$$\sigma_s^b < \sigma_s^* < \pi_x^b = \pi_y^b < \sigma_z^b < \pi_x^* = \pi_y^* < \sigma_z^*$$

This order will be used in Problems 10–16 through 10–18 to *predict* the ground state electronic configurations of a number of simple diatomic species. Electrons are fed into the molecular orbitals according to the *Aufbau* procedure described earlier for atomic orbitals. We must remember, however, that our predictions from the general order of energy levels (Figure 10–3) may not always be compatible with the latest evidence for specific examples. For instance, the best data now available show that there is a crossover of energy levels at O_2 and F_2, and the σ_z^b level is lower in energy than the π_x^b and π_y^b level for these diatomic molecules.

The correct order is **2**. Answer **1** is the order of *decreasing energy.*

SOLUTION 10–16 (4)

Since O_2^- is an A_2 type molecule, we can use the molecular orbitals and the energy sequence discussed in Problem 10–15. There are 17 electrons in O_2^- (eight from each of the two O atoms and one from the net charge). Four of these are not valence electrons and are not involved in bonding. Therefore, we have 13 electrons to feed into our molecular orbitals according to the *Aufbau* procedure. Assuming the order of energy levels shown in Figure 10–3 applies to O_2^-, the resulting ground state electronic configuration is

$$KK\ (\sigma_s^b)^2(\sigma_s^*)^2(\pi_{x,y}^b)^4(\sigma_z^b)^2(\pi_x^*)^2(\pi_y^*)^1$$

The correct answer is **4**.

[*Note:* $(\pi_x^*)^2(\pi_y^*)^1$ could also be written as $(\pi_{x,y}^*)^3$.]

SOLUTION 10–17 (2)

Molecular oxygen (O_2) has one less electron than O_2^- (which was discussed in Problem 10–16). Therefore, the ground state electronic configuration of O_2 in terms of molecular orbital theory (assuming Figure 10–3 applies to O_2) is

$$KK\ (\sigma_s^b)^2(\sigma_s^*)^2(\pi_{x,y}^b)^4(\sigma_z^b)^2(\pi_x^*)^1(\pi_y^*)^1$$

Remember that Hund's principle must be obeyed for molecular orbitals in exactly the same way as it is for atomic orbitals. Hence, there is one electron in each of the two degenerate π^* antibonding molecular orbitals. (Don't forget that these two electrons will have parallel spins.) All the other electrons are paired. So there are only two unpaired electrons in molecular oxygen; thus, O_2 is *paramagnetic.*

Answer **2** is correct. For Answer **5**, you forgot about Hund's rule. [*Note:* On the basis of the Lewis structure for O_2,

$$:\ddot{O}{=}\ddot{O}:$$

one would say that O_2 should be diamagnetic (no unpaired electrons). However, the paramagnetism of O_2 has been confirmed experimentally, and this is regarded as one of the major triumphs of molecular orbital theory over the older theories of bonding.]

SOLUTION 10–18

(3)

In Problems 10–16 and 10–17 we predicted that the ground state electronic configurations of O_2^- and O_2 are

O_2^-: $KK\,(\sigma_s^b)^2(\sigma_s^*)^2(\pi_{x,y}^b)^4(\sigma_z^b)^2(\pi_x^*)^2(\pi_y^*)^1$
O_2: $KK\,(\sigma_s^b)^2(\sigma_s^*)^2(\pi_{x,y}^b)^4(\sigma_z^b)^2(\pi_x^*)^1(\pi_y^*)^1$

O_2^+ has one less electron than O_2. Hence, we would predict its ground state electronic configuration to be

O_2^+: $KK\,(\sigma_s^b)^2(\sigma_s^*)^2(\pi_{x,y}^b)^4(\sigma_z^b)^2(\pi_x^*)^1$

What do these three electronic configurations indicate about the bond strengths in the different species? Consider the σ_s^b bonding molecular orbital. This is filled for all three species as is the corresponding σ_s^* antibonding molecular orbital. Any bonding character derived from the σ_s^b orbital is canceled by the antibonding character of the σ_s^* orbital. Similarly, with the other orbitals, there is either complete or partial cancellation of bonding character. To get some idea of the net bonding character, we define the *bond order* (number of covalent bonds) in a diatomic molecule as the number of electrons in bonding molecular orbitals minus the number of electrons in antibonding molecular orbitals, divided by two. (Remember that we need *two* electrons per covalent bond.) Thus, the bond orders of O_2^-, O_2, and O_2^+ are

O_2^-: $(8-5)/2 = 1\tfrac{1}{2}$
O_2: $(8-4)/2 = 2$
O_2^+: $(8-3)/2 = 2\tfrac{1}{2}$

(The *KK* electrons are excluded from the calculations.)

The ΔH^0's for the three endothermic processes given are the *bond dissociation energies* of the species O_2^-, O_2, and O_2^+. The larger the bond order, the harder it is to break the bond; that is, the larger the bond dissociation energy and the more stable the species. Since O_2^+ has a larger bond order than that in O_2^- or O_2, it will have the largest bond dissociation energy and will be the most stable. Thus, ΔH^0 will be largest for process (c), namely

$$O_2^+(g) \rightarrow O^+(g) + O(g)$$

The correct answer is **3**.

[*Note:* The experimental dissociation energies for the three species are O_2^+, 149 kcal mole^{-1}; O_2, 117 kcal mole^{-1}; O_2^-, 94 kcal mole^{-1}.]

SOLUTION 10–19

(1)

Te_2 is obviously the odd man out because it contains two atoms of the same kind, whereas the other four molecules each contain two different atoms. Te_2 is an example of a *homonuclear* diatomic molecule: two atoms of the same kind bonded together. We already have considered a number of other homonuclear diatomic molecules (e.g., O_2^-, O_2, and O_2^+). S_8 and P_4 are examples of larger homonuclear molecules.

BF, BeH, CO, and NH are all heteronuclear diatomic molecules. From the molecular orbital point of view they can be regarded in much the same way as homonuclear diatomic molecules. We can use the same type of energy level diagram to build up their ground state electronic configurations. The molecular orbitals are filled in the same order and in the same manner as before.

The correct answer is **1**.

SOLUTION 10–20

(2)

Since we must assume that all five positive ions occur in the E region of the ionosphere, the most stable one should be the one with the highest bond dissociation energy, that is, the highest bond order.

The ground state electronic configurations and bond orders for the five positive ions are shown in Table 10–1.

Table 10–1

	Ground state electronic configuration	Bond order
N_2^+:	$KK\ (\sigma_s^b)^2(\sigma_s^*)^2(\pi_{x,y}^b)^4(\sigma_z^b)^1$	$2\frac{1}{2}$
NO^+:	$KK\ (\sigma_s^b)^2(\sigma_s^*)^2(\pi_{x,y}^b)^4(\sigma_z^b)^2$	3
O_2^+:	$KK\ (\sigma_s^b)^2(\sigma_s^*)^2(\pi_{x,y}^b)^4(\sigma_z^b)^2(\pi_x^*)^1$	$2\frac{1}{2}$
Li_2^+:	$KK\ (\sigma_s^b)^1$	$\frac{1}{2}$
Be_2^+:	$KK\ (\sigma_s^b)^2(\sigma_s^*)^1$	$\frac{1}{2}$

NO^+ has the highest bond order and should be the most stable.

The correct answer is **2**.

As might be expected, Li_2^+ and Be_2^+ are not found in the E region of the ionosphere. However, NO^+, N_2^+, and O_2^+ are; NO^+ is the most abundant. The abundance of NO^+ is noteworthy because there is very little NO present in the atmosphere. NO^+ is apparently formed from O^+ and N_2^+ by the following reactions:

$$O^+ + N_2 \rightarrow NO^+ + N$$
$$N_2^+ + O \rightarrow NO^+ + N$$

SOLUTION 10–21

(1)

Since the valence electrons of the atoms in the four species are all from the $n = 2$ shell, it is fair to say that the species with the fewest covalent bonds (i.e., lowest bond order) should have the longest bond. The ground state electronic configurations of the four species and their bond orders are shown in Table 10–2.

Table 10–2

Ground state electronic configuration		Bond order
CN^+:	$KK\,(\sigma_s{}^b)^2(\sigma_s{}^*)^2(\pi_{x,y}{}^b)^4$	2
CN:	$KK\,(\sigma_s{}^b)^2(\sigma_s{}^*)^2(\pi_{x,y}{}^b)^4(\sigma_z{}^b)^1$	$2\frac{1}{2}$
CN^-:	$KK\,(\sigma_s{}^b)^2(\sigma_s{}^*)^2(\pi_{x,y}{}^b)^4(\sigma_z{}^b)^2$	3
NO^+:	$KK\,(\sigma_s{}^b)^2(\sigma_s{}^*)^2(\pi_{x,y}{}^b)^4(\sigma_z{}^b)^2$	3

From this table, it is evident that CN^+ has the lowest bond order and hence should have the longest bond.

The correct answer is **1**.

[*Note:* The observed bond lengths of the four species in order of increasing bond length (decreasing bond order) are NO^+, 1.062 Å; CN^-, 1.14 Å; CN, 1.1719 Å; CN^+, 1.1727 Å. Notice that the bond length of CN^- is longer than that of the isoelectronic NO^+ molecule even though they both have the same bond order. In general, the bond lengths of isoelectronic diatomic molecules increase with increasing negative charge.]

SOLUTION 10–22

(3)

The ground state electronic configuration of N_2 is

$KK\,(\sigma_s{}^b)^2(\sigma_s{}^*)^2(\pi_{x,y}{}^b)^4(\sigma_z{}^b)^2$

The antibonding character of the $\sigma_s{}^*$ molecular orbital cancels the bonding character of the $\sigma_s{}^b$ molecular orbital. This leaves four electrons in the $\pi_{x,y}{}^b$ orbitals and two electrons in the $\sigma_z{}^b$ orbital. Thus, the N_2 molecule has three net bonds (one σ and two π). This triple bond accounts for the unusual stability of N_2. The Lewis structure of N_2 is $:N{\equiv}N:$, with three covalent bonds and two lone pairs. The two lone pairs fill the σ_s^b and $\sigma_s{}^*$ molecular orbitals, while the triple bond results from two π bonds $(\pi_{x,y}{}^b)^4$ and one σ bond $(\sigma_z^b)^2$.

The correct answer is **3**.

[*Note:* In any Lewis structure, a single bond in terms of molecular orbital theory is a σ bond, a double bond is a σ and π bond, and a triple bond is a σ and 2π bonds.]

SOLUTION 10–23 (4)

The Lewis structure of tetracyanoethylene is

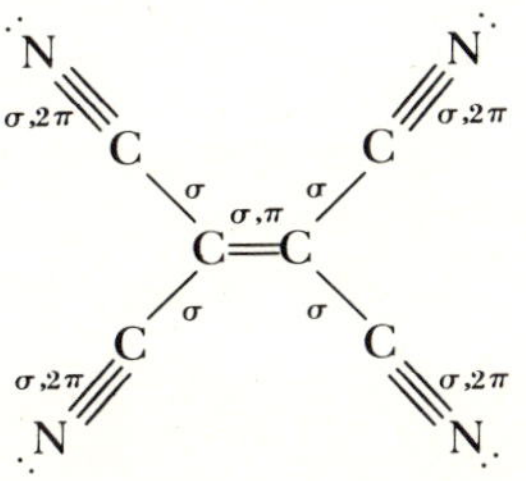

We know that a single bond is a σ bond, a double bond is a σ and a π bond, and a triple bond is a σ and two π bonds. Counting up the bonds in tetracyanoethylene, we have nine σ and nine π bonds. These are indicated on the Lewis structure.

The correct answer is **4**.

SOLUTION 10–24 (5)

The ground state electronic configuration of sulfur is

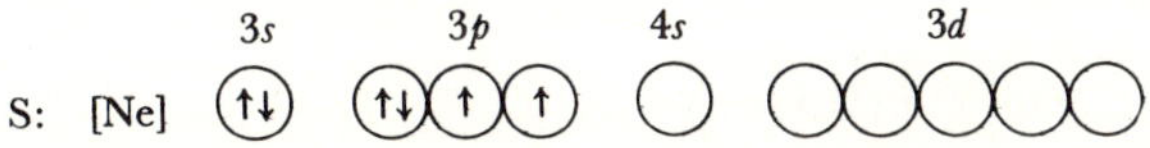

In this configuration, sulfur can form only two covalent bonds. How, then, can SF_6 be accounted for? Sulfur has an empty 4*s* orbital and five empty 3*d* orbitals. Even though in atoms the 4*s* orbital is more stable than the 3*d* orbitals, sulfur uses two of the 3*d* orbitals to increase its bond-forming ability. The stability gained in bond formation more than compensates for the energy necessary to unpair two pairs of electrons and then promote two of the electrons to two of the empty 3*d* orbitals to give the excited state configuration:

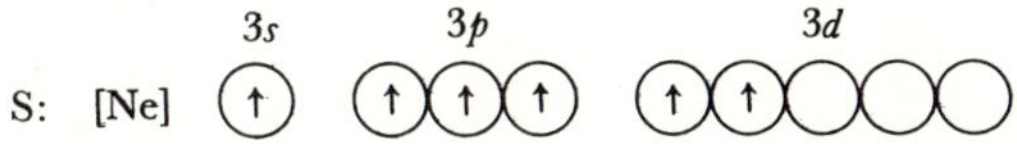

Then the six orbitals (the 3*s*, the three 3*p*'s, and the two 3*d*'s) are hybridized to form a set of six equivalent $3sp^3d^2$ orbitals directed toward the corners of a regular octahedron (Figure 10–5). Then six fluorine–sulfur σ bonds are formed by the overlap of a half-filled *p* orbital on each fluorine with one of the half-filled $3sp^3d^2$ hybridized orbitals on the sulfur atom:

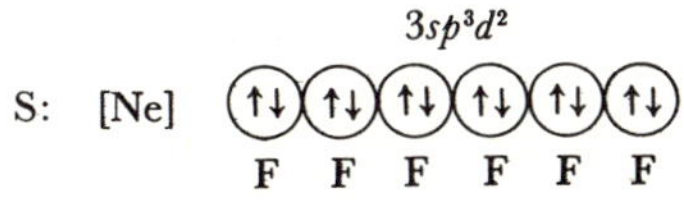

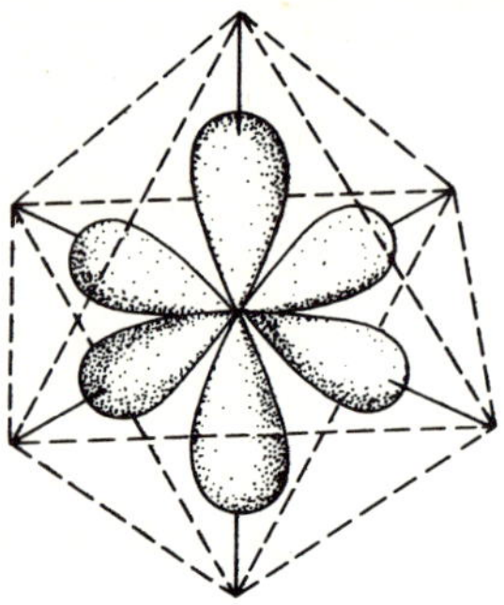

Figure 10–5. sp^3d^2 hybrid orbitals such as those formed by sulfur in SF_6.

The octahedral structure of SF_6 is shown in Figure 10–6.

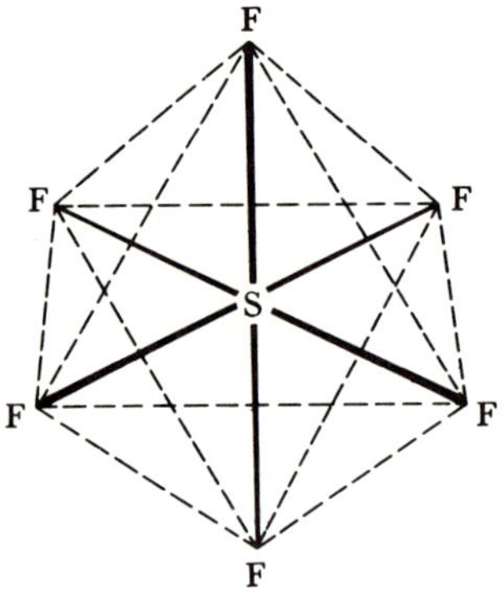

Figure 10–6

Thus, sulfur adopts the sp^3d^2 type of hybridization in SF_6. Answer **5** is correct.

SOLUTION 10–25

(4)

From Problem 10–24 we know that the ground state electronic configuration of sulfur is

```
               3s        3p
S: [Ne]       (↑↓)    (↑↓)(↑)(↑)
```

There are only two unpaired electrons. To form SF_4 we need four unpaired electrons to pair with the unpaired *p* electrons on four fluorine atoms. So one pair of electrons on the sulfur atom is unpaired and one electron is promoted to a vacant $3d$ orbital:

```
               3s        3p              3d
S: [Ne]       (↑↓)    (↑)(↑)(↑)     (↑)( )( )( )( )
```

We now have four unpaired electrons as required and one lone pair of electrons. What hybridization does the S atom adopt: $3p^3d$ or $3sp^3d$? It turns out that, because of the mathematics of hybridization, we always have to include an *s* orbital to account for the common geometries adopted by molecules. In SF_4, the sulfur atom adopts $3sp^3d$ hybridization that affords a set of five equivalent orbitals directed toward the corners of a trigonal bipyramid (Figure 10–7).

```
                   3sp³d
S: [Ne]       (↑↓)(↑)(↑)(↑)(↑)
```

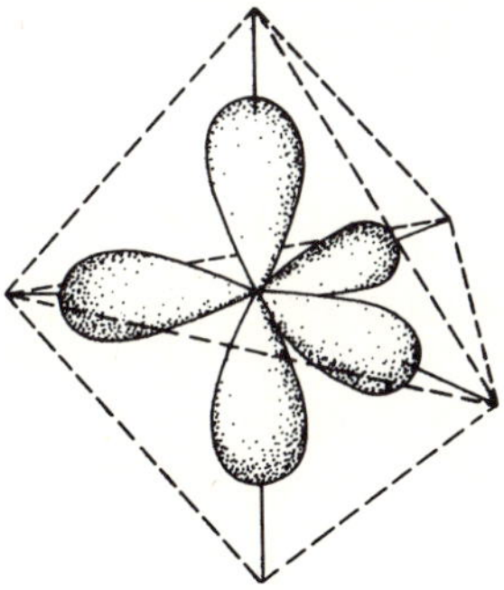

Figure 10–7. sp^3d hybrid orbitals such as those formed by sulfur in SF_4.

Thus, in SF_4, sulfur has the sp^3d type of hybridization. Answer 4 is correct.

SOLUTION 10–26

4

The hybridization of the sulfur atom in SF_4 is $3sp^3d$. The bonding can be represented schematically:

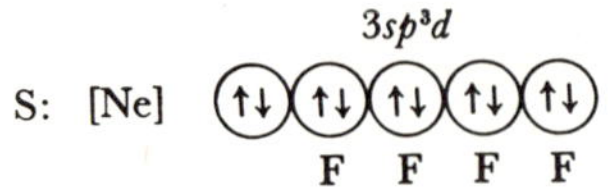

There are four shared electron pairs (bond pairs) and one lone pair arranged in a trigonal bipyramid around the central sulfur atom. Therefore, there are two possible structures for SF_4 based on a trigonal bipyramid. These are shown in Figure 10–8.

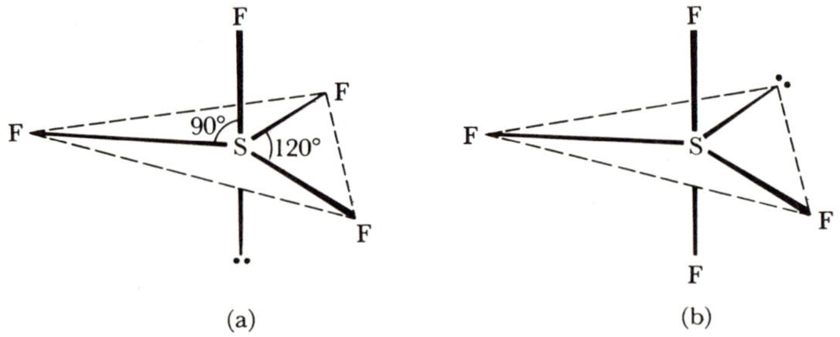

Figure 10–8. The two dots represent a lone pair of electrons; the solid lines between the atoms represent bond pairs of electrons.

One of them must be a more stable structure than the other, but how can we tell which one? To do this we consider all the possible repulsions between the different electron pairs and make use of the fact that the repulsions vary in strength as

lone pair–lone pair > lone pair–bond pair > bond pair–bond pair

Ignoring repulsions between pairs of electrons that lie at angles greater than 90° to each other, we find that structure (a) in Figure 10–8 has

3 lone pair–bond pair repulsions at 90°
3 bond pair–bond pair repulsions at 90°

and structure (b) has

2 lone pair–bond pair repulsions at 90°
4 bond pair–bond pair repulsions at 90°

Structure (b) has one less lone pair–bond pair repulsion at 90° than structure (a), and this outweighs the fact that it has one more bond pair–bond pair repulsion at 90°. Therefore, structure (b) has less electron pair repulsion than structure (a) and thus is more stable. If only the spatial arrangement of the bond pairs is considered, the shape of the SF_4 molecule is that shown in Figure 10–9.

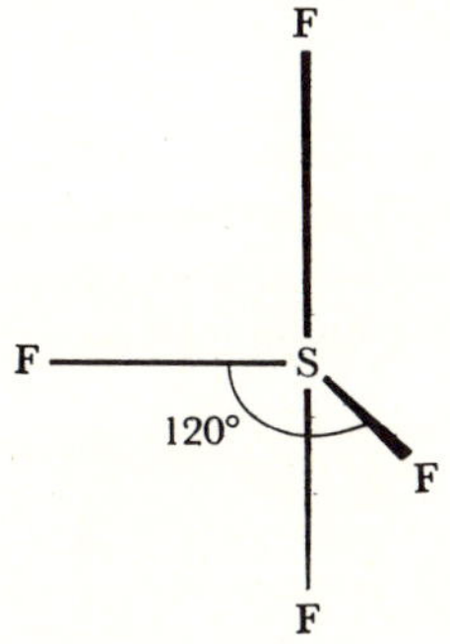

Figure 10–9

The shape is much like that of a "seesaw" in which the two S—F bonds at an angle of 120° to one another act as the support for the linear FSF part of the molecule. This structure is referred to more rigorously as an *irregular tetrahedron* with the S atom at the center of the tetrahedron.

The correct answer is **4**. Answer **2** is almost correct.

SOLUTION 10–27 (3)

Iodine is the central atom in ICl_4^-. The ground state electronic configuration of iodine is

	$5s$	$5p$
I: [Kr]$4d^{10}$	(↑↓)	(↑↓)(↑↓)(↑)

There is only one unpaired electron, yet somehow we must attach three Cl atoms and one Cl^- ion to the iodine atom. First, we must unpair one pair of electrons in I and promote one electron into a vacant $5d$ orbital. This gives us the three unpaired electrons necessary to bond with the unpaired p electrons on three Cl atoms:

	$5s$	$5p$	$5d$
I: [Kr]$4d^{10}$	(↑↓)	(↑↓)(↑)(↑)	(↑)( )( )( )( )

Next, we need another $5d$ orbital to accommodate the lone pair of electrons from the Cl^- ion:

	$5s$	$5p$	$5d$
I: [Kr]$4d^{10}$	(↑↓)	(↑↓)(↑)(↑)	(↑)( )

These six orbitals are hybridized to give six equivalent $5sp^3d^2$ orbitals (such as those in Figure 10–5):

	$5sp^3d^2$
I: [Kr]$4d^{10}$	(↑↓)(↑↓)(↑)(↑)(↑)( )

Thus, the final bonding picture in ICl_4^- is

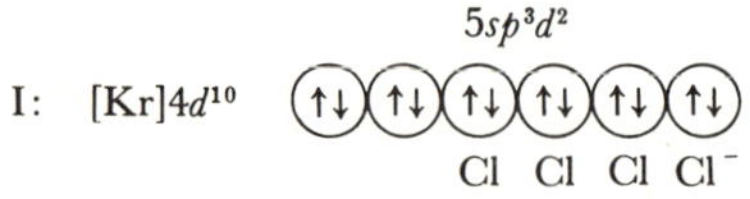

There are four bond pairs and two lone pairs. The two possible structures are shown in Figure 10–10.

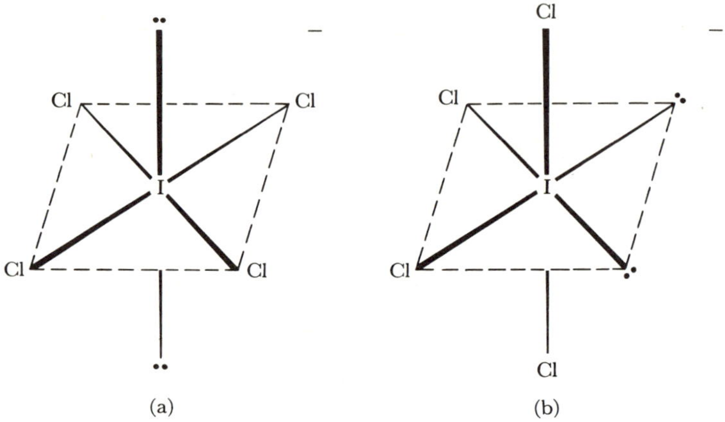

Figure 10–10. The pairs of dots represent lone pairs of electrons.

Neglecting electron pair repulsions at angles greater than 90°, we find that structure (a) has

8 lone pair–bond pair repulsions at 90°
4 bond pair–bond pair repulsions at 90°

and structure (b) has

1 lone pair–lone pair repulsion at 90°
6 lone pair–bond pair repulsions at 90°
5 bond pair–bond pair repulsions at 90°

The lone pair–lone pair repulsion at 90° in (b) is so great that it more than compensates for (a) having two more lone pair–bond pair repulsions at 90° than (b). Hence, structure (a) has the lesser amount of electron pair repulsion and is the more stable. Considering only the bond pairs, we see that the shape of the ICl_4^- ion is square planar (Figure 10–11).

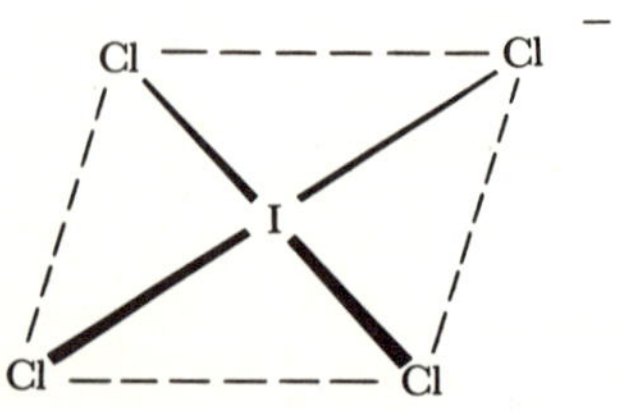

Figure 10–11

The correct answer is **3**.

CHAPTER 11 COORDINATION COMPOUNDS

Coordination chemistry plays a vital role in our very existence. We need hemoglobin to carry the oxygen necessary for us to live; the plants we eat need chlorophyll for their photosynthetic life cycle. Both hemoglobin and chlorophyll are coordination compounds, the former of Fe and the latter of Mg. In this chapter we will consider some of the important aspects of coordination chemistry such as the color of complexes, coordination number, oxidation state, isomerism, chelation, and magnetism.

Nomenclature is important in coordination chemistry because of the existence of *isomers* (compounds with the same chemical composition but different structures). Consequently, it may be necessary for you to learn the internationally accepted rules for naming coordination compounds, given in Section 11–2 of *Chemical Principles* by Dickerson, Gray, and Haight. These rules are

(1) In naming the entire complex, the name of the cation is given first and the anion second (just as for sodium chloride), no matter whether the cation or the anion is the complex species.

(2) In the complex ion, the name of the ligand or ligands precedes that of the central metal atom. Special ligand names are "aquo" for water, "ammine" for NH_3, and "carbonyl" for CO.

(3) Ligand names generally end in "o" if the ligand is negative ("chloro" for Cl^-, "cyano" for CN^-) and "ium" in the rare cases in which the ligand is positive ("hydrazinium" for $NH_2NH_3^+$). The names are unmodified if the ligand is neutral ("methylamine" for CH_3NH_2, "ethylenediamine" for $NH_2CH_2CH_2NH_2$).

(4) A Greek prefix (mono, di, tri, tetra, penta, hexa, etc.) indicates the number of each ligand (mono is often omitted for a single ligand of a given type). If the name of the ligand itself contains the terms mono, di, and so forth (ethylenediamine, en; diethylenetriamine, dien), then the ligand is enclosed in parentheses and its number is given with the alternative prefixes bis and tris, instead of di and tri. For example, $[Pt(en)_3]Br_4$ is tris(ethylenediamine)platinum(IV) bromide.

(5) A Roman numeral or a zero in parentheses indicates the oxidation number of the central metal atom.

(6) If the complex ion is negative, the name of the metal ends in "ate."

(7) If more than one ligand is present in a species, the order of ligands in the name is negative, neutral, and positive.

To answer some of the problems you will need to memorize the following relationships:

$$\text{absorbance} = \log\left(\frac{I_0}{I}\right)$$

$$= \epsilon c \text{ (for a 1.0-cm path-length cell)}$$

in which I_0 and I are the intensities of the incident and transmitted light, respectively.

ϵ = molar absorptivity
c = concentration (moles liter^{-1})

Solutions to problems in Chapter 11 begin on page 313.

PROBLEMS

PROBLEM 11–1

The experimentally determined solubility (13°C) of AgCl in solutions of different $[Cl^-]$ ($K_{sp} = 1 \times 10^{-10}$ at 13°C) is shown on a log–log plot in Figure 11–1. Which statement below is the most reasonable explanation of these data?

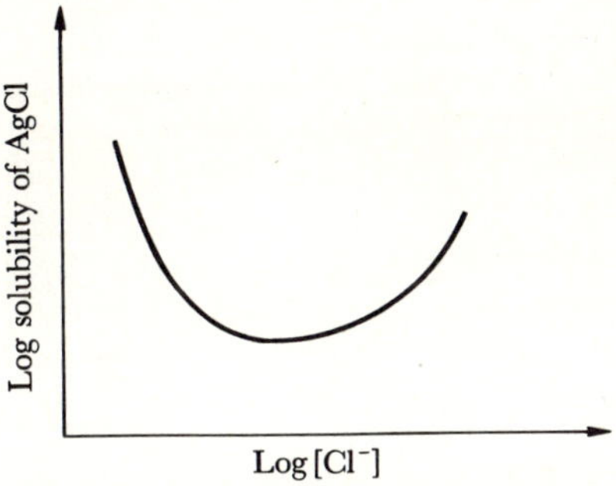

Figure 11–1

1. AgCl is ionized in solution.
2. Cl^- is the anion of a strong acid.
3. Ag^+ forms a complex ion with OH^-
4. AgCl forms a complex ion with Cl^-
5. None of the above.

PROBLEM 11–2

When NH_3 is dissolved in a saturated solution of AgCl in contact with solid AgCl, the solid dissolves. The correct explanation of this phenomenon is that ___________.

Data (25°C)

$K_{sp}(AgCl) = 1.7 \times 10^{-10}$
$K_{sp}(AgOH) = 2 \times 10^{-8}$

1. NH_4OH is a weak base.
2. AgOH is an insoluble base.
3. Ag^+ forms a complex ion with Cl^-.
4. Ag^+ forms a complex ion with NH_3.
5. None of the above.

PROBLEM 11–3

If the absorbance of the tinted glass in a pair of sunglasses is 0.30, what percentage of the light from the sun is stopped by them?

1 30%
2 50%
3 70%
4 100%
5 200%

PROBLEM 11–4

The molar absorptivity of phytochrome (a light-sensitive pigment thought to control the process of flowering in plants) is 76,000. What will be the absorbance of a 5.0×10^{-6} molar aqueous solution of phytochrome if the pathlength of the cell is 1.0 cm?

1 6.6×10^{-11}
2 0.38
3 0.50
4 0.76
5 1.3×10^{10}

PROBLEM 11–5

Color as a property of transition metal complexes is discussed in Section 11–1 of Chemical Principles *by Dickerson, Gray, and Haight.*

The absorption spectrum of a compound is a plot of the amount of light absorbed at each wavelength (or frequency) by the compound versus the wavelength (or frequency) of the light.

The visible absorption spectrum of a biologically important coordination compound is shown in Figure 11–2. The approximate wavelength ranges for the principal colors are shown in Figure 11–3. What is the color of the coordination compound most likely to be?

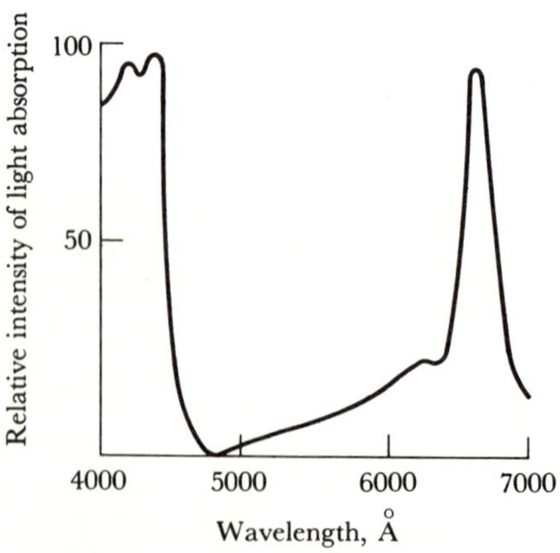

Figure 11–2

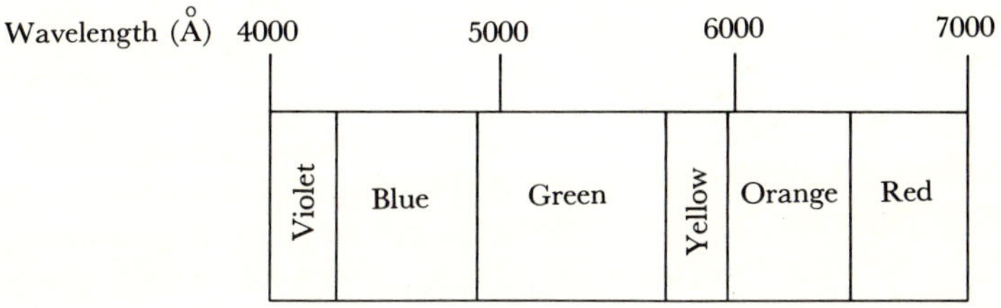

Figure 11–3

1 Violet.
2 Green.
3 Yellow.
4 Orange.
5 Red.

PROBLEM 11–6

Schweitzer's reagent is a deep blue solution containing the complex cation $Cu(NH_3)_4^{2+}$, formed by adding solid cupric hydroxide, $Cu(OH)_2$, to a concentrated aqueous ammonia solution. The reagent dissolves *cellulose,* the chief component of wood and plant fibers. When such a cellulose solution is acidified, the cellulose precipitates. This property has been used in one method for the production of *rayon.* A solution of cellulose in Schweitzer's reagent is squirted through very fine jets into an aqueous solution of sulfuric acid. The cellulose precipitates as silklike threads known as rayon.

What is the *coordination number* of copper in Schweitzer's reagent?

1 1
2 2
3 4
4 5
5 17

PROBLEM 11–7

One of the first octahedral coordination compounds to be investigated by Alfred Werner [one of the founders of modern coordination chemistry and Nobel Laureate (1913)] in the early 1890's was $[Pt(NH_3)_5Cl]Cl_3$. What is the *oxidation state* (or "primary valency" as Werner originally termed it) of platinum in this compound?

Oxidation state and coordination number of transition metal complexes are discussed in Section 11–1 of Chemical Principles.

1 +1
2 +3
3 +4
4 +5
5 +6

PROBLEM 11-8 The biologically important oxygen-carrying molecule in blood, hemoglobin, consists of four sections, each of which contains a heme group [Figure 11-4(a)] bound to a high molecular weight protein molecule (globin) through an imidazole group in the manner shown in Figure 11-4(b). The vacant octahedral coordination position (i.e., empty sp^3d^2 hybrid orbital) on each of the Fe^{2+} ions in the heme groups is thus available for bonding to molecular oxygen, O_2, and other ligands. Which one of the following molecules do you think would *not* bond to the Fe^{2+} ions in hemoglobin?

(H, $Z=1$; B, $Z=5$; C, $Z=6$; N, $Z=7$; O, $Z=8$; F, $Z=9$; P, $Z=15$)

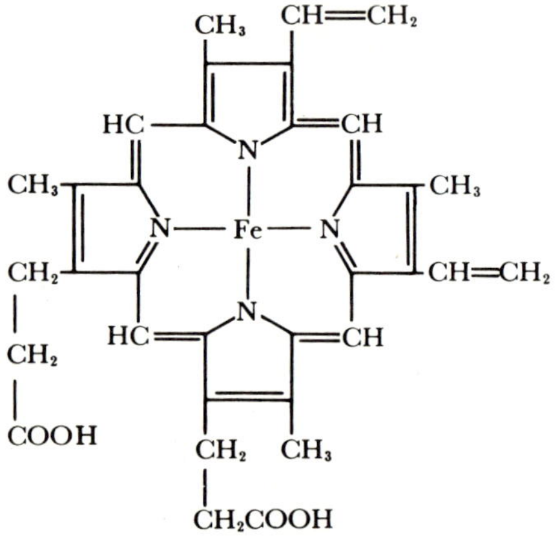

(a) *Heme group*

Figure 11-4

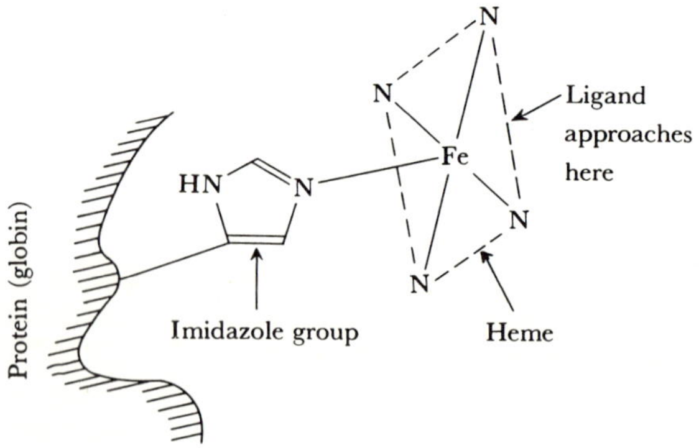

(b)

1 NH_3
2 CO
3 CN^-
4 PF_3
5 BF_3

PROBLEM 11–9

Two of the products emitted during the combustion of coal, petroleum, and diesel fuel are finely divided nickel metal and carbon monoxide. These products readily combine to form the extremely volatile toxic air pollutant nickel carbonyl (bp 43°C). To give you some idea of the magnitude of this environmental hazard, it has been estimated that the annual emission of nickel from the combustion of fossil fuels is equivalent to about 15% (70,000 short tons) of the current commercial production of the metal! What formula would you consider most likely for nickel carbonyl?

[*Hint:* With very few exceptions, the transition metal carbonyls obey the so-called "18-electron rule"; that is, the appropriate *s*, *p*, and *d* valence orbitals are completely filled with electrons (9 orbitals in all; therefore, 18 electrons).] (Ni, $Z = 28$)

1 Ni(CO)
2 $Ni(CO)_2$
3 $Ni(CO)_3$
4 $Ni(CO)_4$
5 $Ni(CO)_6$

PROBLEM 11–10

Iridium forms a complex, $Ir(CO)(\phi_3P)_2Cl$, in which ϕ_3P represents the triphenylphosphine ligand, $(C_6H_5)_3P$, which coordinates to iridium through the lone pair of electrons on the phosphorus atom. [$Ir(CO)(\phi_3P)_2Cl$ is an oxygen carrier and as such is a *model system* for hemoglobin.] What name would you give to the iridium complex? [*Note:* CO is named before ϕ_3P.]

1 Chlorocarbonylbis(triphenylphosphine)iridium(I).
2 Carbonylchlorobis(triphenylphosphine)iridium(I).
3 Bis(triphenylphosphine)chlorocarbonyliridium(I).
4 Chlorocarbonylditriphenylphosphineiridium(I).
5 Chlorocarbonylbis(triphenylphosphine)iridium(III).

PROBLEM 11–11

The square planar structure of the oxygen carrier described in Problem 11–10 is shown in Figure 11–5. What name would you now give to the complex to indicate its stereochemistry as well as its composition?

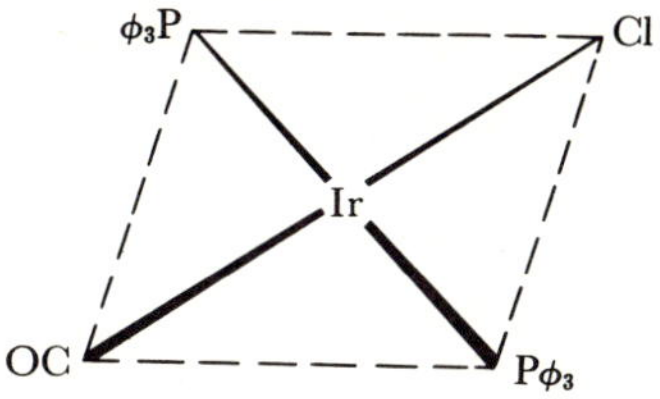

Figure 11–5

1 *trans*-Chlorocarbonyl-*trans*-bis(triphenylphosphine)iridium(I).
2 *cis*-Chlorocarbonyl-*trans*-bis(triphenylphosphine)iridium(I).
3 *cis*-Chlorocarbonylbis(triphenylphosphine)iridium(I).
4 *trans*-Chlorocarbonylbis(triphenylphosphine)iridium(I).
5 Chlorocarbonyl-*cis*-bis(triphenylphosphine)iridium(I).

PROBLEM 11–12

When coal gas is produced it contains hydrogen cyanide (HCN), a highly poisonous gas that must be removed before the coal gas is sold to the public as a fuel. The HCN is removed by passing coal gas through "purifiers" containing ferric oxide (Fe_2O_3). The "spent oxide" is converted into a yellow crystalline solid that is sold as potassium ferrocyanide. Its scientific name is potassium hexacyanoferrate(II). What is the formula of this compound?

1 $K_2[Fe(CN)_4]$
2 $K_3[Fe(CN)_6]$
3 $K_3[Fe(CN)_6]_2$
4 $K_4[Fe(CN)_6]$
5 $K_6[Fe(CN)_8]$

PROBLEM 11–13

Geometrical isomers are described in Section 11–2 of Chemical Principles.

Alfred Werner found that the coordination compound $[Co(NH_3)_4Cl_2]Cl$ exists in two different forms, one green and the other violet. He deduced that the two forms are the result of different octahedral arrangements of the six ligands (four NH_3 and two Cl^-) around the central Co^{3+} ion in the cationic complex, $Co(NH_3)_4Cl_2^+$ (Figure 11–6). The two forms are referred to as *geometrical isomers.* How many different geometrical isomers would you expect for the octahedral manganese carbonyl compound $Mn(CO)_3(\phi_3P)_2Br$?

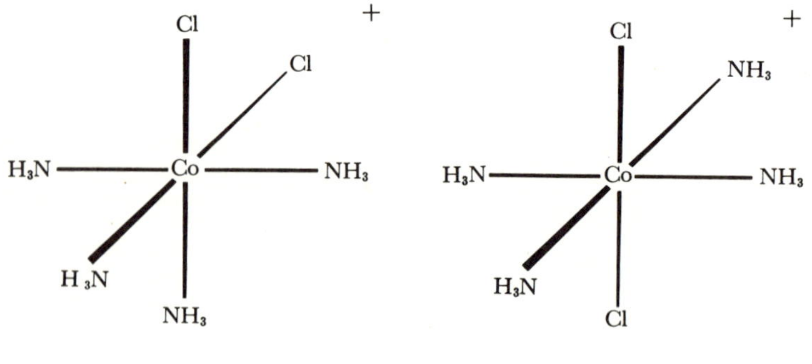

cis-$Co(NH_3)_4Cl_2^+$ *trans*-$Co(NH_3)_4Cl_2^+$

Figure 11–6

1 1
2 2
3 3
4 4
5 5

PROBLEM 11-14

The internationally accepted numbering convention for the ligands in an octahedral complex of an element, M, is shown in Figure 11-7. On the basis of this, what name would you give to isomer (c) of $Mn(CO)_3(\phi_3P)_2Br$ [Figure 11-11(c)] discussed in Solution 11-13? [*Hint:* Refer to Problem 11-10.]

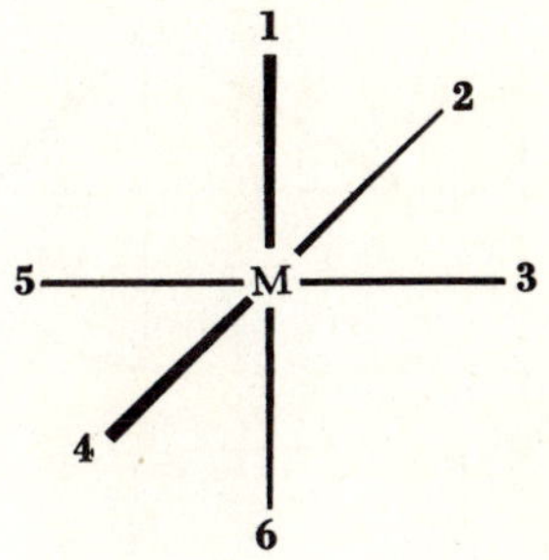

Figure 11-7

1 1-Bromo-2,3-bis(triphenylphosphine)tricarbonylmanganese(I).
2 1-Bromo-2,3,4-tricarbonylbis(triphenylphosphine)manganese(I).
3 1,4-Bis(triphenylphosphine)-2,3,5-tricarbonyl-6-bromomanganese(I).
4 1-Bromo-2,3,4-tris(carbonyl)ditriphenylphosphinemanganese(I).
5 1-Bromo-2,3,4-tricarbonylbis(triphenylphosphine)manganese(III).

PROBLEM 11-15

Optical and stereo-isomerism is discussed in Section 11-2 of Chemical Principles.

Some coordination compounds exhibit a form of isomerism known as *optical isomerism.* This isomerism occurs when the mirror image of a given structure is *not* superimposable on the original structure. Such isomers sometimes can be separated by chemical or physical means. Their solutions rotate a plane of polarized light such that the rotation produced by a solution of one complex is equal in magnitude but opposite in sign to the rotation produced by a solution of the same concentration of the other complex, provided that the light travels the same distance in both solutions. The two optical isomers of the octahedral cationic complex *cis*-$Co(en)_2(NH_3)Cl^{2+}$ (in which en = enthylenediamine, $H_2N{-}CH_2{-}CH_2{-}NH_2$), first resolved (separated) by Werner, are shown in Figure 11-8.

Which one of the following statements concerning the number of geometrical isomers of the *anionic* complex $Co(en)Cl_3Br^-$ and their optical activity is correct?

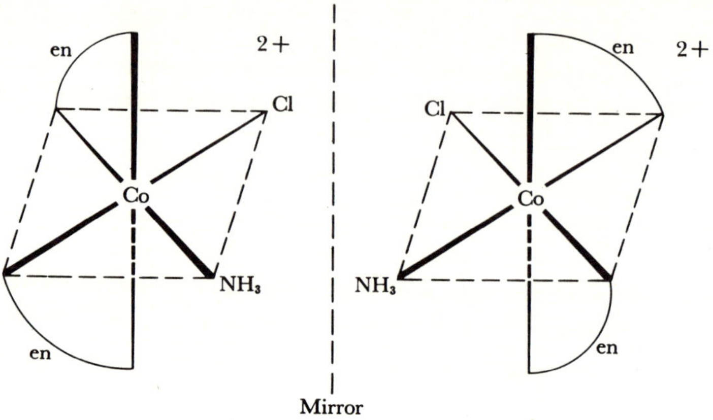

Figure 11–8

1. One geometrical isomer; optically inactive.
2. One geometrical isomer; optically active.
3. Two geometrical isomers; only one optically active.
4. Two geometrical isomers; both optically active.
5. Two geometrical isomers; both optically inactive.

PROBLEM 11–16

A *chelate* is a coordination compound of a metal and another compound (ligand) in which *two* or more of the latter's atoms are coordinated to the metal. For example, in Problem 11–15, ethylenediamine ($H_2N—CH_2—CH_2—NH_2$) is acting as a bidentate (literally, "two-toothed") ligand, chelating through the lone pairs on its two N atoms. Which one of the following molecules would you *not* expect to act as a chelating ligand?

(H, $Z = 1$; C, $Z = 6$; N, $Z = 7$;
O, $Z = 8$; P, $Z = 15$; As, $Z = 33$)

1

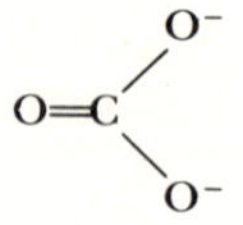

2 $(CH_3)_2P—CH_2—CH_2—P(CH_3)_2$

3

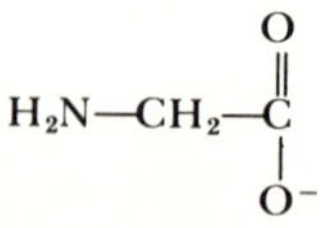

4 ϕ_3As [in which $\phi = (C_6H_5)$]

5

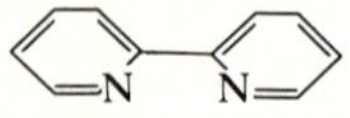

PROBLEM 11–17 When a chelating ligand bonds to a metal, a heterocyclic ring (i.e., a ring containing more than one type of atom) is formed. For example, when ethylenediamine ($H_2N—CH_2—CH_2—NH_2$) reacts with a metal ion (e.g., M^{2+}), a five-membered ring is formed that contains two C's, two N's, and one M:

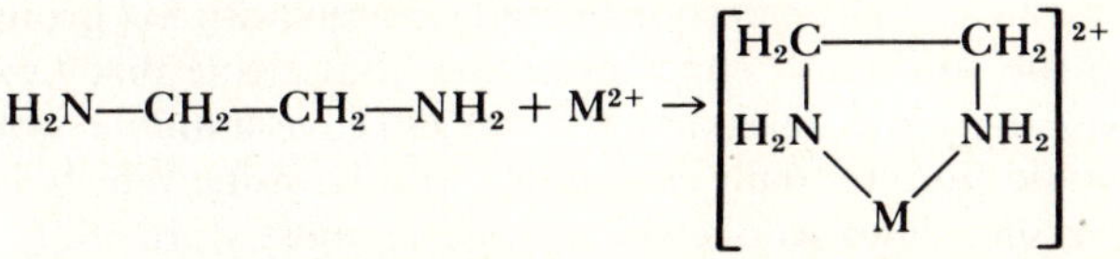

The stability of this chelate far exceeds that of the complex formed from M^{2+} and two monodentate nitrogen ligands such as ammonia (NH_3):

$$2NH_3 + M^{2+} \rightarrow [M(NH_3)_2]^{2+}$$

The stability of a chelate is strongly dependent upon ring size. Five- and six-membered rings are most stable, and stability drops sharply with increased or decreased ring size.

Bearing in mind the points just discussed, determine which one of the following ligands would form the most stable complex with a metal ion M^{2+}.

1 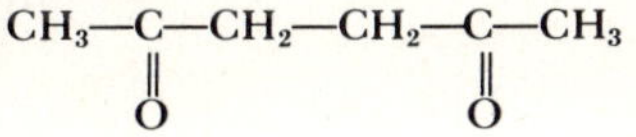

2 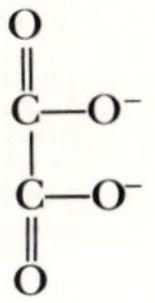

3 $CH_3—O—CH_3$

4 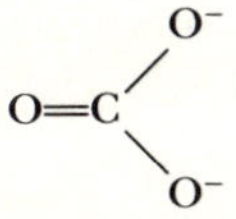

5 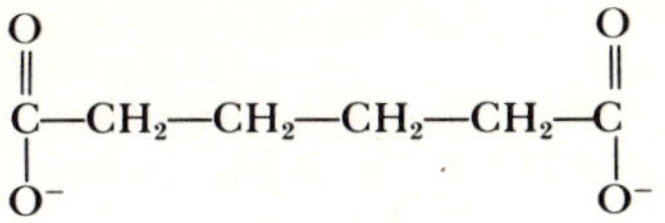

PROBLEM 11–18

At the beginning of World War II there was no guarantee that war gases would not be used, and effective antidotes were immediately sought. For example, one of the gases that might have been used was Lewisite, an arsenic (As^{3+}) containing gas that burns and blisters the skin and damages the bronchial tubes and lungs. It was known that arsenic compounds coordinated strongly with SH groups such as those in the sulfhydryl enzymes of the body tissue that Lewisite attacks. It was therefore reasoned that an effective antidote would be one that would preferentially coordinate with Lewisite. Which one of the following molecules would you choose as most worthwhile for testing as a possible antidote to Lewisite? [*Hint:* Refer to Problem 11–17.]

1 CH_3—SH

2 CH_3—S—S—CH_3

3 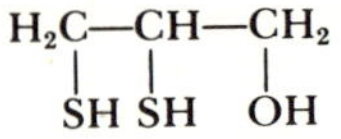

4 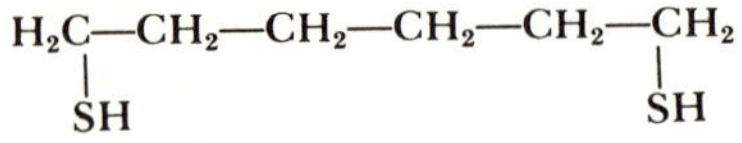

5 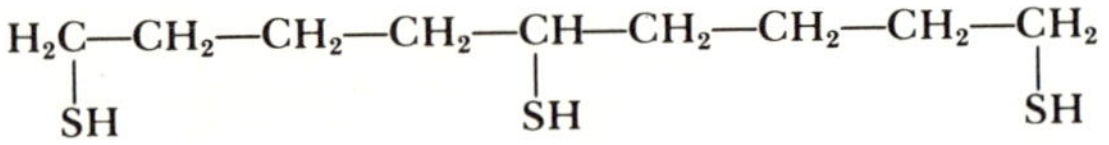

PROBLEM 11–19

Increasing numbers of children in poverty areas are being treated for lead poisoning, caused by eating lead-containing paint that had peeled off the walls of their homes. The treatment for lead poisoning involves an intramuscular injection of a solution of the calcium disodium salt of ethylenediaminetetraacetic acid ($CaNa_2EDTA$), the structure of which is shown in Figure 11–9. The lead located in the soft tissues of the body displaces the calcium from $CaNa_2EDTA$, and the resultant complex is rapidly excreted in the urine.

Which one of the following statements concerning the parent acid EDTA and the use of its salt, $CaNa_2EDTA$, in the treatment of lead poisoning is *incorrect?*

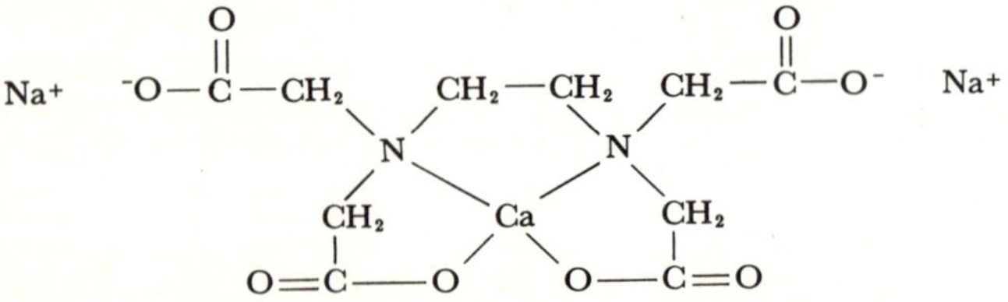

Figure 11–9

1 EDTA forms a more stable complex with lead than with calcium.
2 EDTA acts as a tetradentate ligand in $CaNa_2EDTA$.
3 EDTA could act as a hexadentate ligand.
4 The calcium in $CaNa_2EDTA$ is in the +3 oxidation state.
5 The coordination number of calcium in $CaNa_2EDTA$ is four.

PROBLEM 11–20

The crystal field theory of bonding is discussed in Section 11–3 of Chemical Principles.

In the crystal field theory of bonding in octahedral transition metal complexes the five *d* orbitals of the metal are no longer *degenerate* (of equal energy) because of the electrostatic interaction between the electrons in the *d* orbitals of the metal and the electrons on the six ligands surrounding it. The five *d* orbitals split into two degenerate sets of orbitals (t_{2g} and e_g) that differ in energy by Δ_0, the *crystal field splitting energy*, as shown:

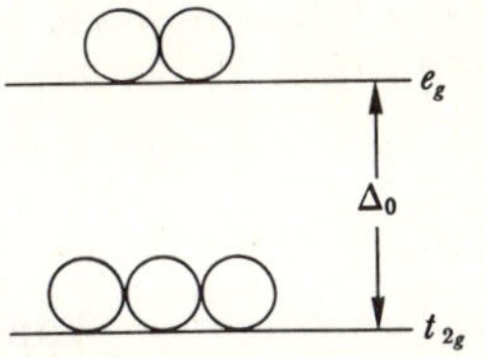

The magnitude of Δ_0 is dependent on the nature of the ligands attached to the metal. If Δ_0 is small (weak crystal field), the *d* electrons of the metal will occupy different orbitals, their spins will be parallel (Hund's principle), and a *high-spin* complex will be formed. If Δ_0 is large (strong crystal field), the electrons will pair in the t_{2g} level before filling the e_g level; a *low-spin* complex will result.

How many unpaired electrons are there in each Fe^{2+} ion of carboxyhemoglobin, assuming that it is a low-spin octahedral complex? [*Hint:* Refer to Problem 9–9.]

(Fe, $Z = 26$)

1 1
2 3
3 4
4 5
5 None; carboxyhemoglobin is diamagnetic.

PROBLEM 11–21 What is the ground state *d* electronic configuration of the Mn^{2+} ion in the octahedral high-spin complex $Mn(H_2O)_6{}^{2+}$?

(Mn, $Z = 25$)

1 $(t_{2g})^3(e_g)^2$
2 $(t_{2g})^4(e_g)^2$
3 $(t_{2g})^5$
4 $(t_{2g})^6$
5 $(t_{2g})^6(e_g)^2$

PROBLEM 11–22 Crystal field theory is extremely useful in explaining the colors of transition metal complexes. Visible light is absorbed by the complexes during the excitation of *d* electrons to higher quantum states. It is customary nowadays to give the energies of these electronic transitions in wave numbers (cm^{-1}), $\bar{\nu}$. Using the following data, predict the color of the octahedral $Ti(H_2O)_6{}^{3+}$ ion if the transition energy is 20,300 cm^{-1}. ($\nu = c\bar{\nu}$)

Data

Color of complex	Main spectral color absorbed	Approximate wavelength of main spectral color absorbed (Å)
Green	Red	6800
Blue	Orange	6100
Violet	Yellow	5600
Red	Blue-green	5000
Yellow	Indigo	4300

1 Yellow
2 Violet
3 Blue
4 Red
5 Green

SOLUTIONS

SOLUTION 11–1

(4)

Answers **1** through **4** are correct statements of fact, but most have nothing to contribute to our understanding of the graph.

At first, the solubility decreases with increasing initial concentration of $[Cl^-]$, as we would expect from the common ion effect:

$$AgCl(s) \rightleftharpoons Ag^+(aq) + Cl^-(aq)$$

But then the solubility starts to increase as the $[Cl^-]$ becomes larger. A second equilibrium with a soluble complex ion

$$AgCl(s) + Cl^-(aq) \rightleftharpoons AgCl_2^-(aq)$$

is a plausible explanation.

Answer **4** is correct.

SOLUTION 11–2

(4)

Answers **1** through **4** are all correct statements of fact, but most of them are irrelevant to the question. Only Answer **4** can supply a reason for the apparent solution of AgCl. The reaction to produce a complex ion of Ag^+ with NH_3 is

$$AgCl(s) + 2NH_3(aq) \rightleftharpoons Ag(NH_3)_2^+(aq) + Cl^-(aq)$$

which actually proceeds in solution by two steps:

$$AgCl(s) \rightarrow Ag^+(aq) + Cl^-(aq)$$
$$Ag^+(aq) + 2NH_3(aq) \rightarrow Ag(NH_3)_2^+(aq) + Cl^-(aq)$$

The second reaction forces the first to the right as the $[Ag^+]$ is lowered until all the solid is dissolved.

Answer **4** is correct. Answer **3** is a correct statement; however, because we are not increasing $[Cl^-]$ by adding NH_3, the answer does not explain why the solid dissolved. Answer **2**: The solubility product for AgOH is 2×10^{-8}, whereas for AgCl it is 1.7×10^{-10}; hence, although the base (actually Ag_2O) is insoluble, AgCl is more insoluble.

SOLUTION 11–3

(2)

$$\text{absorbance} = \log\left(\frac{I_0}{I}\right) = 0.30$$

$$\text{So } \left(\frac{I_0}{I}\right) = \text{antilog } (0.30) = 2.0$$

(Refer to Appendix 2 if this step is difficult.)

The intensity of the light, I, striking the retina of the eye, is therefore $\frac{1}{2}I_0$. The sunglasses effectively stop 50% of the incident light.

2 is correct.
1 is the absorbance.
4 You would not see anything at all!
5 You inverted I_0/I and absorbed more light than was present!

SOLUTION 11-4

(2)

$$\text{absorbance} = \log\left(\frac{I_0}{I}\right) = \epsilon c$$

(for a 1.0-cm path-length cell). So the absorbance of the phytochrome solution is $76{,}000 \times 5.0 \times 10^{-6} = 0.38$.

Answer **2** is correct.

[*Note:* The antilogarithm of 0.38 is 2.5. This means that $I = I_0/2.5$; that is, only 40% of the incident light emerges from the cell, the other 60% is absorbed.]

1 You divided the concentration by the molar absorptivity.
3 is the concentration with some powers of ten missing.
4 is the molar absorptivity with some powers of ten missing.
5 You divided the molar absorptivity by the concentration.

SOLUTION 11-5

(2)

When a coordination compound absorbs visible light, the observed color of the compound is the *complementary* color to that absorbed; it is the part of the visible spectrum remaining after the rest has been absorbed. If a compound absorbs all visible light it will be black; if it absorbs no visible light it will be colorless or white. From the visible absorption spectrum of the compound (Figure 11–2), it is apparent that the compound absorbs all visible light *except* in the region around 5000 Å. Reference to Figure 11–3 shows that this region corresponds to green light; therefore, the compound should appear green.

The correct answer is **2**.

[*Note:* The absorption spectrum in Figure 11–2 is that of chlorophyll *a*, the green pigment in plant leaves that is intimately involved in the process of photosynthesis. Chlorophyll *a* is a coordination compound of magnesium and a molecule (porphyrin) in which there are four nitrogen atoms coordinated to the magnesium atom.]

SOLUTION 11-6

(3)

The participation of d orbitals in the bonding of both nonmetal compounds and transition metal complexes is treated in Review 11 of Programed Reviews of Chemical Principles *by Lassila* et al.

The coordination number of the central atom in a complex is the total number of groups attached to it. Thus, in $Cu(NH_3)_4^{2+}$, there are four neutral NH_3's attached to the central Cu^{2+} ion, and the coordination number of copper is 4.

The correct answer is **3**.

[*Note:* Some examples of metal complexes in which the metal has a coordination number (C.N.) other than 4 are $Ag(NH_3)_2^+$, C.N. = 2; $Au[(C_6H_5)_3P]_2I$, C.N. = 3; $Fe(CO)_5$, C.N. = 5; $Co(NH_3)_6^{3+}$, C.N. = 6; ZrF_7^{3-}, C.N. = 7; $Mo(CN)_8^{4-}$, C.N. = 8; ReH_9^{2-}, C.N. = 9.]

Answer **1** is the number of Cu atoms per stoichiometric unit. Answer **2** is the net charge on the complex ion. Answer **5** is the total number of atoms per stoichiometric unit.

SOLUTION 11-7

(3)

The formula of the compound is $[Pt(NH_3)_5Cl]Cl_3$. The oxidation number of Cl is −1. Therefore, the net charge on the complex cation must be +3 to balance the −3 from the three Cl's; that is, the complex cation is $Pt(NH_3)_5Cl^{3+}$.

Since NH_3 is a neutral molecule and therefore makes no contribution to the charge on the complex cation, the oxidation number (or oxidation state), x, of Pt is

$$x + (-1) = +3$$

so that

$$x = +4$$

Thus, the oxidation state of Pt in the coordination compound is +4.

Answer **3** is correct. Answer **5** is the coordination number of Pt in the compound.

SOLUTION 11-8

(5)

The criterion for a molecule to bond to the Fe^{2+} ions in hemoglobin is that the molecule must have a lone pair of electrons available for donation into the one unoccupied sp^3d^2 hybrid orbital on the Fe^{2+} ion. The Lewis structures of the five molecules are shown below.

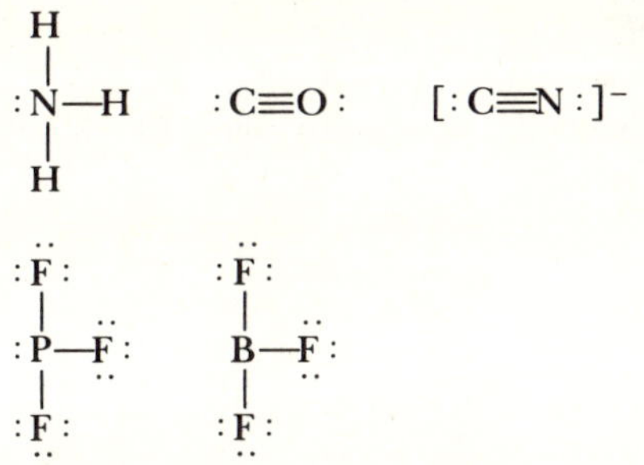

NH_3, CO, CN^-, and PF_3 are all electron pair donors (Lewis bases), but BF_3 is an electron pair acceptor (Lewis acid). The boron atom is electron deficient and needs two electrons to complete its octet; thus, BF_3 would not coordinate to the Fe^{2+} ions in hemoglobin.

The correct answer is **5**.

If you could not draw these Lewis structures, review Chapter 10.

[*Note:* CO, CN^-, PF_3, and, to a lesser extent, NH_3 are toxic as a result of combining with the Fe^{2+} ions in hemoglobin, thereby inhibiting oxygenation. When hemoglobin reacts with CO, carboxyhemoglobin is formed, which is even more intensely red than hemoglobin itself. As a result of the complexing of CN^- with the Fe^{2+} ions in hemoglobin, the blood turns dark blue—a condition known as *cyanosis*.]

SOLUTION 11-9

(4)

From the *Aufbau* process, the ground-state electronic configuration of nickel is

Ni: $[Ar]4s^2 3d^8$

Since carbon monoxide (CO) is a neutral molecule, it follows that nickel carbonyl itself must be uncharged. There are already ten electrons in the outer $4s$ and $3d$ orbitals of the nickel atom; therefore, only eight more are needed to satisfy the "18-electron rule." Since CO is a Lewis base and thus contributes two electrons, the formula of nickel carbonyl must be $Ni(CO)_4$. Answer **4** is correct.

Nickel tetracarbonyl was the first transition-metal carbonyl complex prepared (in South Wales, about 1890). It was discovered accidentally by a chemist who was doing some weighing in a supposedly "inert" atmosphere of CO—his nickel balance pans began to disappear!

On heating to about 150°C, $Ni(CO)_4$ decomposes into nickel metal (≃99.8% pure) and CO:

$$Ni(CO)_4 \rightleftharpoons Ni + 4CO$$

The facile reversibility of this reaction forms the basis of the Mond nickel process for the refinement of nickel.

SOLUTION 11-10

(1)

Nomenclature rules for coordination compounds are given in Section 11-2 of Chemical Principles.

Answer **1** is correct.

In the complex, the names for Cl, CO, and ϕ_3P precede that for Ir, and since the Cl is considered to be an anion, Cl^-, it is named before the neutral ligands. The ligand names are

Cl^- chloro
CO carbonyl
ϕ_3P triphenylphosphine

There are two ϕ_3P's; since the name already has a Greek prefix (tri), we write "bis(triphenylphosphine)" rather than "ditriphenylphosphine." The complex is neutral, and to balance the charge on the Cl^-, iridium must be in the +1 oxidation state. Thus, the name of the complex will end in "iridium(I)." We have only to join the names of the various parts of the complex together, without any hyphens, to obtain the complete name of the complex, which is

chlorocarbonylbis(triphenylphosphine)iridium(I)

Remember that you were told that CO was named before ϕ_3P.

[*Note:* If ever you have occasion to name a fairly complicated molecule that has more than one anionic or neutral ligand, you will need to make use of the two following rules, in addition to the ones that you already know.

(a) Anions are cited in the order H^-, O^{2-}, OH^-, simple monatomic anions (e.g., halides, which are in the order I^-, Br^-, Cl^-, F^-), polyatomic anions, and, last, organic anions in alphabetical order.

(b) Neutral ligands are named in the order H_2O (aquo), NH_3

(ammine), and then in the order in which the coordinating atom falls in the series B, Si, C, Sb, As, P, N, H, Te, Se, S, At, I, Br, Cl, O, F, and, finally, neutral organic ligands in alphabetical order.]

Answer **5** is the correct name but for the oxidation state of Ir.

SOLUTION 11–11

4

The designation "cis" is used for adjacent positions while "trans" is used for opposite (180° apart) positions. Therefore, in Figure 11–5, the Cl is cis to the two ϕ_3P's and trans to the CO. So for a start we can write the name as

trans-chlorocarbonyl . . .

meaning that the Cl and CO are trans to each other. The two ϕ_3P's are trans to each other, so perhaps the full name of the complex will be

trans-chlorocarbonyl-*trans*-bis(triphenylphosphine)iridium(I)

However, the second trans is redundant because we know that the structure is four-coordinate and square planar. Once we have specified that the Cl and CO are trans, the two ϕ_3P's *must* be trans to satisfy the structural requirements. The complete name is therefore

trans-chlorocarbonylbis(triphenylphosphine)iridium(I)

The correct answer is **4**. Answer **1** is *almost* correct. Answer **3** is also almost correct. The name chlorocarbonyl-*trans*-bis(triphenylphosphine)iridium(I) would also convey the same information as Answer **4**. However, convention decrees that we specify stereochemistry as soon as possible in a name, so Answer **4** is the best name.

[*Note:* Sometimes the positions in the structure are referred to by the numbering scheme in Figure 11–10. By this scheme the name of the complex is

1-chloro-3-carbonylbis(triphenylphosphine)iridium(I)

Again, once the Cl and CO are specified to be in Positions 1 and 3, the ϕ_3P's have to be in Positions 2 and 4; thus, there is no need to say so.]

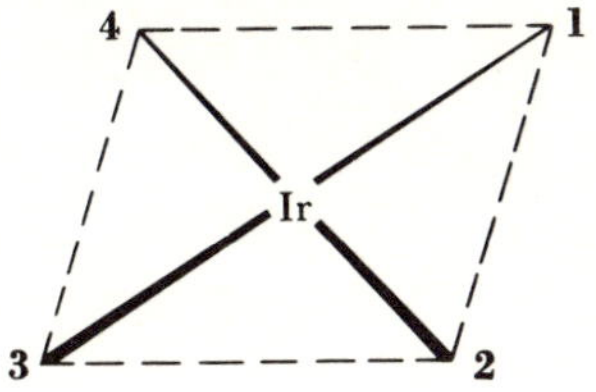

Figure 11–10

SOLUTION 11–12 (4)

The name of the compound is potassium hexacyanoferrate(II). The ending "ferrate(II)" indicates that the complex anion contains iron in the ferrous, Fe(II), or Fe^{2+} state. The name "cyano" indicates the presence of the anionic CN^- ligand in the complex anion. Furthermore, since the prefix "hexa" means six, there are six CN^-'s in the complex anion. Thus, the net charge on the complex anion is given by

$$6(-1) + (+2) = -4$$

The potassium cation is K^+. To balance the −4 charge on the complex anion, we therefore need four K^+'s. The formula of the coordination compound is thus $K_4[Fe(CN)_6]$.

The correct answer is **4**.

[*Note:* Coordination chemistry is usually said to have originated in about 1704 with the discovery by Diesbach, an artists' color maker, of *Prussian Blue,* a compound very closely related to $K_4[Fe(CN)_6]$.]

SOLUTION 11–13 (3)

By trial and error the six ligands (three CO, two ϕ_3P, and one Br^-) can be octahedrally arranged around the Mn^+ ion in only *three* different ways. Thus, there are three different geometrical isomers possible for $Mn(CO)_3(\phi_3P)_2Br$. Their structures are shown in Figure 11–11.

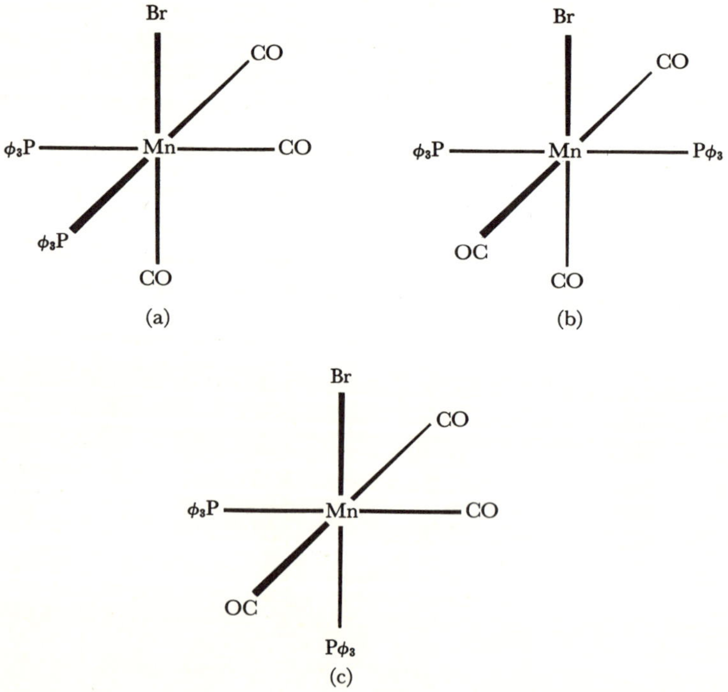

Figure 11–11

In (a), the two ϕ_3P's and the Br are all cis to one another; in (b), the two ϕ_3P's are trans to one another; and in (c), one of the ϕ_3P's is trans to the Br.

[*Note:* Although all three geometrical isomers are possible, only isomers (a) and (b) have been isolated to date.]

The correct answer is **3**.

SOLUTION 11–14 (2)

The numbering system of the ligands in isomer (c) of $Mn(CO)_3(\phi_3P)_2Br$ is shown in Figure 11–12. There are three CO's, two ϕ_3P's, and one Br^- around the central Mn^+ ion. Thus, the proper order of the names of the various parts of the neutral complex is

bromo, tricarbonyl, bis(triphenylphosphine), manganese(I)

The Br is in Position 1; the three CO's are in Positions 2, 3, and 4; and the two ϕ_3P's are in Positions 5 and 6. Once we have specified the ligands in Positions 1, 2, 3, and 4, the two ϕ_3P's must be in Positions 5 and 6. Therefore, there is no need to indicate their positions in the name; the complete name of isomer (c) of $Mn(CO)_3(\phi_3P)_2Br$ is

1-bromo-2,3,4-tricarbonylbis(triphenylphosphine)manganese(I)

The correct answer is **2**.

[*Note:* If position numbers are used to name a complex, the structure of the complex must be oriented so the ligand to be named first, according to nomenclature rules, appears in Position 1, the ligand to be named second is in Position 2, and so on. This was already done for us in this problem.]

Note that the names of the other two isomers of $Mn(CO)_3(\phi_3P)_2Br$, Figure 11–11, are

(a) 1-bromo-2,3,6-tricarbonylbis(triphenylphosphine)manganese(I)
(b) 1-bromo-2,4,6-tricarbonylbis(triphenylphosphine)manganese(I).

Answer **5** is the correct name except for the oxidation state of Mn.

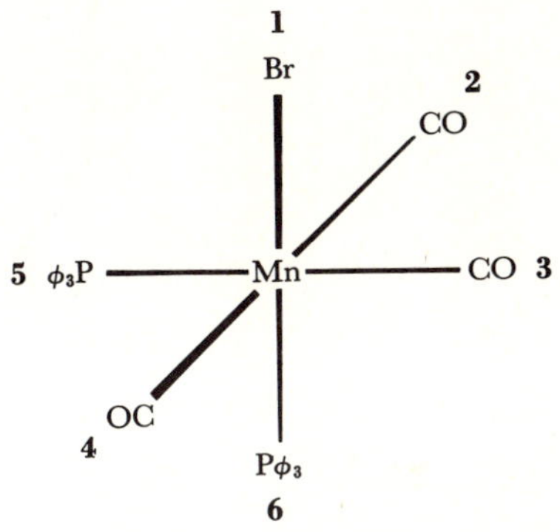

Figure 11–12

SOLUTION 11–15

(5)

There are two possible geometrical isomers of $Co(en)Cl_3Br^-$. The structures of these, (a) and (b), together with their mirror images are shown in Figure 11–13.

Isomer (a) and its mirror image are superimposable; a 90° counterclockwise rotation of the mirror image about the Z axis turns it back into the original structure of isomer (a). What about isomer (b) and its mirror image? These, too, are superimposable, although it is perhaps not so easy to see this at first glance. A 90° counterclockwise rotation of the mirror image about the X axis produces the structure shown in Figure 11–14. A 90° clockwise rotation of this structure about its Z axis produces the original structure of isomer (b). Thus, there are two geometrical isomers, both of which are optically inactive.

The correct answer is **5**.

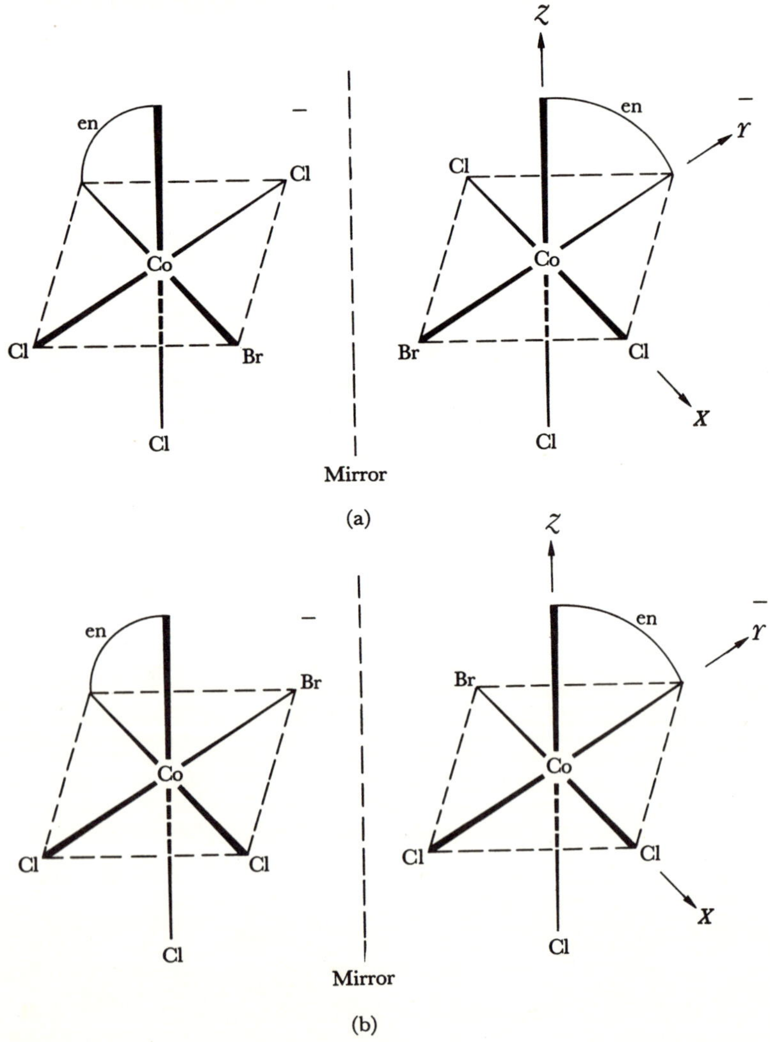

Figure 11–13

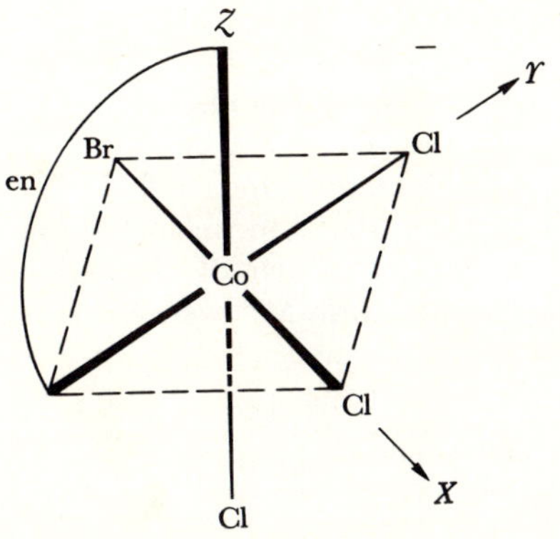

Figure 11–14

SOLUTION 11–16 (4)

The structures of the five molecules can be written as follows:

1 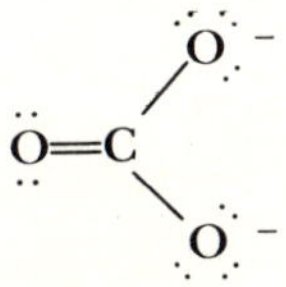

2 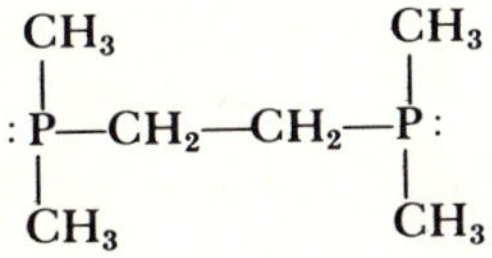

3 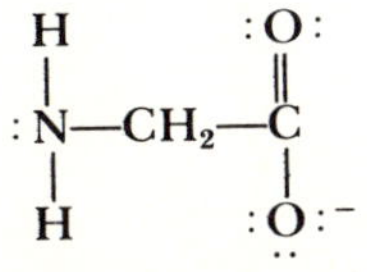

4 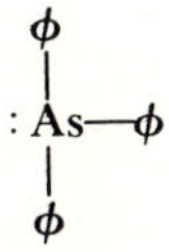

5 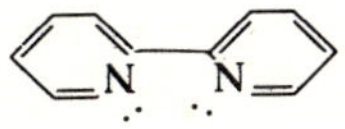

Notice that, in every case except **4**, there are lone pairs of electrons on two or more atoms in the molecules. ϕ_3As has only one lone pair on the As atom and is referred to as a monodentate ligand; thus, it cannot form chelates. The other four molecules are bidentate ligands: **1** coordinates to a metal through the two O^- groups, **2** through the two P atoms, **3** through the N atom and the O^- of the COO^- group, and **5** through the two N atoms.

The correct answer is **4**.

SOLUTION 11–17 (2)

All five ligands will coordinate through oxygen atoms. The complex formed with CH_3—O—CH_3 (Answer **3**) will not be as stable as the complexes formed with the other ligands. CH_3—O—CH_3 is a monodentate ligand, whereas the others are bidentate ligands. The complexes formed with the bidentate ligands will all be chelates and consequently more stable than the nonchelate complex formed with CH_3—O—CH_3. To determine which of the bidentate ligands will form the most stable complex with M^{2+} we must compare the sizes of the chelate rings that will be formed.

1 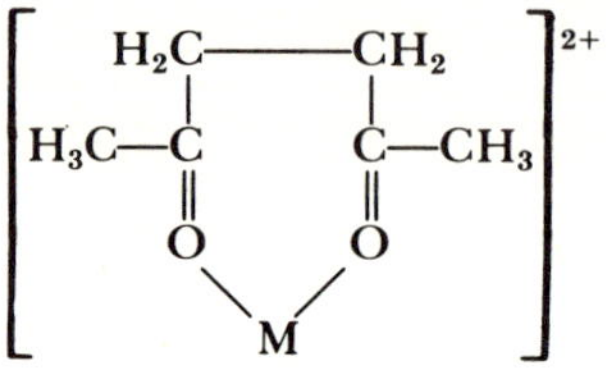

seven-membered chelate ring

2 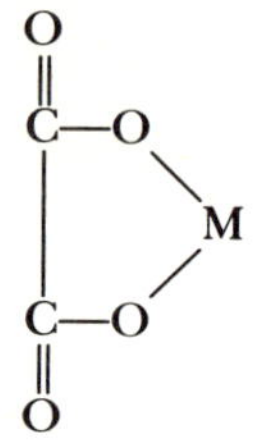

five-membered chelate ring

(Notice that the charge on the M^{2+} ion is neutralized by the charges on the two O^-'s of the $C_2O_4^{2-}$ ion.)

4 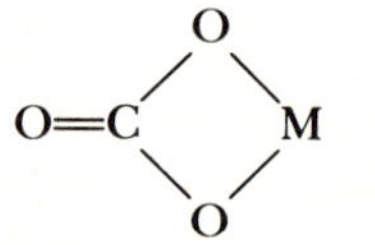

four-membered chelate ring

5 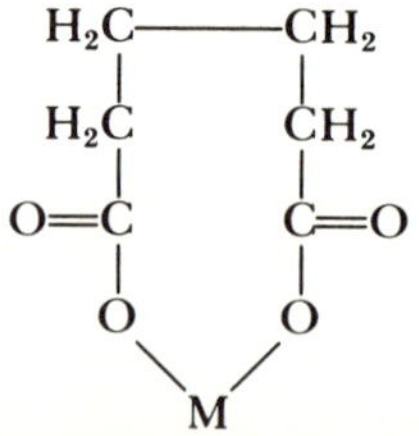

nine-membered chelate ring

The most stable chelate will be the one with the five-membered ring. Thus, Answer **2** is correct.

[*Note:* A polydentate ligand that can form more than one chelate ring when attached to a metal will be more stable than a bidentate ligand that can form only one chelate ring. Thus, a tridentate ligand that can form two five-membered rings will be more stable than a bidentate ligand that can form only one five-membered ring.]

SOLUTION 11–18 (3)

The molecule most worthwhile testing will be the one that contains some SH groups and that forms the most stable complex with arsenic (As^{3+}). We can immediately discard Answer **2** because there are no SH groups. What about the other ligands? Answer **1** is a monodentate ligand, Answer **4** is a bidentate ligand, and Answers **3** and **5** are both tridentate ligands. The polydentate ligands will form chelate complexes that will be more stable than the complex with the monodentate ligand. Thus, Answer **1** can be discarded. To determine which one of the polydentate ligands will form the most stable complex with As^{3+}, we must compare the number and sizes of the chelate rings that will be formed.

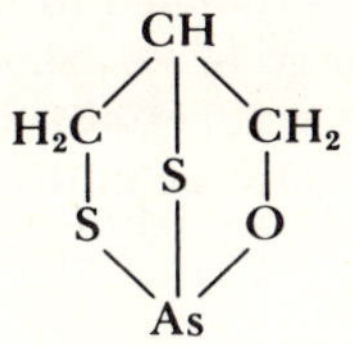

two five-membered chelate rings

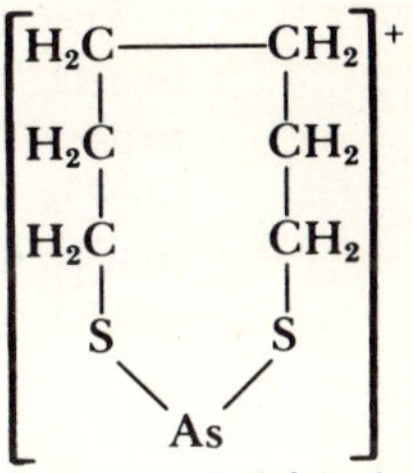

nine-membered chelate ring

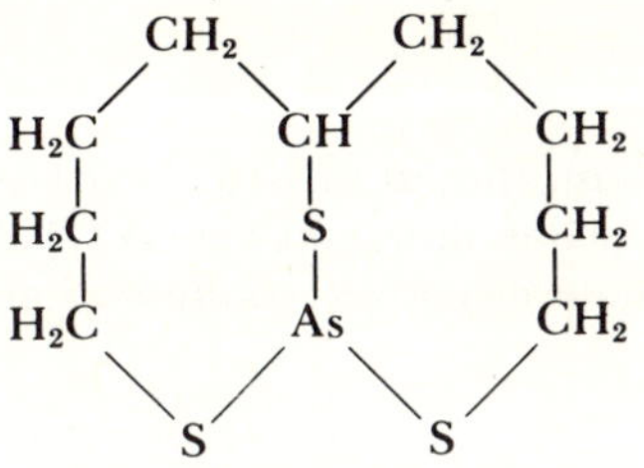

two eight-membered chelate rings

The chelates formed with molecules **3** and **5** will be more stable than the one formed with molecule **4** because they have two rings, whereas the chelate formed with molecule **4** has only one. Furthermore, because **3** leads to two five-membered rings whereas **5** leads to two eight-membered rings, **3** will form the more stable complex.

The correct answer is **3**.

[*Note:* Answer **3** is *dimercaprol* (2,3-dimercaptopropanol) and was the compound developed during World War II as an antidote to arsenical gases. It was known as British Anti-Lewisite (BAL). After the war, dimercaprol was recognized as one of the best antidotes to heavy metal (As, Hg, and Au) poisoning. However, it is useless in the treatment of lead poisoning.]

SOLUTION 11–19

(4)

Since the whole basis for using $CaNa_2EDTA$ in the treatment of lead poisoning is that the lead displaces the calcium from the complex, it must be that EDTA forms a more stable complex with lead than with calcium. So Answer **1** is correct.

In $CaNa_2EDTA$, the EDTA is coordinated to the calcium through two O's and two N's; that is, the EDTA acts as a tetradentate ligand and the calcium is four-coordinated. Therefore, Answers **2** and **5** are correct.

There are two O^-'s in $CaNa_2EDTA$ that are not used to coordinate calcium. These are available for coordination; and so EDTA could (and, in fact, sometimes does) act as a hexadentate ligand. Answer **3** is correct.

The calcium in $CaNa_2EDTA$ is coordinated to two neutral N's and two O^-'s. The EDTA ion has a charge of −4. Since two Na^+ ions are present in $CaNa_2EDTA$, the charge on the CaEDTA part of the complex is −2. Therefore, the calcium must be in the +2 oxidation state, that is, Ca^{2+}. Answer **4** is incorrect.

The required answer is **4**.

SOLUTION 11–20

(5)

The atomic number of Fe is 26, so the ground state electronic configuration is

Fe: $[Ar]4s^23d^6$

From Problem 9–9, we know that in first-row transition metal cations the $3d$ orbitals are more stable (lower in energy) than the $4s$ orbital. Therefore, the ground state electronic configuration of Fe^{2+} in carboxyhemoglobin is

Fe^{2+}: $[Ar]3d^6$

Carboxyhemoglobin is a low-spin octahedral complex (CO is a strong field ligand: Δ_0 large). Thus, the six $3d$ electrons of Fe^{2+} will be allocated in the t_{2g} and e_g orbitals so that there are as few unpaired electrons as possible.

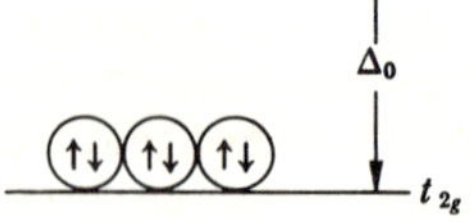

In fact, the t_{2g} level is completely filled and there are no unpaired electrons in carboxyhemoglobin; it is diamagnetic.

Answer **5** is correct.

SOLUTION 11–21 (1)

The atomic number of Mn is 25, so there are 23 electrons in the Mn^{2+} ion. The ground state electronic configuration of Mn^{2+} is

Mn^{2+}: $[Ar]3d^5$

$Mn(H_2O)_6^{2+}$ is a high-spin complex (H_2O is a weak field ligand; Δ_0 small), and each of the five $3d$ electrons will be located in a different $3d$ orbital, with their spins parallel to one another.

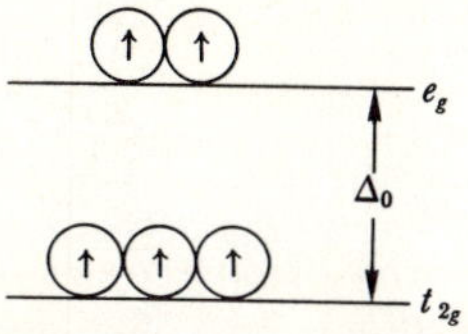

There are three electrons in the t_{2g} level and two in the e_g level. Thus, the ground state d electronic configuration of Mn^{2+} in $Mn(H_2O)_6^{2+}$ is $(t_{2g})^3(e_g)^2$.

Answer **1** is correct.

SOLUTION 11–22 (4)

The energy of the electronic transition is given by

$$E = h\nu = hc\bar{\nu} = \frac{hc}{\lambda}$$

So

$$\bar{\nu} = \frac{1}{\lambda} = 20{,}300 \text{ cm}^{-1}$$

This means that the wavelength of the light absorbed is

$$\frac{1}{20{,}300 \text{ cm}^{-1}} = 4.93 \times 10^{-5} \text{ cm} \times 10^8 \text{ Å cm}^{-1}$$

$$= 4930 \text{ Å}$$

Visible light of 4930 Å is in the blue-green range and the color of the $Ti(H_2O)_6^{3+}$ ion will be the complement, namely, red. Answer **4** is correct.

$Ti(H_2O)_6^{3+}$ is an octahedral complex containing the Ti^{3+} ion with a $3d^1$ ground-state electronic configuration. Thus, in crystal field theory terms, the electronic transition responsible for the color of the complex is

$$(t_{2g})^1(e_g)^0 \rightarrow (t_{2g})^0(e_g)^1$$

and Δ_0 for the transition is 20,300 cm^{-1}. Many other transition-metal complexes are colored because of similar "d–d" electronic transitions.

CHAPTER 12 THE SPECIAL ROLE OF CARBON

The chemistry of carbon compounds, *organic chemistry,* is a unique field, involving practically every aspect of modern life. The complex system of organic chemistry includes a variety of categories of commercial and biological importance such as hormones, insecticides, synthetic fibers, and drugs.

For the problems in this chapter you will need to memorize the following:

(1) The number of carbons in the straight-chain hydrocarbons.

Methane, C_1
Ethane, C_2
Propane, C_3
Butane, C_4
Pentane, C_5
Hexane, C_6

(2) isopropyl $= CH_3—\underset{\textstyle H}{\overset{\textstyle CH_3}{\overset{|}{\underset{|}{C}}}}—$

-ol = hydroxyl = —OH
-ene = —CH=CH—

Cyclohexane = 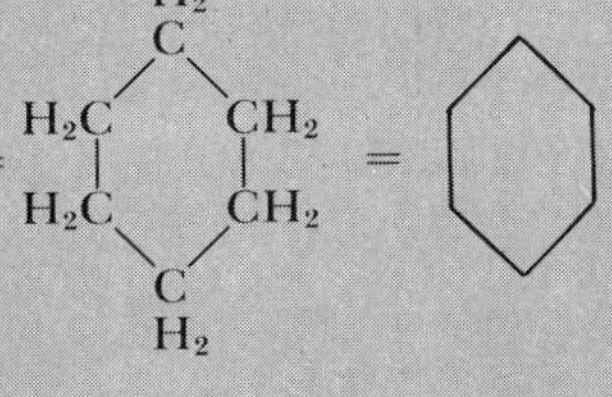

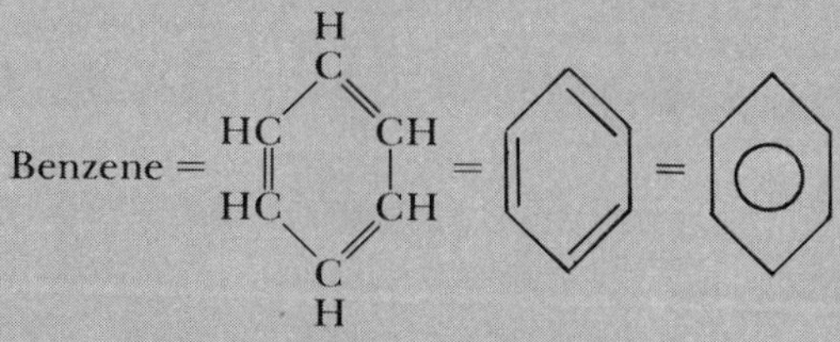

di = 2
tri = 3
tetra = 4

(3) Br_2 plus aromatic or saturated aliphatic compound: generally no reaction.
Br_2 plus unsaturated aliphatic compound: addition reaction across the double bond.

(4) Review the geometries associated with *sp*, *sp*^{2}, and *sp*^{3} hybridization discussed in Chapter 10 of this book.

Solutions to problems in Chapter 12 begin on page 343.

PROBLEMS

PROBLEM 12–1

Which one of the following is the structural formula for 1,2-dichlorotetrafluoroethane?

1
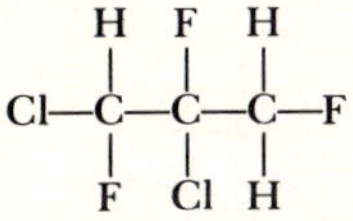

2
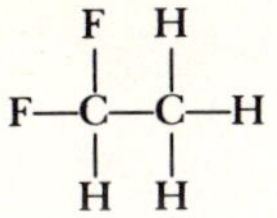

3
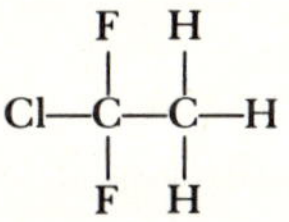

4
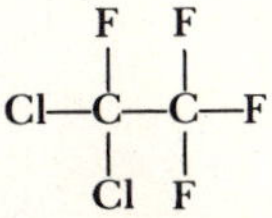

5
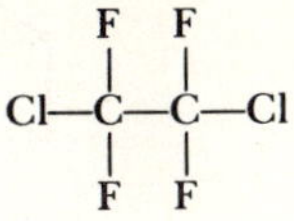

PROBLEM 12–2

The structural formula of 2,2,3-trimethylhexane is ________.

Saturated hydrocarbons are discussed in Section 12–2 of Chemical Principles *by Dickerson, Gray, and Haight.*

1
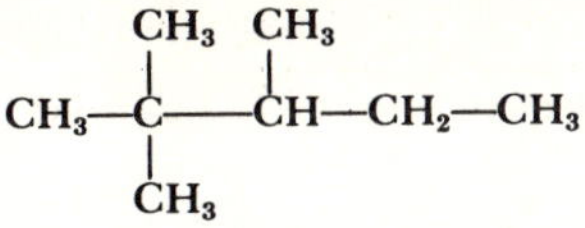

2
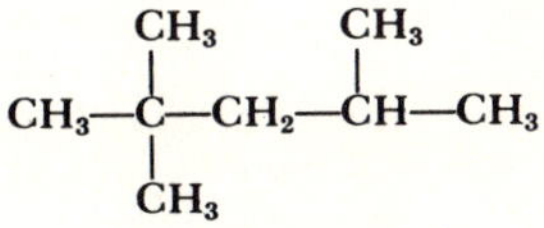

3
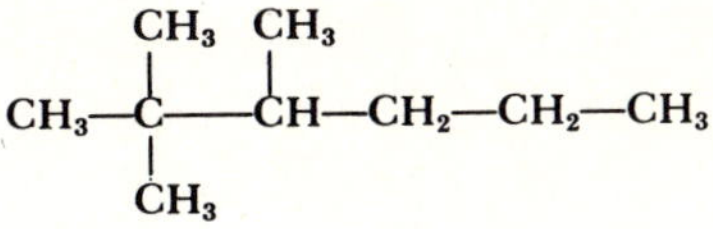

4
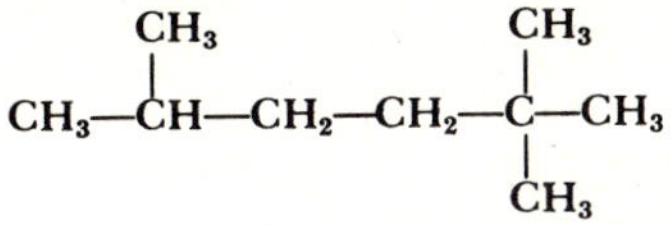

5
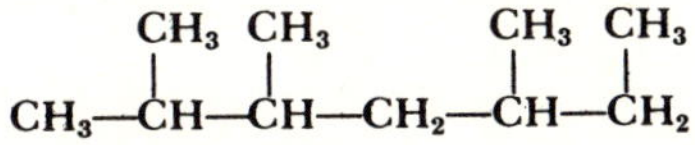

PROBLEM 12–3

The derivatives of hydrocarbons and the properties of functional groups are treated in Section 12–3 of Chemical Principles.

Menthol (2-isopropyl-5-methyl-cyclohexanol) is used in a variety of commercial products such as cough drops, shaving lotion, and mentholated cigarettes. Which one of the following is the structural formula for menthol?

1

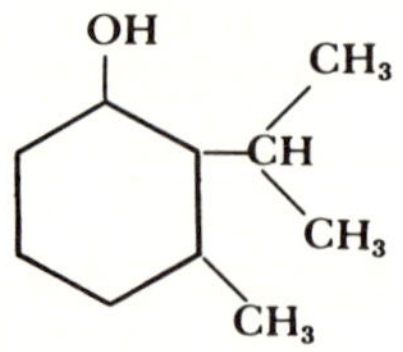

2

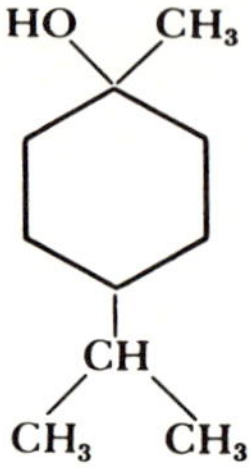

3

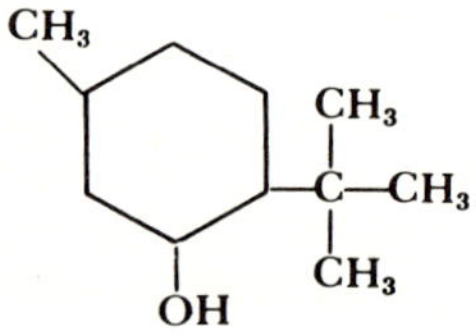

4

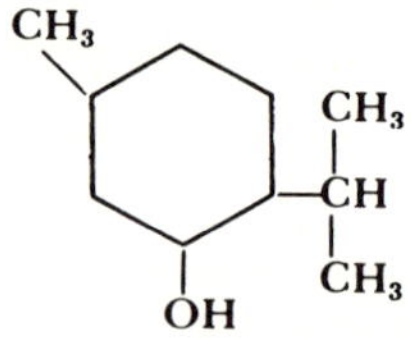

5

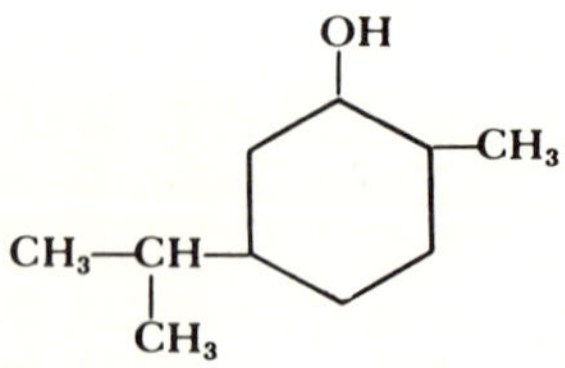

PROBLEM 12–4

The name of the compound below is __________.

$CH_2{=}CH{-}CH{=}CH{-}CH_2CH_3$

1 Hexadiene.
2 1,2,3,4-Hexadiene.
3 3,5-Hexadiene.
4 1,3-Hexadiene.
5 3,4-Hexadiene.

PROBLEM 12–5

In which of the compounds below is there more than one kind of hybridization (sp, sp^2, sp^3) for carbon?

(a) $CH_3—CH_2—CH_2—CH_3$
(b) $CH_3—CH{=}CH—CH_3$
(c) $CH_2{=}CH—CH{=}CH_2$
(d) $H—C{\equiv}C—H$

1 (b) and (c)
2 (b)
3 (c) and (d)
4 (a) and (d)
5 (c)

PROBLEM 12–6

Unsaturated hydrocarbons are described in Section 12–4 of Chemical Principles.

Mycomycin (shown below) is an antibiotic containing a large number of unsaturated linkages (double or triple carbon–carbon bonds).

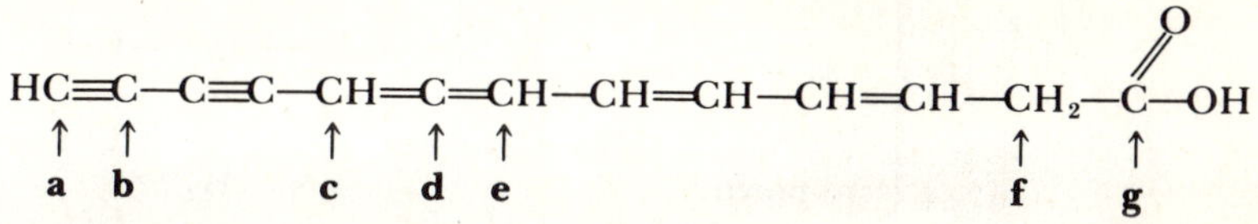

Which of the following correctly describes the bond hybridization at carbon atoms **a–g**?

	a	b	c	d	e	f	g
1	sp^2	sp^2	sp	sp	sp	sp^2	sp
2	sp	sp	sp^2	sp	sp^2	sp^3	sp^2
3	sp^2	sp^2	sp^3	sp^2	sp^3	sp^3	sp^3
4	sp	sp	sp^2	sp^2	sp^2	sp^3	sp^2
5	sp^2	sp^2	sp^2	sp^2	sp^2	sp^2	sp^2

PROBLEM 12–7

On the basis of the information in this chapter, and the known model reactions shown, predict which of the "unknown" reactions would be likely to proceed as written.

(a) Model: $CH_3—CH_2—CH_2—Br + KOH \rightarrow CH_3—CH_2—CH_2—OH + KBr$

Unknown: $CH_3—CH_2—CH_2—Br + KSH \rightarrow CH_3—CH_2—CH_2—SH + KBr$

(b) Model: / Unknown:

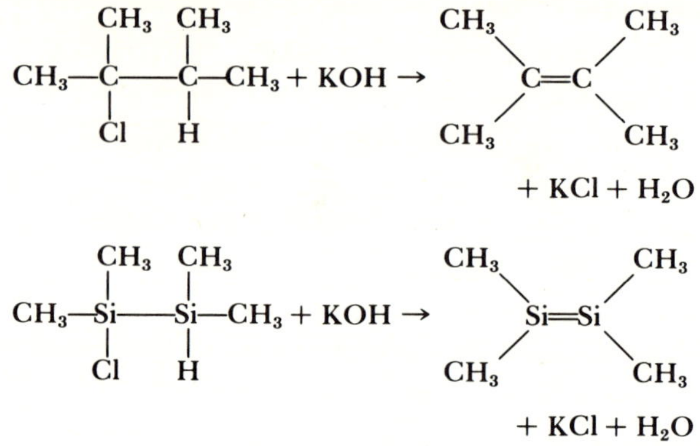

(c) Model: / Unknown:

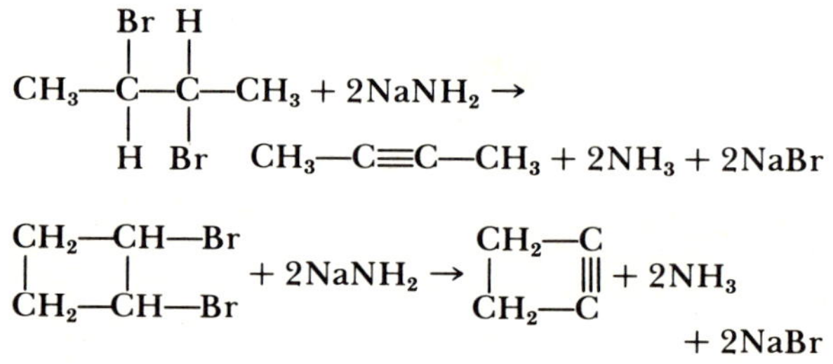

1 (a)
2 (a), (b), (c)
3 (a), (c)
4 (b), (c)
5 (b)

PROBLEM 12–8 How many structural isomers are possible for the formula $C_2H_4Cl_2$?

1 1
2 2
3 3
4 4
5 6

PROBLEM 12–9 How many structural isomers exist for the formula C_6H_{14}?

1 3
2 4
3 5
4 6
5 8

PROBLEM 12–10 The following compound is a tetramer of HCN (hydrocyanic acid, $C_4H_4N_4$) and may have been involved as an intermediate in the formation of nucleic acids from prebiotic chemicals:

$H_2NC(CN){=}C(CN)NH_2$

How many geometrical isomers of this compound are possible? (The rotation about the double bond is restricted.)

1 1
2 2
3 3
4 4
5 6

PROBLEM 12–11 Many butterflies and moths possess extrusible brushlike structures ("hairpencils") that serve an active role during courtship. The compound below was isolated from the hairpencils of 6500 male Monarch butterflies.

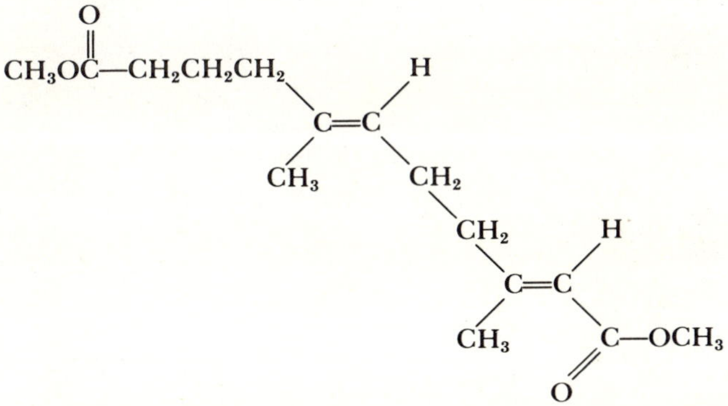

How many geometrical isomers, including the one above, are possible? (Consider rotation about the two carbon–carbon double bonds to be restricted.)

1 1
2 2
3 3
4 4
5 8

PROBLEM 12–12 How many different 1,2-dimethylcyclopropane isomers exist? One such isomer is

Models of optical isomers are shown in Figure 12–12 in Chemical Principles.

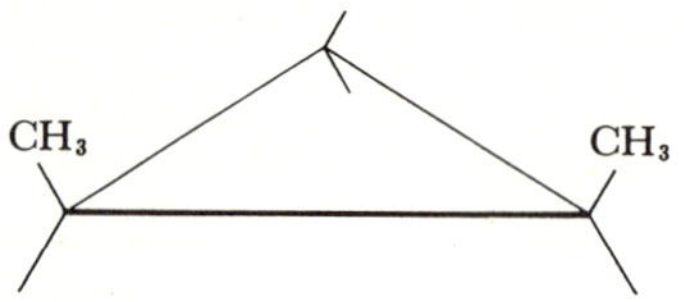

[*Hint:* Consider optical isomerism.]

1 2
2 3
3 4
4 5
5 6

PROBLEM 12–13 Various forms of the compound below are effective medicinal agents.

Note: 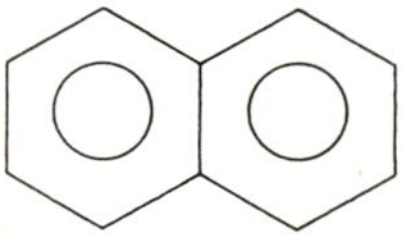 means 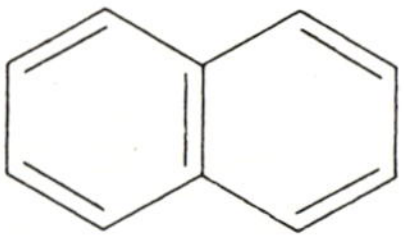

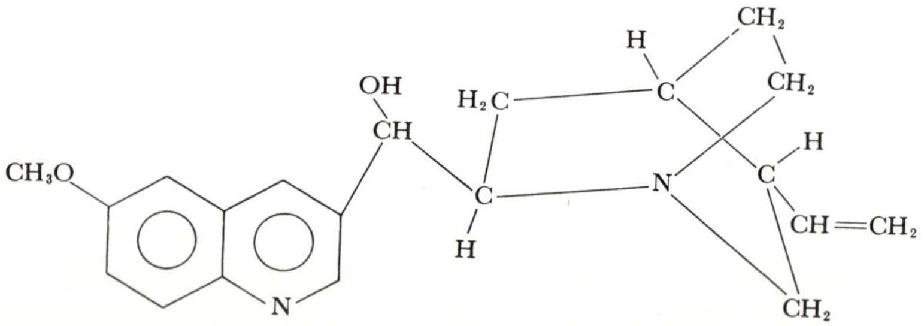

How many stereoisomers are possible for this compound? [*Hint:* Consider only the positions where there are asymmetric atoms (four different kinds of groups attached to the atom). The total number of stereoisomers is 2^n, where n is the number of asymmetric atoms.]

1 2
2 4
3 8
4 16
5 32

PROBLEM 12–14

In the United States, losses attributable to rats amount to over \$800 million annually. Chemists have developed a compound that does not kill the male rats but renders them sterile. Assume that you had developed this compound in the laboratory and found that it possesses two stereoisomers and that 1 mole of it can be converted to 1 mole of a *triol* (a compound with three OH groups) upon treatment with 1 mole of KOH. Which of the following structures is the drug?

1 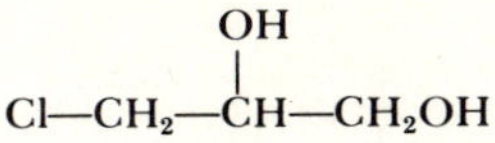

2 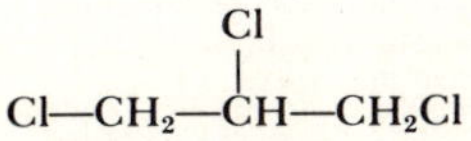

3 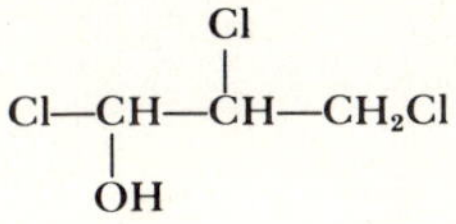

4 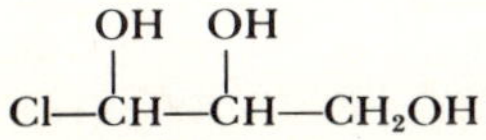

5 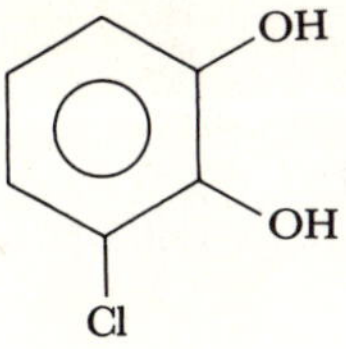

PROBLEM 12–15

The reactions of aromatic compounds are discussed in Section 12–5 of Chemical Principles.

Which of the following structures would be the most reactive against Br_2?

1 

2 $CH_3CH_2CH_2CH_2CH_3$

3 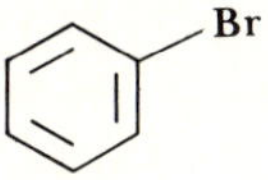

4 

5 $CH_2{=}CH_2$

PROBLEM 12–16 Predict the product ot the following hydrogenation reaction:

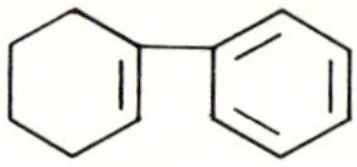 $+ H_2 \xrightarrow[\text{conditions}]{\text{mild}}$

1 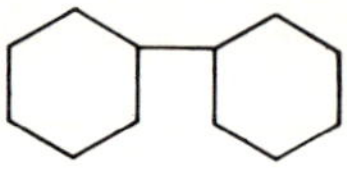

2 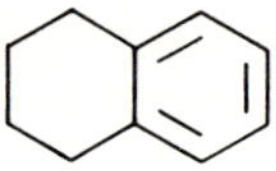

3 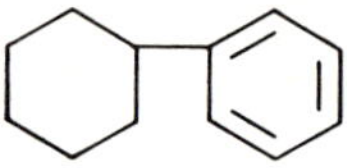

4 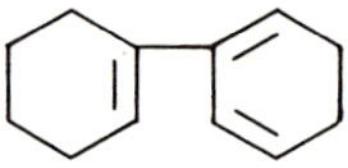

5 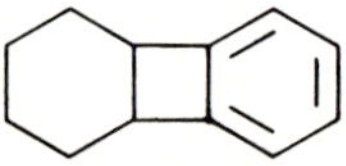

PROBLEM 12–17 Examine the following statements about amino acids:

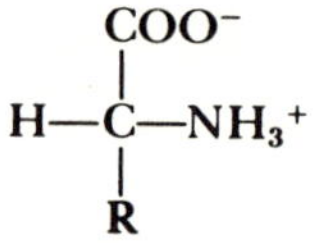

(R = alkyl or aromatic group)

(a) Amino acids are generally water soluble.
(b) Amino acids are generally soluble in organic solvents.
(c) Amino acids are low melting solids.
Which of the above are correct?

1 (a), (c)
2 (a)
3 (b), (c)
4 (b)
5 All of them.

Peptides and polypeptides (a large polypeptide is known as a protein) are molecules containing am.no acids linked together by amide groups:

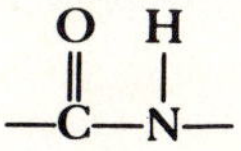

Many polypeptides are found in the body and are physiologically active. On hydrolysis, peptides and polypeptides yield the amino acids comprising them:

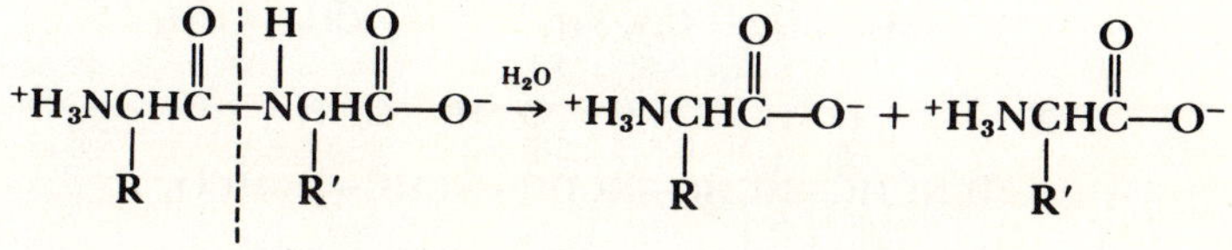

The composition of a polypeptide can be determined by identifying, with an amino acid analyzer, the amino acids formed on hydrolysis. However, no information as to the order of the amino acids in the polypeptide chain is gained from this analysis.

PROBLEM 12-18 For the generalized formula for an amino acid (compound I) the following table differentiates several of the common amino acids.

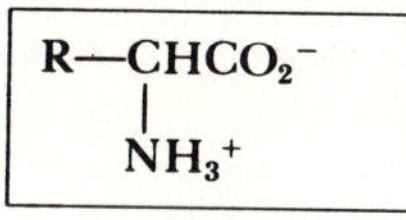

I

R—	Name
CH_3—	Alanine
CH_3 CH— CH_3	Valine
$HSCH_2$—	Cysteine
H—	Glycine

Consider that 1 mole of an unknown polypetide was hydrolyzed so it degraded to the corresponding amino acids. An amino acid analyzer showed that 1 mole of alanine, 2 moles of valine, and 2 moles of cysteine were formed. Which one of the following polypeptides fits the analysis?

1
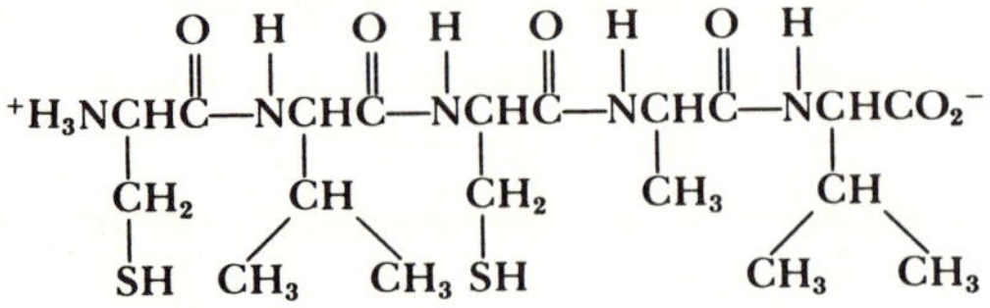

2
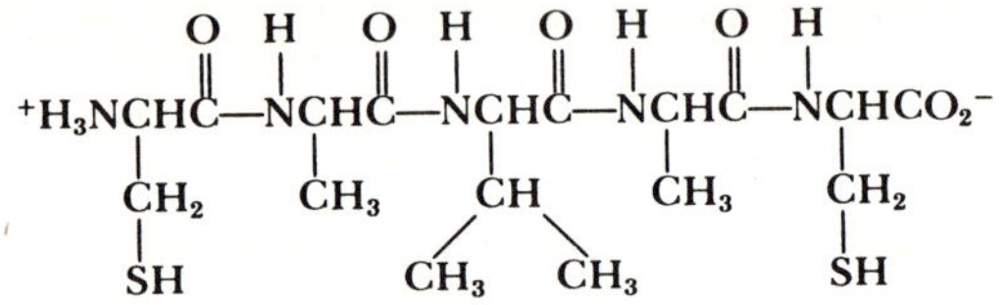

3
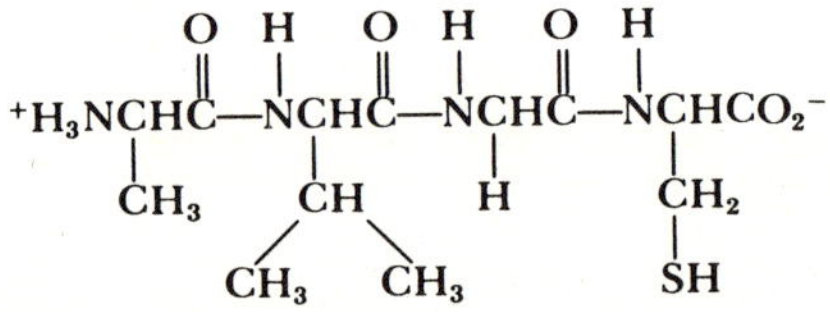

4
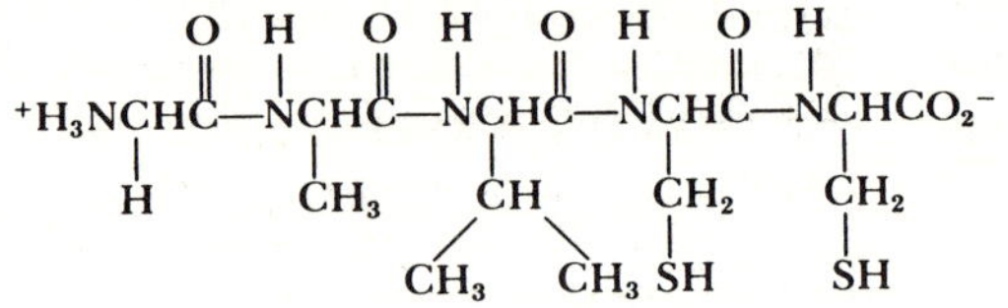

5
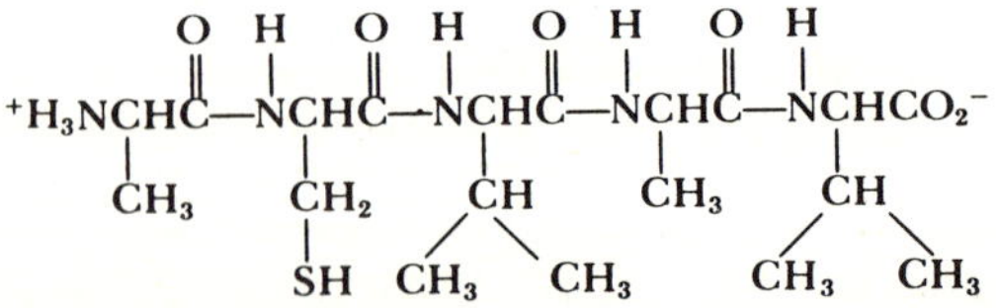

PROBLEM 12–19 This and the following problem lead to the topic of enzyme inhibition. Examine the following common functional groups:

(a)
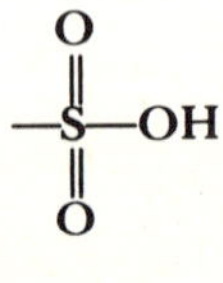

(b) 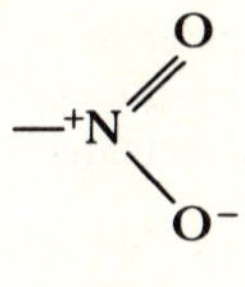

(c) $—NH_2$

(d) $-\overset{\displaystyle O}{\overset{\|}{C}}-OH$

(e) $—CH_2OH$

Which of the above groups contain a π electron system?

1 (a), (b), (d)
2 (a), (d)
3 (a), (c), (e)
4 (c), (e)
5 (b), (c)

PROBLEM 12–20 Examine the following common chemical structures to which simple functional groups (as in Problem 12–19) are often attached:

(a) 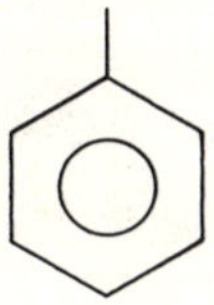

(b) 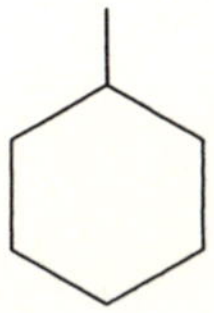

(c) 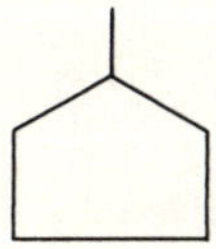

(d) $CH_3CH_2CH_2CH_2CH_2—$

(e) 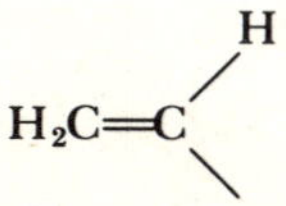

Which of these systems has essentially planar geometry?

1 (b), (c)
2 (b), (c), (d)
3 (d)
4 (a), (b)
5 (a), (e)

PROBLEM 12-21

An important transformation in the citric acid cycle involves the removal of two hydrogen atoms from succinic acid (A) to form fumaric acid (B). The enzyme responsible for this transformation is called succinic dehydrogenase.

For more on the citric acid cycle, metabolism in general, and the catalytic action of enzymes, refer to Sections 12–8 and 12–11 of Chemical Principles.

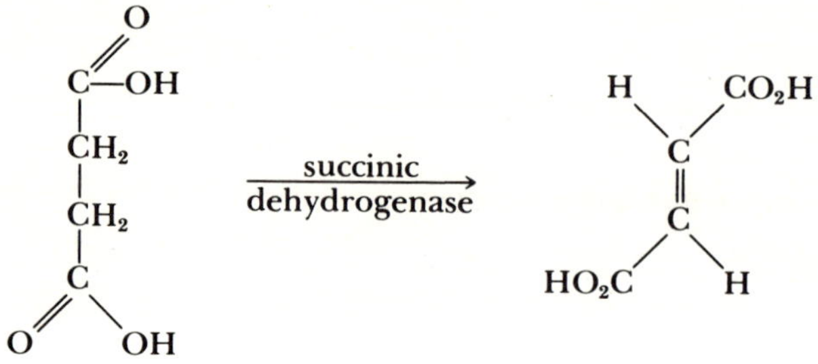

A B

Which of the following compounds would be a likely inhibitor of the above reaction? (Competitive inhibitors act by combining with the enzyme to prevent combination of the enzyme with its *substrate* such as A). Assume that the inhibitor has a chemical structure resembling that of the substrate, A.

1 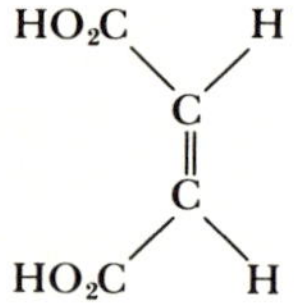

2 $HO_2C{-}C{\equiv}C{-}CO_2H$
3 $HO_2C{-}CH_2CO_2H$
4 $CH_3CH_2CH_2CH_2CH_2NH_2$
5 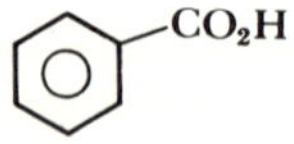

PROBLEM 12-22 *para*-Aminobenzoic acid (shown below) is an essential growth factor (a vitamin) for many bacteria. Which of the following compounds is the most probable candidate as an inhibitor for bacteria growth (i.e., a drug)?

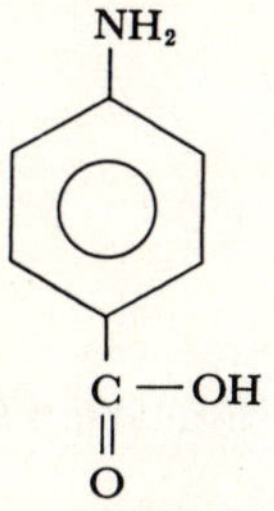

1

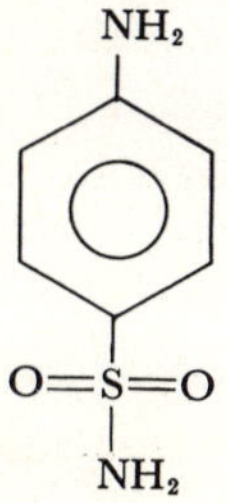

2

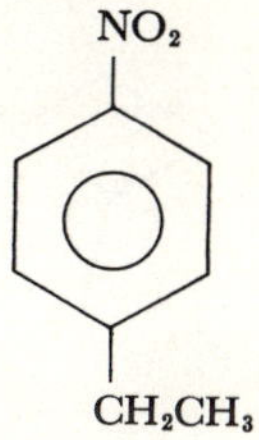

3

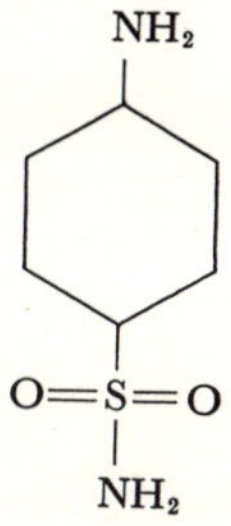

4

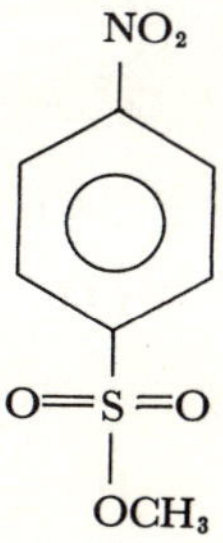

5

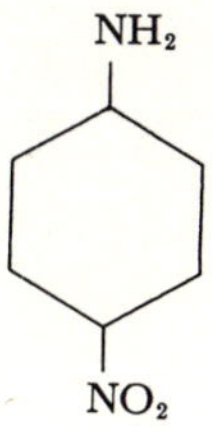

PROBLEM 12–23 Adamantylamine (shown below) is one of the few chemical compounds active against the influenza virus.

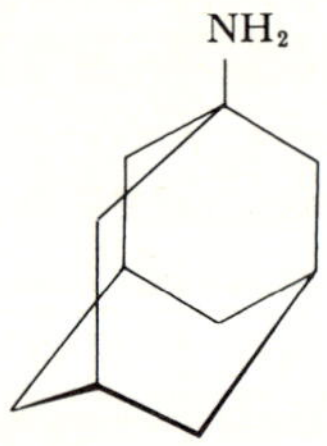

Assume that the following compounds are available for testing.

(a)

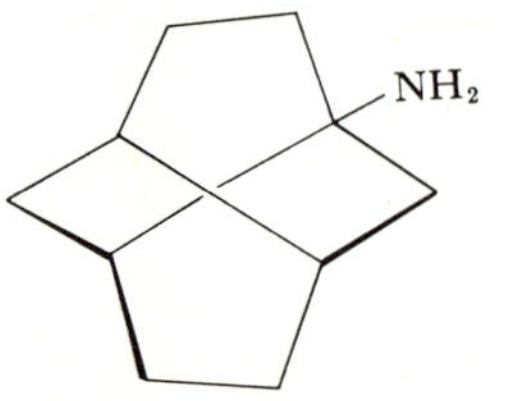

(b)

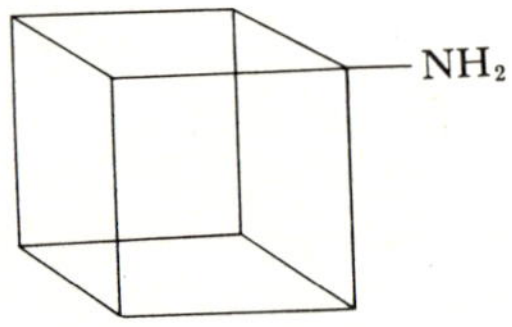

(c) $CH_3CH_2CH_2CH_2CH_2CH_2CH_2NH_2$

(d) $CH_3—C(CO_2^-)(H)—NH_3^+$

What is your guess as to which of the above would be the best candidates for testing against influenza virus?

1 (a), (b)
2 (a), (b), (c)
3 (d)
4 All of them.
5 None of them.

SOLUTIONS

SOLUTION 12–1

(5)

To construct the structural formula from the name, we first write down the carbon skeleton, which is given by the last part of the name, ethane:

C—C

Now we add the substituents that are designated by numbers 1,2-dichloro. And since "di" means two, we attach two Cl's:

1 *2*
Cl—C—C—Cl

where the numbers are assigned to the carbons on the skeleton in order, proceeding from left to right. Finally, "tetra" means four, so we add four fluorines to the available positions:

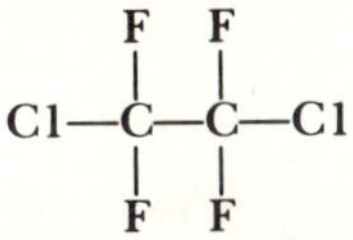

This is compound **5**, which is called Freon-114 and is used as a refrigerant.

Compound **4** is called 1,1-dichlorotetrafluoroethane since both Cl's are on the same C. Here it does not matter to which carbon atom both are attached because

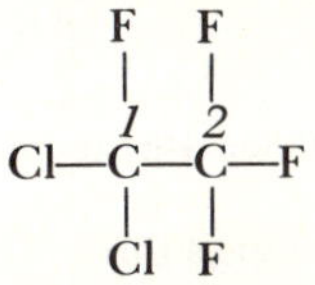

1,1-dichlorotetrafluoroethane

is the same as

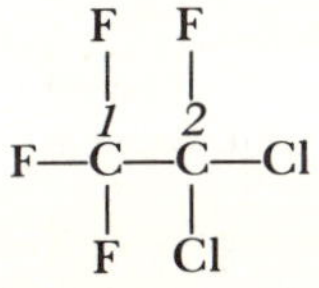

2,2-dichlorotetrafluoroethane

We choose the name with the lowest set of numbers.

Compound **3** is called 1-chloro-1,1-difluoroethane because the one Cl and two F's are on the same C. It is also known as Genetron-142B and is used as a refrigerant.

Compound **2** is called 1,1-difluoroethane. It is known as Genetron-152A and is used as an aerosol propellant.

Answer **1** must be wrong because the skeleton contains one too many carbons and is called 1,2-dichloro-1,2,3-trifluoropropane.

[*Note:* We could write Answer **5** in many ways because the tetrahedrally oriented CF_2Cl groups can rotate freely about the C—C single bond. But the compound and its name are unchanged.]

A structural formula

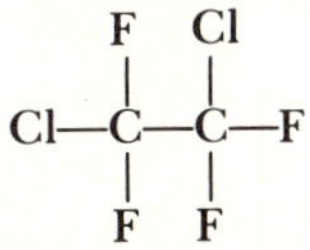

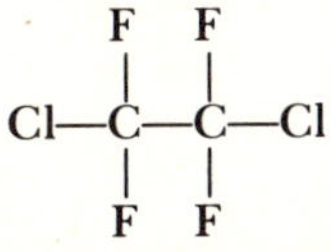

A projection formula

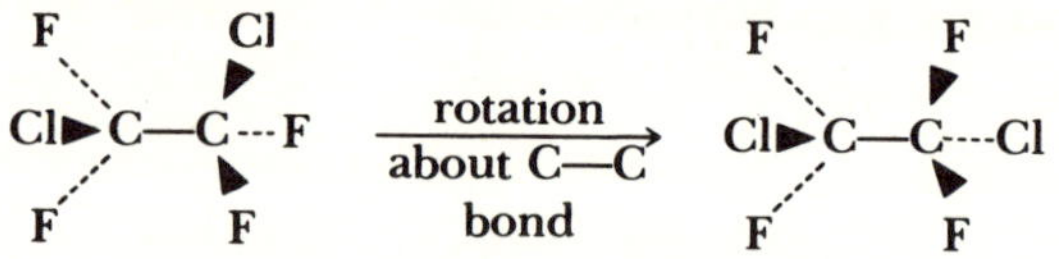

[The designation (◄) is taken to mean that the bond is directed toward the viewer. The designation (----) means that the bond is directed away from the viewer.]

SOLUTION 12–2

(3)

The carbon chain is the hexane skeleton:

1 2 3 4 5 6
C—C—C—C—C—C

Add the three methyl groups at the correct positions:

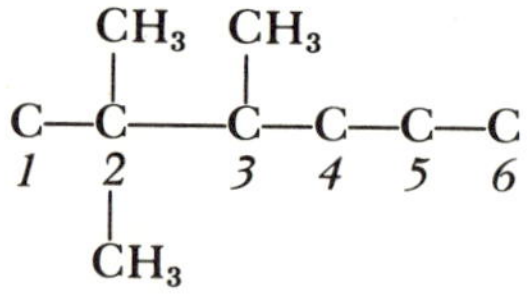

and fill in the missing H's:

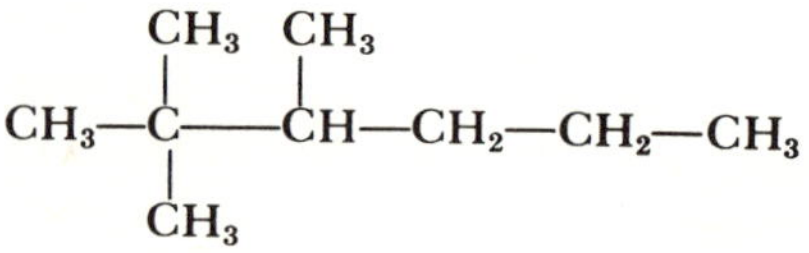

3 is correct.

1 is incorrect because the chain is only five carbons long (pentane skeleton) and is called 2,2,3-trimethylpentane.

2 is also a pentane: 2,2,4-trimethylpentane. (It is not 2,4,4-trimethylpentane, even though this is not an ambiguous name, because international convention dictates that we use the lowest set of prefix numbers.)

4 is numbered from the right-hand side: 2,2,5-trimethylhexane (not 2,5,5 – see preceding sentence).

5 is named 2,3,5-trimethylheptane (not 2,3,5,6-tetramethylhexane). A methyl group (CH_3)

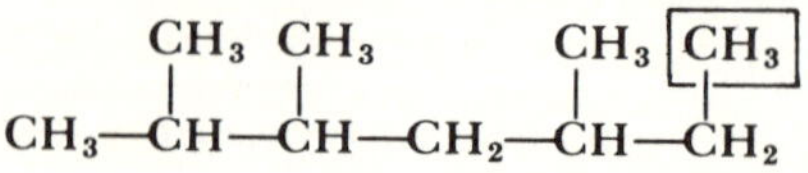

written above the main line of carbons need not mean that it is a substituent group. You must find the maximum chain length regardless of the way the structural formula is written.

SOLUTION 12–3

(4)

The skeleton is a cyclohexane, and the suffix "ol" denotes an alcohol at position *1*:

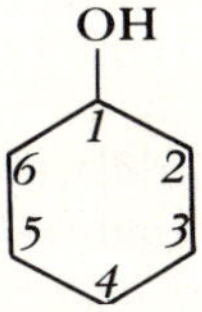

Isopropyl is at position *2*:

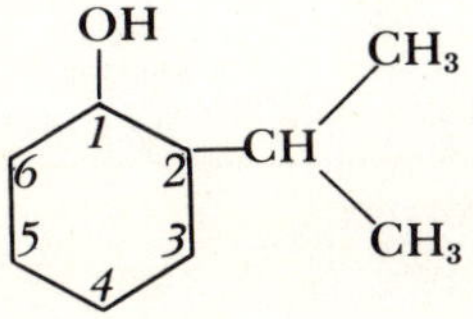

The methyl group is at position *5*:

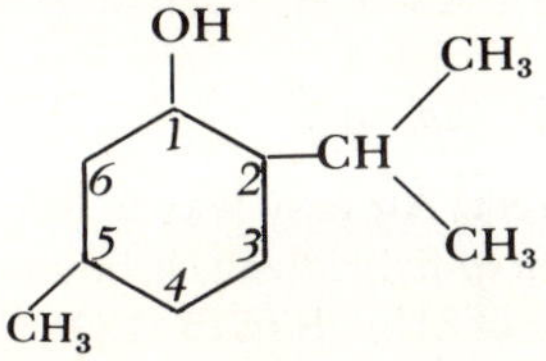

This structure is the same as the one given in Answer **4**. If you chose Answer **1**, you probably numbered clockwise for isopropyl and then counterclockwise for methyl.

SOLUTION 12–4

(4)

If you need help to learn the names, structures, and functional groups of common organic compounds, see Review 12 of Programed Reviews of Chemical Principles *by Lassila* et al.

The skeleton is hexane. The suffix "ene" denotes a double bond, and "diene" means that there are two of them. It is necessary to specify the position of each double bond.

$$\overset{1}{CH_2}{=}\overset{2}{CH}{-}\overset{3}{CH}{=}\overset{4}{CH}{-}\overset{5}{CH_2}{-}\overset{6}{CH_3}$$

The lowest numbered carbon of a double bond designates the position. The carbon chain is numbered to give the lowest set of numbers.

The correct name is Answer **4**: 1,3-hexadiene. Answer **3** corresponds to numbering in the reverse direction.

[*Note:* The biggest customer for a similar compound, 1,3-butadiene, is the synthetic rubber industry. About 2 million short tons of butadiene were sold in 1969.]

SOLUTION 12-5

(2)

In compound (a) every C is sp^3 hybridized (four atoms attached, hence four orbitals involved: one *s* and three *p*). The bonds around each carbon are directed toward the corners of a tetrahedron.

Every C atom in compound (c) is sp^2 hybridized (three atoms attached, one *s* and two *p* orbitals involved).

Both carbons in (d) are *sp* hybridized (two atoms attached, one *s* and one *p* orbital involved). These bonds are collinear.

Compound (b) has two types of hybridization: sp^3 at the terminal carbons and sp^2 at the interior carbons.

$$\underset{sp^3}{\underset{\uparrow}{CH_3}}-\underset{sp^2}{\underset{\uparrow}{CH}}=\underset{sp^2}{\underset{\uparrow}{CH}}-\underset{sp^3}{\underset{\uparrow}{CH_3}}$$

So Answer **2** is correct.

SOLUTION 12-6

(2)

$$\underset{sp}{\underset{\uparrow}{HC}}\equiv\underset{sp}{\underset{\uparrow}{C}}-C\equiv C-\underset{sp^2}{\underset{\uparrow}{CH}}=\underset{sp}{\underset{\uparrow}{C}}=\underset{sp^2}{\underset{\uparrow}{CH}}-CH=CH-CH=CH-\underset{sp^3}{\underset{\uparrow}{CH_2}}-\underset{sp^2}{\underset{\uparrow}{C}}(=O)-OH$$

Answer **2** is correct. An easy way to keep the hybridization clear is to remember that an *sp*-hybridized carbon has two atoms attached. That is, the total number of orbitals in the hybrid is revealed by the designation: one *s* and one *p* orbital that give two *sp* hybrids.

SOLUTION 12-7

(1)

For (a) the behavior of SH^- as a displacing group would qualitatively be expected to parallel that of OH^- since sulfur is immediately below oxygen in the periodic table. [The RSH compounds are called *thiols* (trivial name: *mercaptan*). The lower molecular weight members are among the world's worst smelling substances. One of the constituents of the skunk protectant is 1-butanethiol ($CH_3CH_2CH_2CH_2SH$).]

The unknown reaction in (b) would not be expected to proceed as does the analogous carbon reaction because silicon has not yet been observed to participate in multiple bonding.

The unknown reaction in (c) does not proceed because triple bonds in a four-membered ring would provide too much strain. Remember that *sp* hybridization requires linear geometry for the four carbon atoms:

$$C-C\equiv C-C$$

The correct answer is **1**.

[*Note:* The seemingly dichotomous behavior of OH^- with the alkyl halides in the first two model reactions (substitution versus elimination) is an important area of research in organic chemistry.]

SOLUTION 12–8

(2)

The only ways in which two chlorine atoms can be attached in the formula $C_2H_4Cl_2$ are shown below.

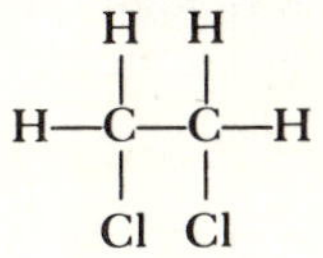

1,2-dichloroethane

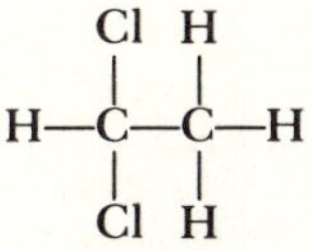

1,1-dichloroethane

Answer **2** is correct. Remember that

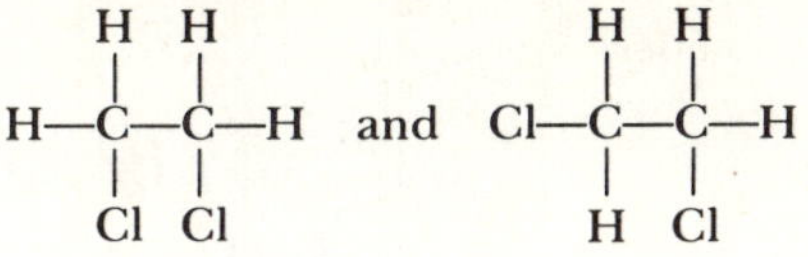

are identical molecules. A good test to see whether two structures are the same is to apply the IUPAC rules for nomenclature. If the name is the same for both structures, then they are not structural isomers. To convert a molecule from one structural isomer to another, it is necessary to break some of the bonds of the first molecule.

SOLUTION 12–9

(3)

Isomerism in organic compounds is discussed in Section 12–4 of Chemical Principles.

The isomers are

$CH_3CH_2CH_2CH_2CH_2CH_3$ *n*-hexane ("*n*" means normal or straight-chained)

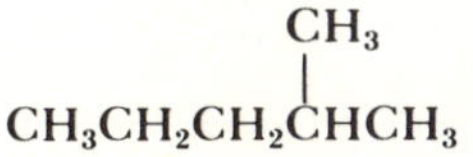

2-methylpentane

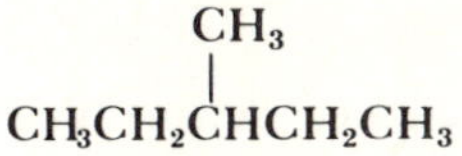

3-methylpentane

There are no additional ways to have five carbons in a chain with one extra CH_3 group.

We construct the next isomers with four carbons as the longest chain:

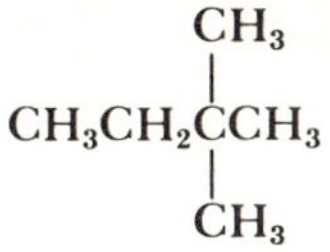

2,2-dimethylbutane

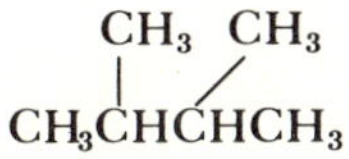

2,3-dimethylbutane

But we cannot construct any isomers with only three carbons as the longest chain. Thus, there is a total of five structural isomers, and Answer **3** is correct.

All of these structural isomers exist. Although their molecular formulas are identical, their physical properties are different. Consider, for instance, the following boiling points (bp).

	bp (°C)
n-hexane	69
2-methylpentane	60
3-methylpentane	63
2,2-dimethylbutane	50
2,3-dimethylbutane	58

SOLUTION 12–10

(2)

There are two possibilities:

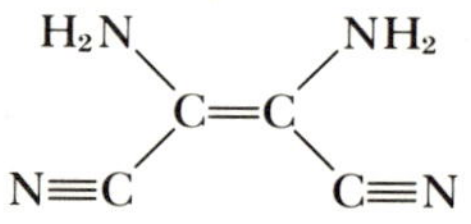

cis isomer

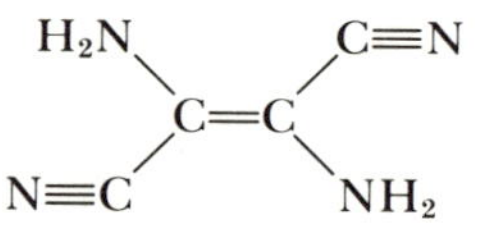

trans isomer

Geometrical isomers can exist only because the rotation about this carbon–carbon double bond is restricted. The NH_2 groups (and the C≡N groups) on the molecule on the left are on the same side of the carbon–carbon double bond. The word "cis" designates this configuration. The molecule on the right has the NH_2 groups on opposite sides of the carbon–carbon double bond and is called "trans". Answer **2** is correct.

There are other ways to write the tetramer; for example,

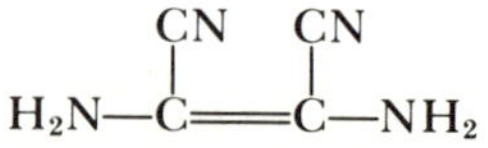

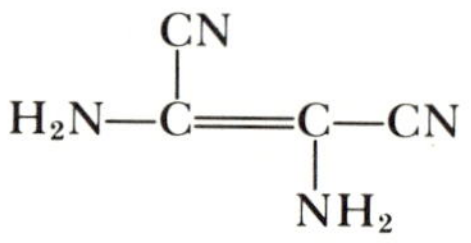

Neither of the above two formulations clearly expresses the correct geometry of the molecule. Alkenes should be written with correct bond angles (120°):

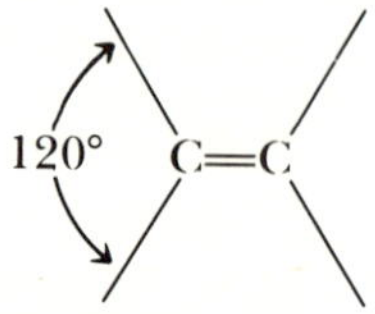

SOLUTION 12-11 (4)

The double bonds about which rotation is restricted are labeled **a** and **b**.

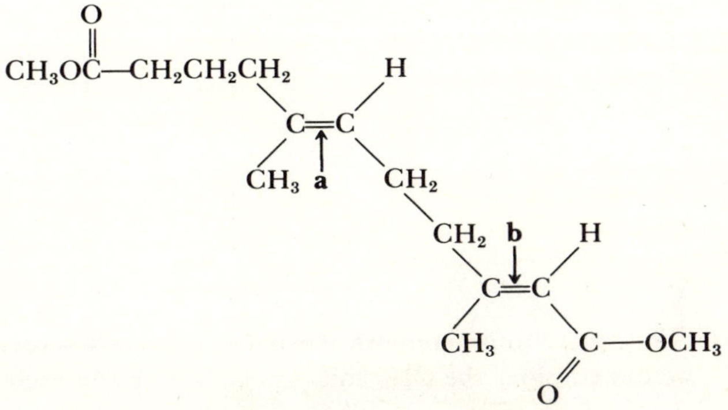

A (already shown)

To write the other isomers we must *imagine* a 180° rotation about the individual bonds **a** and **b**. This is a rotation that each isomer cannot make without breaking bonds. Rotating about bond **b**, we obtain geometrical isomer B.

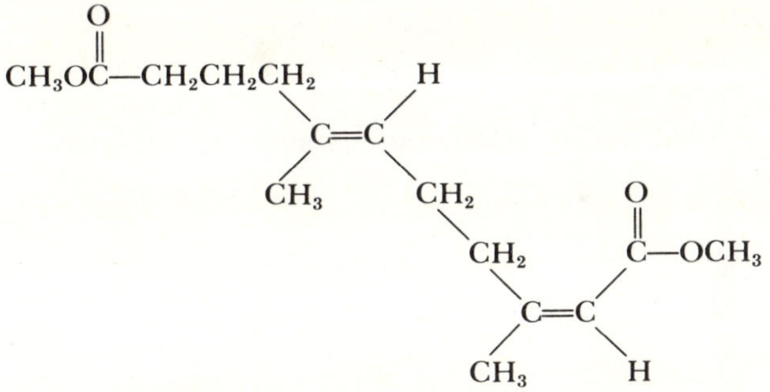

B

Now we take isomer B and rotate about bond **a** to obtain isomer C.

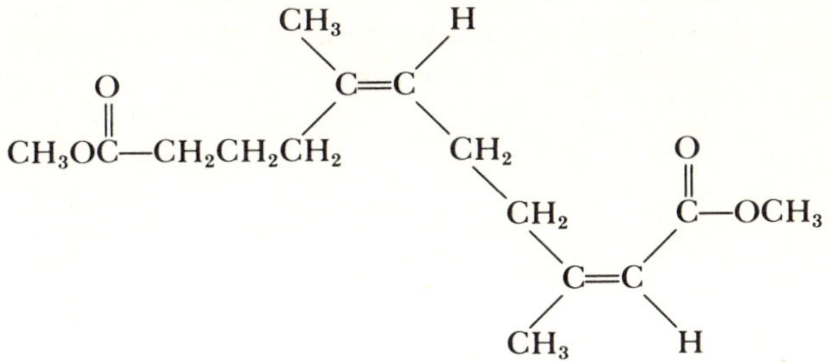

C

Finally, we take isomer C and rotate about bond **b**, thereby giving us isomer D.

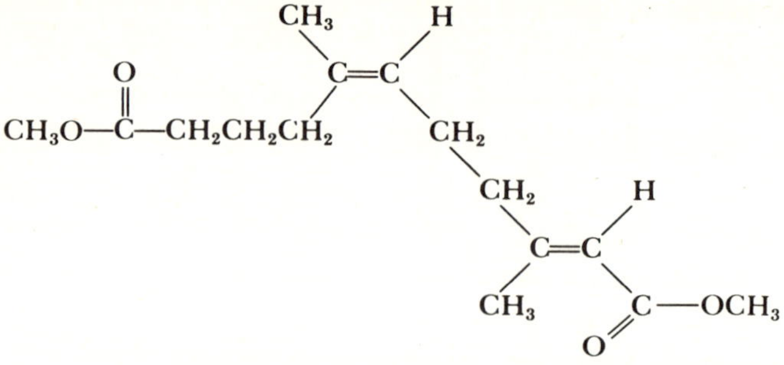

D

There are four geometrical isomers. Answer **4** is correct. As a check, we can consider the CH_3 and H relationships on each double bond.

	a	**b**
A	trans	trans
B	trans	cis
C	cis	cis
D	cis	trans

Since the designations are all different, they are four different geometrical isomers.

SOLUTION 12–12

(2)

First, there are cis and trans isomers.

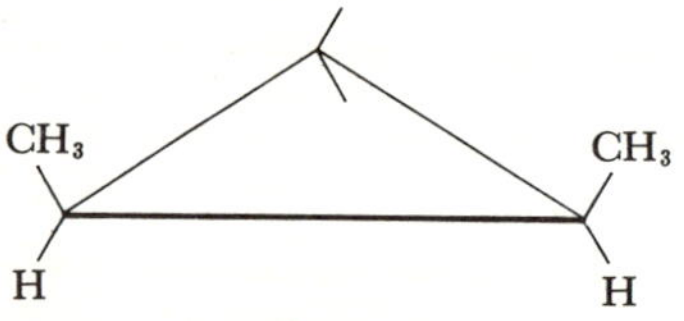

cis

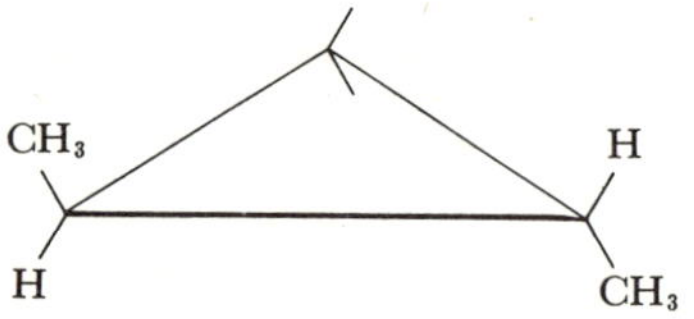

trans

Now we must find out how many optical isomers each one of the above geometrical isomers possesses. *trans*-1,2-Dimethylcyclopropane has a nonsuperimposable mirror image.

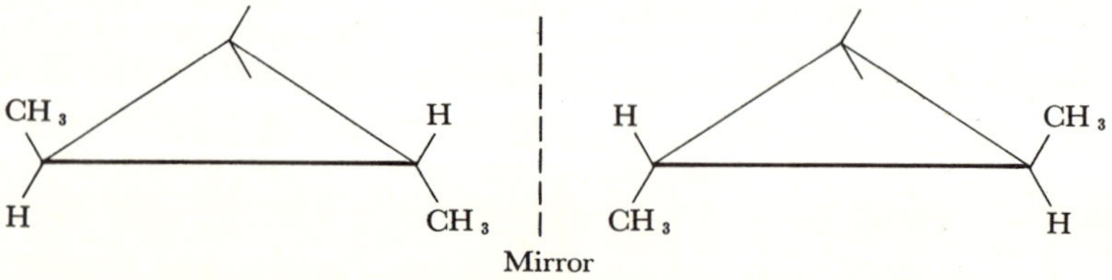

Try to superimpose these two mirror images: They are not the same. But the mirror images of *cis*-1,2-dimethylcyclopropane are superimposable.

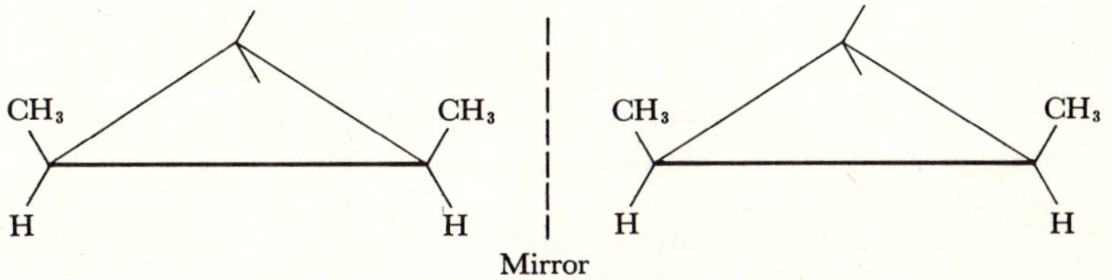

There are three isomers in all; two for trans and one for cis. Answer **2** is correct.

SOLUTION 12–13

(3)

The three asymmetric atoms ("asymmetric" is defined as an atom with four different kinds of groups attached) are numbered in boldface type below. Thus, there are 2^3 or 8 stereoisomers, and Answer **3** is correct.

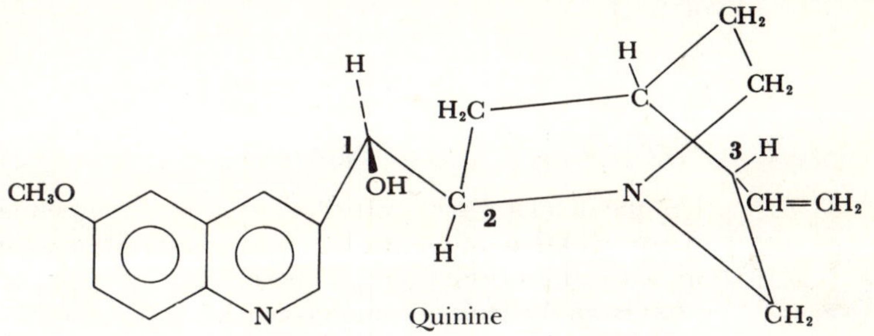

Quinine

The specific isomer shown is quinine, which is useful in the treatment of malaria. In contrast, another isomer (in which the spatial arrangement of the groups at **2** is reversed) is not active against malaria.

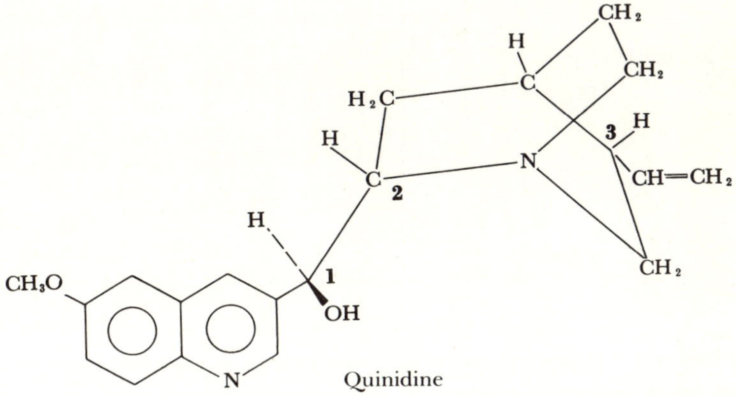

Quinidine

It is called quinidine and is used in the control of auricular fibrillation (heart flutter). Quinine does not possess this property. The point is that a knowledge of stereochemistry is necessary in organic chemistry since molecules as closely related as the preceding two will frequently have quite different physiological properties.

[*Note:* The nonsuperimposable mirror image of *d*-lysergic acid diethylamide (LSD) has limited hallucinogenic properties. In addition, one stereoisomer of monosodium glutamate (MSG) serves as a taste enhancer (also a meat tenderizer), whereas the other stereoisomer is inactive in this role. Recently, companies that produce baby food have stopped using MSG as a flavor enhancer since certain tests indicate that it is harmful (in high doses) to rats.

Active isomer of MSG

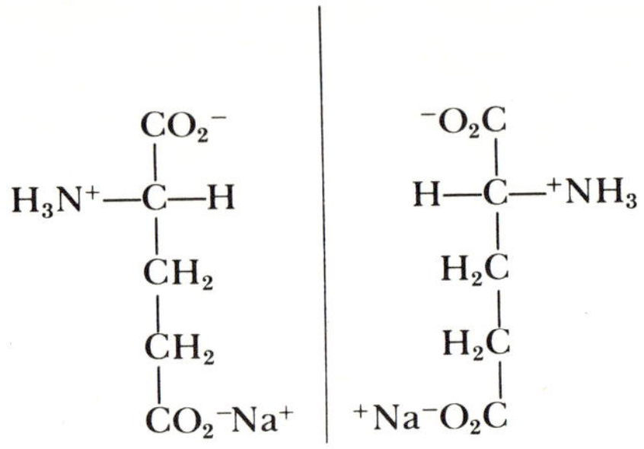

Inactive isomer

Mirror

The physiological activity (or lack of it) of a compound with an asymmetric center is often attributed to whether the compound "fits" in an active site in the body. If there is no fit, presumably there is no effect from the molecule and no activity. A schematic diagram showing how one compound of a mirror pair can fit an active site is in Figure 12–1.]

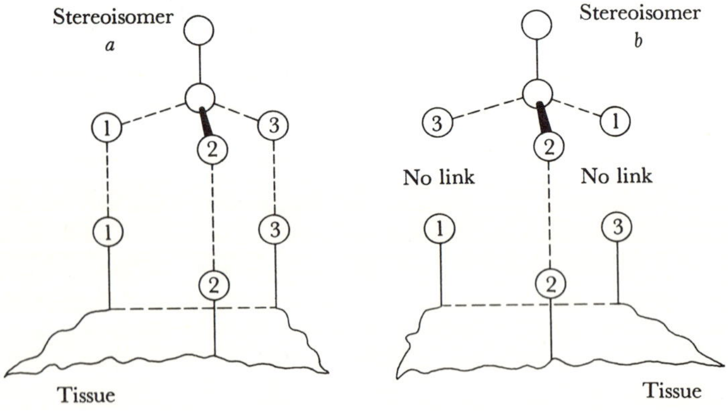

Figure 12–1. Fit, and lack of it, of stereoisomers to tissue.

SOLUTION 12–14

(1)

From the stoichiometry of the KOH reaction, the compound has only one chlorine. Since the product is a triol, however, the drug must have contained two hydroxyl groups. In addition, there must be one asymmetric carbon. Only Structure **1** can satisfy these requirements.

SOLUTION 12–15

(5)

Answer **5** is correct. Answers **1** and **2** are incorrect since the compounds represented lack multiple bonds and would not easily react with Br_2. The compounds in Answers **2** and **4** are aromatic, hence generally unreactive toward reagents such as Br_2 unless an electron-attracting catalyst is present. Ethene (or ethylene), shown in Answer **5**, is the only alkene in the list of compounds and will react with Br_2 to form 1,2-dibromoethane.

SOLUTION 12–16

(3)

Answer **3** is correct. The question illustrates the considerable difference between isolated double bonds and an aromatic (electron delocalized) ring system. The three conjugated (linked sp^2 carbon atoms) double bonds in the right-hand ring have special stability. The ring is resistant to addition reactions. Hydrogen can be added to benzene but only under extreme conditions (very high temperature and pressure).

SOLUTION 12–17

(2)

Enzymes and proteins are treated in Section 12–9 of Chemical Principles.

An amino acid is often written in texts and articles as

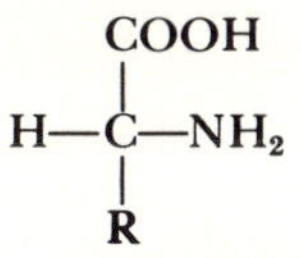

but actually the compound is very *polar* (large separation of charges within the molecule) and will dissolve in polar solvents such as water. Many organic solvents are nonpolar and will not dissolve amino acids. Making the analogy to melting points of ionic crystals, we can see that these amino acids will have high melting points.

Answer **2** is correct.

You should always write the proton of the carboxylic acid on the basic (electron-donating) nitrogen atom. In Problem 5–32 you encountered several charged species that arise from amino acids, namely

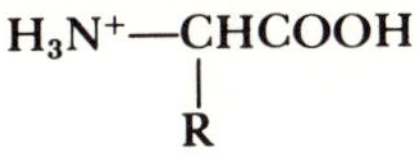

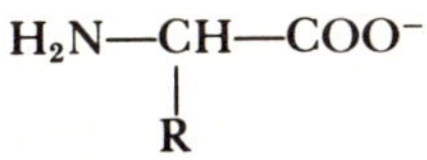

SOLUTION 12–18

(1)

On hydrolysis, 1 mole of the polypeptide in Answer **1** would form 1 mole of alanine, 2 moles of valine, and 2 moles of cysteine.

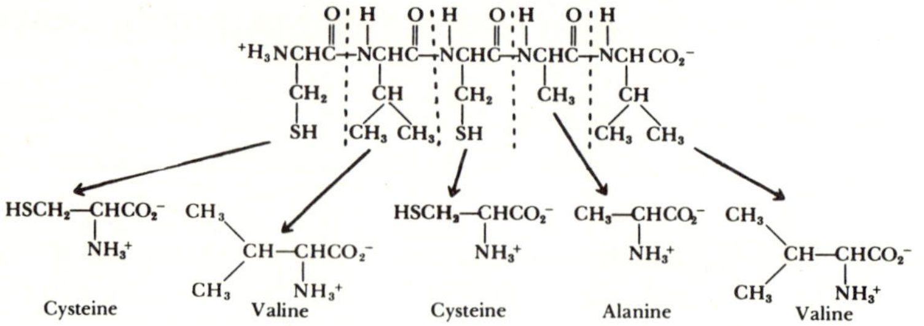

Thus, our unknown polypeptide is Answer **1**.

The number of moles of the different amino acids produced on the hydrolysis of 1 mole of each of the other polypeptides is indicated in Table 12–1.

Table 12–1

Polypeptide hydrolyzed	Number of moles of amino acid produced			
	Alanine	Valine	Cysteine	Glycine
2	2	1	2	0
3	1	1	1	1
4	1	1	2	1
5	2	2	1	0

SOLUTION 12–19

(1)

Only functional groups with double bonds will have π electrons, that is, electrons in a π orbital located above and below the bond axis. (a), (b), and (d) are such functional groups, and Answer **1** is correct.

SOLUTION 12–20

(5)

Structure (b) is best portrayed as

and is far from flat. Each C is sp^3 hybridized, its bonds are tetrahedrally oriented, and it has a bond angle of about 109°. A regular hexagon has 120° angles, so the cyclohexane ring must pucker to relieve the strain.

Structure (c) can be portrayed as

Here again, considerable flexibility exists in the molecule. The structure (d) is freely flexible, and is not constrained to a planar geometry.

The sp^2 hybridization and the π systems in (a) and (e) cause the atoms to lie in a plane.

Answer **5** is correct.

SOLUTION 12–21

(3)

Succinic acid has a nonplanar geometry, so we will reject compounds that are planar, **1**, or linear, **2**. It has a very flexible chain; hence, we eliminate Answer **5** because the ring is planar. Compounds **3** and **4** both have flexible chains, but only compound **3** has the correct kind of functionality (an acidic group, —COOH, rather than a basic group, $—NH_2$).

Answer **3** is the best choice.

[*Note:* This is an oversimplification of the real situation because enzymatic reactions are sometimes inhibited by molecules resembling the product; in such a case, Answer **1** would be a good choice as an inhibitor.]

SOLUTION 12–22

(1)

Answers **3** and **5** can be rejected (at least compared with Answers **1**, **2**, and **4**) since the functional groups are attached to a flexible ring as opposed to a flat benzene ring.

The NO_2 group of Answer **4** is considerably different from the NH_2 of *para*-aminobenzoic acid, and there is no acidic proton in structure **4** (in the $—SO_3CH_3$ function), whereas there is one in the other, —COOH.

For similar reasons, structure **2** can be rejected. Structure **1** resembles *para*-aminobenzoic acid in many respects, and indeed is an effective drug (sulfanilamide, one of the so-called sulfa drugs). It is effective against such cocci infections as pneumonia and gonorrhea. Another antibiotic is sulfadiazine, which has several features similar to *para*-aminobenzoic acid.

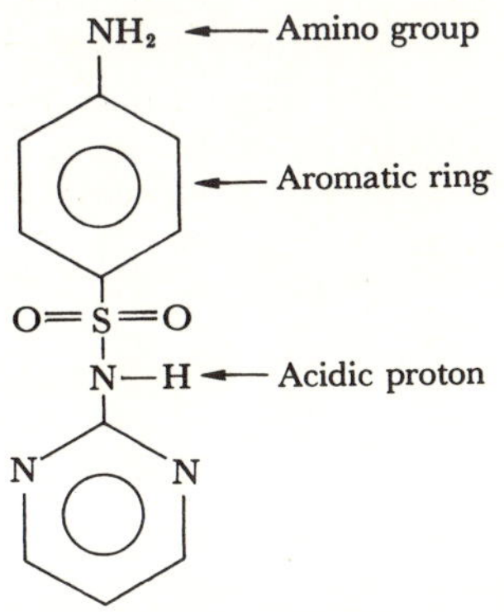

Thus, in drug design it is often necessary to effect minor molecular change (often called molecular roulette) to find drugs with proper therapeutic properties and low toxicity. One disadvantage of some of the sulfa drugs is their low solubility in water; this sometimes leads to an accumulation of the drug in the kidneys. It is generally best to examine all compounds having a reasonable potential for activity even if there are not as many structural similarities (to the vitamin) as above.

[*Note:* Although structural similarities often indicate physiological similarity, it is interesting to note that the two compounds below are both sweet to the taste.

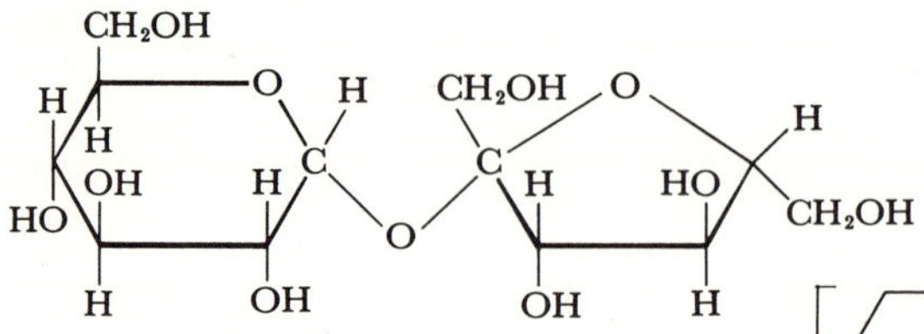

Sucrose

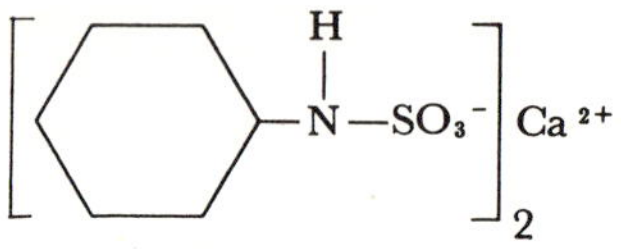

Calcium cyclamate

Sucrose is common table sugar, whereas calcium cyclamate is about 30 times sweeter and has been used as an "artificial" sweetener in many diet drinks and other foods. Large doses of cyclamates fed to rats over a long period have produced bladder cancer. This problem has caused the Canadian and United States governments to place severe restrictions on their use.]

SOLUTION 12–23

(1), (4)

Both (a) and (b) are "cage," or rather spherical or globular, molecules and both bear an NH_2 group. Thus, Answer **1** seems correct. The structures in (c) and (d) would be less likely candidates since (c) is quite different in shape, as is (d). In addition, (d) is a very polar species, whereas (a), (b), and (c) are not. Although it is wise to test practically all compounds for physiological activity, intelligent guesses, such as Answer **1**, often can be made. (Compare, however, the note in the solution of Problem 12–22.) It does happen that compound (a) is an active agent against the Asian flu virus, although it is not possible to predict this kind of activity with precision. Hence, Answer **4** could also be deemed correct.

CHAPTER 13 NUCLEAR CHEMISTRY

It does not matter when we date the beginning of the "atomic era"—the Curies, Ernest Rutherford, Otto Hahn and Lise Meitner, or the Hiroshima bomb. We are in it now and had best understand it so we can make intelligent decisions about bomb testing and nuclear power plants. For if we do not, others will always be willing to make them for us, and perhaps not so intelligently.

To answer some of the problems in this chapter you will need to memorize the following relationships.

^1_0n = neutron

Activity of a radioactive nuclide (rate of its decay) = $k \times N$

in which k is the rate constant and N is the number of radioactive atoms present.

Nuclear reactions are generally not balanced in terms of orbital electrons.

Solutions to problems in Chapter 13 begin on page 366.

PROBLEMS

PROBLEM 13-1

Atomic structure and isotopes are reviewed in Section 13-1 of Chemical Principles *by Dickerson, Gray, and Haight.*

What is the atomic weight of the element neon that consists of three isotopes whose exact masses (relative to $^{12}_{6}C = 12.00000$ amu) and percent natural abundances are as follows.

Isotope	Exact mass (amu)	% Natural abundance
$^{20}_{10}Ne$	19.99244	90.92000
$^{21}_{10}Ne$	20.99395	0.2570000
$^{22}_{10}Ne$	21.99138	8.820000

1 20.17072 amu
2 20.21347 amu
3 20.50000 amu
4 20.94567 amu
5 21.31645 amu

PROBLEM 13-2

The *nuclides* (a nuclide is the general name for any nuclear species) $^{12}_{6}C$, $^{56}_{26}Fe$, and $^{238}_{92}U$ have 12, 56, and 238 *nucleons,* respectively, in their nuclei. The total number of nucleons in a nucleus is equal to ________.

1 The total number of electrons in the atom.
2 The total number of protons in the nucleus.
3 The total number of neutrons in the nucleus.
4 The total number of protons and neutrons in the nucleus.
5 The atomic number.

PROBLEM 13-3

If the nuclides $^{14}_{6}C$, $^{15}_{7}N$, and $^{17}_{9}F$ are *isotones,* how do you think isotones are defined?

1 Nuclides having the same number of electrons but different nuclear charges.
2 Nuclides having different numbers of neutrons and protons but equal mass numbers.
3 Nuclides having the same atomic number but different mass numbers.
4 Nuclides having equal numbers of neutrons but differing in mass number and atomic number.
5 None of the above is correct.

PROBLEM 13–4

Neutron scattering experiments have shown that the radius of the nucleus of an atom is directly proportional to the cube root of the number of nucleons in the nucleus. From $^{7}_{3}Li$ to $^{189}_{76}Os$ the radius is ________.

1 Halved.
2 The same.
3 Doubled.
4 Tripled.
5 Quadrupled.

PROBLEM 13–5

Nuclear decay is described in Section 13–2 of Chemical Principles.

The conversion of a nuclide of one element into that of another is known as *transmutation.* The transmutation of $^{232}_{90}Th$ into $^{228}_{88}Ra$ is an example of α decay. Which one of the transmutations given below does *not* take place by α decay?

1 $^{238}_{92}U \rightarrow ^{234}_{90}Th$
2 $^{227}_{89}Ac \rightarrow ^{223}_{87}Fr$
3 $^{226}_{88}Ra \rightarrow ^{222}_{86}Rn$
4 $^{212}_{84}Po \rightarrow ^{208}_{82}Pb$
5 $^{213}_{83}Bi \rightarrow ^{213}_{84}Po$

PROBLEM 13–6

The transmutation of $^{213}_{83}Bi$ into $^{213}_{84}Po$ (discussed in Problem 13–5) is the result of β^- decay. Which one of the following transmutations does *not* take place by β^- decay?

1 $^{64}_{29}Cu \rightarrow ^{64}_{28}Ni$
2 $^{114}_{49}In \rightarrow ^{114}_{50}Sn$
3 $^{124}_{51}Sb \rightarrow ^{124}_{52}Te$
4 $^{212}_{83}Bi \rightarrow ^{212}_{84}Po$
5 $^{234}_{90}Th \rightarrow ^{234}_{91}Pa$

PROBLEM 13–7

Natural radioactive decay series are discussed in Section 13–3 of Chemical Principles.

In one of the natural radioactive decay series, uranium-235 ($^{235}_{92}U$) decays with the emission of α and β^- particles to form, ultimately, lead-207 ($^{207}_{82}Pb$). How many α and β^- particles are emitted per atom of $^{207}_{82}Pb$ formed?

	α	β^-
1	1	1
2	4	7
3	7	4
4	7	14
5	14	8

PROBLEM 13–8

The actual mass of an atom is always less than the sum of the masses of its constituent particles. The energy equivalent of this mass difference is termed the *binding energy* of the nucleus. The most stable atoms have mass numbers in the region of 55, for example, $^{55}_{25}Mn$. What is the binding energy per nucleon for $^{55}_{25}Mn$?

(isotopic mass: $^{1}_{0}n = 1.00867$ amu, $^{1}_{1}H = 1.00782$ amu,
$^{55}_{25}Mn = 54.9381$ amu;
1.00 amu = 931 MeV)

1 8.2 MeV nucleon^{-1}
2 8.8 MeV nucleon^{-1}
3 9.2 MeV nucleon^{-1}
4 92 MeV nucleon^{-1}
5 482 MeV nucleon^{-1}

PROBLEM 13–9

Phosphorus-32 ($^{32}_{15}P$) is a radioactive nuclide with a half-life of about 14 days. What is the mode of decay of $^{32}_{15}P$ and what is the associated decay energy?

(isotopic mass: $^{32}_{14}Si = 31.9740$ amu, $^{32}_{15}P = 31.9739$ amu,
$^{32}_{16}S = 31.9721$ amu;
1.00 amu = 931 MeV)

	Mode of decay	Decay energy (MeV)
1	β^+	0.55
2	β^-	1.0
3	β^-	1.7
4	β^+	3.0
5	β^-	5.5

PROBLEM 13–10

Nuclear fission and fusion are treated in Section 13–4 of Chemical Principles.

Strontium-90 ($^{90}_{38}Sr$) is one of the fission products present in radioactive "fallout" from the explosion of an atomic bomb. A pure sample of $^{90}_{38}Sr$ gave an activity of 1000 disintegrations min^{-1}. If the activity of this sample after 1.00 year is 975.2 disintegrations min^{-1}, what is the half-life of $^{90}_{38}Sr$? [*Note:* $2.303 \log (N_0/N) = kt$.]

(log 975.2 = 2.9891)

1 1.00 year
2 12.0 years
3 27.6 years
4 38.0 years
5 97.5 years

PROBLEM 13–11

Radioactive materials are often measured in terms of a radiation unit called the *curie*. One curie (C) of a radioactive material is defined as the weight of material that has an activity of 3.70×10^{10} disintegrations sec^{-1}. According to the US Atomic Energy Commission regulations, the maximum permissible dose of $^{90}_{38}Sr$ in the body of an adult is about 1 microcurie (μC). To what weight of $^{90}_{38}Sr$ does this maximum dosage correspond? [*Note:* 2.303 log $(N_0/N) = kt$.]

($t_{1/2}$: $^{90}_{38}Sr = 27.6$ years;
1.00 year $= 3.15 \times 10^7$ sec;
1.00 $\mu C = 1.00 \times 10^{-6}$ C)

1 2.2×10^{-16} g
2 2.2×10^{-10} g
3 6.9×10^{-9} g
4 6.9×10^{-3} g
5 3.7×10^{10} g

PROBLEM 13–12

Suppose that an obese man ingests the 1.00-μC maximum permissible dosage of $^{90}_{38}Sr$ by drinking milk shakes made from milk that was contaminated during a nuclear test. What will be the activity of the $^{90}_{38}Sr$ in the man's body after a 20.0-year period, assuming that all of the $^{90}_{38}Sr$ is concentrated in his bones and none of it is eliminated? [*Note:* 2.303 log$(N_0/N) = kt$.]

($t_{1/2}$: $^{90}_{38}Sr = 27.6$ years;
activity of 1.00 C $= 3.70 \times 10^{10}$ disintegrations sec^{-1};
1.00 $\mu C = 1.00 \times 10^{-6}$ C)

1 2.05×10^4 disintegrations sec^{-1}
2 2.24×10^4 disintegrations sec^{-1}
3 3.70×10^6 disintegrations sec^{-1}
4 2.24×10^{10} disintegrations sec^{-1}
5 2.76×10^{10} disintegrations sec^{-1}

PROBLEM 13–13

Potassium constitutes about 0.35% of the weight of an adult. Naturally occurring potassium contains 0.012% $^{40}_{19}K$, a β^- emitter ($t_{1/2} = 4.5 \times 10^{10}$ years). What will be the activity of the $^{40}_{19}K$ in the body of a 90-kg ($\simeq 200$ lb) man? [*Note:* 2.303 log$(N_0/N) = kt$.]

(1.0 year $= 3.2 \times 10^7$ sec)

1 0.025 disintegration sec^{-1}
2 0.27 disintegration sec^{-1}
3 25 disintegrations sec^{-1}
4 270 disintegrations sec^{-1}
5 4.5×10^{10} disintegrations sec^{-1}

PROBLEM 13–14

Carbon-14 ($^{14}_{6}C$), a β^- emitter with a half-life of 5770 years, is produced in the atmosphere by the interaction of cosmic rays with nitrogen. Because $^{14}_{6}C$ decays at the same rate at which it is formed, there is a constant ratio of $^{14}_{6}C$ to $^{12}_{6}C$ in the CO_2 of the atmosphere. Living organisms establish equilibrium with atmospheric CO_2; thus, the ratio of $^{14}_{6}C/^{12}_{6}C$ in them is identical to that in the atmosphere. On death the exchange between an organism and the atmosphere is halted; hence, the equilibrium is destroyed. No new $^{14}_{6}C$ is incorporated into the organism, and the $^{14}_{6}C/^{12}_{6}C$ ratio decreases in accord with the 5770-year half-life of $^{14}_{6}C$. Therefore, measurement of the $^{14}_{6}C/^{12}_{6}C$ ratio for such a dead organism provides a means of calculating the time of its death.

A sample of wood charcoal from a cypress beam in the tomb of an Egyptian king has a $^{14}_{6}C/^{12}_{6}C$ ratio of 54.2% that of present-day charcoal. Determine when, approximately, the cypress tree was cut down (and presumably when the king was entombed). [*Note:* $2.303 \log(N_0/N) = kt$.]

1 500 A.D.
2 100 A.D.
3 3100 B.C.
4 5100 B.C.
5 5770 B.C.

PROBLEM 13–15

You are at a dinner party and your host brings out a bottle of brandy for which he paid a great deal of money because he was told that it was bottled in the time of Napoleon (1769–1821). Since you are a research chemist, you are asked if there is any way in which the age of the brandy can be determined. You assert that tritium ($^{3}_{1}H$) dating is the way to do it. Ordinary water contains a very small amount of tritium that is formed in the earth's atmosphere by reactions involving cosmic rays. Tritium decomposes by β^- emission at a rate equal to that at which it is formed, and water exposed to the atmosphere will contain a constant tritium content. Since brandy contains water, it should be possible to estimate its age. Because the tritium in the bottle will not be replaced after it decays, the older the brandy, the lower its tritium content. You agree to test a sample of your host's brandy. You find that the sample has a tritium content 30.6% of that of present-day water. How long ago was the brandy bottled? [*Note:* $2.303 \log (N_0/N) = kt$.]

($t_{1/2}$: $^{3}_{1}H = 12.3$ years)

1 12.3 years
2 21.0 years
3 30.6 years
4 120 years
5 170 years

PROBLEM 13–16

Nuclear chemists and physicists abbreviate induced nuclear reactions as follows:

$^{14}_{7}N(\alpha, p)^{17}_{8}O$

where the participants (reading from left to right) are the target nucleus, bombarding particle, ejected particle (p denotes $^{1}_{1}H$), and product nucleus. Which one of the proposed induced nuclear reactions given below would *not* give the product suggested? [*Note:* $d = {}^{2}_{1}H$.]

1 $^{10}_{5}B(\alpha, n)^{13}_{7}N$
2 $^{14}_{7}N(n, p)^{14}_{6}C$
3 $^{28}_{14}Si(d, n)^{29}_{15}P$
4 $^{31}_{15}P(n, 2n)^{30}_{15}P$
5 $^{242}_{96}Cm(\alpha, 2n)^{243}_{97}Bk$

PROBLEM 13–17

The preparation of new elements is treated in Section 13–4 of Chemical Principles.

Plutonium-239 ($^{239}_{94}Pu$) is used as a fissionable material in some nuclear power reactors. But plutonium-239 does not exist naturally and must be produced artificially in "breeder reactors." These reactors not only generate power but also produce the nuclide to supply the plutonium-fueled power reactors. Which of the following reactions might be considered for production of $^{239}_{94}Pu$?

(a) $^{238}_{94}Pu(d, p)$
(b) $^{243}_{95}Am(p, n)$
(c) $^{242}_{96}Cf(n, \alpha)$
(d) $^{238}_{92}U(\alpha, 3n)$
(e) $^{236}_{93}Np(p, \alpha)$

1 Only (d).
2 Only (b) and (e).
3 Only (a), (c), and (d).
4 None of them.
5 All of them.

PROBLEM 13–18

Figure 13–1 shows the energy relationships in the decay of $^{242}_{94}Pu$ to $^{238}_{92}U$. Two alpha particles, α_1 and α_2, and a γ ray are emitted. Predict the energy of the α_1 alpha particle.

($h = 6.626 \times 10^{-27}$ erg sec;
$c = 2.998 \times 10^{10}$ cm sec^{-1};
1 MeV $= 1.602 \times 10^{-6}$ erg)

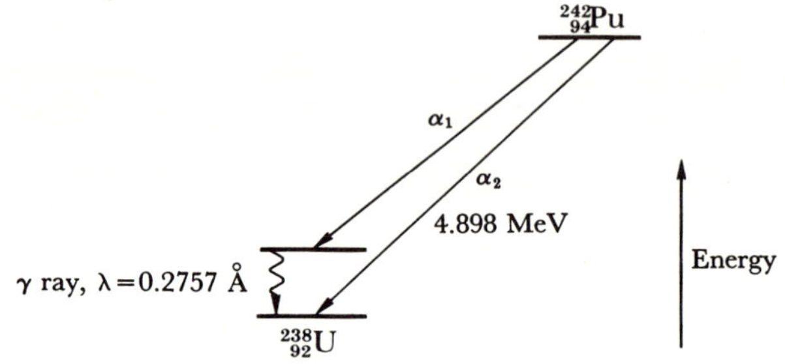

Figure 13–1

1 4.853 MeV
2 4.883 MeV
3 4.898 MeV
4 4.943 MeV
5 6.626 MeV

PROBLEM 13–19

The mass distribution of the fission products from the slow neutron fission of $^{235}_{92}U$ is illustrated in Figure 13–2. Which one of the following nuclides is the most likely to be a product of this induced fission?

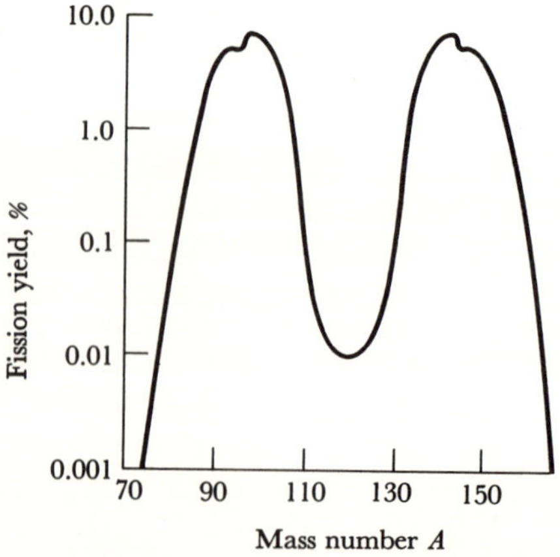

Figure 13–2

1 $^{13}_{7}N$
2 $^{16}_{8}O$
3 $^{95}_{45}Rh$
4 $^{117}_{50}Sn$
5 $^{232}_{90}Th$

PROBLEM 13-20

The overall reaction for one proposed mechanism (the proton–proton cycle) believed to be responsible for the energy of the sun is the transformation of four hydrogen atoms into one helium atom and two positrons:

$$4\,^{1}_{1}H \rightarrow {}^{4}_{2}He + 2\,{}^{0}_{+1}e$$

Estimate the energy released per atom of He formed in this reaction? [*Note:* Neglect the positron mass.]

(isotopic wt: $^{1}_{1}H = 1.00782$ amu, $^{4}_{2}He = 4.00260$ amu; 1 amu = 931 MeV)

1 0.029 MeV (atom He)$^{-1}$
2 27 MeV (atom He)$^{-1}$
3 39 MeV (atom He)$^{-1}$
4 270 MeV (atom He)$^{-1}$
5 931 MeV (atom He)$^{-1}$

PROBLEM 13-21

Several applications of nuclear chemistry are described in Section 13-5 of Chemical Principles.

A proposed fuel for controlled nuclear fusion is lithium deuteride, LiD (where D = $^{2}_{1}H$). Which of the following reactions would possibly contribute to energy production?

(a) $^{6}_{3}Li + {}^{1}_{0}n \rightarrow {}^{4}_{2}He + {}^{3}_{1}H$
(b) $^{6}_{3}Li + {}^{2}_{1}H \rightarrow 2\,{}^{4}_{2}He$
(c) $2\,^{2}_{1}H \rightarrow {}^{3}_{2}He + {}^{1}_{0}n$

(isotopic wt: $^{1}_{0}n = 1.00867$, $^{2}_{1}H = 2.01410$, $^{3}_{1}H = 3.01605$, $^{3}_{2}He = 3.01603$, $^{4}_{2}He = 4.00260$, $^{6}_{3}Li = 6.01513$)

1 (a)
2 (b)
3 (c)
4 All of them.
5 None of them.

SOLUTIONS

SOLUTION 13–1

(1)

The contribution of each isotope to the atomic weight is in proportion to its relative natural abundance. Thus, per gram-atom of neon

$^{20}_{10}$Ne contributes $19.99244 \times 0.9092000 = 18.17713$ amu
$^{21}_{10}$Ne contributes $20.99395 \times 0.002570000 = 0.05395$ amu
$^{22}_{10}$Ne contributes $21.99138 \times 0.08820000 = 1.93964$ amu

Total 20.17072 amu

Thus, the atomic weight of neon (based on $^{12}_{6}$C = 12.00000 amu) is 20.17072 amu.

Answer **1** is correct.

SOLUTION 13–2

(4)

A nucleon is the general name given to a proton or a neutron in a nucleus. Thus, the total number of nucleons in a nucleus is equal to the total number of protons and neutrons in the nucleus.

Answer **4** is correct.

SOLUTION 13–3

(4)

Table 13–1 indicates the mass number, number of protons, atomic number, and number of neutrons for each of the three nuclides.

Table 13–1

Nuclide	Mass number	Number of protons	Atomic number	Number of neutrons
$^{14}_{6}$C	14	6	6	8
$^{15}_{7}$N	15	7	7	8
$^{17}_{9}$F	17	9	9	8

The three nuclides differ in mass number and atomic number but have the same number of neutrons; every other nuclide with eight neutrons will be isotonic with them.

4 is correct.
1 Nuclides do not have any electrons; they are atomic nuclei.
2 is the relationship between *isobars,* such as $^{14}_{6}$C and $^{14}_{7}$N, and $^{110}_{47}$Ag and $^{110}_{49}$In.
3 is the relationship between *isotopes.*

SOLUTION 13–4

(4)

Let R_{Li} and R_{Os} be the radii of the lithium and osmium nuclei, respectively. The radius of the nucleus of an atom is directly proportional to the cube root of the number of nucleons in the nucleus. So

$$R_{Li} = k(7)^{1/3}$$

and

$$R_{Os} = k(189)^{1/3}$$

in which k is a proportionality constant. Hence

$$\frac{R_{Os}}{R_{Li}} = \left(\frac{189}{7}\right)^{1/3} = (27)^{1/3} = 3$$

or

$$R_{Os} = 3R_{Li}$$

The radius is *tripled* in going from ${}^{7}_{3}Li$ to ${}^{189}_{76}Os$.

The correct answer is **4**.

[*Note:* The nucleus of an atom is much smaller than the atom itself; for example, the lithium nucleus has a radius of approximately 2.7×10^{-13} cm, whereas the radius of the lithium atom is 1.23×10^{-8} cm.]

SOLUTION 13–5

(5)

Our example of α decay is

$${}^{232}_{90}Th \rightarrow {}^{228}_{88}Ra$$

The mass number decreases by four units, and the atomic number decreases by two units. Therefore, an α particle is a helium nucleus, ${}^{4}_{2}He$. Let us see which one of the equations for the transmutations cannot be balanced by adding a helium nucleus to the right-hand side.

1 ${}^{238}_{92}U \longrightarrow {}^{234}_{90}Th + {}^{4}_{2}He$; balances.
2 ${}^{227}_{89}Ac \longrightarrow {}^{223}_{87}Fr + {}^{4}_{2}He$; balances.
3 ${}^{226}_{88}Ra \longrightarrow {}^{222}_{86}Rn + {}^{4}_{2}He$; balances.
4 ${}^{212}_{84}Po \longrightarrow {}^{208}_{82}Pb + {}^{4}_{2}He$; balances.
5 ${}^{213}_{83}Bi \not\longrightarrow {}^{213}_{84}Po + {}^{4}_{2}He$; neither the mass numbers nor the atomic numbers balance.

The transmutation of ${}^{213}_{83}Bi$ into ${}^{213}_{84}Po$ does not take place by α decay but by β^- decay (electron emission). In β^- decay, there is no change in mass number, but the atomic number increases by one unit because the electron has an "atomic number" (number of units of positive charge) of -1.

$${}^{213}_{83}Bi \rightarrow {}^{213}_{84}Po + {}^{0}_{-1}e$$

Answer **5** is correct.

SOLUTION 13–6

(1)

β^- decay involves the emission of an electron. Let us see which one of the equations for the five transmutations does not balance when an electron is added to the right-hand side.

1 $^{64}_{29}Cu \not\rightarrow {}^{64}_{28}Ni + {}^{0}_{-1}e$; the mass numbers balance but the atomic numbers do not.
2 $^{114}_{49}In \longrightarrow {}^{114}_{50}Sn + {}^{0}_{-1}e$; balances.
3 $^{124}_{51}Sb \longrightarrow {}^{124}_{52}Te + {}^{0}_{-1}e$; balances.
4 $^{212}_{83}Bi \longrightarrow {}^{212}_{84}Po + {}^{0}_{-1}e$; balances.
5 $^{234}_{90}Th \longrightarrow {}^{234}_{91}Pa + {}^{0}_{-1}e$; balances.

The transmutation of $^{64}_{29}Cu$ into $^{64}_{28}Ni$ does not take place by β^- decay but by β^+ decay [positive electron or *positron* ($^{0}_{+1}e$) emission]. In β^+ decay, there is no change in mass number, but the atomic number *decreases* by one unit because the positron has an "atomic number" of +1.

$$^{64}_{29}Cu \rightarrow {}^{64}_{28}Ni + {}^{0}_{+1}e$$

The correct answer is **1**.

SOLUTION 13–7

(3)

$^{235}_{92}U$ decays through a succession of α and β^- steps to produce $^{207}_{82}Pb$. [The decay takes an extremely long time because $^{235}_{92}U$ and one of its intermediate decay products, protactinium-231 ($^{231}_{91}Pa$), have particularly long half-lives: 8.8×10^8 years and 3.43×10^4 years, respectively.] Let us compare the mass numbers and the atomic numbers of $^{235}_{92}U$ and $^{207}_{82}Pb$. There is a decrease in mass number of 28 units and a decrease in atomic number of 10 units. β^- decay does not involve a change in mass number; hence, the 28-unit mass number change must be the result of $28/4 = 7$ alpha decays (losses of $^{4}_{2}He$). Seven α decays would give us platinum-207 ($^{207}_{78}Pt$), which has an atomic number that is four units lower than $^{207}_{82}Pb$. To raise the atomic number from 78 to 82 we need four β^- decays (losses of $^{0}_{-1}e$). Thus, our final decay equation is

$$^{235}_{92}U \rightarrow {}^{207}_{82}Pb + 7\,{}^{4}_{2}He + 4\,{}^{0}_{-1}e$$

So there are seven α and four β^- decays per atom of $^{207}_{82}Pb$ formed.

The correct answer is **3**. Answer **5** is the formation of two atoms of $^{207}_{82}Pb$.

SOLUTION 13-8

(2)

Binding energy of the nucleus is discussed in Section 13-1 of Chemical Principles.

In terms of its mass, the manganese isotope ($^{55}_{25}Mn$) is composed of 25 hydrogen atoms (i.e., 25 protons plus 25 orbital electrons) and 30 neutrons.

25 hydrogen atoms weigh 25 × 1.00782	= 25.1955 amu
30 neutrons weigh 30 × 1.00867	= 30.2601 amu
mass of constituent particles	= 55.4556 amu
mass of $^{55}_{25}Mn$ isotope	= 54.9381 amu
mass difference	= 0.5175 amu

The energy equivalent of 1.00 amu is 931 MeV. Thus, the binding energy of the $^{55}_{25}Mn$ nucleus is

$$0.5175 \text{ amu} \times 931 \text{ MeV amu}^{-1} \simeq 482 \text{ MeV}$$

There are 55 nucleons (25 protons and 30 neutrons) in the $^{55}_{25}Mn$ nucleus, so the binding energy per nucleon is

$$\frac{482 \text{ MeV}}{55 \text{ nucleons}} \simeq 8.8 \text{ MeV nucleon}^{-1}$$

The correct answer is **2**. Answer **5** is the total energy of all of the nucleons in the nucleus.

SOLUTION 13-9

(3)

For spontaneous, exothermic decay to take place, the mass of the product (or daughter) nuclide must be less than that of the original (or parent) nuclide. This rules out decay of $^{32}_{15}P$ to $^{32}_{14}Si$ but not to $^{32}_{16}S$. The atomic number of $^{32}_{16}S$ is one unit greater than that of $^{32}_{15}P$; therefore, the decay occurs with the emission of a β^- particle and release of energy, Q (decay energy):

$$^{32}_{15}P \rightarrow {}^{32}_{16}S + {}^{0}_{-1}e + Q$$

The isotopic masses of $^{32}_{15}P$ and $^{32}_{16}S$ are 31.9739 amu and 31.9721 amu, respectively. Therefore, Q is given by

$$Q = (31.9739 - 31.9721) \text{ amu} \times 931 \text{ MeV amu}^{-1}$$
$$= 0.0018 \times 931 \text{ MeV}$$
$$\simeq 1.7 \text{ MeV}$$

The correct answer is **3**.

SOLUTION 13–10

(3)

$$2.303 \log\left(\frac{N_0}{N}\right) = kt \qquad \text{(A)}$$

When $N = N_0/2$, $t = t_{1/2}$ and $2.303 \log 2 = kt_{1/2}$, so

$$t_{1/2} = \frac{0.693}{k} \qquad \text{(B)}$$

Therefore, combining Equations (A) and (B) yields

$$t_{1/2} = \frac{0.301\ t}{\log(N_0/N)}$$

Activity of $^{90}_{38}Sr$ at start $= k \times N_0 = 1000$ disintegrations min^{-1}. Activity of $^{90}_{38}Sr$ after 1.00 year $= k \times N = 975.2$ disintegrations min^{-1}. Therefore

$$\frac{N_0}{N} = \frac{1000}{975.2}$$

and

$$t_{1/2} = \frac{0.301 \times 1.00 \text{ year}}{\log(1000/975.2)}$$

$$= \frac{0.301}{3.0000 - 2.9891}$$

$$= 27.6 \text{ years}$$

Answer **3** is correct. Answer **2**: You dropped the factor 2.303 somewhere.

SOLUTION 13–11

(3)

Activity (rate of decay) = number of disintegrations sec^{-1} $= k \times N$, in which k is the rate constant in sec^{-1} and N is the number of atoms present.

$$k = \frac{0.693}{t_{1/2}} = \frac{0.693}{27.6 \text{ years} \times 3.15 \times 10^7 \text{ sec year}^{-1}}$$

(Check back to Problem 13–10 for the derivation of this equation.) The number of atoms (N) present in 1.00 μC of $^{90}_{38}Sr$ is therefore given by

$$N = \frac{\text{activity}}{k}$$

$$= \frac{27.6 \text{ years} \times 3.15 \times 10^7 \text{ sec year}^{-1} \times 1.00 \times 10^{-6} \text{ C} \times 3.70 \times 10^{10} \text{ atoms sec}^{-1} \text{ C}^{-1}}{0.693}$$

$$= 4.64 \times 10^{13} \text{ atoms}$$

The weight of $^{90}_{38}Sr$ corresponding to the 1.00 μC maximum permissible dose is

$$\frac{4.64 \times 10^{13} \text{ atoms} \times 90 \text{ g g-atom}^{-1}}{6.02 \times 10^{23} \text{ atoms g-atom}^{-1}} = 6.9 \times 10^{-9} \text{ g}$$

This is of the order of 10^{-11}% of adult body weight!

The correct answer is **3**. Answer **4**: You did the calculation for 1 C instead of 1 μC. Answer **5** is 1 C with the wrong units.

SOLUTION 13–12

(2)

$$t_{1/2} = \frac{0.301\ t}{\log(N_0/N)}$$

(Refer back to Problem 13–10 for the derivation of this equation.) Therefore

$$\log\left(\frac{N_0}{N}\right) = \frac{0.301 \times 20.0 \text{ years}}{27.6 \text{ years}} = 0.2181$$

whence

$$\frac{N_0}{N} = 1.65$$

Activity of $^{90}_{38}Sr$ at start

$= 1.00 \times 10^{-6} \times 3.70 \times 10^{10}$ disintegrations sec^{-1}
$= 3.70 \times 10^{4}$ disintegrations sec^{-1}
$= k \times N_0$

Activity of $^{90}_{38}Sr$ after 20.0 years $= k \times N$

$$= \frac{N}{N_0} \times 3.70 \times 10^4 \text{ disintegrations sec}^{-1}$$

$$= \frac{3.70 \times 10^4}{1.65} \text{ disintegrations sec}^{-1}$$

$= 2.24 \times 10^4$ disintegrations sec^{-1}

So even after 20.0 years the activity of the $^{90}_{38}Sr$ is about 60% of what it was originally.

The correct answer is **2**. Answer **4**: You did the calculation for 1 C instead of 1 μC.

SOLUTION 13–13

(4)

Weight of $^{40}_{19}K$ in the man's body

$= 90 \text{ kg} \times 1.0 \times 10^3 \text{ g kg}^{-1} \times 0.0035 \times 0.00012$
$= 0.038$ g

The number of atoms (N) in 0.038 g is

$$\frac{0.038 \text{ g} \times 6.0 \times 10^{23} \text{ atoms g-atom}^{-1}}{40 \text{ g g-atom}^{-1}} = 5.7 \times 10^{20} \text{ atoms of } ^{40}_{19}\text{K}$$

Activity $= k \times N$, and

$$t_{1/2} = \frac{0.693}{k}$$

Therefore, the activity of the $^{40}_{19}K$ in the man's body is

$$\frac{0.693 \times N}{t_{1/2}} = \frac{0.693 \times 5.7 \times 10^{20} \text{ atoms}}{4.5 \times 10^{10} \text{ years} \times 3.2 \times 10^7 \text{ sec year}^{-1}}$$

$= 270$ atoms disintegrating per sec

Compare the activity with that of the 1 μC of $^{90}_{38}Sr$ in Problem 13–12. We need not worry too much about the radioactive potassium present in our bodies!

The correct answer is **4**. Answer **2**: You forgot to convert the man's weight to grams.

SOLUTION 13–14

(3)

Radiocarbon dating is described in Section 13–5 of Chemical Principles.

$$t = \frac{\log(N_0/N)\, t_{1/2}}{0.301}$$

(Check Problem 13–10 if you have forgotten how this is derived.) For every 100 atoms of $^{14}_{6}C$ that originally existed, there are 54.2 remaining. Thus

$$t = \frac{\log(100/54.2) \times 5770 \text{ years}}{0.301}$$

$$= \frac{0.266 \times 5770 \text{ years}}{0.301}$$

$$\simeq 5100 \text{ years}$$

$$\simeq 3100 \text{ B.C.}$$

Answer **3** is correct.

[*Note:* Before Professor Willard F. Libby developed the radiocarbon-dating method, archeologists had established that the Egyptian king's tomb was built in 2950 ± 200 B.C. This problem illustrates the utility of the simple radiocarbon-dating method. However, owing to nuclear testing, the $^{14}_{6}CO_2/^{12}_{6}CO_2$ ratio in the atmosphere may have been altered from what it was in bygone days; in the future the results from radiocarbon dating may become suspect.]

Answer **4** is the actual age in years.

SOLUTION 13–15

(2)

$$t = \frac{\log(N_0/N) \times t_{1/2}}{0.301}$$

$$= \frac{\log(100/30.6) \times 12.3 \text{ years}}{0.301}$$

$$= \frac{0.514 \times 12.3 \text{ years}}{0.301}$$

$$= 21.0 \text{ years}$$

Your host was the victim of a fraud!

The correct answer is **2**.

[*Note:* Even if the brandy had been bottled in Napoleon's time, you could not have used tritium dating to definitely establish this. The half-life of tritium is so short that none would be detectable after about 120 years ($10t_{1/2}$). The brandy would exhibit a complete lack of tritium activity; thus, all that you could say is that it was bottled over 120 years ago. You were fortunate that a better chemist was not at the party to point out the inadequacy of your test.]

SOLUTION 13–16 First we must write the shorthand equations in their long form.

(5)

1 $^{10}_{5}B + ^{4}_{2}He \rightarrow ^{1}_{0}n + ^{13}_{7}N$
2 $^{14}_{7}N + ^{1}_{0}n \rightarrow ^{1}_{1}H + ^{14}_{6}C$
3 $^{28}_{14}Si + ^{2}_{1}H \rightarrow ^{1}_{0}n + ^{29}_{15}P$
4 $^{31}_{15}P + ^{1}_{0}n \rightarrow 2\,^{1}_{0}n + ^{30}_{15}P$
5 $^{242}_{96}Cm + ^{4}_{2}He \rightarrow 2\,^{1}_{0}n + ^{243}_{97}Bk$

Next we check to see which one of the equations is improperly balanced. Answer **1** has mass numbers totaling 14 on the left and 14 on the right. It also has atomic numbers totaling 7 on the left and 7 on the right. Thus, Answer **1** is correct. Similarly, Answers **2**, **3**, and **4** all balance. Answer **5** is the incorrect one; both the mass numbers and the atomic numbers lack one unit on the right. To be balanced, the product nucleus would have to be $^{244}_{98}Cf$, not $^{243}_{97}Bk$.

The required answer is **5**.

SOLUTION 13–17 Completing the five equations, we obtain

(3)

(a) $^{238}_{94}Pu + ^{2}_{1}H \rightarrow ^{1}_{1}H + ^{239}_{94}Pu$
(b) $^{243}_{95}Am + ^{1}_{1}H \rightarrow ^{1}_{0}n + ^{243}_{96}Cm$
(c) $^{242}_{96}Cf + ^{1}_{0}n \rightarrow ^{4}_{2}He + ^{239}_{94}Pu$
(d) $^{238}_{92}U + ^{4}_{2}He \rightarrow 3\,^{1}_{0}n + ^{239}_{94}Pu$
(e) $^{236}_{93}Np + ^{1}_{1}H \rightarrow ^{4}_{2}He + ^{233}_{92}U$

Only (a), (c), and (d) result in the production of $^{239}_{94}Pu$.

The correct answer is **3**.

[*Note:* Although reactions (a), (c), and (d) lead to the production of $^{239}_{94}Pu$, none of them is actually used for this purpose. Plutonium-239 is prepared in kilogram amounts in nuclear breeder reactors from $^{238}_{92}U$ by the following sequence of reactions:

$$^{238}_{92}U + ^{1}_{0}n \rightarrow ^{239}_{92}U \xrightarrow[t_{1/2} = 23.5\text{ min}]{\beta^-} ^{239}_{93}Np \xrightarrow[t_{1/2} = 2.35\text{ days}]{\beta^-} ^{239}_{94}Pu]$$

SOLUTION 13–18

(1)

Gamma radiation is electromagnetic radiation (see Figure 8–5). So we can relate the wavelength of our γ ray to the energy of the nuclear transition between the excited and ground states of $^{238}_{92}U$.

$$E = h\nu = \frac{hc}{\lambda}$$

$$= \frac{6.626 \times 10^{-27} \text{ erg sec} \times 2.998 \times 10^{10} \text{ cm sec}^{-1}}{0.2757 \text{ Å} \times 10^{-8} \text{ cm Å}^{-1} \times 1.602 \times 10^{-6} \text{ erg MeV}^{-1}}$$

$$= 0.045 \text{ MeV}$$

Therefore, the energy of the α_1 alpha particle is

$$4.898 - 0.045 = 4.853 \text{ MeV}$$

(The experimental value is 4.854 MeV.)

The correct answer is **1**. Answer **2**: You omitted c and lost a few powers of ten somewhere. For Answer **4**, you *added* 4.898 MeV and 0.045 MeV instead of subtracted them.

SOLUTION 13–19

(3)

In the slow neutron fission of $^{235}_{92}U$, it is believed that a *compound nucleus* $^{236}_{92}U$ is first formed from $^{235}_{92}U$ and $^{1}_{0}n$. This compound nucleus decomposes to yield a variety of fission products. The fission products have mass numbers clustered in the regions of 90 and 140, as shown in the double-humped curve in Figure 13–2. One possible mode of fission could be

$$^{235}_{92}U + ^{1}_{0}n \rightarrow ^{236}_{92}U \rightarrow ^{94}_{38}Sr + ^{140}_{54}Xe + 2\,^{1}_{0}n + \text{energy} + \gamma$$

Of the five nuclides given, only $^{95}_{45}Rh$ would lie anywhere near either of the two maxima in Figure 13–2; hence, it is the most likely of the five to be a fission product.

The correct answer is **3**. Answer **5** has an *atomic number*, not a mass number, of 90.

SOLUTION 13–20

(2)

$$\begin{aligned} \text{weight of 4 } ^{1}_{1}\text{H's} &= 4 \times 1.00782 = 4.03128 \text{ amu} \\ \text{weight of } ^{4}_{2}\text{He} &= 4.00260 \text{ amu} \\ \text{weight difference} &= 0.02868 \text{ amu} \end{aligned}$$

Since 1 amu = 931 MeV, the energy released per atom of He formed

$$= 0.029 \text{ amu (atom He)}^{-1} \times 931 \text{ MeV amu}^{-1}$$
$$= 27 \text{ MeV (atom He)}^{-1}$$

Answer **2** is correct. Answer **1** is the mass difference with the wrong units. Answer **4** has a decimal point error.

SOLUTION 13–21 (4)

When we make our mass difference calculation just as we did in Problem 13–20, we find that the isotopic weight of the products is smaller than that of the reactants for all three cases. So all three reactions are exothermic and are possible energy-producing reactions.

Answer **4** is correct.

[*Note:* The control of the fusion process for power production has excited the imagination of scientists for many years, but has so far eluded them. Fusion has many advantages over fission: The fuels and products are less hazardous, and the fuel is readily available (D_2 can be extracted from D_2O in seawater), but uranium supplies are limited. Recently, a new advance has been announced: Fusion of D atoms, which requires extremely high temperatures (thermonuclear), was initiated by an extremely high energy pulsed laser.]

CHAPTER 14 BONDING IN SOLIDS AND LIQUIDS

We have discussed the forces that hold atoms together to form molecules. But what holds molecules together to form bulk substances—liquids and solids? This question is the topic of this chapter.

No memorization should be necessary for this chapter.

Solutions to problems in Chapter 14 begin on page 388.

PROBLEMS

PROBLEM 14–1

Van der Waals forces are described in Section 14–2 of Chemical Principles *by Dickerson, Gray, and Haight.*

The potential energy curve for the van der Waals interactions between two argon atoms in the gas state is illustrated in Figure 14–1. For comparison the potential energy curve (not drawn to scale) for the H—H covalent bond in the H_2 molecule is also shown. Which one of the following statements is *incorrect?*

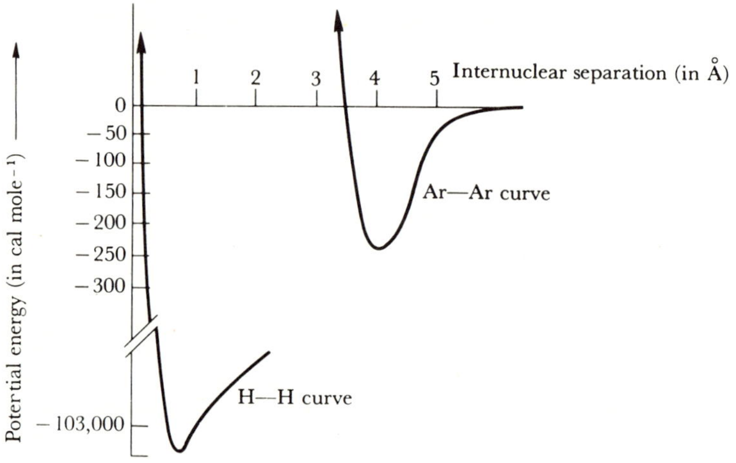

Figure 14–1

1 The equilibrium internuclear separation between Ar atoms is about 4 Å.

2 About 250 cal $mole^{-1}$ of energy are required to separate the Ar atoms to an infinite distance.

3 When the internuclear separation between the Ar atoms is about 4 Å there is a balance between the attractive and repulsive forces between the Ar atoms.

4 Because the H—H potential minimum is so deep, the equilibrium internuclear separation must be greater than that for Ar—Ar.

5 The Ar_2 molecule is much less stable than the H_2 molecule.

PROBLEM 14–2

Which one of the following instantaneous electron cloud diagrams best represents the attraction between two Ar atoms?

1

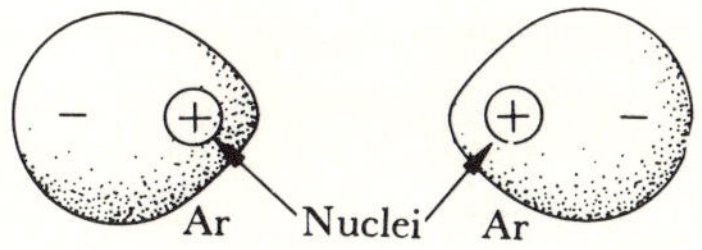

2

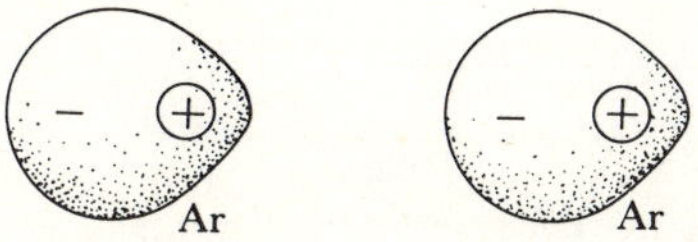

3

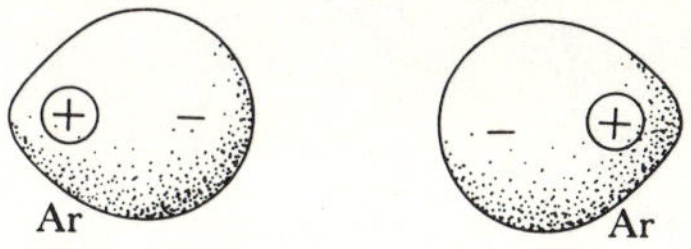

4

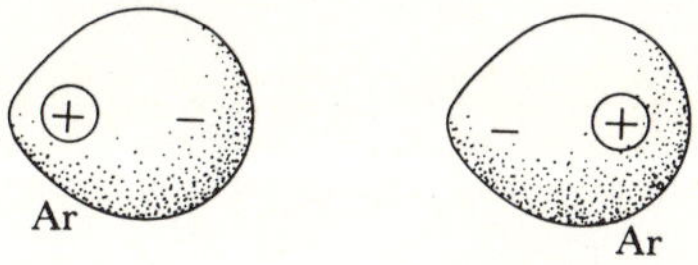

5

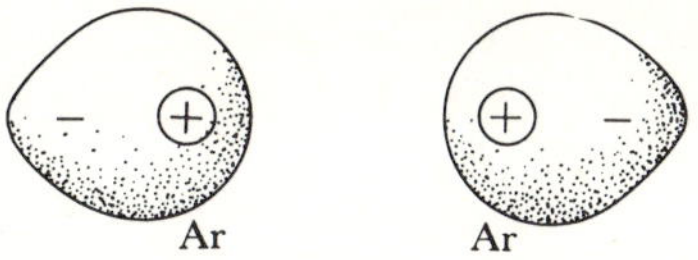

PROBLEM 14–3

Assuming ideal gas behavior for argon, the average kinetic energy of a mole of Ar gas at the melting point of solid Ar should be equal to the energy of van der Waals interaction in solid Ar. If the latter is 250 cal mole^{-1}, predict the melting point of solid Ar. [*Hint:* Refer to Problem 2–16.]

($R = 1.99$ cal mole^{-1} deg^{-1})

1 −189°K
2 83.8°K
3 167°K
4 250°K
5 273°K

PROBLEM 14–4

The influence of permanent and induced dipoles on bonding is discussed in Sections 14–2 and 14–4 of Chemical Principles.

Figure 14–2 is a plot of the boiling points of the binary hydrogen compounds of some of the representative elements. Which one of the following compounds would you expect to exhibit hydrogen bonding?

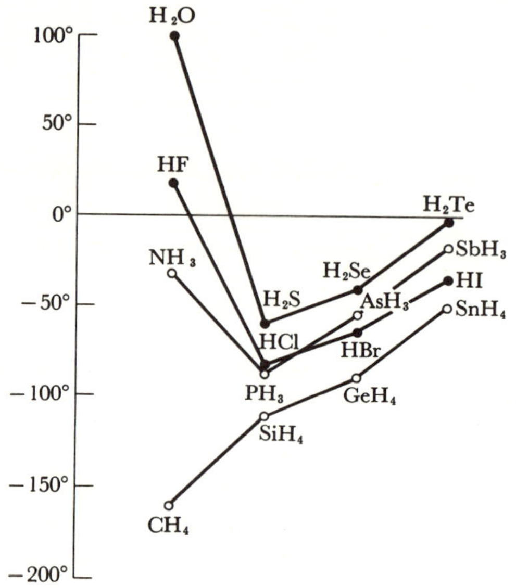

Figure 14–2

1 PH_3
2 CH_4
3 H_2O
4 HBr
5 H_2Se

PROBLEM 14–5

The effect of hydrogen bonding on the physical properties of liquids is discussed in Section 14–4 of Chemical Principles.

Hydrogen fluoride (HF) in aqueous solution is used for etching glass, for example, for making the scales on thermometers.

In the pure state (liquid or vapor), hydrogen fluoride consists of associated HF molecules such as H_2F_2 and H_3F_3. Which one of the following formulas represents the associated H_3F_3 species? [*Note:* The dotted lines represent hydrogen bonds.]

(H, Z = 1; F, Z = 9)

1 H^+—F^-—F^-—F^-—H^+ (with H^+ attached by dotted line below the middle F^-)

2 F^+—H^-···F^+
  | | ⋮
  H^- F^+···H^-

3 F^+—H^-···F^+—H^-···F^+—H^-

4 F^-···H^+···F^-
  | | |
  H^+···F^- H^+

5 F^-—H^+···F^-—H^+···F^-—H^+

PROBLEM 14–6 In the vapor state, methanol (CH_3OH) consists of four methanol molecules (a *tetramer*) held together by hydrogen bonds. Which one of the following structures represents the structure of a methanol tetramer?

1 $CH_3{-}O^-{-}H^+\cdots CH_3{-}O^-{-}H^+\cdots CH_3{-}O^-{-}H^+\cdots CH_3{-}O^-{-}H^+$

2 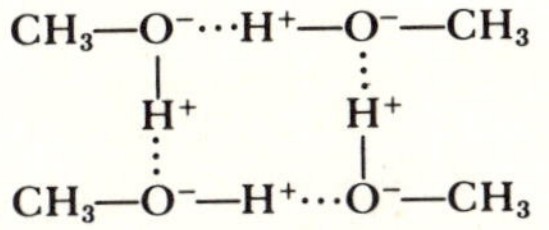

3 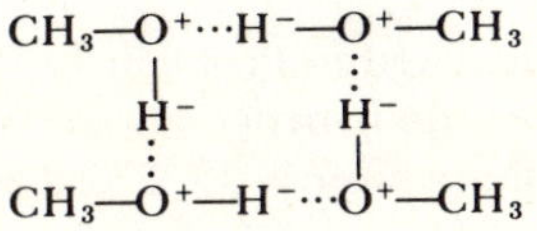

4 $CH_3{-}O^+{-}H^-\cdots CH_3{-}O^+{-}H^-\cdots CH_3{-}O^+{-}H^-\cdots CH_3{-}O^+{-}H^-$

5 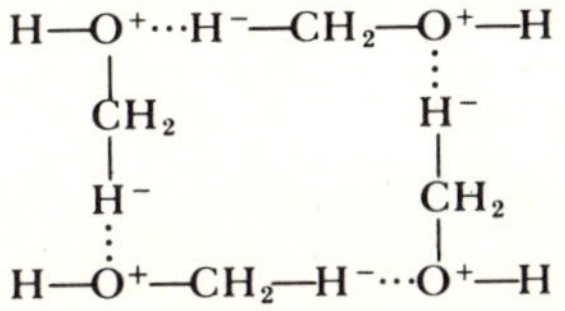

PROBLEM 14–7 GeH_4 and SnH_4 are isoelectronic with H_2Se and H_2Te, respectively. Using the data given below, estimate the heat of sublimation of a *hypothetical* form of ice in which the water molecules are *not* hydrogen bonded. [*Hint:* Refer to Figure 14–2.]

Heats of sublimation (kcal mole^{-1})	
GeH_4	4.0
H_2Se	5.3
SnH_4	5.0
H_2Te	6.6
CH_4	2.0

(H, $Z = 1$; C, $Z = 6$; O, $Z = 8$; Ge, $Z = 32$; Se, $Z = 34$; Sn, $Z = 50$; Te, $Z = 52$)

1 0.38 kcal mole^{-1}
2 1.5 kcal mole^{-1}
3 2.6 kcal mole^{-1}
4 5.2 kcal mole^{-1}
5 6.6 kcal mole^{-1}

PROBLEM 14–8

What is the maximum possible number of hydrogen bonds in which an H_2O molecule can participate? [*Hint:* Refer to Problem 14–4.]

(H, $Z = 1$; O, $Z = 8$)

1 1
2 2
3 3
4 4
5 5

PROBLEM 14–9

The heat of sublimation calculated in Problem 14–7 for a hypothetical form of ice in which there are no hydrogen bonds was 2.6 kcal $mole^{-1}$. If the observed heat of sublimation is 12.2 kcal $mole^{-1}$, what is the energy of a hydrogen bond in ice? [*Hint:* Use the structure for ice given in Solution 14–8.]

1 2.6 kcal $mole^{-1}$
2 4.8 kcal $mole^{-1}$
3 7.4 kcal $mole^{-1}$
4 9.6 kcal $mole^{-1}$
5 12.2 kcal $mole^{-1}$

PROBLEM 14–10

Hydrogen bonding in solids is discussed in Section 14–4 of Chemical Principles.

The heat of fusion of ice is 1.4 kcal $mole^{-1}$, and that of a hypothetical form of ice in which there are no hydrogen bonds is 0.30 kcal $mole^{-1}$. Assuming that the energy of a hydrogen bond is 4.8 kcal $mole^{-1}$, determine what percent of the hydrogen bonds in ice were broken during fusion. [*Note:* Refer to Problem 14–8.]

1 3.6%
2 11%
3 18%
4 23%
5 36%

PROBLEM 14–11 In 1965, a form of water was "discovered" with properties very different from those of normal water: low vapor pressure, high density, solidification at a maximum temperature of −40°C to a glassy form with a much lower expansion than that of ordinary ice, and stability to 500°C. One structure proposed for this "anomalous water" is shown in Figure 14–3. The planar or puckered layer of hexagons is held together by very strong (60–100 kcal $mole^{-1}$) hydrogen bonds. The layer is negatively charged and is held to adjoining layers by protons or hydronium ions (H_3O^+). What is the total charge of the anomalous water layer in Figure 14–3?

[*Note:* After several years of controversy, the staunchest defender of anomalous water now agrees that it does not exist, and that the reported unusual properties are due to impurities in ordinary water.]

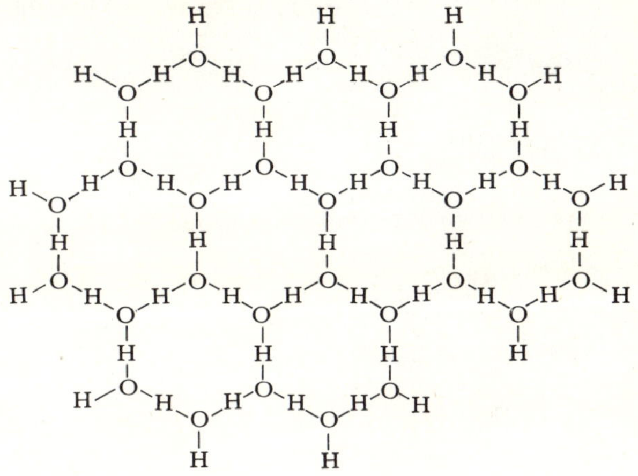

Figure 14–3

1 −112
2 −60
3 −8
4 −2
5 −1

We have been examining many molecules that exhibit intermolecular hydrogen bonding resulting in the formation of molecular aggregates. But hydrogen bonding also occurs between atoms in the same molecule, that is, intramolecular hydrogen bonding, as illustrated in Problems 14–12 and 14–13.

PROBLEM 14–12

In a series of related molecules capable of forming hydrogen bonds, the molecule that undergoes intramolecular rather than intermolecular hydrogen bonding will have the lowest boiling point since it will be less associated than the other molecules. With this in mind, determine which one of the following molecules would have the lowest boiling point.

1 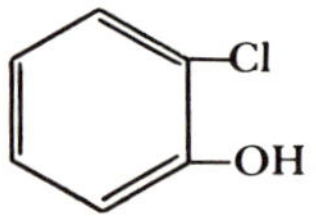

(2-chlorophenol)

2 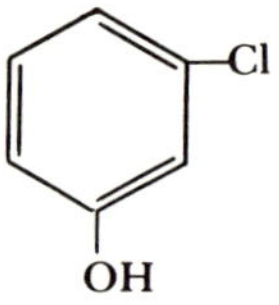

(3-chlorophenol)

3 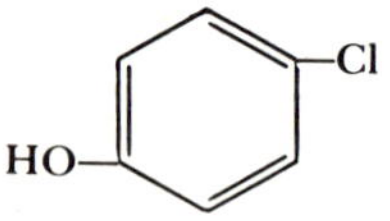

(4-chlorophenol)

4 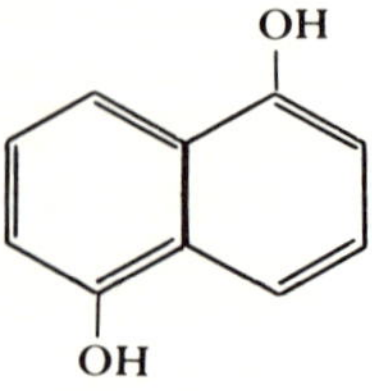

(1,5-dihydroxynaphthalene)

5 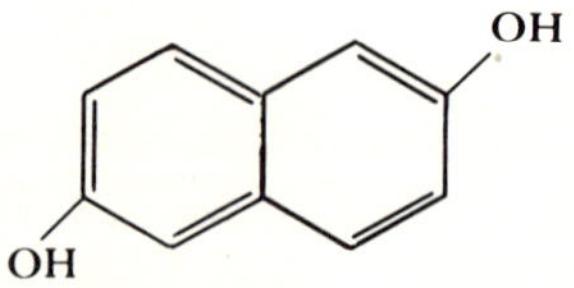

(2,6-dihydroxynaphthalene)

PROBLEM 14–13 Deoxyribonucleic acid (DNA) is an important molecule in cell duplication. DNA consists of a number of units, called *nucleotides*, joined together to form a polynucleotide chain. Each nucleotide is made of three parts: a phosphoric acid molecule, a sugar molecule (deoxyribose), and a "nitrogen base." The nitrogen bases in DNA are *purines* (adenine and guanine) and *pyrimidines* (thymine and cytosine). The planar structures of these molecules are shown below. The double helix structure of DNA is held together by the formation of hydrogen bonds between adenine and thymine and between guanine and cytosine.

If you were proposing a structure for DNA, how many hydrogen bonds would you predict to be between (a) the adenine and thymine molecules and (b) the guanine and cytosine molecules?

(a) (b)

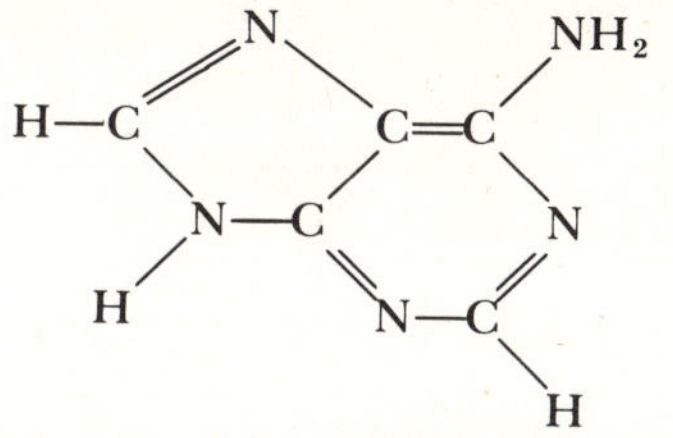

Adenine

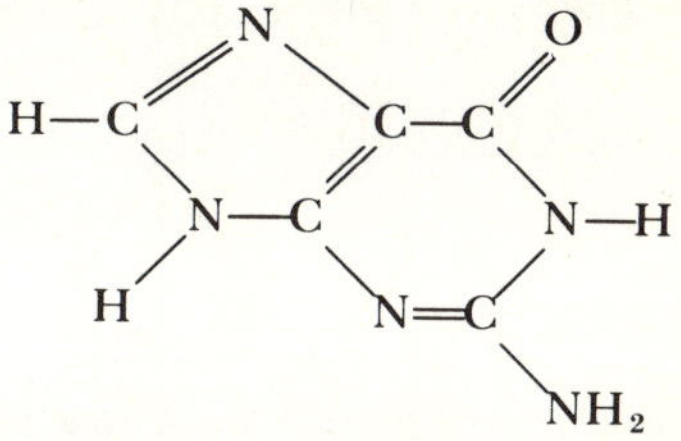

Guanine

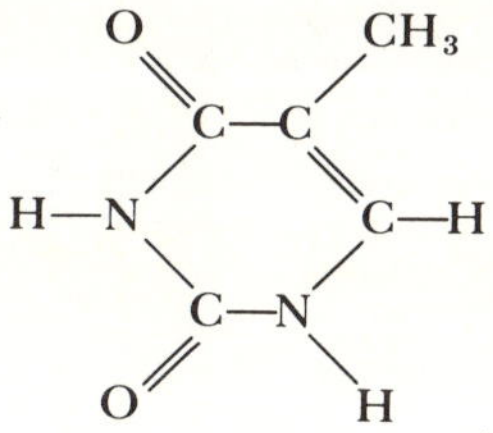

Thymine

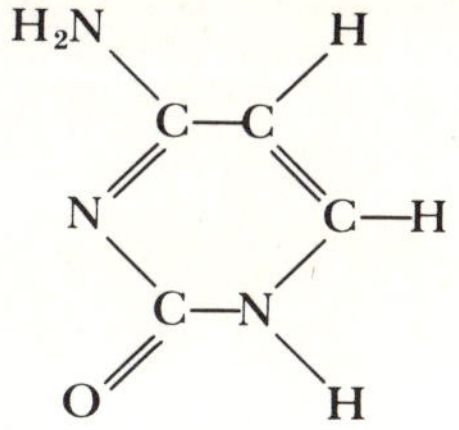

Cytosine

	(a)	(b)
1	1	1
2	1	2
3	2	2
4	2	3
5	3	2

PROBLEM 14–14

The structures of several types of silicates are described in Section 14–5 of Chemical Principles.

Silicon is the second most abundant element in the earth's crust; oxygen is the most abundant. The silicon occurs in the form of silica (SiO_2) and mineral silicates. The fundamental unit for all silicate structures is the SiO_4^{4-} tetrahedron; different structures are possible because of the variety of ways in which the tetrahedra can be joined together by the sharing of oxygen atoms. The structures of the anions in the important mineral silicates *benitoite* and *beryl,* are shown in Figure 14–4. What are the formulas of these two anions?

[*Note:* Closed circles are Si atoms; open circles are O atoms.]

(O, $Z = 8$; Si, $Z = 14$)

I *Benitoite* II *Beryl*

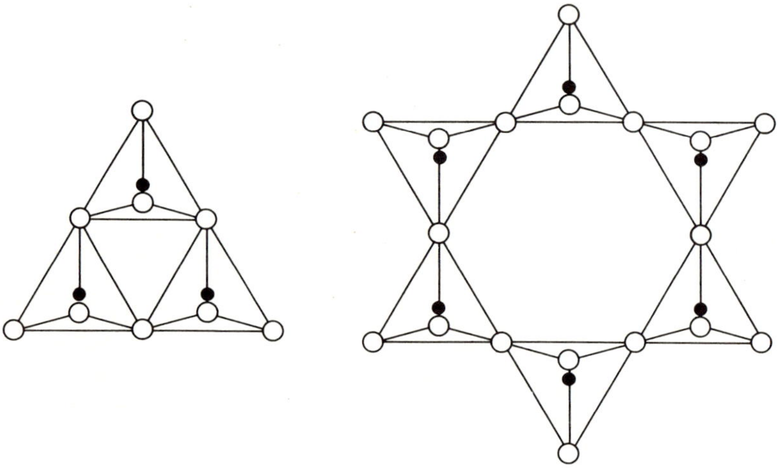

Figure 14–4

	I	II
1	$Si_3O_{10}^{7-}$	$Si_6O_{12}^{10-}$
2	$Si_3O_{18}^{12-}$	$Si_5O_{17}^{6-}$
3	$Si_3O_9^{6-}$	$Si_6O_{18}^{12-}$
4	$Si_2O_7^{6-}$	$Si_{14}O_{10}^{9-}$
5	$Si_3O_9^{6-}$	$Si_6O_{18}^{10-}$

PROBLEM 14–15 *Asbestos* is the name given to many fibrous silicate minerals. The most important asbestos, *chrysotile,* is a hydrated magnesium silicate. About 50% of the world's chrysotile asbestos is mined in Quebec, Canada. The silicate ion in chrysotile consists of double strands of SiO_4 tetrahedra as shown in Figure 14–5. What is the general composition of the silicate ion in chrysotile?

[*Note:* Closed circles are Si atoms; open circles are O atoms.]

(O, $Z = 8$; Si, $Z = 14$)

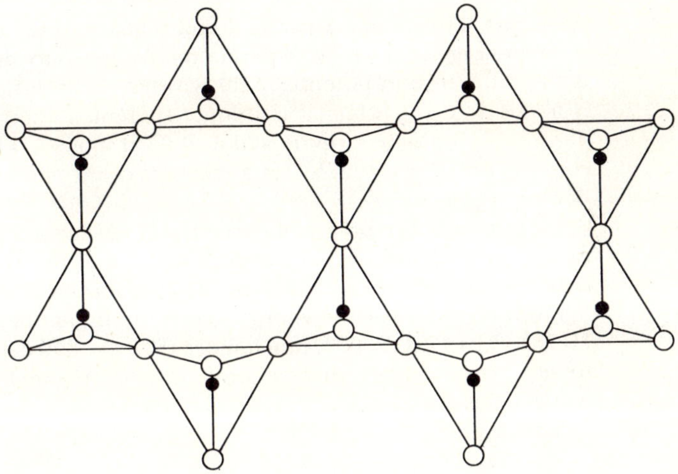

Figure 14–5

1 $(SiO_2)_n$
2 $(SiO_3)_n^{2n-}$
3 $(Si_2O_5)_n^{2n-}$
4 $Si_2O_7^{6-}$
5 $(Si_4O_{11})_n^{6n-}$

SOLUTIONS

SOLUTION 14–1

(4)

When the internuclear separation between Ar atoms is about 4 Å (Figure 14–1) the potential energy is at a minimum and there is an equilibrium between the attractive and repulsive forces between the Ar atoms. Thus, both Answers **1** and **3** are correct.

The energy required to separate the Ar atoms to an infinite distance is the difference in potential energy when the internuclear separation is about 4 Å and when it is infinite (potential energy = zero). From Figure 14–1, this energy is about 250 cal mole^{-1}. Thus, Answer **2** is correct. The 250 cal mole^{-1} is a measure of the van der Waals interaction in argon.

Even though the potential curve for H—H has a deep well, or minimum, the equilibrium internuclear separation for H—H is much *less* than that for Ar—Ar. Therefore, Answer **4** is incorrect.

From Figure 14–1 the energy needed to break the covalent bond in H_2 is 103,240 cal mole^{-1}, much more than the 250 cal mole^{-1} required to separate infinitely the Ar atoms in the Ar_2 molecule. Therefore Answer **5** is correct.

The required answer is **4**.

SOLUTION 14–2

(2)

Answer **2** is correct. The attraction energy between two Ar atoms arises because the electrons on the two atoms avoid each other as much as possible, thereby reducing the repulsion between themselves. At any instant, the Ar atom on the left may have an asymmetric rather than a symmetric electron cloud distribution and thus become a small *dipole.*

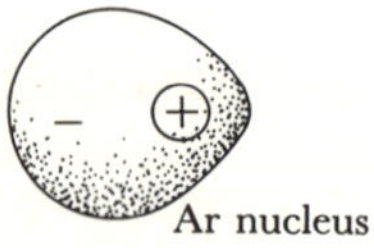

The positive end (the Ar nucleus) of the tiny dipole will attract the electrons on the other Ar atom, thereby causing it to also become a dipole. The second dipole is referred to as an *induced dipole.*

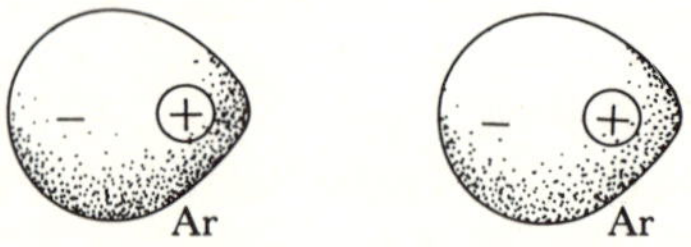

Similarly, at any instant, the Ar atom on the right may act as a dipole and induce the Ar atom on the left to become a dipole.

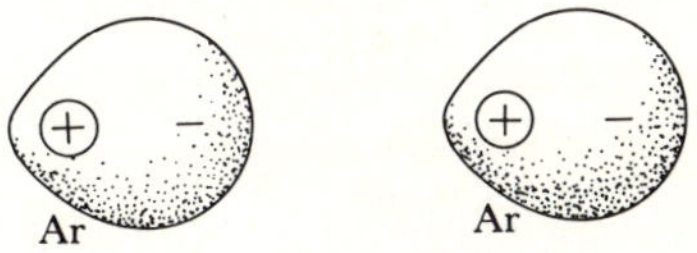

The electron cloud distributions are continually fluctuating, and the net effect is an extremely small attraction between the Ar atoms. This attraction is referred to as an *induced dipole-induced dipole attraction.*

SOLUTION 14–3 (2)

Assuming ideal gas behavior, we say that the average kinetic energy of 1 mole of Ar gas is $\frac{3}{2}RT$. The energy of the van der Waals interaction in solid Ar is 250 cal mole^{-1}. If we set this energy equal to $\frac{3}{2}RT$ (our condition),

$$T = \frac{2 \times 250 \text{ cal mole}^{-1}}{3 \times 1.99 \text{ cal mole}^{-1} \text{ deg}^{-1}}$$

$$= 83.8°\text{K}$$

The observed melting point of solid Ar is 84.0°K. The small difference between the predicted and observed values indicates that the assumption of ideality for Ar gas is reasonable.

The correct answer is **2**. Answer **1** is the melting point in °C not °K. Answer **3**: You forgot to divide by *R* and ignored units.

SOLUTION 14–4 (3)

Answer **3** is correct. In general, boiling points (and also melting points) of the hydrides of the representative elements increase with increasing molecular weight within a group. However, the boiling and melting points of HF, H_2O, and NH_3 are anomalous owing to the presence of *hydrogen bonds* between molecules in both the liquid and solid states. For instance, with H_2O, oxygen is much more electronegative than hydrogen; thus, we can expect ionic structures like

$$\begin{array}{l} H^{+}\text{—}O^{-} \\ \quad\quad\;\, | \\ \quad\quad\;\, H \end{array}$$

In addition to the negative charge, the O atom has two lone pairs. Therefore, it could strongly attract a protonlike H atom of another H_2O molecule, thereby forming

$$\begin{array}{l} H^{+}\text{—}O^{-}\cdots H^{+}\text{—}O^{-} \\ \quad\quad\;\, | \qquad\qquad\quad\; | \\ \quad\quad\;\, H \qquad\qquad\;\;\, H \end{array}$$

where the dotted line represents a hydrogen bond containing two electrons. In this way a large aggregate of H_2O molecules is built up; this results in an increase in molecular weight and consequently a higher boiling point (and melting point) compared to that expected for H_2O in the absence of hydrogen bonding. Because of the electrostatic nature of the hydrogen bond, only the most electronegative atoms form these bonds. So the reason compounds **1**, **2**, **4**, and **5** do not show anomalous behavior is because P, C, Br, and Se atoms are not sufficiently electronegative.

[*Note:* We should not expect a hydrogen (for example, in H_2O) to participate in more than one hydrogen bond. A hydrogen bond involves a protonlike H (e.g., $—H^+$), which interacts with a species that shares a lone pair with it. This reduces the positive charge on the hydrogen (e.g., —H···), and it cannot attract another lone pair.]

SOLUTION 14-5

(5)

In HF, the F atom is much more electronegative than the H atom; consequently, the shared pair of electrons are attracted more toward the F atom than the H atom. This means that the H atom becomes positively charged and protonlike, whereas the F atom becomes negatively charged and fluoridelike:

$H^+—F^-$

Therefore, we have the possibility of hydrogen bond formation by one of the lone pairs on the negatively charged F atom that results in the formation of H_2F_2:

$H^+—F^-\cdots H^+—F^-$

Similarly, H_3F_3 is formed by another HF molecule hydrogen bonding to the H_2F_2 molecule:

$H^+—F^-\cdots H^+—F^-\cdots H^+—F^-$

Molecular species such as H_4F_4, H_5F_5, and H_6F_6 can be formed in a similar manner.

The correct answer is therefore **5**. Answer **1** is incorrect because the F^-'s could not covalently bond to each other. Answer **3** is incorrect because it has the wrong charge distribution, $F^+—H^-$. Answer **4** is incorrect because the central H^+ is involved in two hydrogen bonds at the same time. Answer **2** is incorrect for the reason given for Answer **3** and a reason similar to that given for Answer **4**.

SOLUTION 14-6

(2)

We can immediately rule out structures **3**, **4**, and **5** because oxygen is much more electronegative than hydrogen; so we expect $O^-—H^+$ for the OH groups and not $O^+—H^-$. Answer **5** is further incorrect because C and H have similar electronegativities and the covalent C—H bond would never be $C—H^-$. In structure **1**, we have $O^-—H^+$, as expected, but there is no way for the H^+'s to hydrogen bond to the C's, which have no lone pairs. Structure **2** is the correct one. The O—H groups are correctly polarized, and the lone pairs on the O^-'s hydrogen bond to the H^+'s.

SOLUTION 14–7

(3)

The number of electrons in CH_4 is $6 + 4(1) = 10$. The number of electrons in H_2O is $2(1) + 8 = 10$. Therefore, CH_4 and H_2O are isoelectronic.

The ratio of the heats of sublimation of the isoelectronic pair of molecules, GeH_4 and H_2Se, is $4.0/5.3 = 0.76$, whereas for the isoelectronic pair, SnH_4 and H_2Te, it is $5.0/6.6 = 0.76$. The ratio is the same in both cases. From Figure 14–2, we can see that CH_4, GeH_4, H_2Se, SnH_4, and H_2Te do not exhibit anomalous boiling points and hence are not hydrogen bonded. Therefore, if we assume that the same ratio (0.76) would also apply to the isoelectronic pair, CH_4 and H_2O, if H_2O did not form hydrogen bonds in ice, the heat of sublimation of the hypothetical form of ice would be $2.0/0.76 = 2.6$ kcal mole^{-1}. (Note that we put the H_2X species in the denominator in each case.)

This crude calculation gives an estimate of the van der Waals interactions between the H_2O molecules in ice.

The correct answer is **3**. Answer **1** is the reciprocal of Answer **3**. Answer **2**: You did not observe the convention of keeping the H_2X species in the denominator.

SOLUTION 14–8

(4)

In Problem 14–4, we showed the formation of one hydrogen bond between two H_2O molecules. However, there are two H atoms in one H_2O molecule, and both can participate in hydrogen bonding with lone pairs of electrons on the O atoms of other H_2O molecules:

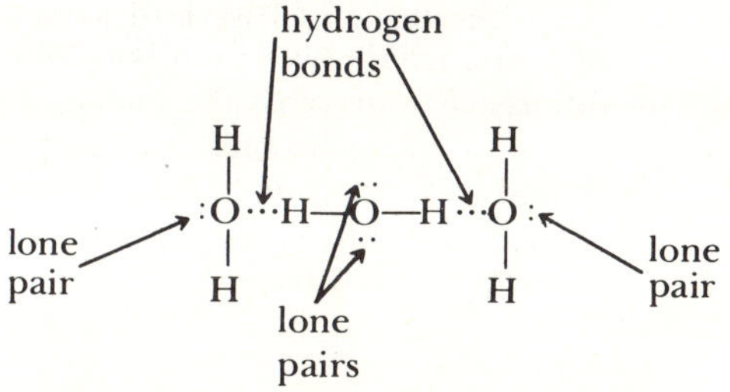

But the two lone pairs on the central O atom can also form hydrogen bonds with H atoms on two other H_2O molecules:

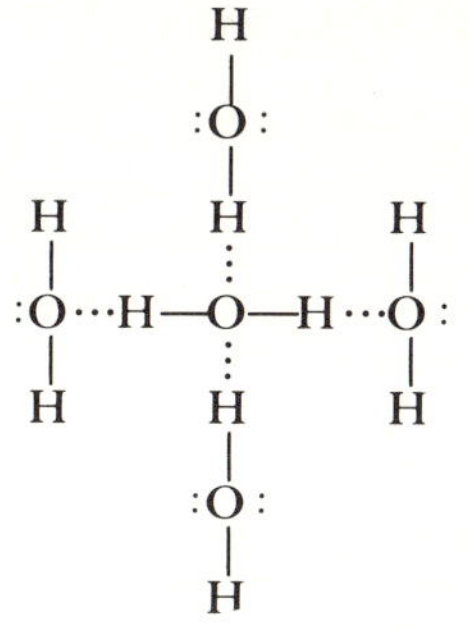

If we look at the central H—O—H, we can see that it participates in *four* hydrogen bonds.

Answer **4** is correct.

[*Note:* The structure is not planar; each oxygen is tetrahedrally surrounded by four hydrogens. This is a structure for ice.]

SOLUTION 14–9

(2)

The difference between the observed heat of sublimation of ice and the value calculated for a hypothetical form of ice in which there are no hydrogen bonds is $12.2 - 2.6 = 9.6$ kcal mole^{-1}. This difference must be attributed to the energy of the hydrogen bonds in ice. From Problem 14–8 we know that in ice each water molecule participates in four hydrogen bonds. But each hydrogen bond is shared by two H_2O molecules. So the *net* number of hydrogen bonds broken per molecule when a mole of ice sublimes is two. And the energy per hydrogen bond must be $9.6/2 = 4.8$ kcal mole^{-1} for a mole of hydrogen bonds.

2 is the correct answer.
1 is the calculated heat of sublimation of the hypothetical form of ice.
3 You added 12.2 and 2.6 kcal mole^{-1} instead of subtracted them.
4 You forgot that there are two hydrogen bonds per H_2O molecule in ice.
5 is the observed heat of sublimation of ice.

SOLUTION 14–10

(2)

The difference between the heats of fusion of ice and a hypothetical form of ice in which there are no hydrogen bonds is $1.4 - 0.30 = 1.1$ kcal mole^{-1}. This difference represents the energy required to break some fraction of the two hydrogen bonds for each H_2O molecule in ice. The energy required to break 2 moles of hydrogen bonds in ice is $2 \times 4.8 = 9.6$ kcal mole^{-1}. Therefore, the fraction of hydrogen bonds broken per H_2O molecule in the process of fusion is $1.1/9.6 = 0.11$ or 11%.

2 is correct.
3 You added 1.4 and 0.30 kcal mole^{-1} instead of subtracted them.
4 You forgot that there are two hydrogen bonds per H_2O molecule in ice.
5 Both mistakes in **3** and **4**.

SOLUTION 14–11

(3)

All we have to do is to count the number of oxygen and hydrogen atoms present in the polywater layer and then determine the net charge from the oxidation numbers of O and H.

There are 30 oxygen atoms and 52 hydrogen atoms; hence, the net charge of the polywater layer is $30(-2) + 52(+1) = -8$.

The correct answer is **3**. Answer **1**: You assigned the H atoms an oxidation number of −1 instead of +1. Answer **2** is the charge of the 30 O atoms.

SOLUTION 14–12 (1) 1,5- and 2,6-dihydroxynaphthalene (molecules **4** and **5**, respectively) can associate through *intermolecular* hydrogen bonds involving the hydroxyl groups:

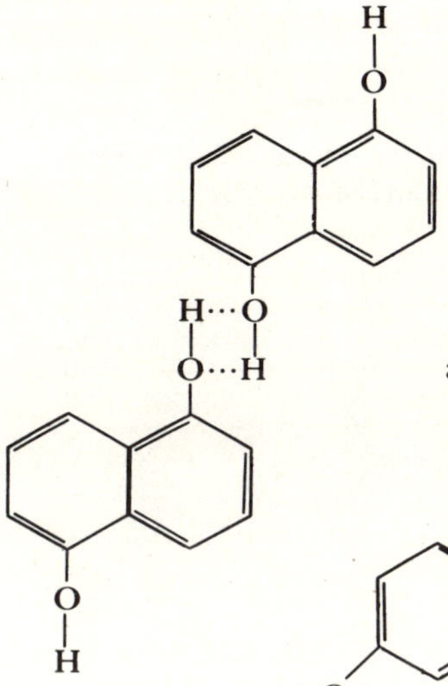

and

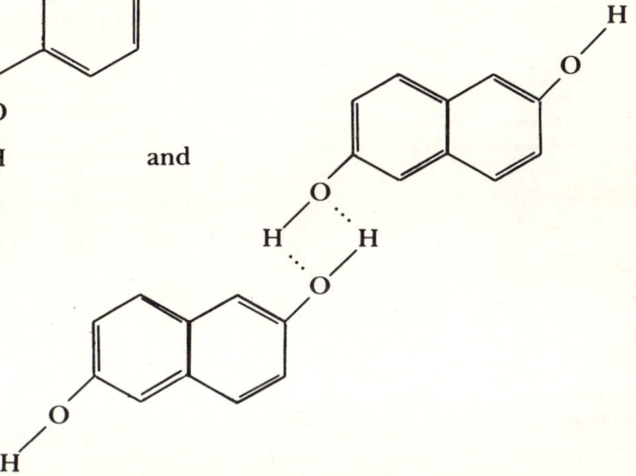

However, in both cases, the hydroxyl groups are too far apart to participate in *intramolecular* hydrogen bonds. All three of the chlorophenols (molecules **1**, **2**, and **3**) can undergo intermolecular hydrogen bonding, but only 2-chlorophenol (molecule **1**) can form *intramolecular* hydrogen bonds. Some examples of the associations that can occur for the three molecules are shown below.

A. Intermolecular

Molecule **1**

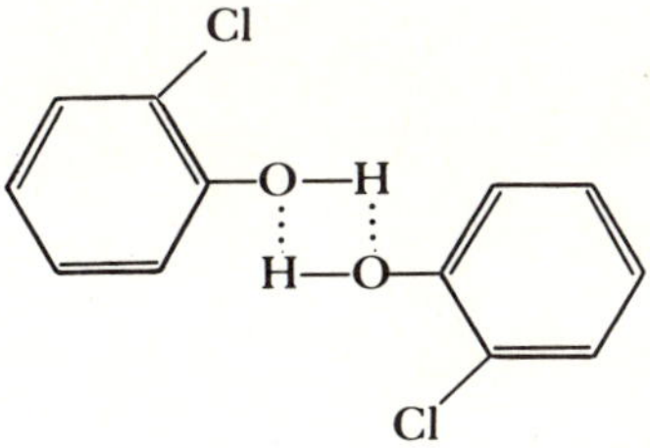

and

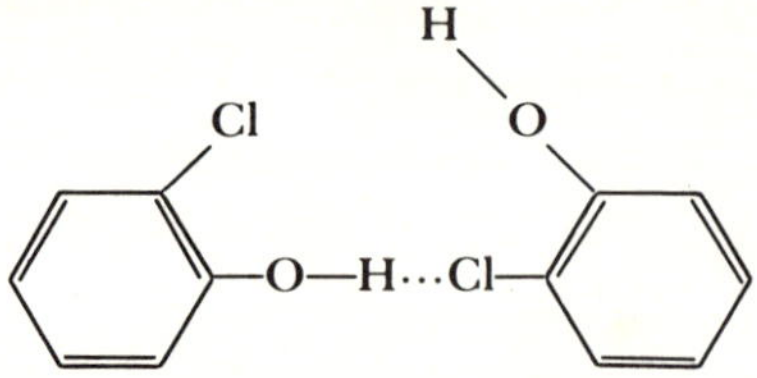

[*Remember:* The Cl atoms have lone pairs that can readily form bonds with the protonlike H atoms of the OH groups.]

Molecule **2**

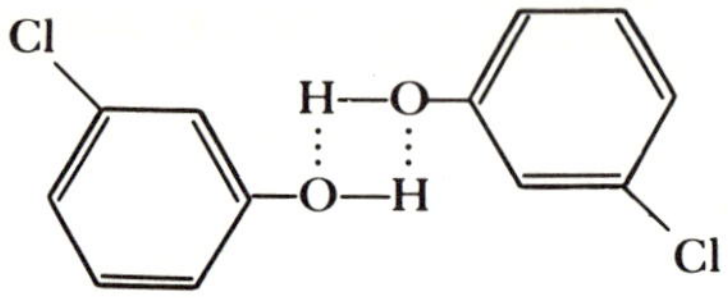

and

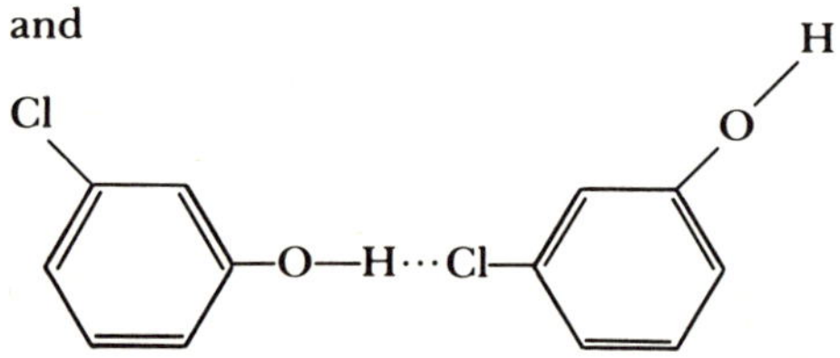

Molecule **3**

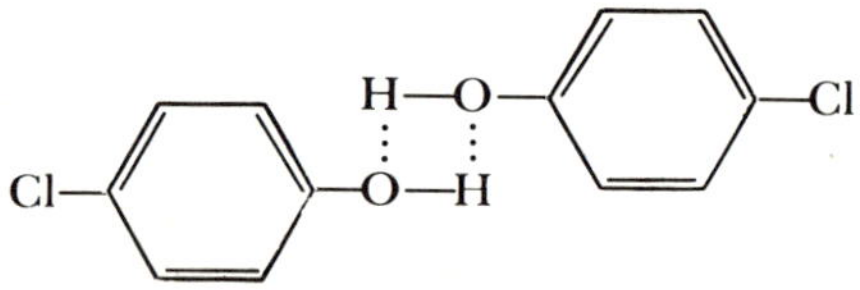

and

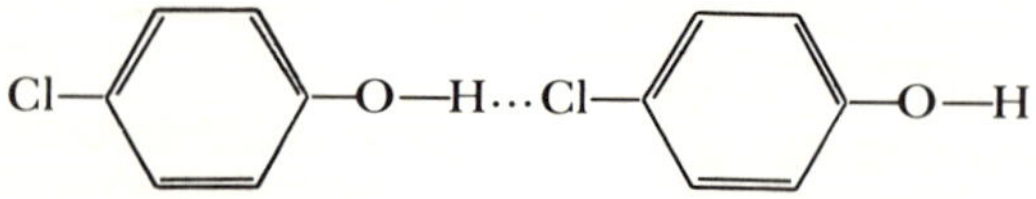

B. Intramolecular

Molecule **1**

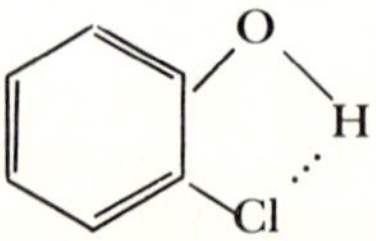

Since it can associate intramolecularly, 2-chlorophenol (molecule **1**) will have fewer hydroxyl hydrogens available for intermolecular association than the other two chlorophenols (molecules **2** and **3**); therefore, it will have a lower boiling point. The dihydroxynaphthalene molecules (**4** and **5**) will also have higher boiling points than 2-chlorophenol because they are of higher molecular weight and will be intermolecularly associated. Hence, 2-chlorophenol will have the lowest boiling point of the five molecules concerned.

The correct answer is **1**.

[*Note:* The observed boiling points are as follow: **1**, 176°C; **2**, 214°C; **3**, 217°C. Molecules **4** and **5** are high melting solids; melting points are as follow: **4**, 258°C; **5**, 218°C. The boiling points are unknown because **4** decomposes above its melting point and **5** sublimes.]

SOLUTION 14–13

(4)

The X—H groups in the molecules are N—H and C—H. N—H will undergo hydrogen bonding because the N atom is much more electronegative than the H atoms; C—H will not because the electronegativities of C and H are comparable.

First we must draw the structures of each pair of molecules and then, remembering that the molecules are planar, juggle them to form as many hydrogen bonds as possible between the two molecules. For the adenine–thymine pair (a), we obtain

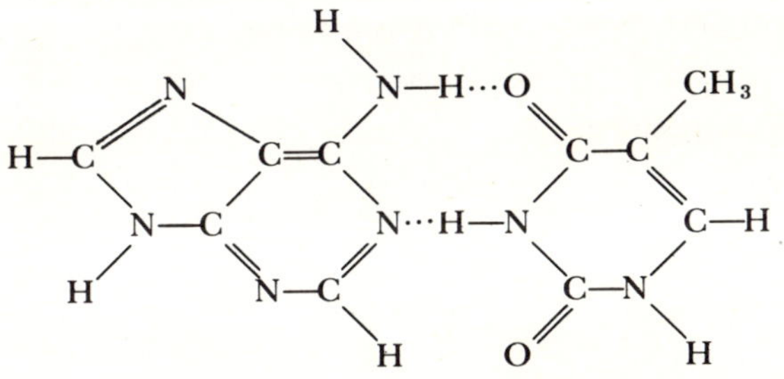

Adenine   Thymine

For this pair of molecules there are two possible hydrogen bonds.

For the guanine–cytosine pair (b), we obtain

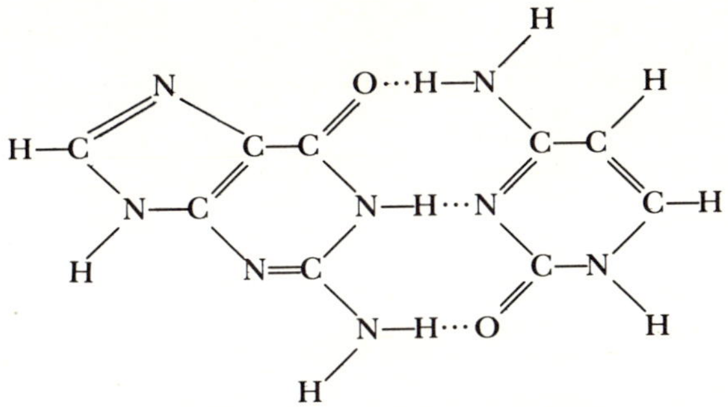

Guanine   Cytosine

In this case there are three hydrogen bonds possible.

The correct answer is **4**.

[*Note:* The two strands in DNA are believed to be held together by the hydrogen bonding system that you have just considered. Since we consider DNA a molecule, the distinction between intermolecular and intramolecular bonding is somewhat artificial.]

SOLUTION 14–14

(3)

First you must count the number of silicon and oxygen atoms in each structure.

I: 3 Si atoms and 9 O atoms
II: 6 Si atoms and 18 O atoms

The fundamental unit in both structures is the SiO_4^{4-} tetrahedron. Each oxygen atom in the SiO_4^{4-} unit has an oxidation number of −2; therefore, the oxidation number of the silicon atom must be +4. Now we can calculate the net charge on each of the two anions.

I	3 Si atoms at +4 per atom	= +12
	9 O atoms at −2 per atom	= −18
	Net charge	= −6
II	6 Si atoms at +4 per atom	= +24
	18 O atoms at −2 per atom	= −36
	Net charge	= −12

Hence, the formulas of the two anions must be

I: $Si_3O_9^{6-}$ II: $Si_6O_{18}^{12-}$

This is answer **3**.

[*Note: Benitoite* is $BaTiSi_3O_9$ and *beryl* is $Be_3Al_2Si_6O_{18}$.]

SOLUTION 14–15

(5)

The simplest repeating unit of the double stranded silicate structure in Figure 14–5 is shown in Figure 14–6. There are four Si atoms and eleven O atoms, thereby giving the repeating unit a net charge of $4(+4) + 11(-2) = -6$. Therefore, the general composition of the silicate ion in chrysotile is $(Si_4O_{11})_n^{6n-}$, in which n is the number of times the repeating unit $Si_4O_{11}^{6-}$ occurs.

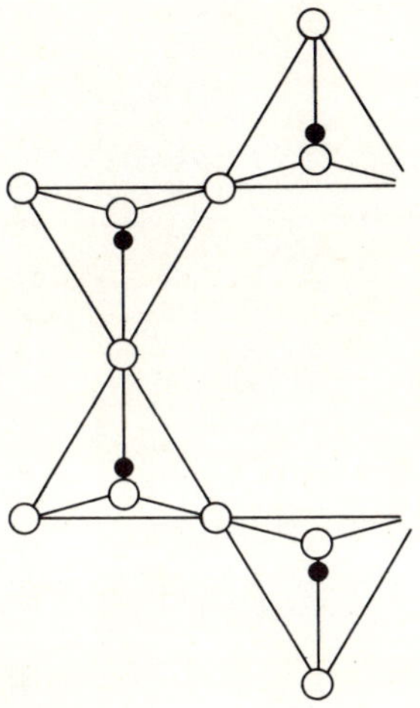

Figure 14–6

The correct answer is **5**.

1 is the general formula of silica.
2 is the general composition of the silicate ion in pyroxene minerals such as *spodumene*, $LiAl(SiO_3)_2$.
3 is the general composition of the silicate ion in micas and clays.
4 is the pyrosilicate ion found in minerals such as *thortveitite*, $ScSi_2O_7$.

CHAPTER 15 THERMODYNAMICS

Practically all chemical reactions either absorb or give off heat. In this chapter we will relate the heat behavior of reactions to their chemical behavior and touch on their relation to the properties of molecules.

You will need to memorize the following relationships to solve the problems in this chapter.

ΔH_f^0 = standard heat of formation
ΔG_f^0 = standard free energy of formation
ΔH_f^0 (elements) = ΔG_f^0 (elements) = 0
(if the elements are in their standard states)

$\Delta E = q - w$

in which

q = heat into system, w = work by system

$w = P\Delta V$ = mechanical work at constant P
$\Delta H = \Delta E + \Delta(PV)$

$\Delta S = q_{rev}/T$

in which

q_{rev} = reversible heat

$G = H - TS$
$\Delta T_b = k_b m_A$ for the boiling point elevation
$\Delta T_f = k_f m_A$ for the freezing point depression

Unless otherwise specified, all thermochemical data are at 25°C.

Solutions to problems in Chapter 15 begin on page 411.

PROBLEMS

PROBLEM 15–1

If the average adult burns 3×10^3 kcal of food per day, what weight of fat alone would be necessary to furnish this daily caloric requirement? (Refer to Problem 4–22 for data.)

1 $\frac{1}{3}$ g
2 9 g
3 $\frac{1}{3}$ kg
4 3 kg
5 $\frac{1}{3} \times 10^3$ kg

PROBLEM 15–2

When an "inch of rain" falls on New York City, it results in a rainfall of 19.8×10^9 liters [5.2 billion gallons (US)] over the city's area (about 300 square miles). If we assume a density for H_2O of 1.00 g cm^{-3}, how much heat was released when this quantity of water condensed?

Data

fusion: $\Delta H_{fus}(H_2O) = 1.4$ kcal $mole^{-1}$
vaporization: $\Delta H_{vap}(H_2O) = 9.7$ kcal $mole^{-1}$
(mol wt: $H_2O = 18.0$)

1 19.8×10^6 kcal
2 1.5×10^{12} kcal
3 11×10^{12} kcal
4 27×10^{12} kcal
5 20×10^{13} kcal

PROBLEM 15–3

Part of the temperature-regulatory mechanism of the human body is the dissipation of heat by the vaporization of sweat. Let's say an athlete generates about 1400 watts during a race. At what rate, in g min^{-1}, must water be vaporized to dissipate this energy?

Data

$\Delta H_{vap}(H_2O) = 9.7$ kcal $mole^{-1}$
$\Delta H_{fus}(H_2O) = 1.4$ kcal $mole^{-1}$
$D\,[H_2O(l)] = 1.00$ g cm^{-3}

(mol wt: $H_2O = 18.0$;
1 watt = 0.0143 kcal min^{-1})

1 0.63
2 2.1
3 11.7
4 38
5 None of the above.

PROBLEM 15–4

A sample of an ideal gas is allowed to escape reversibly and isothermally into a vacuum. Which statement is correct? [*Hint:* What is the average kinetic energy of an ideal gas?]

1 $w = 0$, $\Delta E = q =$ a positive number
2 $w = P\Delta V + V\Delta P$
3 $w = P\Delta V$, $\Delta E = 0$, $q = -P\Delta V$ and is a negative number
4 $\Delta E = 0$, $q = -w =$ a positive number
5 $w = 0$, $q = 0$, $\Delta E = 0$

PROBLEM 15–5

Steam engines may come back into favor in the near future as an alternative to internal combustion engines to power automobiles. When 18.0 g of liquid water are vaporized at 100°C against a constant external pressure of 1.00 atm, how much work is done?

Data

$\Delta H_{vap}(H_2O) = 9.7$ kcal mole^{-1};
$\Delta H_{fus}(H_2O) = 1.4$ kcal mole^{-1};
$R = 0.0820$ liter atm mole^{-1} deg^{-1}
$= 1.99$ cal mole^{-1} deg^{-1};
$D\ [H_2O(l)] = 1.00$ g cm^{-3};
(mol wt: $H_2O = 18.0$)]

1 No work is done.
2 740 cal
3 1400 cal
4 9700 cal
5 11,100 cal

PROBLEM 15–6

Using the data and solution of Problem 15–5, calculate what percentage of the heat that vaporized the water was consumed in useful work.

1 0%
2 7.6%
3 50%
4 92%
5 100%

PROBLEM 15–7

The relationship between energy and work is discussed in Section 15–2 of Chemical Principles.

In Problem 15–6 we found the percentage of the energy that was used to do work in a model steam engine. The efficiency was rather low. Which of the following modifications would improve the efficiency; that is, which would increase the percentage? (Assume that the ΔH's do not change with T.)

1 Start with ice rather than liquid water.
2 Use twice as much water.
3 Let the water vaporize at a higher T at the same external pressure.
4 Increase the external pressure.
5 There is no way to increase the efficiency.

PROBLEM 15–8

The term *adiabatic* describes a process in which there is no heat transfer between the system and the surroundings. This is either because the system is well insulated or because the process occurs very rapidly. For such a process, it follows that ________.

1 $q = +w$
2 $q = 0$
3 $\Delta E = q$
4 $\Delta E = w$
5 $P\Delta V = 0$

PROBLEM 15–9

A chinook results when atmospheric air masses pass over the Rocky Mountains and are suddenly compressed on the lee (eastern) side of the mountain range. Regarding this as an adiabatic compression, which one of the following statements regarding the chinook is correct? [*Hint:* Use the first law of thermodynamics; assume ideality.]

1 The temperature of the air is increased.
2 The temperature of the air is decreased.
3 The temperature of the air does not change.
4 The internal energy of the air remains constant.
5 Insufficient information; need to know the average elevation of the Rocky Mountains.

PROBLEM 15–10

Find the heat of formation of $CH_4(g)$ in the reaction

$$C(\text{graphite}) + 2H_2(g) = CH_4(g)$$

Data

$$CH_4(g) + 2O_2(g) = CO_2(g) + 2H_2O(l) \qquad \Delta H^0 = -212.80 \text{ kcal mole}^{-1}$$
$$C(\text{graphite}) + O_2(g) = CO_2(g) \qquad \Delta H^0 = -94.05 \text{ kcal mole}^{-1}$$
$$H_2(g) + \tfrac{1}{2}O_2(g) = H_2O(l) \qquad \Delta H^0 = -68.32 \text{ kcal mole}^{-1}$$

(All the ΔH^0 are in kcal mole^{-1} taken at 298°K.)

1 $\Delta H_f^0 = -375.2$ kcal mole^{-1}
2 $\Delta H_f^0 = -17.89$ kcal mole^{-1}
3 $\Delta H_f^0 = +17.89$ kcal mole^{-1}
4 $\Delta H_f^0 = +50.43$ kcal mole^{-1}
5 None of these.

PROBLEM 15–11

What is the standard heat of formation of 1 mole of the *interhalogen* (a molecule containing two or more different halogens) compound ICl(g), ΔH_f^0 [ICl(g)]?

The heat of formation of compounds is discussed in Section 15-3 of Chemical Principles.

Data	ΔH^0 (kcal mole^{-1})
$Cl_2(g) = 2Cl(g)$	57.9
$I_2(g) = 2I(g)$	36.1
$ICl(g) = I(g) + Cl(g)$	50.5
$\Delta H^0_{sublimation}$ [$I_2(s)$]	15.0

The standard states are

$I_2(s)$
$Cl_2(g)$

1 -50.5 kcal mole^{-1}
2 -3.5 kcal mole^{-1}
3 $+4.0$ kcal mole^{-1}
4 $+58.5$ kcal mole^{-1}
5 $+105$ kcal mole^{-1}

PROBLEM 15–12

Ethylene is the starting material (*monomer*) in the manufacture of polyethylene, one of the most widely used plastics. Calculate the standard heat of formation of gaseous ethylene, ΔH_f^0 [$C_2H_4(g)$].

Data	ΔH_f^0 (kcal mole^{-1})
$CO_2(g)$	-94.1
$H_2O(l)$	-68.3

Heat of combustion of $C_2H_4(g) = -337.3$ kcal mole^{-1}.

1 -175 kcal mole^{-1}
2 -12.5 kcal mole^{-1}
3 $+12.5$ kcal mole^{-1}
4 $+175$ kcal mole^{-1}
5 $+490$ kcal mole^{-1}

PROBLEM 15-13

The *bond energy* is the energy necessary to sever one bond in a molecule. By calculating ΔH^0 for the reaction to produce one carbon atom and four hydrogen atoms, calculate the average bond energy for the C—H bond in methane, $CH_4(g)$.

Compound	ΔH_f^0 (kcal mole^{-1})
$CH_4(g)$	−17.9
$H(g)$	52.10
$C(g)$	171.7

1 18 kcal mole^{-1}
2 60 kcal mole^{-1}
3 73 kcal mole^{-1}
4 99 kcal mole^{-1}
5 398 kcal mole^{-1}

PROBLEM 15-14

In Problem 15–13 we calculated the bond energy as though all the bonds were of equal strength, even if we broke them one after the other rather than simultaneously. Now instead of calculating the average C—H bond energy, we will use some more sophisticated data to obtain a better estimate, namely, the energy necessary to break the *first* bond. The C—H bond energy, from the data below, is ________.

Compound	ΔH_f^0 (kcal mole^{-1})
$CH_4(g)$	−17.9
$CH_3(g)$	34.0
$H(g)$	52.1

1 34 kcal mole^{-1}
2 52 kcal mole^{-1}
3 99 kcal mole^{-1}
4 104 kcal mole^{-1}
5 200 kcal mole^{-1}

PROBLEM 15-15

If we accept the results of Problems 15–13 and 15–14, the average C—H bond energy in $CH_4(g)$ is 99 kcal mole^{-1}, and the energy necessary to break the first C—H bond is 104 kcal mole^{-1}. Which one of the following inferences from these data is correct?

1 It is impossible to remove more than one H atom from methane.
2 Some of the other C—H bonds are more difficult to break after the first H has been removed.
3 Some of the other C—H bonds are easier to break after the first H has been removed.
4 All of the other C—H bonds are easier to break after the first H has been removed.
5 None of the above.

PROBLEM 15–16 The amount of heat absorbed when CO_2 gas reacts with solid CaO to form solid $CaCO_3$ is measured in a bomb calorimeter (i.e., at constant volume). The data obtained give a direct measure of

____________.

1 ΔE
2 ΔH
3 $V\Delta P$
4 ΔG
5 The work.

PROBLEM 15–17 Which one of the following statements is correct?

1 The standard heat of formation of an element is negative.
2 Since $\Delta H = \Delta E + \Delta(PV)$, ΔH can never equal ΔE.
3 q and w are state functions.
4 $q - w$ is a state function.
5 S is not a state function since it is always increasing.

PROBLEM 15–18 The reaction

$$CO(g) + \tfrac{1}{2}O_2(g) \rightleftharpoons CO_2(g)$$

is extremely important in efforts to reduce noxious emissions from automobile exhausts. At constant T and P, which one of the following statements is correct for this reaction?

$(R = 1.99 \text{ cal mole}^{-1} \text{ deg}^{-1})$

1 $\Delta H = \Delta E$
2 $\Delta H < \Delta E$
3 $\Delta H > \Delta E$
4 Insufficient data; T must be known.
5 ΔH is independent of the physical states of reactants.

PROBLEM 15–19 Which of the following five-card poker hands is characterized by the *lowest* "entropy"?

Entropy is discussed in Section 15–7 of Chemical Principles

1 Four aces
2 No pairs
3 Royal flush
4 Three of a kind
5 Two pairs

PROBLEM 15–20

Suppose that we are examining a biological polymer that undergoes a reaction

polymer → product

and we are able to determine that at a given temperature the reaction is spontaneous and endothermic. What does this information imply about the *structure* of the product?

1 Nothing.
2 The product is a less ordered structure than the polymer.
3 The product is a more ordered structure than the polymer.
4 The heat of formation of the polymer is more positive than that of the product.
5 Since the process is endothermic, the reaction cannot occur and is not spontaneous. The data are suspect.

PROBLEM 15–21

We consider a growing plant, limiting our system to the plant itself, to be an example of decreasing entropy. Small molecules, like CO_2 and H_2O, are used to build complex but orderly arrangements of macromolecules. Which one of the following statements applies?

1 The second law of thermodynamics—that the entropy of the universe is increasing—is being violated.
2 Plant growth is so complex that the laws of thermodynamics cannot be applied successfully.
3 The second law of thermodynamics is not being violated because the entropy of the plant's surroundings is increasing.
4 The second law of thermodynamics is not being violated because the entropy of the plant's surroundings is decreasing.
5 None of the above.

PROBLEM 15–22

The spontaneity of chemical reactions is discussed in Section 15–5 of Chemical Principles.

Processes can be characterized by the sign of ΔH and ΔS, as in Table 15–1.

Table 15–1

Process	ΔH	ΔS
(a)	−	+
(b)	+	−
(c)	−	−
(d)	+	+

Which of the processes are *definitely* spontaneous, and which are *possibly* spontaneous for processes at constant T and P?

1 Definitely (a), possibly (b).
2 Definitely (b) and (c), possibly (d).
3 Definitely (c), possibly (a) and (b).
4 Definitely (d), possibly (a).
5 Definitely (a), possibly (c) and (d).

PROBLEM 15–23

For the process $C(s) + O_2(g) = CO_2(g)$, it is found that ΔG at a given temperature is -10 kcal mole^{-1}. Which statement is correct at this T?

1 The system is at equilibrium.
2 CO_2 will be formed spontaneously.
3 CO_2 will decompose spontaneously.
4 The process will proceed reversibly.
5 The process is not possible.

PROBLEM 15–24

What is the free energy change, ΔG, when 1.0 mole of water is converted, at 100°C and 1 atm, to steam at 100°C and 1 atm in a reversible manner?

[ΔH_{vap} (H_2O) = 9.7 kcal mole^{-1}]

1 −(9700/373) cal = −26 cal
2 −970 cal
3 0 cal
4 970 cal
5 9700 cal

PROBLEM 15–25

The following equation represents the oxidation of glucose:

$$C_6H_{12}O_6(s) + 6O_2(g) = 6CO_2(g) + 6H_2O(l)$$

Given the following data, what is the standard free energy of reaction?

Substance	ΔG_f^0 (kcal mole^{-1})
$C_6H_{12}O_6(s)$	−218
$CO_2(g)$	−94.3
$H_2O(l)$	−57.0

1 +66.7 kcal
2 −66.7 kcal
3 −690 kcal
4 +690 kcal
5 Insufficient data; we must have ΔG_f^0 for $O_2(g)$.

PROBLEM 15-26

Blood plasma has a normal hydrogen ion concentration of 4×10^{-8} molar; in gastric juice the concentration is about 2×10^{-1} molar. The acid in gastric juice is obtained from the blood. In supplying the usual daily 3 liters of gastric juice secreted into an adult's stomach, the body is ________.

1 Storing energy for later use.
2 Doing work to effect this concentration.
3 Violating the second law of thermodynamics.
4 Violating the first law of thermodynamics.
5 None of the above.

PROBLEM 15-27

The fundamental features of equilibrium are discussed in Sections 5-2 and 16-1 of Chemical Principles *by Dickerson, Gray, and Haight.*

When a liquid is in dynamic equilibrium with its vapor, the following conditions could exist.

(a) There is no transfer of molecules between liquid and vapor.
(b) The vapor pressure has a unique value.
(c) The opposing processes (liquid to vapor and vapor to liquid) proceed at equal rates.
(d) The concentration of vapor is dependent on time.

Which of the above choices are applicable?

1 (a)
2 (b) and (c)
3 (a), (b), and (c)
4 (b) and (d)
5 (a) and (b)

PROBLEM 15-28

The vapor pressure of all liquids ________.

1 Is the same at 100°C.
2 Is the same at their freezing points.
3 Increases with the volume of liquid present.
4 Decreases with the increasing volume of the container.
5 Increases with temperature.

PROBLEM 15-29 Figure 15-1 is an example of a *phase diagram* for a pure substance. What is the correct identification of the areas A, B, and C?

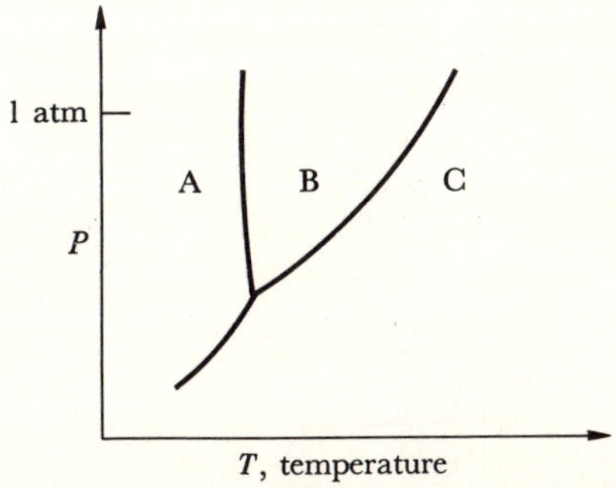

Figure 15-1. P denotes vapor pressure except *for the line between areas A and B, for which it denotes the external pressure.*

	A	B	C
1	Vapor	Liquid	Solid
2	Vapor	Solid	Liquid
3	Solid	Liquid	Vapor
4	Liquid	Vapor	Solid
5	None of the above.		

PROBLEM 15-30 The quantity indicated by X in Figure 15-2 is ________.

If you do not understand the calculations involving colligative properties (i.e., freezing point lowering, boiling point elevation, and osmotic pressure), refer to Review 13 of Programed Reviews of Chemical Principles *by Lassila* et al.

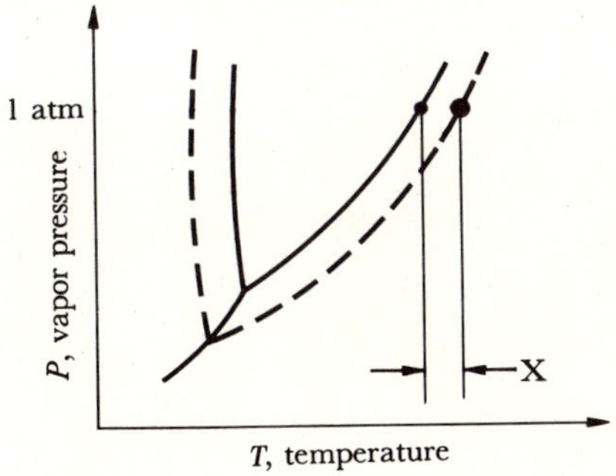

Figure 15-2. Solid black lines represent solvent, B. Dashed black lines represent solution (nonvolatile solute, A).

1 The heat of fusion.
2 ΔT_f
3 $k_f m_A$
4 $k_b m_A$
5 ΔP

PROBLEM 15-31

You have been asked to develop a model for the behavior of sap in trees that is based on osmosis. For a tree in which the sap rises to a height of 60 feet, we will approximate the sap to be a 0.10-molar sucrose solution. What must be the approximate concentration of the ground water outside the tree expressed as an equivalent sucrose solution? (Take T to be 27°C.)

[$\Pi = c_A RT$ = osmotic pressure;
$R = 0.082$ liter atm mole^{-1} deg^{-1};
1 atm $\simeq$ 30 ft of water;
assume density of sap $\simeq$ 1.0 g cm^{-3};
mol wt: sucrose = 342.3]

1 0.0 molar
2 0.02 molar
3 0.4 molar
4 0.10 molar
5 0.12 molar

PROBLEM 15-32

A proposed method for desalinating sea water is to submerge a long tube, its bottom closed by a semipermeable membrane, to a depth such that the hydrostatic pressure overbalances the osmotic pressure and forces pure water through the membrane.

Taking a model for sea water to be a 0.50M aqueous solution of NaCl at 17°C, what is the pressure at which this process will start?

($\Pi = c_A RT$;
$R = 0.082$ liter atm mole^{-1} deg^{-1})

1 1.4 atm
2 6.0 atm
3 12 atm
4 24 atm
5 None of the above.

SOLUTIONS

SOLUTION 15–1 (3)

The number that we need from Problem 4–22 is the caloric value of fat, which we shall take to be 9 kcal g^{-1}. The weight of fat that corresponds to 3×10^3 kcal is

$$\frac{3 \times 10^3 \text{ kcal}}{9 \text{ kcal g}^{-1}} = \tfrac{1}{3} \times 10^3 \text{ g} = \tfrac{1}{3} \text{ kg (almost } \tfrac{3}{4} \text{ lb)}$$

The correct answer is **3**. This is not an appetizing menu.

[*Note:* In nutrition, the Calorie (capital "C") usually denotes a kilocalorie.]

SOLUTION 15–2 (3)

How many moles of H_2O?

$$\begin{aligned} 19.8 \times 10^9 \text{ liters} &= 19.8 \times 10^9 \text{ liters} \times 10^3 \text{ cm}^3 \text{ liter}^{-1} \times 1.00 \text{ g cm}^{-3} \\ &= 19.8 \times 10^{12} \text{ g} \\ &= \frac{19.8 \times 10^{12} \text{ g}}{18.0 \text{ g mole}^{-1}} = 1.1 \times 10^{12} \text{ moles} \end{aligned}$$

At this point it is obviously important to distinguish between ΔH_{fus}, which is the heat of the solid to liquid process, and ΔH_{vap}, which is the heat of the liquid to vapor process. It is clear that the latter is the number we shall use.

$$\begin{aligned} \Delta H_{vap} &= 9.7 \text{ kcal mole}^{-1} \\ \text{heat} &= 1.1 \times 10^{12} \text{ moles} \times 9.7 \text{ kcal mole}^{-1} \\ &= 11 \times 10^{12} \text{ kcal} \end{aligned}$$

Answer **3** is correct. Answer **2** results from using ΔH_{fus}.

SOLUTION 15–3 (4)

$$\begin{aligned} 1400 \text{ watts} &= 1400 \times 1.43 \times 10^{-2} \text{ kcal min}^{-1} \\ &= 20 \text{ kcal min}^{-1} \end{aligned}$$

The latent heat of vaporization is 9.7 kcal $mole^{-1}$. Therefore, water must vaporize at the rate of

$$\begin{aligned} \frac{20 \text{ kcal min}^{-1}}{9.7 \text{ kcal mole}^{-1}} &= 2.1 \text{ mole min}^{-1} \\ &= 2.1 \text{ mole} \times 18 \text{ g mole}^{-1} \text{ min}^{-1} \\ &= 38 \text{ g min}^{-1} \end{aligned}$$

4 is correct.
1 This is the rate of g sec^{-1}.
2 is the moles of H_2O vaporized per minute.
3 You divided the molecular weight by the moles vaporized per minute.

Of course, the body also dissipates heat by radiation and convection.

SOLUTION 15-4

(5)

The expression for mechanical work (at constant pressure) is $P\Delta V$. But even when the pressure is not constant, w must be zero when $P = 0$. The P that is important is the P that the system is pushing against; this will determine the work that the system performs. When the external pressure is zero, as in a vacuum, there is nothing to push against and no work can be done.

We recall that, for an ideal gas, $\bar{E}_k = \frac{3}{2}RT$; hence, it is determined only by T. In an isothermal expansion, T does not change, and ΔE must be zero.

5 is correct.
1 ΔE must be zero. There was no transfer of heat.
2 is an incorrect expression for work. It is an expression for $\Delta(PV) = P\Delta V + V\Delta P = w + V\Delta P$.
3 and **4** have finite values for w.

SOLUTION 15-5

(2)

Expansion or PV work is discussed in Section 15-2 of Chemical Principles *by Dickerson, Gray, and Haight.*

We must calculate $P\Delta V$ for our 1.00-mole sample.

$$V(g) = \frac{373}{273} \times 22.4 \text{ liters}$$

$$= 30.6 \text{ liters} \qquad (\text{since } P = 1.00 \text{ atm})$$

$$V(l) = 18.0 \text{ cm}^3 \text{ (for an 18.0-g sample whose density is 1.00 g cm}^{-3}\text{)}$$

$$\Delta V = 30.6 \text{ liters} - 0.018 \text{ liter}$$

$$\simeq 30.6 \text{ liters}$$

$$w = P\Delta V = 30.6 \text{ liter atm} \qquad (\text{since } P = 1.00 \text{ atm})$$

We need only convert these energy units to calories. Since R is given in both units, we can use those numbers for this conversion.

$$1.99 \text{ cal} = 0.0820 \text{ liter atm}$$

$$\frac{1.99 \text{ cal}}{0.0820 \text{ liter atm}} \text{ is our conversion factor}$$

$$w = 30.6 \text{ liter atm} \frac{1.99 \text{ cal}}{0.0820 \text{ liter atm}}$$

$$= 740 \text{ cal}$$

Answer **2** is correct.

SOLUTION 15-6

(2)

A quantity of 740 cal went into useful work. A quantity of 9700 cal was used to vaporize 1 mole of H_2O.

$$100 \times \frac{740}{9700} \simeq 7.6\%$$

SOLUTION 15–7

(4)

Answer **1**: Starting with ice rather than water will decrease rather than increase the efficiency since the work will be the same but additional energy (ΔH_{fus}) will be used.

Answer **2**: The efficiency does not depend on how much material we use.

Answer **3**: Water will vaporize at 1 atm pressure *only* at 100°C under equilibrium conditions.

Answer **4**:

$$w = P\Delta V = P(V_{\text{final}} - V_{\text{init}})$$

If we neglect the initial volume

$$w \simeq PV_{\text{final}} = nRT_{\text{final}} \qquad \text{(from the ideal gas law)}$$

At a higher pressure, the boiling point is increased and T_{final} is higher; so w is larger. The process converts liquid to vapor at this new higher T_{final} with about the same heat input (ΔH_{vap}), but the amount of useful work is greater.

[*Note:* Our approximation (that the initial volume is inconsequential compared to the final volume) can be considered quantitatively. For 1 mole of H_2O

$V_{\text{init}} = 18.0\ \text{cm}^3$ if we assume that the density is $1\ \text{g cm}^{-3}$
$V_{\text{final}} = 22{,}400\ \text{cm}^3$ at STP, and is even greater under our conditions

So our approximation gives less than 0.1% error.]

SOLUTION 15–8

(2)

For an adiabatic process, $q = 0$ since there is no heat flow into or out of the system.

2 is correct.
1 q will equal the work only when $\Delta E = 0$ ($\Delta E = 0 = q - w$); for instance, during the isothermal expansion of an ideal gas.
3 ΔE will equal q only when $w = 0$, as when a gas expands into a vacuum.
4 ΔE will equal w only in the special case of $q = 2w$.
5 This is the same situation as Answer **3**.

SOLUTION 15–9

(1)

$$\Delta E = q - w$$

in which

q = heat absorbed by system,
w = work done by system.

Since the process is adiabatic, $q = 0$; the system has not done work, rather work has been done on it, so w is a negative number.

$$\Delta E = -w = \text{positive number} = E_{\text{final}} - E_{\text{initial}}$$

The internal energy has increased. Recall that E for an ideal gas must change when T does (from Chapter 2, $\bar{E}_k = \frac{3}{2}RT$). So Answer **1** is correct. Answers **3** and **4** are equivalent and incorrect.

SOLUTION 15–10

(2)

	Equation	ΔH^0 (kcal mole^{-1})
(a)	$CH_4(g) + 2O_2(g) = CO_2(g) + 2H_2O(l)$	−212.80
(b)	$C(\text{graphite}) + O_2(g) = CO_2(g)$	−94.05
(c)	$H_2(g) + \frac{1}{2}O_2(g) = H_2O(l)$	−68.32
(d)	$C(\text{graphite}) + 2H_2(g) = CH_4(g)$	?

Since $CH_4(g)$ must be on the right, we shall reverse Equation (a).

(−a) $CO_2(g) + 2H_2O(l) = 2O_2(g) + CH_4(g)$ +212.80

On the left-hand side we need one C(graphite); we shall add Equation (b).

(b) $C(\text{graphite}) + O_2 = CO_2(g)$ −94.05

Now on the left we need $2H_2(g)$, so we shall add 2 × (c).

2 × (c) $2H_2(g) + O_2 = 2H_2O(l)$ −136.64

Sum: $C(\text{graphite}) + 2H_2(g) = CH_4(g)$ 212.80

(−a) + (b) + (2c) −230.69

$\Delta H^0 = \Delta H_f^0[CH_4(g)] =$ −17.89

Answer **2** is correct.

SOLUTION 15–11

(3)

The reaction for which the heat of reaction is also the heat of formation is

$\frac{1}{2}I_2(s) + \frac{1}{2}Cl_2(g) = ICl(g)$

since the reactants are elements and in their standard states. Hence, we want to take the reactions for which we have data and shuffle them so, when summed, the result is the above equation.

The compound $ICl(g)$ must be a product, so we shall reverse the third equation

		ΔH^0 (kcal mole^{-1})
Note the sign change	$I(g) + Cl(g) = ICl(g)$	−50.5
Note division of ΔH^0 in all cases when the stoichiometry is halved	$\frac{1}{2}Cl_2(g) = Cl(g)$	29.0
	$\frac{1}{2}I_2(g) = I(g)$	18.0
One half the heat of sublimation	$\frac{1}{2}I_2(s) = \frac{1}{2}I_2(g)$	7.5
Sum:	$\frac{1}{2}I_2(s) + \frac{1}{2}Cl_2(g) = ICl\ (g)$	4.0

$\Delta H_f^0\ [ICl(g)] = 4.0$ kcal mole^{-1}

Answer **3** is correct.

1 $I(g) + Cl(g) \rightarrow ICl(g)$ is not a heat of formation reaction because the reactants are not elements in their standard states.
2 You did not use *solid* I_2.
4 You just forgot the stoichiometric factors.
5 You did not reverse the sign on ΔH^0 when you reversed the first equation above.

SOLUTION 15–12

(3)

The heat of combustion of a hydrocarbon is the heat released when the compound is burned in $O_2(g)$ to give $CO_2(g)$ and $H_2O(l)$.

	ΔH^0 (kcal mole^{-1})
$C_2H_4(g) + 3O_2(g) = 2CO_2(g) + 2H_2O(l)$	-337.3

The negative sign is used because such a reaction is exothermic. But the process that we are interested in is

$$2C(s) + 2H_2(g) = C_2H_4(g)$$

since ΔH^0 for this reaction is also the heat of formation of ethylene. So we shall reverse our combustion equation (changing the sign of ΔH^0) and use the other data to eliminate O_2, CO_2, and H_2O.

	$2CO_2(g) + 2H_2O(l) = 3O_2(g) + C_2H_4(g)$	$+337.3$
Double the ΔH_f^0 (CO_2) equation	$2C(s) + 2O_2(g) = 2CO_2(g)$	-188.2
Double the ΔH_f^0 (H_2O) equation	$2H_2(g) + O_2(g) = 2H_2O(l)$	-136.6
Sum:	$2C(s) + 2H_2(g) = C_2H_4(g)$	$+12.5$

$$\Delta H_f^0\ [C_2H_4(g)] = 12.5 \text{ kcal mole}^{-1}$$

Answer **3** is correct.

You could do this problem in a shorter way.

$$C_2H_4(g) + 3O_2(g) = 2CO_2(g) + 2H_2O(g) \qquad \Delta H^0 = -337.3 \text{ kcal mole}^{-1}$$
$$\Delta H^0 = \Delta H_f^0 \text{ (products)} - \Delta H_f^0 \text{ (reactants)}$$
$$-337.3 = [2(-94.1) + 2(-68.3)] - \{3(0) + \Delta H_f^0\ [C_2H_4(g)]\}$$
$$\Delta H_f^0\ [C_2H_4(g)] = 12.5 \text{ kcal mole}^{-1}$$

For Answer **4**, you forgot the stoichiometry. For Answer **5**, you changed signs incorrectly.

SOLUTION 15-13

(4)

	$CH_4(g)$ →	$C(g)$ +	$4H(g)$	
ΔH_f^0:	−17.9	171.7	52.10	kcal mole^{-1}
$\Delta H_f^0 \times$ stoich:	−17.9	171.7	208.4	kcal mole^{-1}

The calculation of bond energies is described in Section 15-4 of Chemical Principles.

The standard enthalpy change for the reaction is

$\Delta H^0 = (171.7 + 208.4) - (-17.9) = +398$ kcal mole^{-1}

This is the energy necessary to break four C—H bonds. To break one bond, about 99 kcal mole^{-1} are necessary.

4 is correct.
2 You forgot the stoichiometry.
3 You thought that the H atoms were H molecules?
5 is the energy required to break all the C—H bonds.

SOLUTION 15-14

(4)

	$CH_4(g)$ →	$CH_3(g)$ +	$H(g)$	
ΔH_f^0:	−17.9	34.0	52.1	kcal mole^{-1}

The standard enthalpy change for the reaction is

$\Delta H^0 = (34.0 + 52.1) + 17.9$
$\Delta H^0 = 104$ kcal mole^{-1}

This is the bond energy.
Answer **4** is correct.

SOLUTION 15-15

(3)

The energy input necessary to break the first C—H bond is 104 kcal mole^{-1}, 5 kcal mole^{-1} more than the average for breaking all the C—H bonds. So *some* of the bonds must be easier to break, and will require an energy input of less than 99 kcal mole^{-1}. But it does not follow from these data that all of the bonds must be weaker, because we can postulate a sequence 104, 105, 94, 93 for the bond energies that would still give an average of 99. The correct answer is **3**. Actually, the best estimates of the bond energies are

CH_3—H	104.0 kcal mole^{-1}
CH_2—H	104 kcal mole^{-1}
CH—H	108 kcal mole^{-1}
C—H	80.9 kcal mole^{-1}

SOLUTION 15-16

(1)

$\Delta E = q - w$
$\Delta E = q - P\Delta V$

In a bomb calorimeter, the process is at constant volume, so $\Delta V = 0$. $\Delta E = q_v$, the heat absorbed at constant volume.
Answer **1** is correct.

SOLUTION 15–17

(4)

If you do not know what is meant by "state function," see Section 15–2 of Chemical Principles.

Answer **1**: ΔH_f^0 (element) is defined to be zero. (Remember, though, that the element must be in its standard state.)

Answer **2**: $\Delta H = \Delta E$ when $\Delta(PV) = 0$. This will be approximately true for a reaction in which the number of moles of gas reactants and gas products is the same ($\Delta n = 0$), or in any reaction in which volume changes are negligible.

Answer **3**: The work, w, is not a state function; that is, w depends on *how* a system changed from state A to state B, and not simply on the characteristics of the two states. The same is true of q.

Remember that the reason we employ the function E is that it is a state function. q and w are not state functions, but $q - w$ is.

Answer **4**: $q - w$ is ΔE, and is a state function. This is the correct answer.

Answer **5**: S is a state function. The criterion is not how the quantity behaves, but that it is determined by the characteristics of the initial and final states and not by the path used to connect the states. Moreover, S does not increase in *all* cases.

SOLUTION 15–18

(2)

$$\begin{aligned} \Delta H &= \Delta E + \Delta(PV) \\ PV &= nRT \\ \Delta(PV) &= \Delta(nRT) \\ &= (\Delta n)RT \quad \text{(when } T \text{ is constant)} \\ \Delta H &= \Delta E + (\Delta n)RT \end{aligned}$$

Remember that Δ usually means "final minus initial." There is, in the equation, 1 mole of gas product and 1.5 moles of gas reactants; so Δn is -0.5.

$$\Delta H = \Delta E - 0.5RT$$

Since R and T are positive,

$$\Delta H < \Delta E$$

Answer **2** is correct. For Answer **5**, ΔH is *not* independent of the physical states of the reactants. For instance, the heat of formation of $H_2O(g)$ is -57.8 kcal mole^{-1}, but for $H_2O(l)$ it is -68.3 kcal mole^{-1}. The difference is the heat necessary to condense a mole of water (or, conversely, to vaporize it).

SOLUTION 15–19

(3)

The entropy is related to the probability of an event. The event here is the occurrence of a poker hand. The higher the probability (i.e., the more likely an event), the larger is the entropy. The winning hand of the five is Answer **3**, the royal flush. There are only four such hands possible, one in each suit. It beats the four of a kind (**1**) because Answer **1** is a more probable configuration. There are 48 such hands, with only a different fifth card.

The lowest entropy system, the least probable configuration, and therefore the most difficult to obtain, is the winning hand.

Answer **3** is correct.

[*Note:* The three laws of thermodynamics have been proposed in poker oriented language: (1) the odds are against you; (2) you can't win; (3) you can't get out of the game.]

If the answers had been given as

1	A♤	A♧	A♡	A◇	K♡
2	A♤	10◇	8◇	3♧	2♡
3	A◇	K◇	Q◇	J◇	10◇
4	10◇	10♡	10♧	7♤	4♡
5	6♤	5◇	5♧	3♡	2◇

all of the poker hands would have the same entropy, because we have the same amount of information—we know the identity of each card—for all the hands.

SOLUTION 15–20

(2)

$\Delta G = \Delta H - T\Delta S$

Spontaneous: ΔG is negative
Endothermic: ΔH is positive

Therefore, $T\Delta S$ (hence ΔS) must be positive. ΔS positive means that the entropy of the final state (product) is greater than that of the initial state (reactant) and that the product is a more random, less ordered structure. Thus, Answer **2** is correct. For instance, we might imagine that the polymer is like a coiled spring, and that the product is just a loose, floppy fiber, which is the result of the uncoiling of the spring.

[*Note:* This is the kind of process that occurs when we boil an egg: The ordered protein is denatured, or broken down to a disordered insoluble protein.]

Answer **4**:

$$\begin{aligned}\Delta H &= \Delta H_f^0 \text{ (product)} - \Delta H_f^0 \text{ (polymer)} \\ &= \text{(number)} - \text{(more positive number)} \\ &= \text{negative number}\end{aligned}$$

But this implies that the reaction is exothermic, and we are told that it is endothermic.

Answer **5**: ΔG is the term that determines the spontaneity of reactions, not ΔH. Even if a reaction is energetically not favored (endothermic), the chemical driving force can be supplied by a favorable change in entropy for the reaction (ΔS positive).

SOLUTION 15–21

(3)

A correct statement of the second law is

$\Delta S_{\text{universe}}$ is positive

$$\underbrace{\Delta S_{\text{plant}}}_{\text{system}} + \underbrace{\Delta S_{\text{surroundings}}}_{\text{everything in universe except the plant}} = \Delta S_{\text{universe}}$$

If the first term is negative (disorder proceeding to order) the second term must be larger and positive for the law to hold. [*Remember:* Large entropy means large randomness or disorder; small entropy means order and low probability.]

We have to ask what processes are occurring outside our system that contribute to the plant's growth. There are many (such as the production of solar energy necessary for the photosynthesis reaction, the discharge of molecules like O_2 into the atmosphere by the plant, etc.) These processes are characterized by an increase in entropy; so $\Delta S_{\text{surroundings}}$ is positive and larger than ΔS_{plant}, and the second law holds.

Answer **3** is correct.

SOLUTION 15–22

(5)

The criterion for spontaneity at constant T and P is

$\Delta G < 0$

where

$\Delta G = \Delta H - T\Delta S$

(a): When ΔH is negative and ΔS positive, ΔG must be negative; the process is then definitely spontaneous.

(b): When ΔH is positive and ΔS negative, ΔG must be positive; the process is then definitely nonspontaneous. Here both tendencies are in the same (wrong) direction. The enthalpy is increasing; so the process is going, if at all, against energy considerations, and the entropy is decreasing as well.

(c) and (d): If both ΔH and ΔS are negative (or both positive), ΔG *may* be negative, but it will depend on the relative magnitudes of ΔH and ΔS. So these processes are *possibly* spontaneous.

The correct answer is **5**.

SOLUTION 15–23

(2)

ΔG negative means that the reaction will proceed spontaneously as written. $CO_2(g)$ will be formed spontaneously.

Answer **2** is correct.

SOLUTION 15-24

(3)

$$\Delta G = \Delta H - T\Delta S$$
$$\Delta G_{vap} = \Delta H_{vap} - T_{vap}\Delta S_{vap}$$

$$\frac{q_{rev}}{T} = \Delta S_{vap} = \frac{\Delta H_{vap}}{T_{vap}}$$ (because the vaporization of water at 100°C and 1 atm is an equilibrium and reversible process)

$$T\Delta S_{vap} = \Delta H_{vap}$$
$$\Delta G_{vap} = 0$$

Answer **3** is correct. But we should have known this because $\Delta G = 0$ is the criterion for equilibrium, and the vaporization of H_2O at 100°C and 1 atm is an equilibrium process.

SOLUTION 15-25

(3)

For more on free energy and free energy change, see Section 15–8 of Chemical Principles.

The standard free energy of reaction is given by

$$\Delta G^0 = \Sigma\Delta G_f^0 \text{ (products)} - \Sigma\Delta G_f^0 \text{ (reactants)}$$

[*Note:* ΣA means "the sum of the A quantities."]

Products / Reactants

$$\Delta G^0 = [6(-94.3) + 6(-57.0)] - [-218 + 0]$$

(the coefficients 6: remember the stoichiometry; the 0: O_2, an element in its standard state)

$$\Delta G^0 = -690 \text{ kcal mole}^{-1}$$

3 is correct.
1 If you simply forgot the stoichiometric coefficients, your answer would be $\Delta G^0 = +66.7$ kcal mole^{-1}.
2 If you made the above error and also thought that

$$\Delta G^0 = \Sigma\Delta G_f^0 \text{ (reactants)} - \Sigma\Delta G_f^0 \text{ (products)}$$

then the sign of the answer would be reversed.
4 is almost correct, but there is a mistake, as in the second part of Answer **2**.
5 $\Delta G_f^0\ [O_2(g)]$ is, by definition, zero since the most stable form of O_2 at 25°C is a diatomic gas.

SOLUTION 15-26

(2)

A spontaneous change implies ΔG negative. But a reaction that takes a dilute solution and separates it into a concentrated solution and an even more dilute solution contradicts our experience with spontaneous reactions. (If it did not, it would be quite easy to obtain potable water from the sea.) So the production of gastric juice is not a spontaneous reaction, ΔG is positive, and the body supplies this energy by ion transport across the membranes of the stomach's parietal cells. Answer **2** is correct.

[Quantitatively

$$\Delta G = n_2RT \times 2.3 \log\left(\frac{c_{final}}{c_{init}}\right)$$

in which n_2 is the number of moles of solute taken from c_{init} to c_{final}.]

SOLUTION 15–27

(2)

(a): There is always molecular motion, and exchange of molecules between liquid and vapor phases, but there is no *net* transfer of mass from one phase to another.

(b): At equilibrium, the vapor pressure is characterized by a unique value. At a given T, it does not fluctuate, and the equilibrium *constant* for the vaporization reaction, pure liquid → vapor, is $K_p = p_{\text{vapor}}$.

(c): This is true as well, and is the condition for equilibrium and the observation of *stasis* in the macroscopic properties.

(d): It may take some time to reach equilibrium, but once there the system will not fluctuate unless we change the temperature.

(b) and (c) are correct, and therefore **2** is the appropriate answer.

SOLUTION 15–28

(5)

Liquid → vapor

is an endothermic process, according to our experience with water. We have to supply heat to things to make them boil.

$K_p = p_{\text{vapor}}$

and since the reaction is endothermic, K_p will increase with T. *Or* our experience is that the vapor pressure will increase with temperature.

Answer **5** is correct. Answers **3** and **4**: The vapor pressure is independent of the amount of liquid present so long as *some* liquid is there. [*Note:* Vapor pressure is an *intensive* property since it does not depend on the quantity of material present.]

SOLUTION 15–29

(3)

Let us start at low temperature and 1 atm vapor pressure and slowly increase the temperature. This corresponds to following the horizontal dashed line in Figure 15–3 from left to right.

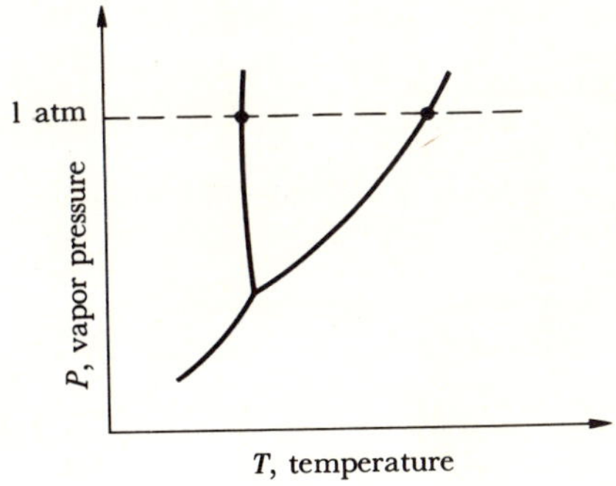

Figure 15–3

Since we are proceeding from low to high temperature, we expect the sequence solid, liquid, vapor to be followed; thus, Answer **3** is correct. The dashed line makes two intersections with lines of our phase diagram which separate the different phases. The intersections are the freezing point and the boiling point of the pure substance.

SOLUTION 15-30 (4)

We expect the vapor pressure of a solvent to be *lowered* by addition of nonvolatile solute, and the dashed curves for the resulting solution lie below the solid curves for the solvent on our phase diagram. Now we can see why the freezing point is depressed but the boiling point is elevated. As we trace the horizontal dashed line at 1 atm in Figure 15-4, we encounter the freezing point of the solution (large dot), which is at a lower T than that of the pure solvent (small dot). As we continue to higher temperatures we find the boiling points, first of the solvent (small dot) and then of the solution (large dot). So X must be the elevation in the boiling point, ΔT_b. But this is not given as one of the choices. At this point we must remember that

$$\Delta T_b = k_b m_A$$

The correct answer is **4**.

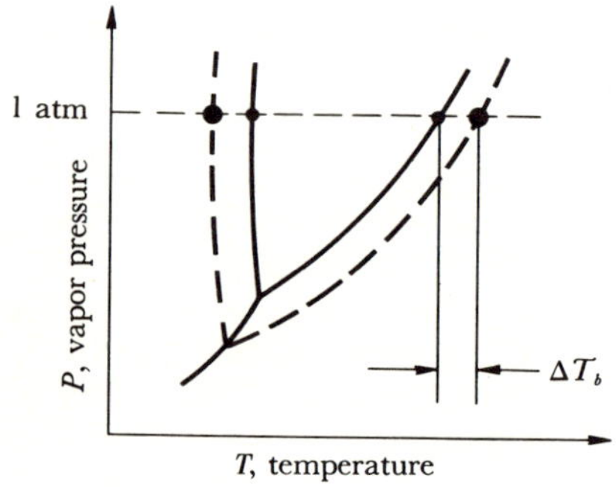

Figure 15-4

SOLUTION 15-31 (2)

The osmotic pressure on the sap will be the osmotic pressure calculated for the sap minus that for the ground water:

$$\Pi = (c_{\text{sap}} - c_{\text{water}})RT$$
$$\Pi = 60 \text{ ft}/30 \text{ ft atm}^{-1} = 2.0 \text{ atm}$$

since we can accept, for this rough calculation the sap density to be that of water.

$$\frac{2.0 \text{ atm}}{(0.082)(300) \text{ liter atm mole}^{-1}} = (0.10 - c_{\text{water}}) \text{ molar}$$

$$c_{\text{water}} = -\frac{2.0}{24.6} + 0.10 \simeq 0.02 \text{ molar}$$

Answer **2** is correct.

SOLUTION 15–32

(4)

To calculate the osmotic pressure of the solution, we must remember that c_A refers to the total concentration of all dissolved species:

$$c_A = [Na^+] + [Cl^-] = 0.50 + 0.50 = 1.0 \text{ molar}$$
$$\Pi = 1.0 \text{ mole liter}^{-1} \times 0.082 \text{ liter atm mole}^{-1} \text{ deg}^{-1} \times 290 \text{ deg} = 24 \text{ atm}$$

4 is correct.
3 you forgot to take account of *all* the concentrations.
1 You did not convert the temperature to degrees Kelvin.
2 You may have used $c_A = 0.50^2$.

CHAPTER 16 FREE ENERGY AND EQUILIBRIUM

In this chapter we will review equilibrium calculations and extend them to the case when solids are present, just as we did in Chapter 5 for K_{sp}. We will make the connection between ΔG^0 and K.

Memorize: $-\Delta G^0 = RT \ln K$
Review: Chapters 5 and 15

Solutions to problems in Chapter 16 begin on page 430.

PROBLEMS

PROBLEM 16–1

The thermodynamic properties of equilibrium are discussed in Section 16–1 of Chemical Principles.

Dynamic equilibrium implies ________.

1 A tendency toward minimum enthalpy only.
2 A tendency toward molecular chaos or maximum entropy only.
3 A balance between the tendencies toward minimum enthalpy and maximum entropy.
4 That ΔH is negative.
5 A balance between the tendencies toward maximum heat and minimum work.

PROBLEM 16–2

For the reaction $C(s) + CO_2(g) \rightleftharpoons 2CO(g)$, an equilibrium mixture exhibited partial pressures for CO and CO_2 of 8.0 atm and 4.0 atm, respectively. What is K_p?

1 0.25 atm
2 0.50 atm
3 2.0 atm
4 4.0 atm
5 16 atm

PROBLEM 16–3

Which one of the following equilibria is *unaffected* by a pressure change, that is, will not result in a net increase of reactants or products?

1 $2NaCl(s) \rightleftharpoons 2Na(s) + Cl_2(g)$
2 $2NO_2(g) \rightleftharpoons N_2O_4(g)$
3 $CO_2(g) + H_2O(l) \rightleftharpoons$ solution of CO_2
4 $H_2(g) + I_2(g) \rightleftharpoons 2HI(g)$
5 $O_3(g) \rightleftharpoons 3O(g)$

PROBLEM 16–4

See Section 16–2 of Chemical Principles *for a review of calculating equilibrium constants for gases.*

The equilibrium constant for the reaction

$C(s) + CO_2(g) \rightleftharpoons 2CO(g)$ at 1000°C

is ________ atm.

Data

$C(s) + 2H_2O(g) \rightleftharpoons CO_2(g) + 2H_2(g) \quad K_p\ (1000°C) = 3.85$ atm
$H_2(g) + CO_2(g) \rightleftharpoons H_2O(g) + CO(g) \quad K_p\ (1000°C) = 0.71$ atm

PROBLEM 16–5

1 $1/(3.85)^2(0.71) \simeq 0.095$
2 $0.71/(3.85)^2 \simeq 0.48$
3 $(0.71)^2(3.85) \simeq 1.9$
4 $2(0.71) + 3.85 \simeq 5.3$
5 $(3.85)^2(0.71) \simeq 21$

The equilibrium constant, K_p, for the reaction

$$NH_4HS(s) \rightleftharpoons NH_3(g) + H_2S(g)$$

is equal to 12 when the pressures are expressed in atmospheres. Some solid NH_4HS is placed in a flask containing 1.0 atm of NH_3. What is the pressure of NH_3 at equilibrium?
[*Hint:* Use trial-and-error approach instead of solving the quadratic equation.]

1 1.0 atm
2 2.0 atm
3 3.0 atm
4 4.0 atm
5 12 atm

PROBLEM 16–6

The macromolecule that is probably most important in the metabolic process is ATP (adenosine-5′-triphosphate). It is a molecule whose reactions provide energy for the cell. Sometimes it can store its energy in the form of creatine phosphate (creatine-P^{2-}) by the reaction

$$\text{ATP} + \text{creatine} \rightleftharpoons \text{ADP} + \text{creatine-}P^{2-} + H^+$$

in which ADP is adenosine-5′-diphosphate. Under standard conditions in the body (pH = 7.4, 30°C, 0.01-molar Mg^{2+}), ΔG is +3 kcal $mole^{-1}$. When will the forward reaction become biologically important?

1 Never.
2 When [creatine-P^{2-}] increases.
3 When pH decreases.
4 When [ADP] increases.
5 When [ATP] increases.

PROBLEM 16–7

Carbon monoxide is dangerous because human hemoglobin has a greater affinity for CO than for O_2. The equilibrium constants for the reactions of human hemoglobin molecules with O_2 and CO (K_1 and K_2, respectively) are in the ratio 1:200 at human body temperature, 37°C.

$$\text{Heme} + O_2 \rightleftharpoons \text{Heme}\cdot O_2 \qquad \text{Heme} + CO \rightleftharpoons \text{Heme}\cdot CO \qquad \frac{K_1}{K_2} = \frac{1}{200}$$

What is the *difference* in the ΔG^0's for the two equilibria?

($\ln x \approx 2.30 \log x$;
$R = 0.082$ liter atm mole^{-1} deg^{-1} or 1.98 cal mole^{-1} deg^{-1})

1 5.3 cal mole^{-1}
2 11 cal mole^{-1}
3 130 cal mole^{-1}
4 1.4 kcal mole^{-1}
5 3.3 kcal mole^{-1}

PROBLEM 16–8

Given the following data for the temperature dependence of the pH of pure water, which one of the statements below is correct? [*Hint:* Distinguish between ΔG^0 and ΔG.]

pH	T, °C
7.47	0
7.27	10
7.00	25
6.77	40
6.55	60

1 ΔG^0 for the dissociation of water at equilibrium cannot be calculated from the data.
2 ΔG for the dissociation of water at equilibrium is zero.
3 Water is acidic above 25°C.
4 The dissociation of water is exothermic.
5 Since ΔG^0 can be calculated from standard free energies of formation, ΔG^0 is temperature-independent.

PROBLEM 16–9

From the data of Problem 16–8, calculate ΔG^0 for the following reaction at 25°C:

$$H_2O(l) \rightleftharpoons H^+(aq) + OH^-(aq)$$

($\ln x = 2.303 \log x$;
$R = 1.98$ cal mole^{-1} deg^{-1})

1 8.26 kcal mole^{-1}
2 8.61 kcal mole^{-1}
3 9.61 kcal mole^{-1}
4 19.0 kcal mole^{-1}
5 None of the above. ΔG^0 must be negative.

PROBLEM 16–10 Using the result of the calculation of ΔG^0 for $H_2O \rightleftharpoons H^+ + OH^-$ at 25°C in Problem 16–9, calculate the standard free energy of formation of $OH^-(aq)$ at this temperature.

Data	ΔG_f^0, kcal mole^{-1} (25°C)
$H_2O(l)$	−56.7
$H^+(aq)$	0

1 −94.4 kcal mole^{-1}
2 −75.7 kcal mole^{-1}
3 −37.7 kcal mole^{-1}
4 +37.7 kcal mole^{-1}
5 None of the above.

SOLUTIONS

SOLUTION 16–1

(3)

Answer **3** is correct. Recall that ΔG, the quantity that indicates whether a reaction is spontaneous, is the result of such a balance between ΔH and ΔS. Answer **4**: ΔH negative simply means that the reaction is exothermic, but indicates nothing about the equilibrium condition.

SOLUTION 16–2

(5)

K_p means that the equilibrium constant is expressed in partial pressures, usually atmospheres:

$$K_p = \frac{p_{CO}^2}{p_{CO_2}}$$

Note that C does not enter into the expression because it is a pure solid.

$$p_{CO} = 8.0 \text{ atm}$$
$$p_{CO_2} = 4.0 \text{ atm}$$

$$K_p = \frac{(8.0)^2}{4.0} = 16 \text{ atm}$$

5 is correct.
2 $K_p \neq p_{CO_2}^2/p_{CO}^2 = 0.50$
1 $K_p \neq p_{CO_2}^2/p_{CO}^2 = 0.25$
3 $K_p \neq p_{CO}/p_{CO_2} = 2$ or $K_p \neq p_{CO_2}^2/p_{CO} = 2$
4 $K_p \neq 2p_{CO}/p_{CO_2} = 4$

SOLUTION 16–3

(4)

An equilibrium will be affected by a pressure change only if the volume of reactants differs from the volume of products. Liquids and solids have densities of the same order of magnitude, and the pressure effect on the equilibrium will be minimal. Reactions that involve gases (perhaps together with liquids or solids) in which the number of moles of gaseous reactants and products is not the same will show a marked pressure dependence. And all the reactions except Answer **4** will be affected by pressure.

SOLUTION 16–4

(3)

$$3.85 = \frac{[H_2]^2[CO_2]}{[H_2O]^2} \quad \text{(a)}$$

$$0.71 = \frac{[H_2O][CO]}{[H_2][CO_2]} \quad \text{(b)}$$

$$K = ? = \frac{[CO]^2}{[CO_2]}$$

To obtain a $[CO]^2$ term we shall square Equation (b):

$$0.71^2 = \frac{[H_2O]^2[CO]^2}{[H_2]^2[CO_2]^2}$$

and to remove the $[H_2O]^2$ term from the numerator, we shall multiply by Equation (a):

$$0.71^2 \times 3.85 = \frac{[H_2O]^2[CO]^2}{[H_2]^2[CO_2]^2} \times \frac{[H_2]^2[CO_2]}{[H_2O]^2}$$

$$= \frac{[CO]^2}{[CO_2]} = K$$

Answer **3** is correct.

This solution shows that K's cannot be summed like ΔH's but are multiplicative.

Equation	Reaction	K
(b)	$H_2(g) + CO_2(g) \rightleftharpoons H_2O(g) + CO(g)$	0.71
2 × (b)	$2H_2(g) + 2CO_2(g) \rightleftharpoons 2H_2O(g) + 2CO(g)$	$(0.71)^2$
(a)	$C(s) + 2H_2O(g) \rightleftharpoons CO_2(g) + 2H_2(g)$	3.85
2 × (b) + (a)	$C(s) + CO_2(g) \rightleftharpoons 2CO(g)$	$3.85 \times (0.71)^2$

SOLUTION 16–5

(4)

Let x be the amount of NH_4HS that dissociates, when expressed in atmospheres.

	$NH_4HS(s) \rightleftharpoons$	$NH_3(g)$ +	$H_2S(g)$	
init:		1.0	0	atm
equil:		$1.0 + x$	x	atm

$$K_p = p_{NH_3}p_{H_2S}$$
$$= (1.0 + x)x = 12 \text{ atm}^2$$

By trial and error $x = 3.0$ satisfies this equation.

$$p_{NH_3} = 4.0 \text{ atm}$$
$$p_{H_2S} = 3.0 \text{ atm}$$

Answer **4** is correct. This problem can also be solved with the quadratic formula,

$$x = \frac{-b \pm \sqrt{b^2 - 4ac}}{2a}$$

for the equation

$$ax^2 + bx + c = 0$$

If you are unfamiliar with this formula, refer to Appendix 2.

Our equation is

$$x^2 + x - 12 = 0$$

$$x = \frac{-1 \pm \sqrt{1+48}}{2} = \frac{-1 \pm 7}{2} = +3.0 \text{ atm}$$

since the positive root is the only one with physical significance. [x cannot be negative because unless some NH_4HS dissociates (x positive), there will be no H_2S to react with the NH_3.]

SOLUTION 16–6 (5)

Under standard conditions, the reaction is not spontaneous. But as ATP accumulates (in a resting muscle), the reaction will be forced to the right; that is, it will become spontaneous and will store the energy as creatine phosphate until it is needed.

So Answer **5** is correct. Answer **3**: pH decreasing means $[H^+]$ increasing, which forces the reaction to the left. Answer **2**: Reaction is forced to the left.

SOLUTION 16–7 (5)

We can write the relationship between K and ΔG^0 for the two equilibria:

$$-RT \ln K_1 = \Delta G_1^0$$
$$-RT \ln K_2 = \Delta G_2^0$$

We are given a ratio of the K's and are asked for the quantity $\Delta G_1^0 - \Delta G_2^0$. Subtracting the second equation from the first, we get

$$-RT(\ln K_1 - \ln K_2) = \Delta G_1^0 - \Delta G_2^0$$

But

$$\ln K_1 - \ln K_2 = \ln(K_1/K_2)$$

$$-RT \ln \frac{K_1}{K_2} = -RT\ 2.30 \log \frac{K_1}{K_2} = +RT \times 2.30 \log \frac{K_2}{K_1}$$

$$= 1.98 \text{ cal mole}^{-1} \text{ deg}^{-1} \times 310 \text{ deg} \times 2.30 \times \log 200$$

$$= 3.25 \times 10^3 \text{ cal mole}^{-1}$$

So

$$\Delta G_1^0 - \Delta G_2^0 \simeq 3.3 \text{ kcal mole}^{-1}$$

5 is correct.
1 This is just ln 200 with some units tacked on. You forgot RT.
2 You left out T.
3 You used the wrong value of R. Had you checked the units of your calculation you would have seen they didn't cancel to give cal mole^{-1}.
4 You forgot the conversion factor between ln and log.

SOLUTION 16–8

(2)

Pure water is neutral at all temperatures (**3** is incorrect), $[H^+] = [OH^-]$, and the data give us the variation of K_w with temperature. The H^+ and OH^- concentrations increase with temperature so the reaction must be endothermic; heat input drives the reaction to the right. **4** is incorrect.

From K_w, ΔG^0 can be calculated:

$$\Delta G^0 = -RT \ln K_w;$$

1 is incorrect. It is not true that ΔG^0 is temperature-independent—the free energies of formation are themselves temperature-dependent, as is shown by the fact that we must specify the temperature for the standard free energy of formation.

But ΔG is a different quantity. At equilibrium and constant temperature and pressure, ΔG must be zero. **2** is correct.

SOLUTION 16–9

(4)

Activity is reviewed in Section 16–1 of Chemical Principles.

Since we know the pH, we also know K_w. At 25°C, K_w is $(1.0 \times 10^{-7})^2$ or 1.0×10^{-14}. And since this is the equilibrium constant for the dissociation equilibrium,

$$\Delta G^0 = -RT \ln K_w$$

At 25°C,

$$\begin{aligned}\Delta G^0 &= -1.98 \text{ cal mole}^{-1} \text{ deg}^{-1} \times 298 \text{ deg} \times \ln(1.0 \times 10^{-14}) \\ &= -1.98 \times 298 \times 2.303 \log(1.0 \times 10^{-14}) \text{ cal mole}^{-1} \\ &= 19.0 \text{ kcal mole}^{-1}\end{aligned}$$

4 is correct.
1 You did not use the conversion factor between ln and log.
2 You forgot R.
3 You had K_w equal to 1.0×10^{-7}.
5 ΔG^0 need not be negative. When it is positive, it indicates that the equilibrium constant is small, that is, that $\ln K_w$ is negative.

[*Note:* K_w is still the equilibrium constant for this reaction even though it does not contain $[H_2O]$ explicitly because the *activity* of pure water is unity, while its *molarity* is 55.]

SOLUTION 16–10

(3)

Since ΔG^0 for the reaction is 19.0 kcal mole^{-1},

$$\begin{aligned}19.0 \text{ kcal mole}^{-1} &= \Delta G_f^0 \text{ (products)} - \Delta G_f^0 \text{ (reactants)} \\ &= [0 + \Delta G_f^0(OH^-) - (-56.7 \text{ kcal mole}^{-1})]\end{aligned}$$

$$\Delta G_f^0(OH^-) = 19.0 - 56.7 = -37.7 \text{ kcal mole}^{-1}$$

3 is correct.
1 You doubled the H_2O value.
2 Sign errors.
4 You subtracted reactants from products.

CHAPTER 17 OXIDATION-REDUCTION EQUILIBRIA AND ELECTROCHEMISTRY

This chapter will examine the applications of electrochemical principles to batteries, cells, and direction of oxidation-reduction reactions.

Memorize the following relationships.

(1) $\Delta G = -n\mathscr{F}\mathscr{E}$ (T, P constant; reversible process; only electrical work)

n = number of electrons transferred in the cell reaction

$\mathscr{E}$ = cell potential

$\mathscr{F}$ = faraday

(2) $\mathscr{E} = \mathscr{E}^0 - \dfrac{RT}{n\mathscr{F}} \ln Q$

where for the reaction

$a\text{A} + b\text{B} = c\text{C} + d\text{D}$

$Q = \dfrac{[\text{C}]^c[\text{D}]^d}{[\text{A}]^a[\text{B}]^b}$

but the concentrations are *not necessarily* measured at equilibrium.

$\mathscr{E}^0$ = standard cell potential

(3) cathode implies reduction

(4) watt = joule sec^{-1} = volt coulomb sec^{-1}

(5) kilowatt = 10^3 watts

Unless otherwise specified, all systems are at 25°C and all concentrations are unity.

Solutions to problems in Chapter 17 begin on page 443.

PROBLEMS

PROBLEM 17-1

Given the following (unbalanced) equation:

$H^+ + MnO_4^{2-} \rightarrow MnO_2 + MnO_4^- + H_2O$

the MnO_2 species is ________.

1 An oxidizing agent.
2 A reducing agent.
3 Disproportionating.
4 Amphoteric.
5 None of the above.

PROBLEM 17-2

Half-reactions are discussed in Section 17-4 of Chemical Principles *by Dickerson, Gray, and Haight.*

For the reaction

$H_2O + PbO_2 + Cl^- + OH^- \rightarrow ClO^- + Pb(OH)_3^-$

the two half-reactions involved are ________.

(a) $Pb^{2+} + 2e^- \rightarrow Pb$
(b) $Pb^{2+}SO_4^{2-}(s) + 2e^- \rightarrow Pb + SO_4^{2-}$
(c) $H_2 + 2e^- \rightarrow 2H^-$
(d) $2H^+ + 2e^- \rightarrow H_2$
(e) $O_2 + 4H^+ + 4e^- \rightarrow 2H_2O$
(f) $PbO_2 + 4H^+ + 2e^- \rightarrow Pb^{2+} + 2H_2O$
(g) $ClO^- + H_2O + 2e^- \rightarrow Cl^- + 2OH^-$
(h) $2H_2O + PbO_2 + 2e^- \rightarrow Pb(OH)_3^- + OH^-$

1 (a) and (c)
2 (a) and (f)
3 (f) and (g)
4 (g) and (h)
5 None of the above.

PROBLEM 17-3

A reaction proceeds as written (assuming all concentrations are unity) when ________.

1 $\mathscr{E}$ is negative.
2 $\mathscr{E}^0$ is negative.
3 $\mathscr{E}$ is zero.
4 $\mathscr{E}^0$ is zero.
5 None of the above.

PROBLEM 17-4

The cell in Figure 17-1 is set up. The reaction that occurs and the cell voltage is given by which answer?

Half-reaction	Standard reduction potential, $\mathscr{E}^0$ (volts)
$Cu^{2+} + 2e^- \leftrightarrows Cu$	0.34
$Zn^{2+} + 2e^- \leftrightarrows Zn$	−0.76

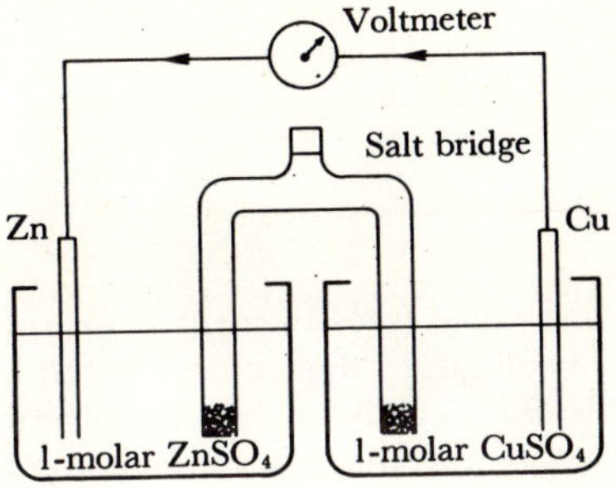

Figure 17-1

		Cell voltage, $\mathscr{E}$
1	$Cu + Zn^{2+} \leftrightarrows Cu^{2+} + Zn$	1.10
2	$Zn + Cu^{2+} \leftrightarrows Zn^{2+} + Cu$	0.42
3	$Cu^{2+} + Zn^{2+} \leftrightarrows Zn + Cu$	0.42
4	$Zn + Cu^{2+} \leftrightarrows Zn^{2+} + Cu$	1.10
5	No reaction occurs.	0.0

PROBLEM 17-5

Electrical energy and cell potentials are discussed in Section 17-2 of Chemical Principles.

A lead storage battery involves the two half-reactions:

	Standard reduction potential, $\mathscr{E}^0$ (volts)
$PbSO_4(s) + 2e^- \leftrightarrows Pb + SO_4^{2-}$	−0.36
$PbO_2 + 4H^+ + SO_4^{2-} + 2e^- \leftrightarrows PbSO_4 + 2H_2O$	1.69

For a cell that duplicates the lead storage battery discharge reaction at unit concentrations, what is the overall reaction and potential in volts?

1. $2PbSO_4(s) + 2H_2O \leftrightarrows Pb + PbO_2 + 2SO_4^{2-} + 4H^+$ $\mathscr{E}^0 = -2.05$
2. $PbO_2 + 4H^+ + 4e^- \leftrightarrows Pb + 2H_2O$ $\mathscr{E}^0 = 1.33$
3. $Pb + 2H_2O \leftrightarrows PbO_2 + 4H^+ + 4e^-$ $\mathscr{E}^0 = 1.33$
4. $Pb + PbO_2 + 2SO_4^{2-} + 4H^+ \leftrightarrows 2PbSO_4(s) + 2H_2O$ $\mathscr{E}^0 = 2.05$
5. None of the above.

PROBLEM 17-6 The cathode reaction in the cell

$Pt|Br_2(l)|HBr(aq)|H_2(g)|Pt$

is ________.

		Standard reduction potential, $\mathscr{E}^0$ (volts)
1	$Pt \rightarrow Pt^{2+} + 2e^-$	−0.98
2	$Br_2(l) + 2e^- \rightarrow 2Br^-$	1.06
3	$2Br^- \rightarrow Br_2(l) + 2e^-$	−1.06
4	$2H^+ + 2e^- \rightarrow H_2(g)$	0
5	$H_2(g) \rightarrow 2H^+ + 2e^-$	0

PROBLEM 17-7 One method of cleaning coins (mostly copper and partially oxidized) at the site of an archaeological "dig" is to suspend the object on a copper wire attached to the negative pole of a battery, dip it in a 2.5% NaOH solution, and insert into the solution a graphite electrode connected to the positive terminal. What is the reaction that occurs at the coin?

1 $Cu \rightarrow Cu^{2+} + 2e^-$
2 $Cu \rightarrow Cu^+ + e^-$
3 $Cu^{2+} + 2e^- \rightarrow Cu$
4 $2H_2O \rightarrow O_2 + 4H^+ + 4e^-$
5 $O_2 + 4H^+ + 4e^- \rightarrow 2H_2O$

PROBLEM 17-8 The Nernst equation,

$$\mathscr{E} = \mathscr{E}^0 - \frac{RT}{n\mathscr{F}} \ln Q$$

gives the relation between $\mathscr{E}$, the electromotive force (emf), and concentration. It indicates that the equilibrium constant, K_c, equals Q (i.e., that the system is at equilibrium and the concentrations are equilibrium concentrations) when ________.

1 $\mathscr{E} = \mathscr{E}^0$
2 $RT/n\mathscr{F} = 1$
3 $\mathscr{E} = 0$
4 $\mathscr{E}^0 = 0$
5 K_c cannot equal Q.

PROBLEM 17–9 Suppose that you forgot the sign in the Nernst equation,

$$\mathscr{E} = \mathscr{E}^0 \pm \frac{RT}{n\mathscr{F}} \ln Q$$

Which of the following arguments would lead to a correct choice of the sign between the two terms on the right-hand side of the equation? Use the reaction $aA + bB \rightleftharpoons cC + dD$.

1 It must be positive because any concentration greatei than unity will drive the equilibrium farther to the right and increase the cell voltage.
2 It must be positive because

$$Q = \frac{[A]^a[B]^b}{[C]^c[D]^d}$$

and when $Q > K_c$, the reaction proceeds to the right.
3 It must be negative because any concentration greater than unity interferes with the mobility of the electrons and decreases the cell's voltage.
4 It must be negative because

$$Q = \frac{[C]^c[D]^d}{[A]^a[B]^b}$$

and $Q > K_c$ means that the reaction proceeds to the left.
5 None of the above arguments is convincing.

PROBLEM 17–10 What is the chemical form of the equilibrium constant for the following reaction?

Equilibrium constants and reduction potentials are related in Section 17–5 of Chemical Principles.

$$2Cr(s) + 3Cu^{2+}(aq) \leftrightarrows 2Cr^{3+}(aq) + 3Cu(s)$$

1 $\dfrac{[Cr^{3+}]^2[Cu]^3}{[Cr]^2[Cu^{2+}]^3}$

2 $\dfrac{[Cr^{3+}]^{3/2}}{[Cu^{2+}]^{2/3}}$

3 $\dfrac{2[Cr^{3+}]}{3[Cu^{2+}]}$

4 $\dfrac{[Cr^{3+}]^2}{[Cu^{2+}]^3}$

5 $\dfrac{\frac{1}{2}[Cr^{3+}]}{\frac{1}{3}[Cu^{2+}]}$

PROBLEM 17–11

What is the numerical value of the *standard* cell potential for the reaction in Problem 17–10?

$2Cr(s) + 3Cu^{2+}(aq) \rightarrow 2Cr^{3+}(aq) + 3Cu(s)$

	Half-reaction	Standard reduction potential, $\mathscr{E}^0$ (volts)
(a)	$Cr^{3+} + 3e^- \rightleftarrows Cr$	−0.74
(b)	$Cr^{3+} + e^- \rightleftarrows Cr^{2+}$	−0.41
(c)	$Cr_2O_7^{2-} + 14H^+ + 6e^- \rightleftharpoons 2Cr^{3+} + 7H_2O$	1.33
(d)	$Cu^+ + e^- \rightleftarrows Cu$	0.52
(e)	$Cu^{2+} + 2e^- \rightleftarrows Cu$	0.34
(f)	$Cu^{2+} + e^- \rightleftarrows Cu^+$	0.16

1 −1.08 volts
2 −0.40 volt
3 0.40 volt
4 1.08 volts
5 2.50 volts

PROBLEM 17–12

For the reaction of Problems 17–10 and 17–11

$2Cr(s) + 3Cu^{2+}(aq) \rightleftarrows 2Cr^{3+}(aq) + 3Cu(s)$

$\mathscr{E}^0 = 1.08$ volts. What is the numerical value of the equilibrium constant, K_c?

[ln $a = 2.303 \log a$ (see Appendix 2)
$2.303RT/\mathscr{F} = 0.0592$ volt at 25°C]

1 $10^{-[6(1.08)/0.0592]} = 10^{-109}$
2 $10^{0.0592/6(1.08)} = 10^{0.00915}$
3 $10^{0.0592/1.08} = 10^{0.0549}$
4 $10^{1.08/0.0592} = 10^{18.2}$
5 $10^{6(1.08)/0.0592} = 10^{109}$

PROBLEM 17–13

Suppose that you want to keep a solution containing Fe^{2+} (ferrous) ions pure and, in particular, free of Fe^{3+} (ferric) ions usually formed by atmospheric oxidation. Which procedure is recommended? [*Note:* Treat this problem by considering unit concentrations.]

Half-reaction	Standard reduction potential, $\mathscr{E}^0$ (volts)
$Fe^{2+} + 2e^- \leftrightarrows Fe$	−0.41
$Fe^{3+} + e^- \leftrightarrows Fe^{2+}$	0.77
$Ag^+ + e^- \leftrightarrows Ag$	0.80
$Zn^{2+} + 2e^- \leftrightarrows Zn$	−0.76

1 Add Ag^+.
2 Add Zn^{2+}.
3 Add Ag.
4 Add Zn.
5 Add Fe.

PROBLEM 17–14

You are asked to select anode, cathode, and electrolyte material for a high performance electrochemical cell that can generate high power for minimum weight. Which set of characteristics is best?

	Cathode		Electrolyte
	Electronegativity	Equivalent wt	conductivity
1	Low	Low	Low
2	Low	Low	High
3	Low	High	Low
4	High	Low	High
5	High	High	High

PROBLEM 17–15

You are to design an electrochemical cell for automobile propulsion by using a cell whose materials reflect our optimization in Problem 17–14.

$Li|LiCl$ (fused salt)$|Cl_2$

What is the optimum ratio of weights of Li to Cl_2, that is, wt Li/wt Cl_2?

(at. wt: Li = 6.94, Cl = 35.5)

1 $6.94/35.5 \simeq 0.195$
2 $6.94/71.0 \simeq 0.098$
3 $6.94/(35.5 + 6.94) \simeq 0.16$
4 $2(6.94)/35.5 \simeq 0.39$
5 $(6.94 + 35.5)/35.5 \simeq 1.2$

PROBLEM 17–16

Electrical work is discussed in Section 17–2 of Chemical Principles.

At 500°C the $Li|LiCl|Cl_2$ cell has a voltage of 3.4 volts. How much energy is released when a battery whose electrode material weighs a total of 1.00 kg is run until it is completely discharged? (Assume that there are no internal power losses in the battery, and use the results of Problem 17–15.)

($\mathcal{F} \simeq 9.6 \times 10^4$ coulombs)

1 3.4 joules kg^{-1}
2 23.6 joules kg^{-1}
3 $3.4 \times 23.6 \simeq 80$ joules kg^{-1}
4 9.6×10^4 joules kg^{-1}
5 $3.4 \times 23.6 \times 9.6 \times 10^4 \simeq 7.7 \times 10^6$ joules kg^{-1}

PROBLEM 17–17 We have calculated the maximum theoretical specific energy for the

$Li|LiCl|Cl_2$

cell to be about 2.1×10^3 watts h kg^{-1}. Actually, we can expect the specific energies for this cell to be only 21% of this value, or 4.4×10^2 watts h kg^{-1}. If a 180-kg cell is used, and the energy requirement for an urban automobile is estimated to be 0.27 kwatt h $mile^{-1}$, what is the range of the automobile?

1 1.6 miles
2 160 miles
3 290 miles
4 29,000 miles
5 It will never work.

SOLUTION 17–1

(2)

Oxidation numbers

$$\overset{+6}{H^+ + \mathbf{MnO_4^{2-}}} \rightarrow \overset{+4}{\mathbf{MnO_2}} + \overset{+7}{\mathbf{MnO_4^-}} + H_2O$$

(reduction: MnO_4^{2-} (+6) → MnO_2 (+4); oxidation: MnO_4^{2-} (+6) → MnO_4^- (+7))

In the *reverse* reaction, MnO_2 is oxidized (+4 → +6) to MnO_4^{2-} and is therefore a *reducing agent* since it is the "agent" by which MnO_4^- is reduced (+7 → +6) to MnO_4^{2-}.

Answer **2** is correct.

If we want to break down the original reaction into half-reactions,

$MnO_4^{2-} \rightarrow MnO_4^- + e^-$ *oxidation*

$MnO_4^{2-} + 4H^+ + 2e^- \rightarrow MnO_2 + 2H_2O$ *reduction*

If we reverse the second half-reaction, it becomes an oxidation half-reaction. In this reaction MnO_2 is oxidized to MnO_4^{2-}

$$MnO_2 + 2H_2O \rightarrow MnO_4^{2-} + 4H^+ + 2e^-$$

and MnO_2 therefore behaves as a reducing agent.

SOLUTION 17–2

(4)

One half-reaction must connect PbO_2 and $Pb(OH)_3^-$; the other must connect Cl^- and ClO^- since these are the only species whose oxidation numbers are being changed. Only reactions (g) and (h) involve these species, and we can see that by reversing (g) and summing with (h) we can obtain the overall reaction:

$$Cl^- + 2OH^- \rightarrow ClO^- + H_2O + 2e^-$$
$$2H_2O + PbO_2 + 2e^- \rightarrow Pb(OH)_3^- + OH^-$$

Sum: $2H_2O + PbO_2 + Cl^- + 2OH^- \rightarrow ClO^- + Pb(OH)_3^- + H_2O + OH^-$

If we subtract H_2O and OH^- from each side,

$$H_2O + PbO_2 + Cl^- + OH^- \rightarrow ClO^- + Pb(OH)_3^-$$

Therefore, Answer **4** is correct.

SOLUTION 17–3 (5)

We must remember that

$$\mathscr{E} = \mathscr{E}^0 - \frac{RT}{n\mathscr{F}} \ln Q$$

and $Q = 1$ when all concentrations are unity. In this case

$$\mathscr{E} = \mathscr{E}^0$$

and

$$\Delta G = -n\mathscr{F}\mathscr{E} = -n\mathscr{F}\mathscr{E}^0$$

Since n and $\mathscr{F}$ are positive numbers, a positive $\mathscr{E}^0$ implies a negative ΔG. This means that the process proceeds spontaneously as written. Thus, a reaction proceeds as written when $\mathscr{E}$ (or $\mathscr{E}^0$, if all the concentrations are unity) is positive.

Answer **5** is the only correct choice.

SOLUTION 17–4 (4)

We can do this by trial and error. If we reverse the second equation, the electrons cancel from the sum of the two equations.

	$\mathscr{E}^0$	
$Cu^{2+} + 2e^- \leftrightarrows Cu$	0.34	
$Zn \leftrightarrows Zn^{2+} + 2e^-$	0.76	(note sign change)
$Cu^{2+} + Zn \leftrightarrows Cu + Zn^{2+}$	1.10	

Since the concentrations are equal, $Q = 1$, and $\mathscr{E} = \mathscr{E}^0 = 1.10$ volts.

There are now two possibilities: Either the reaction proceeds as written or it occurs in the reverse sense, namely, Answers **4** and **1**, respectively. But a positive $\mathscr{E}$ indicates that the process is spontaneous and goes in the direction as written.

Answer **4** is correct.

[*Note:* Our thermodynamic criterion for spontaneous reaction is ΔG negative. But since $\Delta G = -n\mathscr{F}\mathscr{E}$, the same criterion means $\mathscr{E}$ must be positive. (See Problem 17–3.)]

SOLUTION 17–5 (4)

Since this is the discharge reaction, the reaction must be spontaneous, and the cell potential, $\mathscr{E}^0$, must be positive. If we reverse the first half-reaction and sum,

	$\mathscr{E}^0$
$Pb + SO_4^{2-} \rightleftarrows PbSO_4 + 2e^-$	0.36
$PbO_2 + 4H^+ + SO_4^{2-} + 2e^- \rightleftarrows PbSO_4 + 2H_2O$	1.69
Sum: $Pb + PbO_2 + 2SO_4^{2-} + 4H^+ \leftrightarrows 2PbSO_4 + 2H_2O$	2.05

$\mathscr{E}^0$ is positive and the reaction proceeds as written.

Answer **4** is correct.

SOLUTION 17–6

(2)

Reduction occurs at the cathode. A reduction half-reaction is one in which electrons appear on the left-hand side, so only Answers **2** and **4** are possibilities. The cell either has the reaction

$$Br_2(l) + H_2(g) \rightleftarrows 2H^+ + 2Br^-$$

which is the sum of the half-reactions **2** and **5**, or it is the reverse

$$2H^+ + 2Br^- \rightarrow H_2(g) + Br_2(l)$$

which is the sum of half-reactions **3** and **4**. $\mathscr{E}^0$ for the first reaction is 1.06 and for the second is −1.06. A positive potential indicates that the reaction proceeds as written.

Therefore, Answer **2** is the cathode reaction.

SOLUTION 17–7

(3)

Reduction occurs at the cathode, so electrons must appear on the left-hand side of this half-cell reaction. Our only choices are Answers **3** and **5**. We can dismiss Answer **5** because it occurs in acidic solution, whereas ours is basic. Answer **3** is correct. This process attempts to reconstruct the coin by reducing the metal at the place where it was oxidized (corroded) to Cu^{2+}.

SOLUTION 17–8

(3)

The use of the Nernst equation is discussed in Section 17–5 of Chemical Principles.

$$\Delta G = -n\mathscr{F}\mathscr{E}$$

The equilibrium condition is

$$\Delta G = 0$$

This means that the equilibrium condition in terms of cell potential is

$$\mathscr{E} = 0$$

Answer **3** is correct. This means that $\mathscr{E}^0 = (RT/n\mathscr{F}) \ln K_c$.

Answer **1**: $\mathscr{E} = \mathscr{E}^0$ when $Q = 1$. For Answer **5**, K_c equals Q when the equilibrium condition is attained.

SOLUTION 17–9

(4)

For a reaction

$$aA + bB \rightleftharpoons cC + dD$$

$$Q = \frac{[C]^c[D]^d}{[A]^a[B]^b}$$

Suppose Q is greater than K_c. This means that the reaction must proceed to the left to reach equilibrium. Our criteria of spontaneity were as follows: $\mathscr{E}$ positive means spontaneous forward; $\mathscr{E}$ zero means equilibrium; and $\mathscr{E}$ negative means spontaneous reverse, which is the present situation under our supposition. So when we *increase* Q we must *decrease* $\mathscr{E}$, and the correct sign in the Nernst equation is a minus. Answer **4** is our choice. Answer **2** is wrong because Q is inverted.

SOLUTION 17–10 (4)

The solid species do not enter into the equilibrium constant. This equilibrium constant is constructed exactly like all the others:

$$K_c = \frac{[Cr^{3+}]^2}{[Cu^{2+}]^3}$$

Answer **4** is correct.

Answer **1** is almost correct, but the concentrations of the solid species cancel since, if they are pure, they are both unity.

SOLUTION 17–11 (4)

We must pick out the correct half-reactions, ones which connect Cr with Cr^{3+} and Cu^{2+} with Cu. These are (a), reversed, and (e).

	$\mathscr{E}^0$
$Cr \rightleftarrows Cr^{3+} + 3e^-$	+0.74
$Cu^{2+} + 2e^- \rightleftarrows Cu$	+0.34

To make the electrons cancel, we must multiply the first equation by two and the second by three. But the $\mathscr{E}^0$ values are *unaffected*.

		$\mathscr{E}^0$
	$2Cr \rightleftarrows 2Cr^{3+} + 6e^-$	+0.74
	$3Cu^{2+} + 6e^- \leftrightarrows 3Cu$	+0.34
Sum:	$3Cu^{2+} + 2Cr \leftrightarrows 2Cr^{3+} + 3Cu$	+1.08

4 is correct.
1 This is the potential for the reverse reaction.
2 You forgot to change the sign when you reversed the equation.
5 You multiplied the $\mathscr{E}^0$'s by the stoichiometric coefficients.

SOLUTION 17–12 (5)

From Problems 17–8 and 17–9

$$\mathscr{E}^0 = \frac{RT}{n\mathscr{F}} \ln K_c$$

$$\ln K_c = \frac{n\mathscr{F}\mathscr{E}^0}{RT}$$

$$\log K_c = \frac{n\mathscr{F}\mathscr{E}^0}{2.303RT}$$

$$= \frac{n\mathscr{E}^0}{0.0592} \text{ at } 25°C$$

Half-reaction	$\mathscr{E}^0$
$2Cr^{3+} + 6e^- \leftrightarrows 2Cr$	−0.74
$3Cu^{2+} + 6e^- \leftrightarrows 3Cu$	0.34

Reversing the first reaction and summing, we obtain

$$3Cu^{2+} + 2Cr \leftrightarrows 2Cr^{3+} + 3Cu \qquad \mathscr{E}^0 = 1.08$$

$n = 6$ because there are $6e^-$ involved in each half-reaction.

$$\log K_c = 6 \times \frac{1.08}{0.0592}$$

$$K_c = 10^{6(1.08)/0.0592}$$
$$K_c = 10^{109}$$

Answer **5** is correct. Answer **1**: This is $1/K_c$. For Answer **4**, you forgot that $n = 6$.

SOLUTION 17–13

(5)

Half-reaction	Standard reduction potential, $\mathscr{E}^0$ (volts)
$Fe \leftrightarrows Fe^{2+} + 2e^-$	+0.41
$2Fe^{3+} + 2e^- \leftrightarrows 2Fe^{2+}$	+0.77
$Fe + 2Fe^{3+} \leftrightarrows 3Fe^{2+}$	1.18

When $\mathscr{E}^0$ is large, the equilibrium constant, K_c, is large. (Even an $\mathscr{E}^0$ of 0.1 volt gives a large K_c. See Problem 17–12, for instance.) This indicates that by adding Fe metal, we will immediately reduce any Fe^{3+} that forms and the solution will remain pure. Answer **5** is correct.

Answers **1** and **2**: Adding Ag^+ or Zn^{2+} will do nothing but contaminate the solution. These species can only oxidize something and, in the process, return to their metallic state. Answer **3**: Ag will not reduce Fe^{3+} to Fe^{2+} since $\mathscr{E}^0$ for that process is $\mathscr{E}^0 = -0.80 + 0.77 = -0.03$. Answer **4**: Zn will reduce Fe^{3+} to Fe^{2+}, but then it will also reduce Fe^{2+} to Fe.

	$\mathscr{E}^0$
$Zn \leftrightarrows Zn^{2+} + 2e^-$	+0.76
$Fe^{2+} + 2e^- \leftrightarrows Fe$	−0.41
$Fe^{2+} + Zn \leftrightarrows Zn^{2+} + Fe$	+0.35

Hence, this reagent will destroy our Fe^{2+} solution and convert it to a Zn^{2+} solution.

[*Note:* The actual K_c for the $Fe + 2Fe^{3+} \rightarrow 3Fe^{2+}$ reaction is

$$K_c = \frac{[Fe^{2+}]^3}{[Fe^{3+}]^2}$$

and

$$\log K_c = \frac{n\mathscr{F}\mathscr{E}^0}{2.303RT}$$

$$= \frac{n\mathscr{E}^0}{0.0592} \text{ at } 25°C$$

$$= 2 \times \frac{1.18}{0.0592} \simeq 40$$

$K_c \simeq 10^{40}$, so at equilibrium there is almost *no* ferric ion present.]

SOLUTION 17–14

(4)

High power means high voltage. So the cell's cathode (the electrode at which reduction occurs) should be an element with a high electronegativity (Groups VIA and VIIA) because these elements have the greatest affinity for electrons. Minimum weight means that we should receive maximum current for minimum weight, so low equivalent weights are necessary. High performance means that we do not lose power by heating the cell itself. If the electrolyte is a good conductor, it will not present a barrier to the passage of charge.

The only set of conditions that fits is **4**.

[*Note:* To obtain high cell voltage the difference between the cathode and anode electronegativity should be large; hence, that of the anode should be low. This requirement suggests the elements of Groups IA and IIA.]

SOLUTION 17–15

(1)

The reaction is

$$2Li + Cl_2 \rightarrow 2LiCl$$

Thus, an optimum ratio would be one equivalent weight of Li (= at. wt of Li = 6.94) to one equivalent weight of Cl (= at. wt of Cl = 35.5) or 6.94/35.5 = 0.195.

Answer **1** is correct. Answer **2**: 71.0 is not the *equivalent* weight of Cl. Try the reaction

$$H_2 + Cl_2 \rightarrow 2HCl$$
1 mole of $H_2 \simeq 2$ g
1 mole of $Cl_2 = 71.0$ g

But the gram equivalent weight is the weight of that material that will react with 1 g of hydrogen.

SOLUTION 17–16

(5)

We obtain 1.00 $\mathscr{F}$ from 6.9 g of Li plus 35.5 g of Cl_2, or 1.00 $\mathscr{F}$/42.4 g of electrode material ($Li + Cl_2$).

$$\frac{1.00\ \mathscr{F}}{42.4\ \text{g}} = 23.6\ \mathscr{F}\ \text{kg}^{-1}$$

The energy in joules is volt × coulomb, and the energy per kilogram is

$$3.4\ \text{volts} \times 23.6\ \mathscr{F}\ \text{kg}^{-1} \times 9.6 \times 10^4\ \text{coulombs}\ \mathscr{F}^{-1}$$
$$\simeq 7.7 \times 10^6\ \text{joules kg}^{-1}$$

Answer **5** is correct.

This is called the *maximum theoretical specific energy* and is equal to about 2.1×10^3 watts h kg^{-1}. [*Remember:* 1 joule = 1 watt sec = (1/3600) watt h.]

SOLUTION 17-17

(3)

$$\frac{4.4 \times 10^2 \text{ watts h kg}^{-1} \times 180 \text{ kg}}{0.27 \text{ kwatt h mile}^{-1} \times 10^3 \text{ watts kwatt}^{-1}} \simeq 290 \text{ miles}$$

3 is correct.
1 You forgot that the cell weighed 180 kg.
4 You forgot to convert kilowatts to watts.
2 Both the above errors.
5 "That is the biggest fool thing we have ever done. The [atomic] bomb will never go off, and I speak as an expert in explosives."

Admiral William Leahy, 1945

If you chose Answer **5**, you may be making an error of the same kind, but of smaller magnitude.

CHAPTER 18 CHEMICAL DYNAMICS

We can understand everything that we have studied thus far and still not understand what happens between molecules at the instant when they undergo reaction. Chemical dynamics is the field that may provide the answers on a molecular level.

To do the problems in this chapter, memorize the following relations.

For the reaction

$$aA + bB \rightarrow cC + dD$$

if the rate is given by

$$\text{rate} = R = \frac{-\Delta[A]}{\Delta t} = k[A]^x[B]^y$$

in which t is the time, then

k = rate constant
$x + y$ = order of the reaction
x = order with respect to A
y = order with respect to B

The temperature dependence of the rate constant is

$$k = A \exp(-E_a/RT)$$

in which E_a is the activation energy, and A is the preexponential factor.

In the following questions, assume that the rate has a simple dependence on concentration (as above).

Solutions to problems in Chapter 18 begin on page 463.

PROBLEMS

PROBLEM 18–1

A pressure cooker reduces cooking time because ________.

1 The heat is more evenly distributed.
2 The higher pressure tenderizes the food.
3 The cooking juices marinate the meat.
4 The boiling point is elevated.
5 A larger flame must be used.

PROBLEM 18–2

The rate of the reaction $2H_2 + O_2 \rightarrow 2H_2O$ is equal to

$$k[H_2]^2[H_2O]$$

under certain conditions. The rate equation is ________.

The order of chemical reactions is discussed in Section 18–2 of Chemical Principles *by Dickerson, Gray, and Haight.*

1 Zero order.
2 First order.
3 Second order.
4 Third order.
5 Incorrect because of the stoichiometry.

PROBLEM 18–3

The gas reaction $CCl_3CHO + NO \rightarrow CHCl_3 + CO + NO$ has a rate equal to

$$k[CCl_3CHO][NO]$$

If concentrations are expressed in moles per liter, the units of the rate constant, k, are ________.

1 sec
2 sec^{-1}
3 liter $mole^{-1}$ sec^{-1}
4 mole $liter^{-1}$ sec^{-1}
5 $liter^{2}$ $mole^{-2}$ sec^{-1}

PROBLEM 18–4

Photochemical smogs, such as that over Los Angeles, are thought to be initiated, at least partially, by the *photolysis* (breakup after absorption of light) of NO_2 to form NO and O. You decide to study the subsequent gas reaction, $2NO + O_2 \rightarrow 2NO_2$, and you find: (a) When the initial concentration of O_2 is doubled and that of NO held constant, the initial reaction rate doubles; (b) When the initial concentration of NO is doubled and that of O_2 held constant, the initial reaction rate quadruples. On the basis of these data only, which one of the following statements is true?

1 The reaction is second order in O_2.
2 The rate constant could have units of mole $liter^{-1}$ sec^{-1}.
3 The reaction is third order overall.
4 The reaction is third order in NO.
5 The activation energy of the forward reaction is greater than that of the reverse reaction.

PROBLEM 18–5

Stoichiometry and rate expressions of chemical reactions are discussed in Section 18–2 of Chemical Principles.

Consider the reaction

$A + B \rightarrow C$

The following are some experimental results giving initial concentrations of reactants and the initial reaction rates.

	Initial concentrations		
	$[A]_0$	$[B]_0$	Initial rate (mole $liter^{-1}$ min^{-1})$_{(t=0)}$
(a)	1.0	1.0	0.15
(b)	2.0	1.0	0.30
(c)	3.0	1.0	0.45
(d)	1.0	2.0	0.15
(e)	1.0	3.0	0.15

What is the rate expression?

1 rate = $k[B]$
2 rate = $k[A][B]$
3 rate = $k[A]^2[B]$
4 rate = $k[A]$
5 rate = $k[A]^2$

PROBLEM 18–6

For a reaction such as $A + 2B \rightarrow$ products, we have the data shown in Table 18–1.

Table 18–1

Initial rate of reaction (mole $liter^{-1}$ sec^{-1})$_{(t=0)}$	Initial concentration (mole $liter^{-1}$) $[A]_0$	$[B]_0$
5.7×10^{-7}	2.0×10^{-3}	4.0×10^{-3}
11.4×10^{-7}	2.0×10^{-3}	8.0×10^{-3}
22.8×10^{-7}	4.0×10^{-3}	4.0×10^{-3}

What is the rate law?

1 rate = k
2 rate = $k[A][B]$
3 rate = $k[A][B]^2$
4 rate = $k[A]^2[B]$
5 rate = $k[A]^2[B]^2$

PROBLEM 18–7

The isotope $^{42}_{19}K$ has a half-life of about 12 h. What is the fraction of the initial concentration of $^{42}_{19}K$ remaining after 48 h?

1 $\frac{1}{16}$
2 $\frac{1}{8}$
3 $\frac{1}{4}$
4 $\frac{1}{2}$
5 2

PROBLEM 18–8

Radioactive tritium ($^{3}_{1}H$) is sometimes used as a tracer in medical diagnosis. Its radioactive half-life is 12.5 years. However, a human is not a closed system since one must periodically ingest and excrete. The turnover time for H_2O in the body may be expressed as its "biological half-life," which is 12 days. The effective half-life of $^{3}_{1}H$ in water, in a medical diagnosis experiment that includes both these factors, is ________.

(365.25 days per sidereal year)

1 6 days
2 12 days
3 6.15 years
4 12.3 years
5 24.6 years

PROBLEM 18–9

The rate of a certain biochemical reaction at body temperature in the absence of enzyme has been measured in the laboratory. The rate of the *same* reaction, when enzyme-catalyzed in the human body, is 10^6 times faster. The activation energy for this reaction ________.

Arrhenius' activation energy is discussed in Section 18–3 of Chemical Principles.

1 Is $6/RT$ kcal mole^{-1}.
2 Can be determined from the information above.
3 Can be determined from the information above if the temperature of the above reactions are given.
4 Is different in the two cases.
5 Is a concept that cannot be applied to biochemical systems.

PROBLEM 18–10

Campers on a mountain pass at 3 km (9840 ft) altitude experience an atmospheric pressure of 0.69 atm. Their pure water boils at 90°C, and they discover that it takes 300 min of cooking to make a "three-minute" egg. What is the ratio of rate constants, $k_{100°C}/k_{90°C}$?

1 −1/300
2 1/300
3 1
4 100
5 300

PROBLEM 18–11 Using the data and results of Problems 18–9 and 18–10, determine the activation energy for the reaction that occurs when the egg is cooked (denaturation of protein).

($R = 1.99$ cal deg^{-1} mole^{-1};
$2.30 \log x = \ln x$)

1 $(\ln 100)\, R\left(\frac{363 \times 373}{-10}\right) \simeq -120$ kcal mole^{-1}

2 $(\ln 2)\, R\left(\frac{373 \times 363}{10}\right) \simeq 19$ kcal mole^{-1}

3 $R\left(\frac{373 \times 363}{10}\right) \simeq 27$ kcal mole^{-1}

4 $2(\ln 2)\, R\left(\frac{373 \times 363}{10}\right) \simeq 37$ kcal mole^{-1}

5 $(\ln 100)\, R\left(\frac{373 \times 363}{10}\right) \simeq 120$ kcal mole^{-1}

PROBLEM 18–12 Ozone (O_3) in the atmosphere shields the earth from harmful ultraviolet radiation by absorbing it, and reemitting the energy at a different wavelength. Professor Harold Johnston (Berkeley, Calif.) has proposed the following mechanism for the destruction of ozone, which involves nitric oxide (NO) from supersonic transport (SST) exhaust:

$NO + O_3 \rightarrow NO_2 + O_2$ (Step 1)
$NO_2 + O \rightarrow NO + O_2$ (Step 2)

In this scheme NO is
1 A product of the overall reaction.
2 An inhibitor.
3 A reactant in the overall reaction.
4 A catalyst.
5 None of the above.

PROBLEM 18–13 Collision theory sometimes gives the rate in the form

$$\text{rate} = pZ \exp(-E_a/RT)$$

in which Z is the collision frequency, and the exponential is the probability that the energy of collision is equal to or greater than E_a. The steric factor, p, accounts for __________.

1 Molecular collisions.
2 Collisions with insufficient energy.
3 Collisions with sufficient energy.
4 Collisions with unfavorable orientation.
5 The reverse reaction.

PROBLEM 18–14 The collision theory of chemical reactions maintains that ________.

1 A chemical reaction occurs with every molecular collision.
2 Reactions in the gas phase are always first order.
3 Only energetic collisions occur.
4 Reaction rates are of the order of molecular speeds.
5 Rate is directly proportional to the number of collisions per second.

PROBLEM 18–15 Figure 18–1 shows a diagram of a potential-energy surface for the reaction AB + C → BC + A. The indicated configuration (*x*) corresponds to ________.

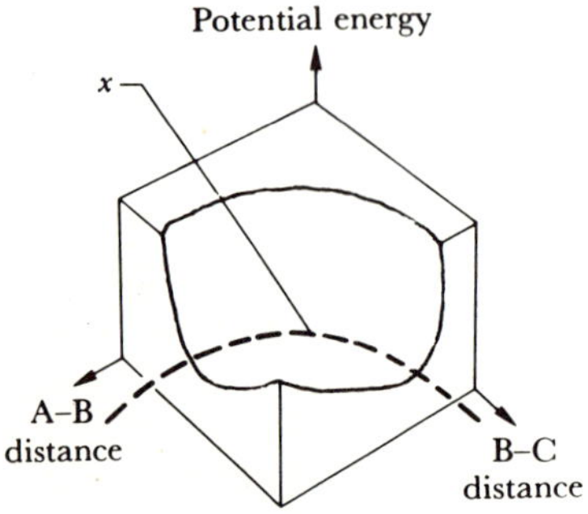

Figure 18–1

1 Reactants AB and C.
2 Products BC and A.
3 The activated complex, ABC.
4 The dissociated atoms, A, B, and C.
5 The ionized separated atoms A^+, B^+, and C^+.

PROBLEM 18–16 The contour map in Figure 18–2 is a representation of a potential-energy surface. Essentially, you are looking down on the surface. The lines are contour lines and denote regions of the same potential energy. Suppose the potential energy surface was made of soft white cake. Now looking down on it as it sits on the table you make a series of horizontal passes through the cake with a chocolate fudge covered knife. The result will look like a contour map.

Which one of the following is the correct identification of the region X for the reaction A + BC → AB + C? The dashed line represents the reaction trajectory.

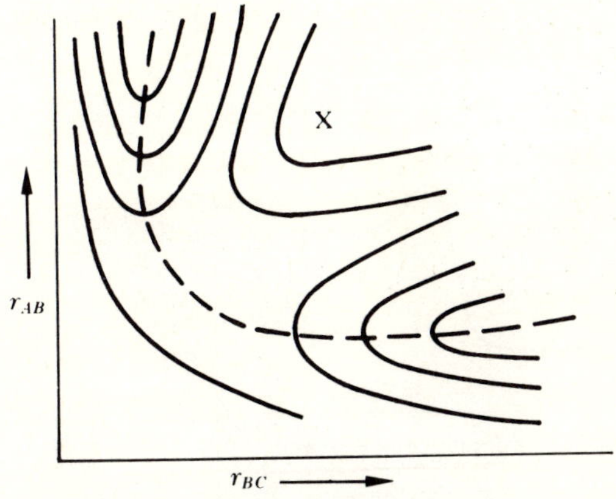

Figure 18–2

1 The reactants A and BC.
2 The products AB and C.
3 The activated complex ABC.
4 The separated atoms A, B, and C.
5 The separated ions A^+, B^+, and C^+.

PROBLEM 18–17 In Figure 18–3 the quantity x is which of the five choices?

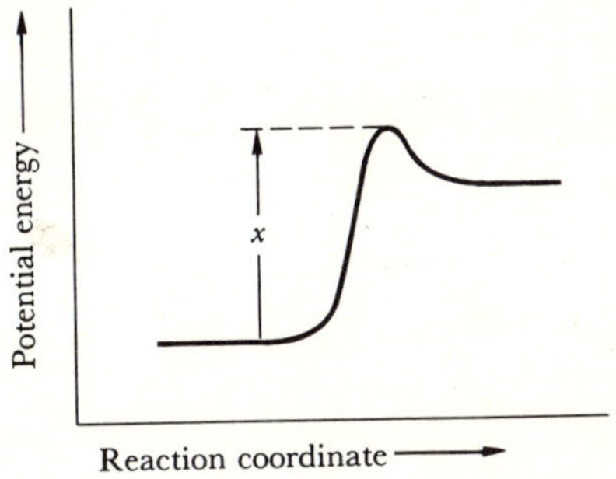

Figure 18–3

1 ΔE
2 E_a(forward)
3 E_a(reverse)
4 The collision energy.
5 The steric factor, p.

PROBLEM 18–18 For the system described by Figure 18–4, which statement is correct?

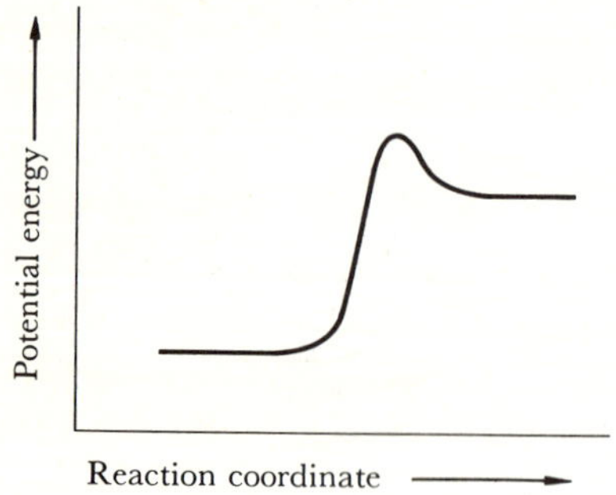

Figure 18–4

1 The forward reaction is exothermic.
2 The forward reaction is endothermic.
3 E_a for the reverse reaction is greater than for the forward reaction.
4 E_a for the forward reaction = 0.
5 E_a for the reverse reaction = 0.

PROBLEM 18–19 From Figure 18–5, which of the statements below is true?

Activated complexes are discussed in Section 18–3 of Chemical Principles.

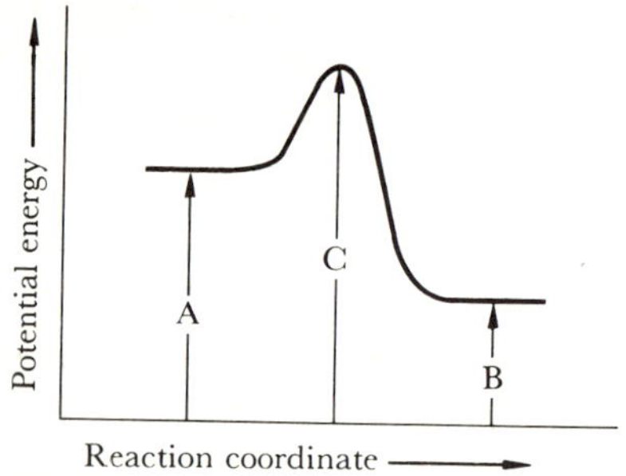

Figure 18–5

1 ΔE of the forward reaction is C − B.
2 ΔE of the forward reaction is B − A.
3 E_a(forward) > E_a(reverse).
4 E_a(reverse) = C − A.
5 The potential energy of the activated complex is greater for the forward reaction than for the reverse reaction.

PROBLEM 18–20 From Figure 18–6, which set of identifications is correct for the total forward reaction?

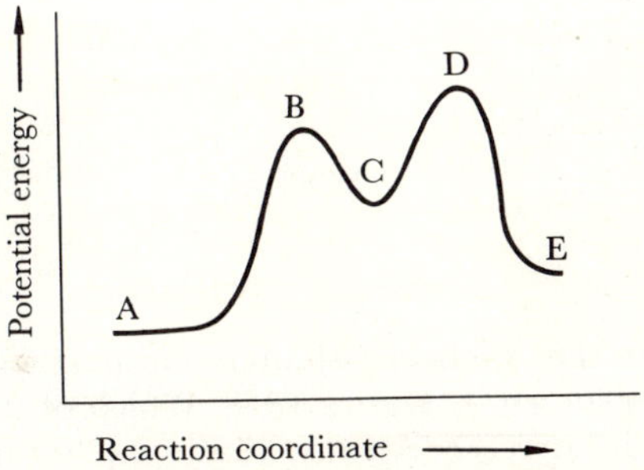

Figure 18–6

	Reactant	Activated complex(es)	Intermediate(s)	Product
1	E	B, D	C	A
2	A	B	None	C
3	C	D	None	E
4	A	C	B, D	E
5	A	B, D	C	E

PROBLEM 18–21 For a simple two-step elementary reaction

$$\text{reactants} \xrightarrow{\text{step 1}} \text{intermediates} \xrightarrow{\text{step 2}} \text{products}$$

the potential energy profile is given in Figure 18–7. Which one of the following statements applies?

Enzyme catalysis is discussed in Section 18–5 of Chemical Principles.

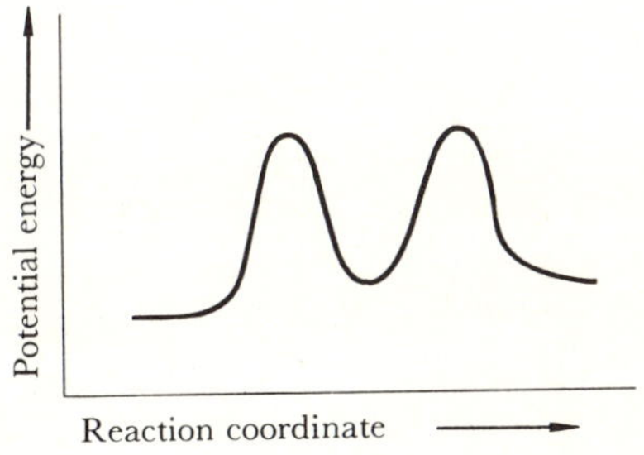

Figure 18–7

1 The intermediates accumulate in measurable quantities, and the reaction is easily resolved into two steps.
2 The reactants quickly form intermediates that slowly form products.
3 The presence of intermediates is difficult to determine kinetically although the two-step reaction is not ultrafast.
4 The intermediates are the final products.
5 None of the above.

PROBLEM 18–22 A simple model for enzyme-catalyzed reactions is given by the following set of equations:

$$\underset{\text{(enzyme)}}{E} + \underset{\text{(reactant)}}{S} \overset{\text{fast}}{\rightleftharpoons} \underset{\text{(intermediate 1)}}{ES}$$

$$\underset{\text{(intermediate 1)}}{ES} \rightleftharpoons \underset{\text{(intermediate 2)}}{EP}$$

$$\underset{\text{(intermediate 2)}}{EP} \rightleftharpoons \underset{\text{(enzyme)}}{E} + \underset{\text{(product)}}{P}$$

This is known as the Michaelis-Menten mechanism. The potential energy diagram is shown in Figure 18–8. Which of the following sets of identifications is correct? (Assume that temperature and pressure are constant.)

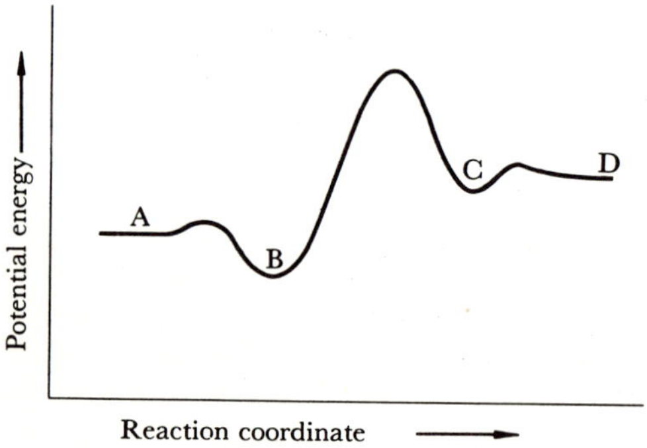

Figure 18–8

	(A)	(B)	(C)	(D)
1	$E + P$	EP	ES	$E + S$
2	ES	Activated complex	EP	Activated complex
3	EP	Activated complex	ES	Activated complex
4	$E + S$	Activated complex	Activated complex	$E + P$
5	$E + S$	ES	EP	$E + P$

PROBLEM 18–23 For the same mechanism as Problem 18–22, refer to Figure 18–9 and choose the correct set of identifications.

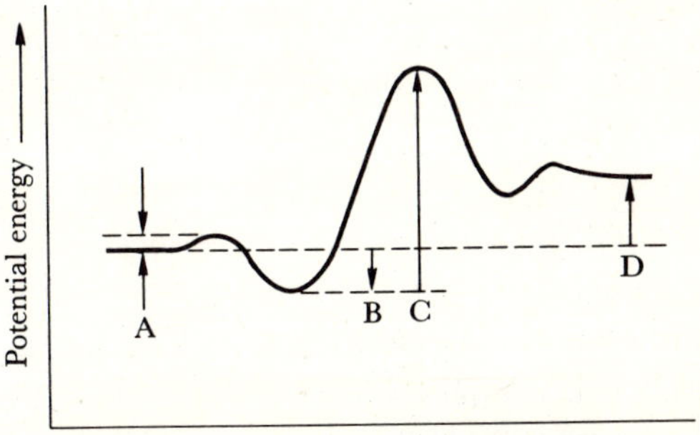

Figure 18–9

	(A)	(B)	(C)	(D)
1	ΔE for $E + S \rightarrow ES$	E_a for $ES \rightarrow EP$	$\Delta E_{\text{overall}}$ for $S \rightarrow P$	E_a for $EP \rightarrow E + P$
2	E_a for $E + S \rightarrow ES$	ΔE for $E + S \rightarrow ES$	E_a for $ES \rightarrow EP$	$\Delta E_{\text{overall}}$ for $S \rightarrow P$
3	E_a for $ES \rightarrow EP$	E_a for $EP \rightarrow E + P$	$\Delta E_{\text{overall}}$ for $S \rightarrow P$	ΔE for $EP \rightarrow E + P$
4	E_a for $E + S \rightarrow ES$	E_a for $ES \rightarrow EP$	E_a for $EP \rightarrow E + P$	$\Delta E_{\text{overall}}$ for $S \rightarrow P$
5	ΔE for $E + S \rightarrow ES$	$\Delta E_{\text{overall}}$ for $S \rightarrow P$	ΔE for $EP \rightarrow E + P$	E_a for $EP \rightarrow E + P$

PROBLEM 18-24

The citric acid cycle is discussed in Section 12–8 of Chemical Principles.

The fumarase-catalyzed hydrolysis of fumarate to malate is one of the reactions developed in the Krebs cycle (also called the citric acid cycle). This cycle is responsible for the oxidation of carbohydrates, fatty acids, and amino acids in the cell. Given the following mechanism for the fumarate to malate reaction, choose the most correct potential energy diagram for the reaction. (Neglect the difference between ΔH and ΔE.)

Reaction	ΔH (kcal mole^{-1})	E_a (kcal mole^{-1})
fumarate + H_2O → malate	−3.6	
fumarate + fumarase → fumarate–fumarase complex	+4.2	
fumarate-fumarase complex ⇄ malate–fumarase complex		6.1 (forward) 15 (reverse)
malate–fumarase complex → malate + fumarase	+1.2	

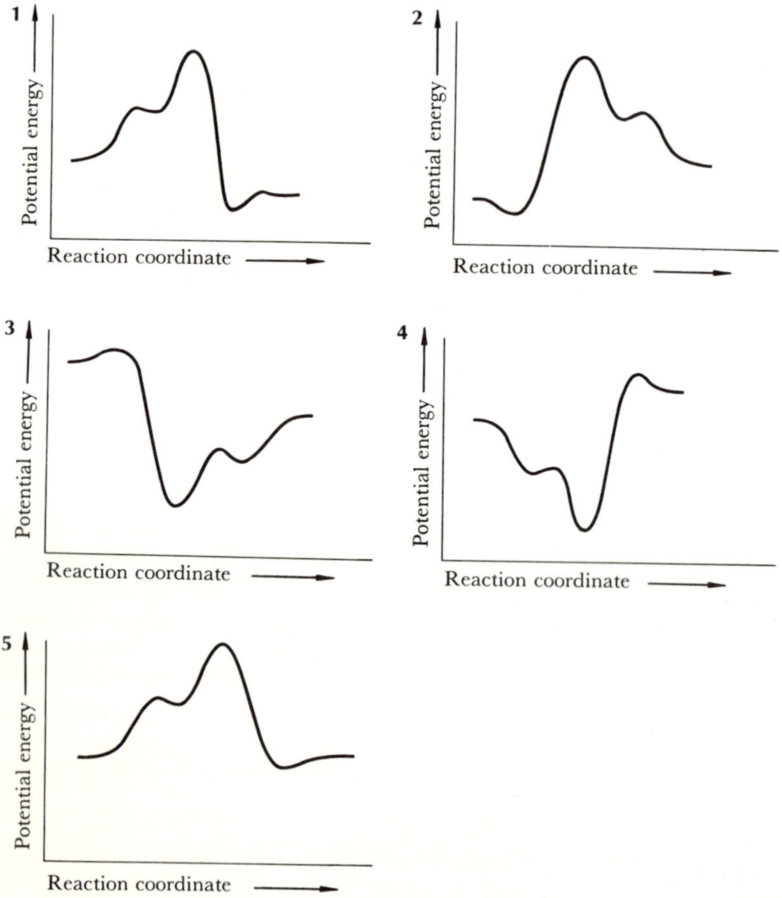

SOLUTION 18–1

(4)

As P increases, the boiling point for H_2O increases. In a pressure cooker, therefore, the reactions involved in cooking occur at the new and higher temperature of the boiling point, and the reactions proceed faster.

Answer **4** is correct. Answer **5**: A larger flame does not help when we are boiling something since the temperature, if equilibrium is preserved and $P = 1$ atm, remains at 100°C. Answers **1** and **3** may be correct, but they have nothing to do with the special function of a pressure cooker.

[*Note:* We can see how the boiling point is raised when the pressure is increased by considering the phase diagram for water.]

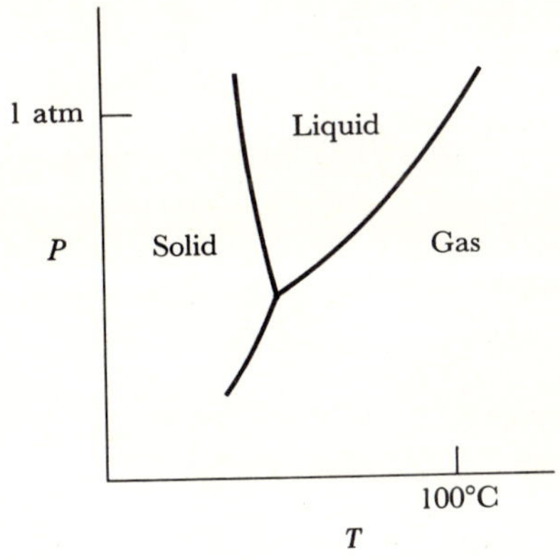

SOLUTION 18–2

(4)

The order of the reaction is the sum of the exponents on the concentration terms of the rate equation:

order = 2 + 1 = 3

4 is correct.
1 Zero is the order with respect to O_2, which does not appear in the rate equation.
2 One is the order with respect to H_2O, a product of reaction. As the concentration of product increases as the reaction proceeds, the rate goes up. This is known as *autocatalysis.*
3 Two is the order with respect to H_2.
5 *Remember:* The stoichiometry of a reaction tells us nothing about the order of a rate equation unless you know that the reaction is an elementary reaction.

SOLUTION 18–3

(3)

If you want to test your understanding of the rates of chemical reactions, see Review 14 of Programed Reviews of Chemical Principles *by Lassila* et al.

$$R = \frac{\Delta \text{conc}}{\Delta \text{time}} = \frac{\text{mole liter}^{-1}}{\text{sec}} = k[\text{conc}]^2 = k \text{ mole}^2 \text{ liter}^{-2}$$

$$k = \frac{1}{\text{mole liter}^{-1} \text{ sec}} = \text{liter mole}^{-1} \text{ sec}^{-1}$$

Answer **3** is correct.

[*Note:* CCl_3CHO is called *chloral*, and when added to water forms *chloral hydrate*, the compound in "knockout drops."]

SOLUTION 18–4

(3)

We shall suppose that the rate law has a simple form

$$\text{rate} = [O_2]^x[NO]^y$$

in which we are to find x and y.

(a) $\begin{cases} 2R = [2O_2]^x[NO]^y \\ R = [O_2]^x[NO]^y \end{cases}$

If we divide, we obtain

$$2 = [2]^x \qquad x = 1$$

(b) $\begin{cases} 4R = [O_2]^1[2NO]^y \\ R = [O_2]^1[NO]^y \end{cases}$

If we divide, we obtain

$$4 = [2]^y \qquad y = 2$$

The rate equation is

$$\text{rate} = k[O_2][NO]^2$$

Answer **3** is correct. Answer **2**: If the rate constant has such units, the reaction must be zero order:

$$\begin{aligned} R &= \text{mole liter}^{-1} \text{ sec}^{-1} \\ &= (k, \text{ mole liter}^{-1} \text{ sec}^{-1}) \times (\text{conc})^0 \end{aligned}$$

Answer **5**: All this means is that the reaction is endothermic.

SOLUTION 18–5

(4)

Again we shall hope that the rate equation has a simple form

$$\text{rate} = R = k[A]^x[B]^y$$

(a) and (b): As $[A]_0$ doubles with $[B]_0$ constant, R doubles. In fact, considering (a), (b), and (c), we see that R is directly proportional to $[A]_0$; $x = 1$.

(a), (d), and (e) are experiments in which $[B]_0$ varies, but $[A]_0$ is constant. We see that R does not change, so y must be zero.

$$R = k[A]$$

Answer **4** is correct.

SOLUTION 18–6 (4)

$R = k[A]^x[B]^y$

The first two experiments are at constant initial concentration of A.

$$11.4 \times 10^{-7} = k[2.0 \times 10^{-3}]^x[8.0 \times 10^{-3}]^y$$
$$5.7 \times 10^{-7} = k[2.0 \times 10^{-3}]^x[4.0 \times 10^{-3}]^y$$

If we divide, we obtain

$$2 = \left[\frac{8.0 \times 10^{-3}}{4.0 \times 10^{-3}}\right]^y = (2)^y$$

$y = 1$

The first and third experiments are at constant initial concentration of B.

$$22.8 \times 10^{-7} = k[4.0 \times 10^{-3}]^x[4.0 \times 10^{-3}]^1$$
$$5.7 \times 10^{-7} = k[2.0 \times 10^{-3}]^x[4.0 \times 10^{-3}]^1$$

If we divide, we obtain

$4 = (2)^x \qquad x = 2$

The rate equation is

$R = k[A]^2[B]$

Answer **4** is correct.

SOLUTION 18–7 (1)

It is not necessary to remember any complicated exponential decay law. But we must recall that half of the radioisotope remains after the first 12-h period, half of that after the second 12-h period, and so forth.

$$1 \xrightarrow{12\text{ h}} \frac{1}{2} \xrightarrow{12\text{ h}} \frac{1}{4} \xrightarrow{12\text{ h}} \frac{1}{8} \xrightarrow{12\text{ h}} \frac{1}{16}$$

Answer **1** is correct.

A first order rate equation and the decay of $_{14}C$ is discussed in Section 18–2 of Chemical Principles.

SOLUTION 18–8 (2)

Because the turnover time for any water in the body has a biological half-life of only 12 days, we cannot expect the effective half-life to be longer than 12 days. It might be shorter if the radioactive half-life is comparable to 12 days. But this is not the case. During the 12 days only a very small fraction of the tritium decays according to its 12.5-year half-life. So the biological half-life is the determining factor, and the effective half-life is 12 days.

Answer **2** is correct.

SOLUTION 18–9 (4)

The two activation energies are different. When enzyme-catalyzed, the activation energy is reduced so the rate constant (and therefore the rate of reaction) is increased.

Answer **4** is correct.

[*Note:* Quantitatively, we can use the Arrhenius equations

$$k_{\text{lab}} = A\ \exp(-E_a^{\text{lab}}/RT)$$
$$k_{\text{body}} = A\ \exp(-E_a^{\text{body}}/RT)$$

Dividing these equations [*if the A's are the same* (a fair approximation) and if the T's are the same], we obtain

$$\frac{k_{\text{lab}}}{k_{\text{body}}} = \exp\left(\frac{-(E_a^{\text{lab}} - E_a^{\text{body}})}{RT}\right)$$

This ratio is equal to 10^{-6}. So we can only tell the *difference* between the two activation energies if the T's are the same and we are given their value.]

Answer **3** is incorrect because the E_a's are different. Thus, even if the T's were known, there is not enough information (after eliminating A there is only one equation) to find the two unknowns, E_a^{lab} and E_a^{body}.

SOLUTION 18–10 (4)

We don't know the rate equation for the reaction, but we can assume for the moment that it has a simple form

$$R = k[\text{conc}]^x$$

At the two different temperatures

$$R_{100°\text{C}} = k_{100°\text{C}}\ [\text{conc}]^x$$
$$R_{90°\text{C}} = k_{90°\text{C}}\ [\text{conc}]^x$$

Dividing these two equations, we find that

$$\frac{R_{100°\text{C}}}{R_{90°\text{C}}} = \frac{k_{100°\text{C}}}{k_{90°\text{C}}}$$

The rate at 90°C is 100 times *slower* (300/3) than at 100°C, so

$$\frac{R_{100°\text{C}}}{R_{90°\text{C}}} = 100 = \frac{k_{100°\text{C}}}{k_{90°\text{C}}}$$

Answer **4** is correct.

SOLUTION 18–11 (5)

This problem is different from Problem 18–9 because here the temperatures are different, but the activation energy does not change. (The chemical system is the same; no catalyst has been added.) But we can use the same kinds of mathematical manipulations of the Arrhenius equations.

$$k_1 = A \exp(-E_a/RT_1)$$
$$k_2 = A \exp(-E_a/RT_2)$$

Dividing these equations and assuming that the A's are equal, we obtain

$$\frac{k_1}{k_2} = \exp\left(\frac{-E_a}{RT_1} + \frac{E_a}{RT_2}\right)$$
$$= \exp\left[\frac{E_a}{R}\left(\frac{1}{T_2} - \frac{1}{T_1}\right)\right]$$
$$= \exp\left[\frac{E_a}{R}\left(\frac{T_1 - T_2}{T_1 T_2}\right)\right]$$

Now take the natural (or Naperian) logarithm of the equation

$$\ln \frac{k_1}{k_2} = \frac{E_a}{R}\left(\frac{T_1 - T_2}{T_1 T_2}\right)$$

and solve for E_a:

$$E_a = \left(\ln \frac{k_1}{k_2}\right) R\left(\frac{T_1 T_2}{T_1 - T_2}\right)$$

Insert the values from Problem 18–10, and remember to change T to °K.

$$E_a = (\ln 100)\ 1.99 \text{ cal deg}^{-1} \text{ mole}^{-1} \left(\frac{373 \times 363}{10}\right) \text{deg}$$

$$= 2.30\ (\log 100)\ 1.99\ (3.73 \times 3.63 \times 10^3) \text{ cal mole}^{-1}$$
$$= 124 \text{ kcal mole}^{-1}$$
$$\simeq 120 \text{ kcal mole}^{-1}$$

Answer **5** is correct. (We considered this process *thermodynamically* in Problem 15–20.)

[*Note:* If we wished to speed up this egg-cooking process, we might employ the usual procedure of dissolving common salt (NaCl) in the cooking water to raise the boiling point (Chapter 15). You should now be able to calculate what concentration is necessary and if heterogeneous equilibria considerations (Chapter 5) will allow such a solution. Or you might accept the solution of the Indian army, which equips its soldiers with pressure cookers when on maneuvers in the Himalayas.]

SOLUTION 18–12 (4)

If we add the two steps of the mechanism, we obtain the overall reaction:

$O_3 + O \rightarrow 2O_2$

that is, the destruction of ozone by oxygen atoms. **1** and **3** must be incorrect because NO does not appear in the overall reaction. NO is the agent, unconsumed by the reaction (it is regenerated in step 2), by which this reaction proceeds, and is therefore a catalyst. **4** is correct.

SOLUTION 18–13 (4)

Collision theory is described in Section 18–3 of Chemical Principles.

Answer **4** is correct. The steric factor is our way of including molecular geometry in our reaction probability calculations. Answers **2** and **3**: The activation energy is the term that distinguishes between collisions with sufficient and insufficient energy.

SOLUTION 18–14 (5)

1 If a reaction occurred with every collision, the rates of all gas reactions would be about the same and extremely fast, but this is not the case.
2 Gas phase reactions may have any order.
3 Collisions with all energies can occur because all molecular velocities are present.
4 Same objection as for Answer **1**.
5 Correct. The rate is *proportional,* but not necessarily equal, to the number of collisions per second since this number is modified by the fraction that is sufficiently energetic and has a favorable orientation.

SOLUTION 18–15 (3)

At point x the A–B and B–C distances are about equal; in other words, A is no further from B than is C. This corresponds to the activated complex ABC.

3 is correct.
1 AB and C: This is the system configuration when the B–C distance is large and the A–B distance is small, at the right-hand end of the dashed line.
2 BC and A: This is at the left-hand end of the dashed line where the A–B distance is large and the B–C distance is small.
4 The separated atoms, A + B + C, are located where the A–B and the B–C distances are both large, at the "near" corner (the one projecting out of the book) of the potential energy surface.
5 The ionized species would correspond to the same region as **4** on another (different) potential energy surface.

SOLUTION 18–16 (4)

If you compare the contour diagram with Figure 18–1 in Problem 18–15, the region at X is the corner of the potential energy diagram that projects itself closest to the reader. It is the plateau region where both r_{AB} and r_{BC} are large, so all the atoms are distant from each other. **4** is correct. The other choices are shown on the contour map in Figure 18–10. The ionic species (**5**) would lie on a different potential energy surface.

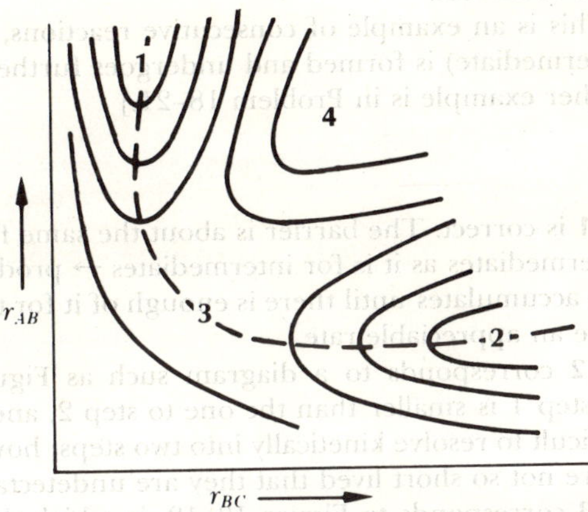

Figure 18–10

SOLUTION 18–17 (2)

x is the barrier height for the forward reaction, E_a(forward).

Answer **2** is correct. Answer **4**: There are many collision energies possible when two molecules collide. x is only the collision energy necessary for reaction (steric considerations aside).

SOLUTION 18–18 (2)

The potential energy increases in going forward from reactants to products, so there must have been a net energy input to effect the reaction. That means the reaction is endothermic.

Answer **2** is correct. Answer **3**: The barrier is greater in the forward direction than in the reverse direction: So $E_a(\text{forward}) > E_a(\text{reverse})$. Answers **4** and **5**: Both are greater than zero.

SOLUTION 18–19 (2)

ΔE of the forward reaction is B − A since it is the difference between the final and initial states.

2 is correct.
1 and **4** E_a for the reverse reaction is C − B.
3 The forward activation energy (C − A) is smaller than the reverse activation energy (C − B).
5 The potential energy of the activated complex is C, and is the same for reactions in either direction.

SOLUTION 18–20

(5)

It should be clear that A corresponds to the reactants that pass over a barrier at B, which must be an activated complex, and into a valley at C. This would correspond to product formation in a usual reaction (Answer **2**). But in this case the species at C continue reacting to produce products at E by passing over another barrier and by forming another activated complex at D; C corresponds to reaction intermediates.

Answer **5** is correct.

[*Note:* This is an example of consecutive reactions, in which a substance (intermediate) is formed and undergoes further chemical reaction. Another example is in Problem 18–21.]

SOLUTION 18–21

(1)

Answer **1** is correct. The barrier is about the same for step 1: reactants → intermediates as it is for intermediates → products. Hence, intermediate accumulates until there is enough of it for the second reaction to have an appreciable rate.

Answer **2** corresponds to a diagram such as Figure 18–11. The barrier to step 1 is smaller than the one to step 2, and the reaction is usually difficult to resolve kinetically into two steps; however, the intermediates are not so short lived that they are undetectable.

Answer **3** corresponds to Figure 18–12, in which the potential well for the intermediates is very shallow. The barrier to step 2 is so small that the intermediates decompose to products as fast as they are formed.

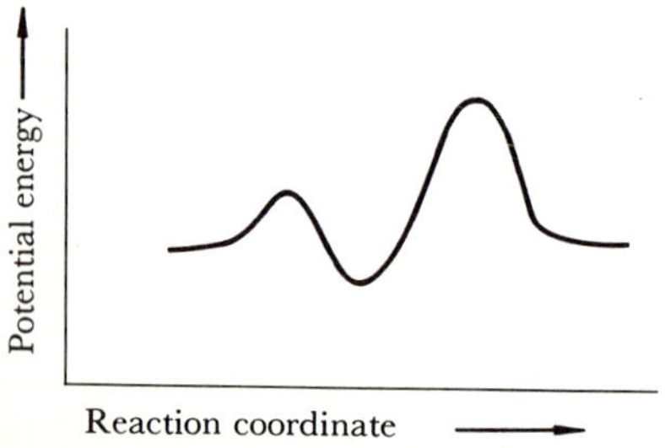

Figure 18–11

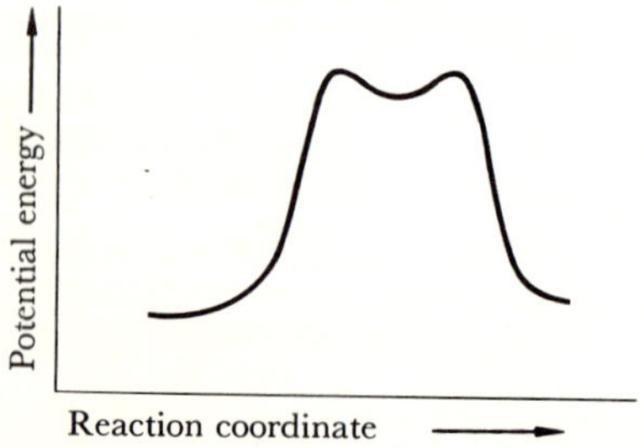

Figure 18–12

SOLUTION 18–22

(5)

The two intermediates are located in the two valleys (B) and (C). The first intermediate is *ES* at (B); the second, *EP* at (C).

(A) and (D) are the reactants and products, and are therefore $E + S$ and $E + P$, respectively.

Answer **5** is correct.

SOLUTION 18–23

(2)

The identifications that we made in Problem 18–22 will serve us here. The first hump in the curve is the barrier between $E + S$ and *ES*, so (A) is E_a for $E + S \rightarrow ES$. Hence, (B) is the energy difference between $E + S$ and *ES*, and is ΔE for $E + S \rightarrow ES$. The second barrier (C) is between *ES* and *EP*; therefore, (C) is E_a for $ES \rightarrow EP$, whereas (D) is the energy difference between $E + S$ and $E + P$, and is ΔE for the overall enzyme-catalyzed reaction $S \rightarrow P$.

Answer **2** is correct.

SOLUTION 18–24

(1)

The first line of the mechanism gives the overall ΔH for the reaction fumarate → malate and shows that it is exothermic by 3.6 kcal. This immediately eliminates Answers **2**, **4**, and **5**. Answers **2** and **4** are endothermic overall, and Answer **5** is thermoneutral ($\Delta H_{\text{overall}} = 0$).

The E_a information indicates that the valley to the left is shallower than the valley to the right. Note that these are two intermediates and correspond to the valleys in the potential energy curve. Both Answers **1** and **3** have two valleys, but only in Answer **1** is the valley on the right deeper.

We will have to see if the other information also conforms to Answer **1**. ΔH for the reaction malate–fumarase complex → malate + fumarase shows that it is slightly endothermic, and Answer **1** shows a slight rise from the second valley to the plateau on the right, which corresponds to the product malate and the enzyme fumarase. ΔH for the reaction fumarate + fumarase → fumarate–fumarase complex shows that it is a bit more endothermic. And Answer **1** indicates a larger rise from the left-hand plateau (fumarate + fumarase) to the first valley (complex).

Answer **1** is correct.

APPENDIXES

APPENDIX 1 EXAMINATIONS

EXAMINATION I

This examination will test your progress in mastering the material in Chapters 1 through 7. Allow yourself three hours to complete the 43 questions of the examination. You may refer to the periodic table on the inside front cover. Assume ideal gas behavior unless otherwise specified. For the purposes of this examination, ignore the difference between ΔH and ΔE.

1–1 The molecular weight of glucose is 180, and its density is 1.56 g cm^{-3}. What is the volume of 1 mole of glucose?

1 $180/(1.56 \times 6.0 \times 10^{23}) = 1.9 \times 10^{-22}$ cm^3
2 $180/(6.0 \times 10^{23}) = 3.0 \times 10^{-22}$ cm^3
3 $180/1.56 = 115$ cm^3
4 $180 \times 1.56 = 281$ cm^3
5 22.4 liters

1–2 What is the *weight* of HF produced by the reaction of 1.5×10^{23} molecules of H_2 according to the reaction

$H_2 + F_2 \rightarrow 2HF$

if the reaction goes to completion?

(at. wt: H = 1.0, F = 19)

1 0.50 g
2 2.0 g
3 5.0 g
4 10 g
5 40 g

1–3 Bauxite is the chief ore of aluminum. It consists of a mixture of aluminum oxide (Al_2O_3), silica (SiO_2), and oxides of iron and titanium. In 1967 the total production of aluminum in the United States was 3,300,000 short tons that sold at approximately 25¢ per pound, thereby grossing about $1,700,000,000. The bauxite used in the aluminum-refining process contains about 50% alumina (Al_2O_3) by weight. Assuming that bauxite costs about $6.00 per short ton and that it was the only factor contributing to the cost of aluminum production, estimate the material cost to the aluminum industry in 1967.

(at. wt: O = 16, Al = 27)

1 $5,200,000
2 $19,000,000
3 $37,000,000
4 $60,000,000
5 $75,000,000

1–4 When excess NH_3 gas reacts with 1.36 g of BF_3 gas, they form a white solid ($H_3N—BF_3$). How many grams of solid are formed?

(at. wt: H = 1.01, B = 10.8, N = 14.0, F = 19.0)

1 0.020 g
2 0.34 g
3 1.36 g
4 1.70 g
5 2.72 g

1-5 Barium is precipitated as the sulfate, $BaSO_4$. In one experiment, 4.66 g of $BaSO_4$ were obtained. How much barium was present initially? (The reaction is complete.)

$Ba^{2+} + SO_4^{2-} \rightarrow BaSO_4$

(at. wt: O = 16.0, S = 32.1, Ba = 137)

1 1.37 g
2 2.33 g
3 2.74 g
4 72.0 g
5 137 g

1-6 A pure sample of the important dyestuff *indigo* gave an analysis of 73.3% carbon, 3.86% hydrogen, 10.7% nitrogen, and 12.2% oxygen. What is its empirical formula?

(at. wt: H = 1.01, C = 12.0, N = 14.0, O = 16.0)

1 CHNO
2 C_2N_2HO
3 C_6H_4NO
4 C_7NOH_3
5 C_8H_5NO

1-7 The dirigible *Hindenburg* had a gas volume of 1.40×10^8 liters (about 5 million cubic feet). What was the difference between the weight of hydrogen (H_2) gas that filled the dirigible and the weight of the 1.40×10^8 liters of air that it displaced? (Assume that the average molecular weight of air is 29.0, that the temperature was 27°C, and that the gas pressure in the airship was 1.00 atm.) This difference is the lifting capacity of the dirigible.

($R = 0.0820$ liter atm mole^{-1} deg^{-1};
at. wt: H = 1.01)

1 1.14×10^7 g
2 2.28×10^7 g
3 1.54×10^8 g
4 1.65×10^8 g
5 17.1×10^8 g

1–8 How many grams of argon (Ar) gas will occupy 82 cm^3 at 0.30 atm pressure and 27°C?

($R = 0.082$ liter atm $mole^{-1}$ deg^{-1};
at. wt: Ar = 40)

1 0.0010 g
2 0.040 g
3 0.060 g
4 0.12 g
5 0.20 g

1–9 A flask contains 1 mole of He gas at 27°C. Then 2 moles of H_2 gas are added at the same temperature. The average kinetic energy per mole for He is changed in what way?

($R = 1.99$ cal $mole^{-1}$ deg^{-1};
at. wt: H = 1, He = 4)

1 It is quartered.
2 It is reduced to one third its original value.
3 It is halved.
4 It is doubled.
5 It remains the same.

1–10 Flask A of volume V contains 3 moles He at T_1°K. Flask B of volume V contains 3 moles Ar at T_2°K. If the root mean square (rms) speeds of He and Ar are the same, that is, $v_{rms}(\text{He}) = v_{rms}(\text{Ar})$, then ______.

(at. wt: He = 4.0, Ar = 40.0)

1 The pressure in A is less than that in B.
2 T_1 equals T_2.
3 The pressure in A is equal to that in B.
4 T_2 is less than T_1.
5 Insufficient data; the densities must be given.

1–11 The partial pressure of O_2 in a flask containing 28 g of N_2 and 16 g of O_2 at a total pressure of 6.0 atm (and 0°C) is ______.

(at. wt: N = 14, O = 16)

1 2.0 atm
2 2.4 atm
3 3.6 atm
4 4.0 atm
5 None of the above.

1–12 When a solution of sodium acetate ($Na^+CH_3COO^-$) is electrolyzed, the anode reaction is

$$2CH_3COO^- \rightarrow \underset{\text{ethane}}{C_2H_6} + 2CO_2 + 2e^-$$

What volume of CO_2 gas (at STP) would be produced if a current of 0.268 amp were passed through a sodium acetate solution for 10.0 h?

($R = 0.082$ liter atm mole^{-1} deg^{-1}; 1 $\mathscr{F} = 96{,}500$ coulombs)

1 0.187 liter
2 0.373 liter
3 1.12 liters
4 2.24 liters
5 224 liters

1–13 The mass spectrum of BrCl in the parent molecular ion region is shown in Figure E1–1. If chlorine has two isotopes of mass numbers 35 and 37, what are the isotopes of bromine? [*Note:* The relative intensities of the peaks in the mass spectrum are not drawn to scale.]

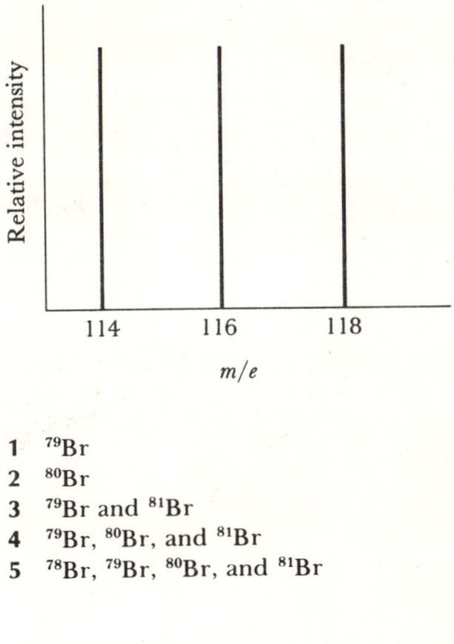

Figure E1–1

1 ^{79}Br
2 ^{80}Br
3 ^{79}Br and ^{81}Br
4 ^{79}Br, ^{80}Br, and ^{81}Br
5 ^{78}Br, ^{79}Br, ^{80}Br, and ^{81}Br

1–14 Consider the mass spectrum of an unknown compound in Figure El–2. What is the unknown compound? [*Note:* For the purposes of this problem assume that the only isotopes present in the compound are ^{1}H, ^{12}C, ^{14}N, ^{16}O, ^{19}F, and ^{35}Cl. The relative intensities of the peaks in the mass spectrum are not to scale.]

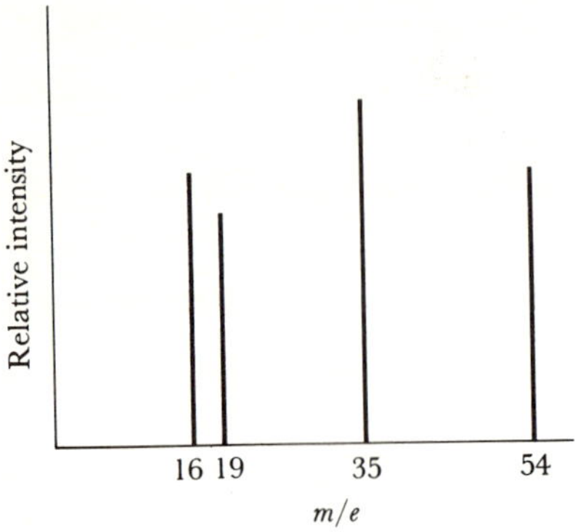

Figure E1–2

1 FCl
2 F_2O
3 C_2N_2
4 C_4H_6
5 C_4H_8

1–15 What are the coordination numbers of Zn^{2+} and S^{2-} in the zinc blende structure shown in Figure E1–3?

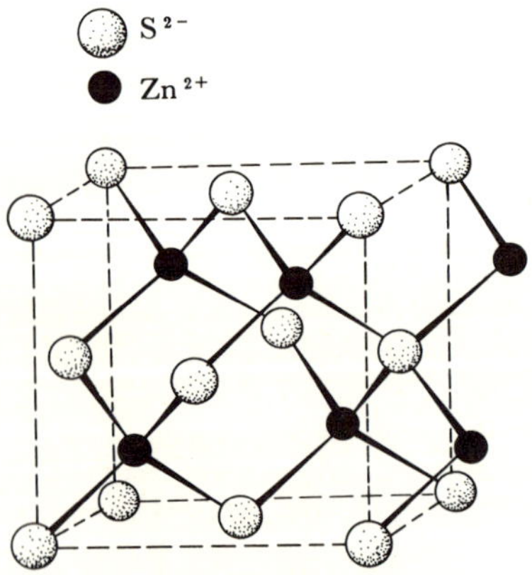

Figure E1–3

Coordination numbers

	Zn^{2+}	S^{2-}
1	1	1
2	4	4
3	6	3
4	6	6
5	8	8

1–16 Crystalline cadmium oxide (CdO) has a density of 8.2 g cm^{-3}. If the edge of the unit cell is 4.7 Å, how many Cd^{2+} and O^{2-} ions are there in the unit cell?

(mol wt: CdO = 128)

1 $0.8Cd^{2+}$ and $0.8O^{2-}$
2 $1Cd^{2+}$ and $1O^{2-}$
3 $4Cd^{2+}$ and $4O^{2-}$
4 $6Cd^{2+}$ and $6O^{2-}$
5 $8Cd^{2+}$ and $8O^{2-}$

1–17 What volume of hydrogen (H_2) gas, measured at STP, is liberated when 6.5 g of zinc (Zn) dissolve in excess dilute hydrochloric acid (HCl)?

(R = 0.082 liter atm $mole^{-1}$ deg^{-1};
at. wt: Zn ≃ 65)

$$Zn + 2HCl \rightarrow ZnCl_2 + H_2$$

1 0.0224 liter
2 0.112 liter
3 0.336 liter
4 2.24 liters
5 22.4 liters

1–18 The average concentration of potassium ions (K^+) in blood serum is 0.195 g $liter^{-1}$. What is the molarity?

(at. wt: K = 39)

1 0.0050
2 0.050
3 0.50
4 $0.195 \times 39 = 7.6$
5 19.5

1–19 In the titration of a 42.0-ml sample of commercial vinegar (density = 1.02 g cm^{-3}) with 1.20-normal NaOH solution, it required 56.0 ml of the base to reach an end point. What is the percent by weight of acetic acid (mol wt = 60.0) in the vinegar? [*Note:* Acetic acid is a monoprotic acid.]

1 1.6%
2 3.2%
3 5.4%
4 9.4%
5 9.6%

1-20 Find ΔH^0 for the reaction

$$CH_4(g) + 2O_2(g) = 2H_2O(l) + CO_2(g)$$

Substance	ΔH_f^0 (kcal mole^{-1})
$CH_4(g)$	−18
$H_2O(l)$	−68
$CO_2(g)$	−94

1 −212 kcal mole^{-1}
2 −180 kcal mole^{-1}
3 0 kcal mole^{-1}
4 +180 kcal mole^{-1}
5 +212 kcal mole^{-1}

1-21 The following equilibrium was established for an organic reaction (hydrolysis):

	(ester) R'COOR	+ H_2O ⇌	(organic acid) R'COOH	+ (alcohol) ROH	
equil:	1.0	1.0	0.50	0.50	no. of moles

The number of moles present at equilibrium for each liquid substance is given beneath each term. What is the value of the equilibrium constant, K_c, if the total volume of liquid is 2.0 liters?

1 0.25
2 0.50
3 1.0
4 2.0
5 4.0

1-22 The equilibrium constant, K_p, of the reaction

$$H_2(g) + I_2(g) \rightleftharpoons 2HI(g)$$

is 66.5 at 633°K and 50.7 at 713°K. Which of the statements below are true?

(a) $K_p = \dfrac{2[HI]}{[H_2][I_2]}$

(b) The reaction is exothermic.
(c) The reaction is endothermic.
(d) K_p is dimensionless; that is, it has no units.

1 (a) and (b)
2 (a) and (c)
3 (c) and (d)
4 (b) and (d)
5 (b) alone

1–23 The equilibrium reaction

$2NOCl(g) \rightleftharpoons 2NO(g) + Cl_2(g)$

was studied at 462°C in a 1.0-liter vessel. Initially, 2.00 moles of NOCl were placed in a container and at equilibrium 33% of the NOCl had dissociated. K_c is (in units of mole liter^{-1}) ________.

1 0.083
2 0.17
3 0.33
4 0.66
5 1.0

1–24 At 27°C and 1.0 atm total pressure, 20% of the molecules of $N_2O_4(g)$ have dissociated into $NO_2(g)$ molecules. What is K_p for the following dissociation?

$N_2O_4(g) \rightleftharpoons 2NO_2(g)$

1 0.05 atm
2 0.17 atm
3 0.25 atm
4 0.93 atm
5 4.0 atm

1–25 The equilibrium partial pressures for the reaction below were

	$C(s) + CO_2(g)$	$\rightleftharpoons 2CO(g)$	
equil:	4.0	8.0	atm

If 4.0 atm of $CO(g)$ are removed from the reaction mixture, what is the new partial pressure of $CO_2(g)$ when equilibrium is reached?

1 2.1 atm
2 4.0 atm
3 5.3 atm
4 6.6 atm
5 7.9 atm

1–26 In the reaction

$$HCO_3^- + HO^- \rightarrow CO_3^{2-} + H_2O$$

CO_3^{2-} is ________.

1 An acid.
2 A base.
3 An oxidizing agent.
4 Conjugate with H_2O.
5 A reducing agent.

1–27 A solution is prepared by dissolving 1.0×10^{-3} mole of ammonia (NH_3) in enough water to make 10 ml of solution. It is determined that 1.3% of the hydrated ammonia (NH_4OH, ammonium hydroxide) has dissociated according to the equation

$$NH_4OH \rightleftharpoons NH_4^+ + OH^-$$

What is K_b for NH_4OH? (Use concentrations of mole liter^{-1}.)

1 1.3×10^{-2}
2 1.3×10^{-3}
3 1.7×10^{-3}
4 1.3×10^{-5}
5 1.7×10^{-5}

1–28 A solution has been prepared by dissolving 0.40 mole of NaCN in enough water to make 1.0 liter of solution. What is the pH?

[$K_a(HCN) = 4.0 \times 10^{-10}$ mole liter^{-1};
$K_w = 1.0 \times 10^{-14}$ mole^2 liter^{-2}]

1 3.2×10^{-3}
2 2.5
3 7.0
4 10
5 11.5

1–29 What is the approximate pH of a solution that was prepared by dissolving 0.010 mole of NaCN (a strong electrolyte) and 0.10 mole of HCN in enough water to form 1.0 liter of solution?

[$pK_a(HCN) = 9.4$]

1 1.0
2 7.0
3 8.4
4 9.4
5 10.4

1–30 The titration curve in Figure E1–4 shows the titration of 1.0 ml of an unknown solution with 0.01-normal NaOH solution. Which one of the following statements about the unknown is correct?

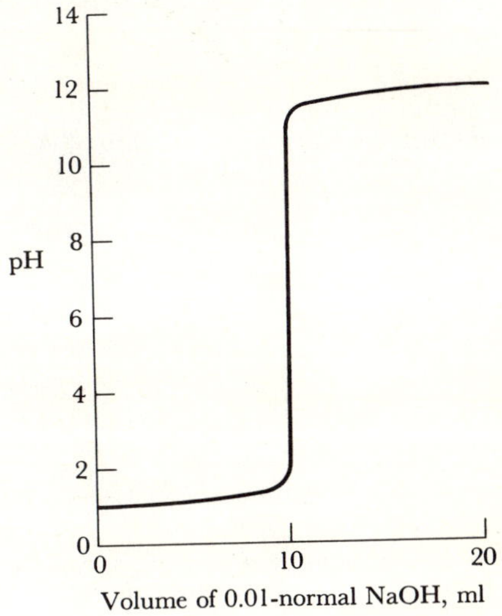

Figure E1–4

1 The unknown is 0.1-normal weak base.
2 The unknown is 0.1-normal weak acid.
3 The unknown is 0.01-normal strong acid.
4 The unknown is 0.01-normal strong base.
5 The unknown is 0.1-normal strong acid.

1–31 The solubility of $La(IO_3)_3$ in pure water is 1.0×10^{-3} mole liter^{-1} at a given temperature. What is K_{sp}? (Use units of mole liter^{-1}.)

1 1.0×10^{-12}
2 3.0×10^{-12}
3 27×10^{-12}
4 1.0×10^{-3}
5 4.0×10^{-3}

1–32 K_{sp} of AgCl is 1.7×10^{-10} mole² liter⁻² at 20°C. What is the solubility of AgCl in a solution whose $[Cl^-]$ is 3.4×10^{-2} molar?

($K_w = 1 \times 10^{-14}$ mole² liter⁻²)

1 $1.7 \times 10^{-10} \times 3.4 \times 10^{-2} \simeq 5.8 \times 10^{-12}$ molar
2 1.7×10^{-10} molar
3 $\dfrac{1.7 \times 10^{-10}}{3.4 \times 10^{-2}} = 5.0 \times 10^{-9}$ molar
4 $\dfrac{1.7 \times 10^{-10}}{(1.0 \times 10^{-14})^{1/2}} = 1.7 \times 10^{-3}$ molar
5 3.4×10^{-2} molar

1–33 At a certain temperature, what is the solubility of AgOH in a buffer of pH = 10?

[$K_w = 1 \times 10^{-14}$ mole² liter⁻²; $K_{sp}(AgOH) = 1 \times 10^{-8}$]

1 1×10^{-14} molar
2 1×10^{-10} molar
3 1×10^{-7} molar
4 1×10^{-4} molar
5 1×10^{-2} molar

1–34 Moseley found that the frequency of the K_α x ray emitted by an element is related to the atomic number (Z) of the element by the following equation:

$$\nu = A(Z - 1)^2$$

If the frequency of the K_α x ray emitted by nickel (Ni) is 1.8×10^{18} sec⁻¹, what is the frequency of the one emitted by calcium (Ca)?

(Ca, Z = 20; Ni, Z = 28)

1 0.89×10^{18} sec⁻¹
2 1.3×10^{18} sec⁻¹
3 1.8×10^{18} sec⁻¹
4 2.6×10^{18} sec⁻¹
5 3.7×10^{8} sec⁻¹

1–35 Which one of the following elements does not belong to the family indicated?

1 Fluorine (F, $Z = 9$)—halogen
2 Krypton (Kr, $Z = 36$)—noble gas
3 Ruthenium (Ru, $Z = 44$)—transition metal
4 Astatine (At, $Z = 85$)—alkali metal
5 Radium (Ra, $Z = 88$)—alkaline earth metal

1–36 ΔH^0 for the process

$$X(solid) \rightarrow X(liquid)$$

in which X is a neutral atom, is called the ________.

1 Electron affinity of X.
2 First ionization energy of X.
3 Heat of sublimation of X.
4 Solvation energy of X.
5 Heat of fusion of X.

1–37 If ΔH^0 for the process

$$S(g) + 2e^- \rightarrow S^{2-}(g)$$

is +94.5 kcal $mole^{-1}$ and the electron affinity of S is −48.0 kcal $mole^{-1}$, what is the electron affinity of S^-?

1 −48.0 kcal $mole^{-1}$
2 −143 kcal $mole^{-1}$
3 +46.5 kcal $mole^{-1}$
4 +94.5 kcal $mole^{-1}$
5 +143 kcal $mole^{-1}$

1–38 Which one of the following isoelectronic ions has been assigned too large an ionic radius?

Data	
Br^-:	1.96 Å
Sr^{2+}:	1.71 Å
Rb^+:	1.47 Å
Y^{3+}:	0.89 Å
Zr^{4+}:	0.79 Å

1 Br^-
2 Rb^+
3 Sr^{2+}
4 Y^{3+}
5 Zr^{4+}

1–39 A compound of xenon and fluorine is found on analysis to contain 53.5% xenon. What is the oxidation number of xenon in the compound?

(at. wt: F = 19.0, Xe = 131)

1 −4
2 0
3 +2
4 +4
5 +6

1–40 The following redox reaction is balanced by which set of coefficients?

$$(?)CuO + (?)NH_3 \rightarrow (?)Cu + (?)N_2 + (?)H_2O$$

	CuO	NH₃	Cu	N₂	H₂O
1	4	3	4	$\frac{3}{2}$	4
2	4	6	4	3	4
3	3	2	3	1	3
4	6	4	6	1	6
5	None of the above.				

1–41 Choose the set of stoichiometric coefficients that correctly balances the equation.

$$(?)K_2Cr_2O_7 + (?)HI + (?)HClO_4 \rightarrow (?)KClO_4 + (?)Cr(ClO_4)_3 + (?)I_2 + (?)H_2O$$

	$K_2Cr_2O_7$	HI	$HClO_4$	$KClO_4$	$Cr(ClO_4)_3$	I_2	H_2O
1	1	6	1	2	2	3	1
2	1	6	8	2	2	3	1
3	1	6	1	1	2	3	1
4	1	6	8	2	2	3	7
5	None of the above.						

1–42 Choose the set of stoichiometric coefficients that correctly balances the equation.

$$(?)MnO_4^- + (?)SnO_2^{2-} + (?)H_2O \rightarrow (?)MnO_2 + (?)SnO_3^{2-} + (?)OH^-$$

	MnO_4^-	SnO_2^{2-}	H_2O	MnO_2	SnO_3^{2-}	OH^-
1	1	1	1	1	1	1
2	1	1	0	1	1	1
3	1	3	1	1	3	1
4	1	3	2	1	3	1
5	2	3	1	2	3	2

1–43 The amount of copper (Cu) in a 2.00-g sample of the mineral *cuprite* was determined by dissolving the sample in nitric acid (HNO_3) and adding an excess of iodide (I^-) solution to the resultant copper nitrate [$Cu(NO_3)_2$] solution. The iodine (I_2) liberated required 15.7 ml of 0.200-normal sodium thiosulfate ($Na_2S_2O_3$) solution to be titrated to an end point. What is the percentage of Cu in the mineral? The essential reactions are

$$2Cu^{2+} + 4I^- \rightarrow 2CuI + I_2$$
$$I_2 + 2S_2O_3^{2-} \rightarrow S_4O_6^{2-} + 2I^-$$

(at. wt: Cu = 63.5)

1 5.0%
2 10%
3 20%
4 40%
5 Insufficient information; normality of nitric acid must be known.

Examination I Answers

1–1	(3)	1–16	(3)	1–30	(5)
1–2	(4)	1–17	(4)	1–31	(3)
1–3	(5)	1–18	(1)	1–32	(3)
1–4	(4)	1–19	(4)	1–33	(4)
1–5	(3)	1–20	(1)	1–34	(1)
1–6	(5)	1–21	(1)	1–35	(4)
1–7	(3)	1–22	(4)	1–36	(5)
1–8	(2)	1–23	(1)	1–37	(5)
1–9	(5)	1–24	(2)	1–38	(3)
1–10	(1)	1–25	(4)	1–39	(5)
1–11	(1)	1–26	(2)	1–40	(3)
1–12	(4)	1–27	(5)	1–41	(4)
1–13	(3)	1–28	(5)	1–42	(5)
1–14	(2)	1–29	(3)	1–43	(2)
1–15	(2)				

EXAMINATION II

This examination will test your understanding of the material in Chapters 8 through 18. Allow yourself three hours to complete the 49 questions of the examination. You may refer to the periodic table on the inside front cover.

2–1 The wavelength ranges for the different colors in the visible spectrum are shown in Figure E2–1. What color flame would you expect to see when you heat sodium chloride in a bunsen burner flame if the frequency of the light emitted is approximately 5.1×10^{14} sec^{-1}?

($c = 3.0 \times 10^{10}$ cm sec^{-1})

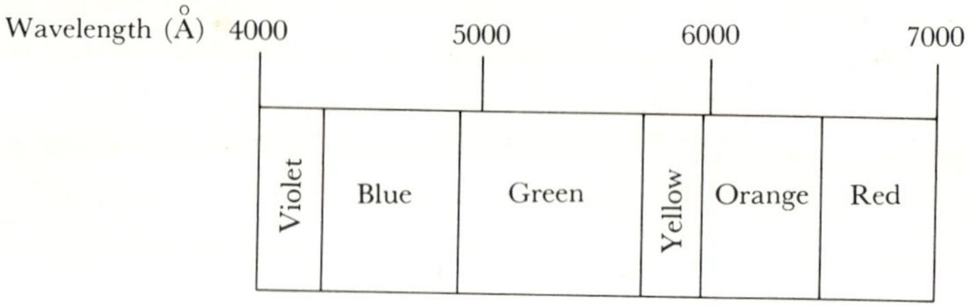

Figure E2–1

1 Blue.
2 Green.
3 Red.
4 Violet.
5 Yellow.

2–2 The work function for sodium is approximately 2.0 eV. What will be the kinetic energy of a photoelectric electron ejected from sodium by ultraviolet light of wavelength 3000 Å? [*Note:* $h\nu = E + \phi_M$.]

($c = 3.0 \times 10^{10}$ cm sec^{-1};
$h = 6.6 \times 10^{-27}$ erg sec;
1.0 eV $= 1.6 \times 10^{-12}$ erg)

1 6.6×10^{-27} erg
2 3.4×10^{-12} erg
3 9.8×10^{-12} erg
4 3.0×10^{10} ergs
5 An electron would not be ejected photoelectrically from sodium by light of wavelength 3000 Å.

2–3 When vanadium metal is bombarded with a beam of high energy electrons, x rays of wavelength 2.5Å are emitted. What is the approximate energy of these x rays?

($h = 6.6 \times 10^{-27}$ erg sec;
$c = 3.0 \times 10^{10}$ cm sec^{-1};
1.0 eV $= 1.6 \times 10^{-12}$ erg)

1 7.9×10^{-9} eV
2 1600 eV
3 5000 eV
4 6600 eV
5 2.0×10^{4} eV

2–4 What is the de Broglie wavelength of a 1.0-g snail traveling at 5.0 cm h^{-1}? [*Note:* $\lambda = h/\text{momentum}$.]

($h = 6.6 \times 10^{-27}$ erg sec)

1 1.3×10^{-27} Å
2 7.9×10^{-26} Å
3 4.8×10^{-24} Å
4 4.8×10^{-16} Å
5 7.9×10^{-16} Å

2–5 The element with the ground state electronic configuration $[Ar]4s^23d^6$ is the fourth most abundant (5.00% by weight) element in the earth's crust. What is this element?

1 Magnesium (Mg, Z = 12).
2 Aluminum (Al, Z = 13).
3 Scandium (Sc, Z = 21).
4 Iron (Fe, Z = 26).
5 Nickel (Ni, Z = 28).

2–6 Aqueous solutions of chromic (Cr^{3+}) salts are used in chromium plating. What is the ground state electronic configuration of the Cr^{3+} ions in the electroplating solutions?

(Cr, Z = 24)

1 [Ar]
2 $[Ar]4s^13d^2$
3 $[Ar]4s^23d^1$
4 $[Kr]5s^14d^2$
5 $[Ar]3d^3$

2–7 Which one of the following elements would you expect to have the *lowest* first ionization energy?

1 Strontium (Sr, $Z = 38$).
2 Arsenic (As, $Z = 33$).
3 Xenon (Xe, $Z = 54$).
4 Sulfur (S, $Z = 16$).
5 Fluorine (F, $Z = 9$).

2–8 Which one of the following elements would you expect to have the *highest* electronegativity?

1 Magnesium (Mg, $Z = 12$).
2 Sulfur (S, $Z = 16$).
3 Boron (B, $Z = 5$).
4 Tellurium (Te, $Z = 52$).
5 Barium (Ba, $Z = 56$).

2–9 When molten sulfur is reacted with chlorine gas, an orange liquid with a vile smell is formed. The liquid is analyzed to have the empirical formula SCl. Its Lewis structure is __________.

(S, $Z = 16$; Cl, $Z = 17$)

1 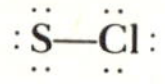

2 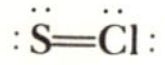

3 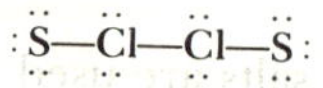

4 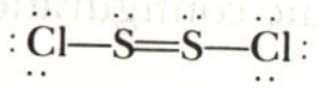

5 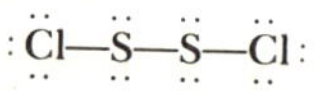

2–10 The structure of the azide ion (N_3^-) in the explosive detonator, lead azide, $Pb(N_3)_2$, is formulated as a resonance hybrid. Which of the following structures could contribute to resonance hybrid structure of the azide ion?

(N, Z = 7)

(a) $[:\ddot{N}=N=\ddot{N}:]^-$

(b) $[:\underset{\cdot\cdot}{\ddot{N}}-N\equiv N:]^-$

(c) $[:\underset{\cdot\cdot}{\ddot{N}}=N=\underset{\cdot\cdot}{\ddot{N}}:]^-$

(d) $[:N\equiv N-\underset{\cdot\cdot}{\ddot{N}}:]^-$

(e) $[:\underset{\cdot\cdot}{\ddot{N}}-N=\underset{\cdot\cdot}{\ddot{N}}:]^-$

1 (a) and (d)
2 (c) and (e)
3 (a), (b), and (d)
4 (c), (d), and (e)
5 All of them.

2–11 Which of the following molecules would *not* be regarded as Lewis acids?

(H, Z = 1; C, Z = 6; N, Z = 7; F, Z = 9;
Cl, Z = 17; Co, Z = 27; Ge, Z = 32;
Sb, Z = 51; I, Z = 53)

(a) I^-
(b) GeF_4
(c) Co^{3+}
(d) $H_2N-CH_2-CH_2-NH_2$
(e) $SbCl_3$

1 (a)
2 (a) and (d)
3 (b) and (e)
4 (c) and (d)
5 (b), (c), and (e)

2–12 The ground state electronic configuration of CN^- in terms of molecular orbital theory is ________.

(C, Z = 6; N, Z = 7)

1 $KK\ (\sigma_s^b)^2(\sigma_s^*)^2(\pi_{x,y}^b)^4(\sigma_z^b)^2$
2 $KK\ (\sigma_s^b)^2(\sigma_s^*)^2(\pi_{x,y}^b)^4$
3 $KK\ (\sigma_s^*)^2(\pi_{x,y}^b)^4(\sigma_s^b)^2(\sigma_z^b)^2$
4 $KK\ (\sigma_s^*)^2(\pi_{x,y}^b)^4(\sigma_s^b)^2(\sigma_z^*)^2$
5 $KK\ (\sigma_s^b)^2(\sigma_s^*)^2(\pi_{x,y}^b)^4(\sigma_z^b)^2(\pi_{x,y}^*)^3$

2–13 Carbon suboxide (C_3O_2), a foul-smelling gas, is prepared by dehydrating malonic acid (HOOC—CH_2—COOH). How many σ and π bonds are in the structure of carbon suboxide?

(C, $Z = 6$; O, $Z = 8$)

1 Three σ and two π.
2 Four σ and four π.
3 Four σ and no π.
4 Five σ and four π.
5 Five σ and three π.

2–14 In the triiodide ion (I_3^-), which is linear, the hybridization of the central I is ________.

[I: $[Kr]5s^2 4d^{10} 5p^5$]

1 sp
2 sp^2
3 sp^3
4 sp^3d
5 sp^3d^2

2–15 An unusual complex of molybdenum is $[Mo(CN)_8]^{4-}$. What is the coordination number of molybdenum in this complex?

1 −4
2 1
3 2
4 8
5 12

2–16 What is the oxidation state of nickel in the oxalate complex, $K_2[Ni(C_2O_4)_2]$? [*Hint:* Oxalic acid is $H_2C_2O_4$.]

1 −4
2 2
3 3
4 4
5 6

2–17 Which is the correct structure for the 1-bromo-6-aquotetraamminecobalt(III) cation?

1

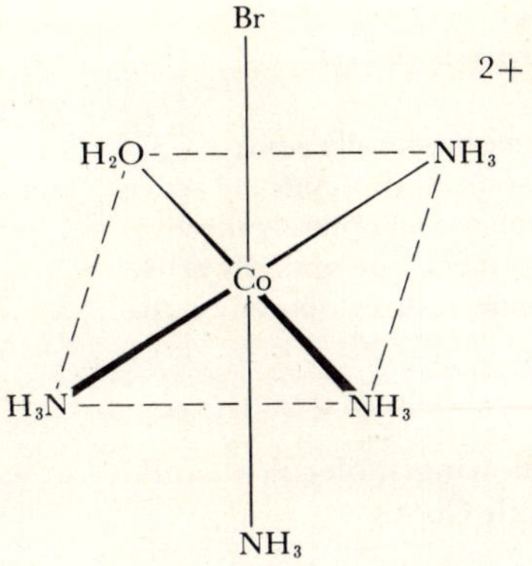

2

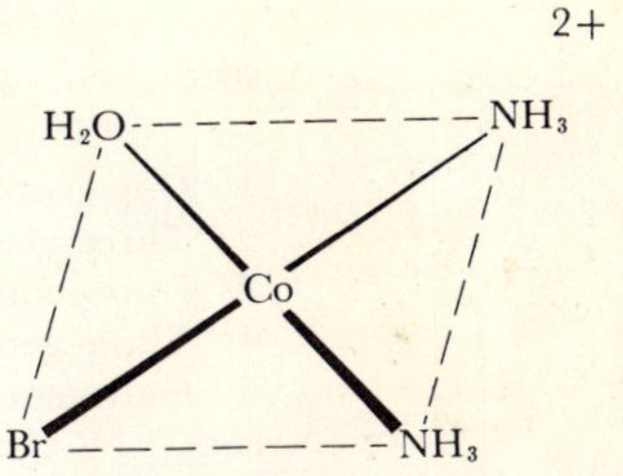

3

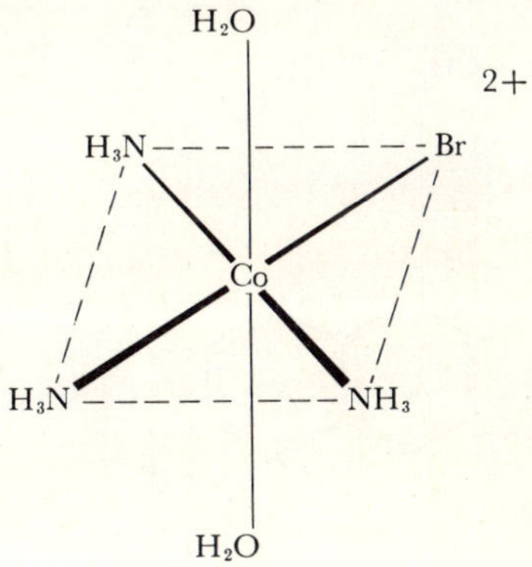

4

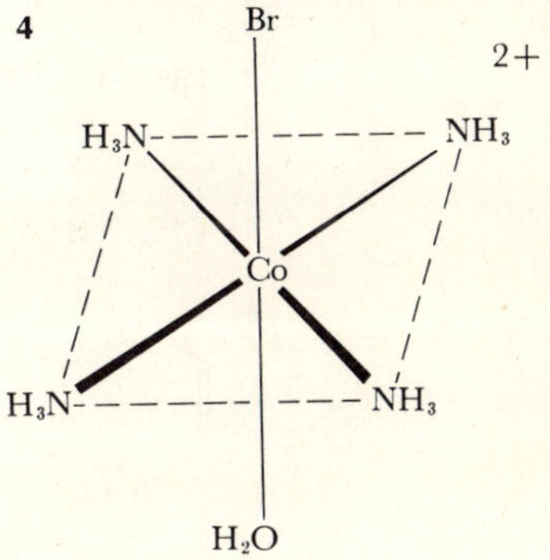

5

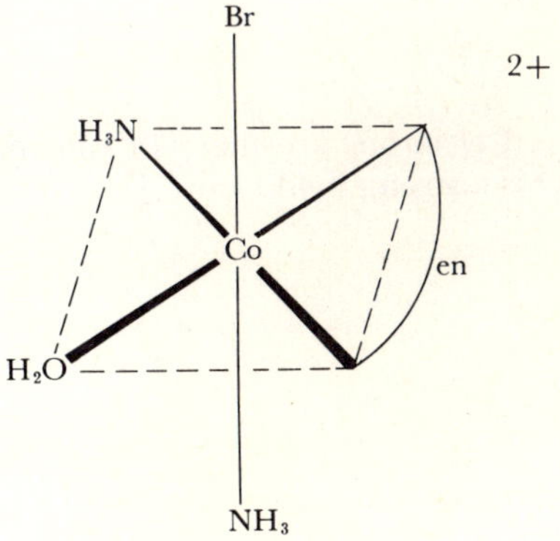

2–18 Which one of the following statements concerning the number of geometrical isomers of the trigonal bipyramidal complex

$Mn(CO)_3(\phi_3P)NO$

and their optical activity is *correct?*

1 One geometrical isomer; optically active.
2 Three geometrical isomers; two optically active.
3 Four geometrical isomers; all optically inactive.
4 Three geometrical isomers; one optically active.
5 Four geometrical isomers; three optically active.

2–19 Which one of the following molecules would you expect to form the most stable chelate with Cu^{2+}?

1 $H_2N—CH_2—CH_2—NH_2$

2 $CH_3—\underset{\underset{O}{\|}}{C}—CH_2—CH_2—CH_2—\underset{\underset{O}{\|}}{C}—CH_3$

3 $CH_3—NH_2$

4

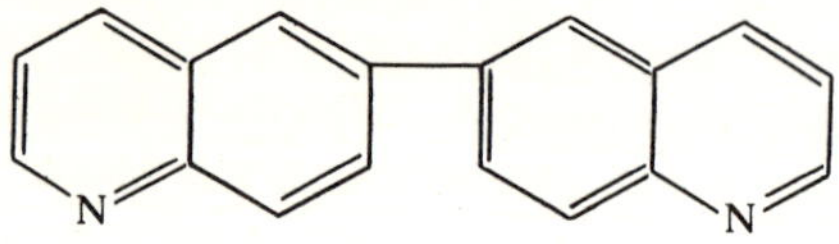

5 $CH_3—\underset{\underset{O}{\|}}{C}—CH_3$

2–20 How many unpaired electrons are there in vanadium hexacarbonyl, $V(CO)_6$? [*Hint:* CO is a strong field ligand.]

(V, $Z = 23$)

1 1
2 2
3 3
4 5
5 None; vanadium hexacarbonyl is diamagnetic.

2–21 What is the correct structural formula for 2,2,3-trimethylpentane?

1

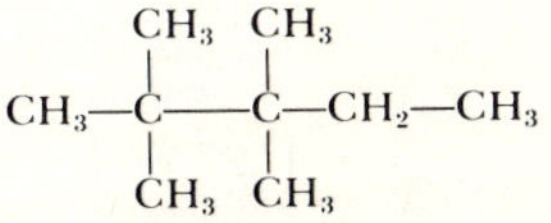

2

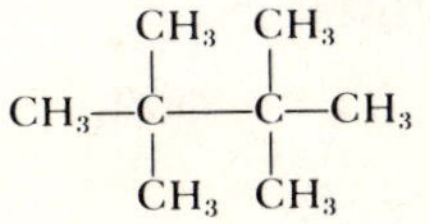

3

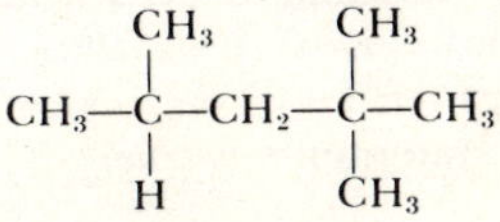

4

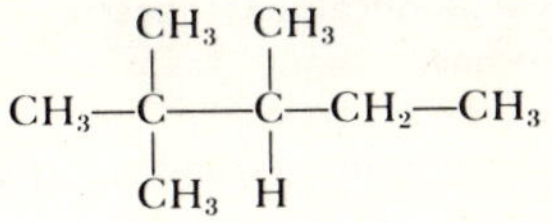

5

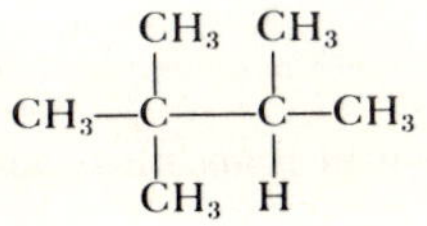

2–22 Examine the structural formula of the following polypeptide:

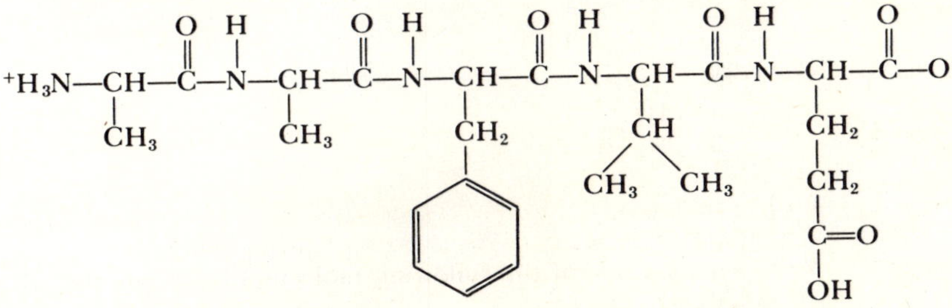

How many different amino acids would be detected on an amino acid analyzer?

1 2
2 3
3 4
4 5
5 6

2–23 Examine the following compounds:

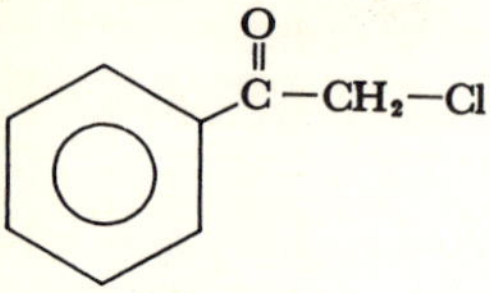

Mace: an eye irritant, used in crowd control

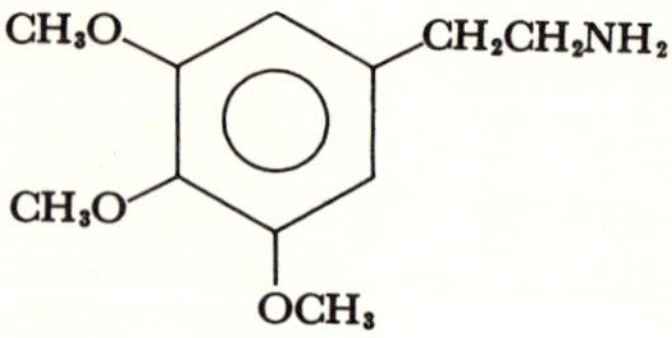

Mescaline: a hallucinatory drug

Which one of the following statements is *true?*

1 Both compounds would give identical mass spectra.
2 Mace has two geometrical isomers.
3 Mescaline has two optical isomers.
4 All substituent groups on the ring in mescaline are planar.
5 Both compounds have aromatic rings.

2–24 Which compound listed below would react most easily with Br_2?

1 $CH_3CH_2CH_2CH_3$
2 $CH_3CH{=}CH{-}CH{=}CH{-}CH_3$
3 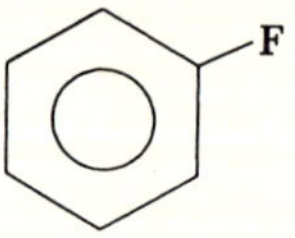

4 CH_4
5 $CH_3{-}Br$

2–25 Which one of the following molecules has a cylindrically shaped structure?

1 $H{-}C{\equiv}C{-}C{\equiv}C{-}C{\equiv}C{-}H$
2 $CH_3{-}CH_2{-}CH_2{-}CH_2{-}CH_2{-}CH_2{-}CH_3$
3 $CH_2{=}CH{-}CH{=}CH{-}CH{=}CH_2$
4 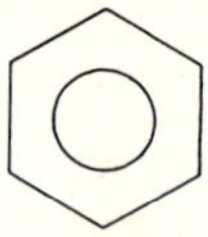
5 $CH_3{-}Cl$

2–26 In 1940, the first isotope of a transuranium element was discovered. The element was named neptunium, and the isotope was neptunium-239 ($^{239}_{93}Np$). How many nucleons are there in $^{239}_{93}Np$?

1 −93
2 93
3 146
4 239
5 332

2–27 One of the seven naturally occurring isotopes of the rare earth element, ytterbium (Yb, $Z = 70$), is $^{170}_{70}Yb$. What is the binding energy per nucleon for $^{170}_{70}Yb$?

(isotopic mass: $^{1}_{0}n = 1.00867$ amu, $^{1}_{1}H = 1.00782$ amu,
$^{170}_{70}Yb = 169.9349$ amu;
1.00 amu = 931 MeV)

1 4.9 MeV nucleon^{-1}
2 7.5 MeV nucleon^{-1}
3 8.1 MeV nucleon^{-1}
4 8.7 MeV nucleon^{-1}
5 1.4×10^3 MeV nucleon^{-1}

2–28 A soil sample taken from the underground nuclear testing site in the Aleutian Islands has an activity of 7.8×10^{-9} C (where "C" means "curie"). Assuming that this activity is solely caused by $^{90}_{38}Sr$, calculate the number of grams of $^{90}_{38}Sr$ that are present in the soil sample.

(activity of 1.0 C = 3.7×10^{10} disintegrations sec^{-1};
rate constant for $^{90}_{38}Sr = 7.8 \times 10^{-10}$ sec^{-1})

1 1.8×10^{-18} g
2 3.9×10^{-17} g
3 7.8×10^{-12} g
4 3.9×10^{-11} g
5 5.6×10^{-11} g

2–29 A cobalt-60 ($^{60}_{27}Co$) γ source has an activity of 10,000 counts per minute (cpm). If the half-life of cobalt-60 is 5.20 years, after how long will the activity of the γ source be 500 cpm? [*Note:* $2.303 \log(N_0/N) = kt$.]

1 5.2 years
2 10 years
3 22 years
4 30 years
5 350 years

2-30 Which of the following proposed nuclear reactions would yield $^{243}_{96}Cm$ as the product?

($d = ^{2}_{1}H$)

(a) $^{242}_{96}Cm$ (d, p)
(b) $^{247}_{97}Bk$ (electron capture)
(c) $^{246}_{98}Cf(n, \alpha)$
(d) $^{242}_{94}Pu$ (α, 3n)
(e) $^{240}_{95}Am$ (p, α)

1 (d)
2 (c) and (e)
3 (a) and (b)
4 (a), (c), and (d)
5 All of them.

2-31 Which one of the following molecules would you *not* expect to be intermolecularly hydrogen bonded?

1 HF
2 $CH_3—CH_2—OH$
3 $CH_3—NH_2$

4 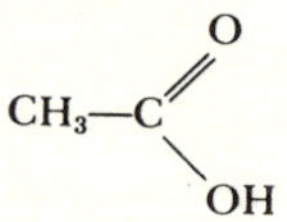

5 H_2S

2-32 Which of the following molecules would you expect to have the lowest boiling point?

1

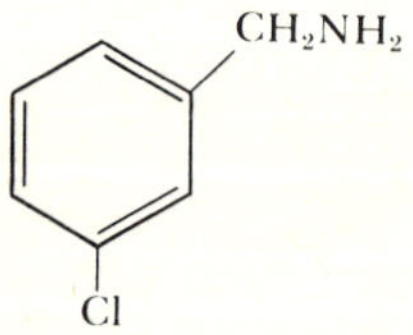

2

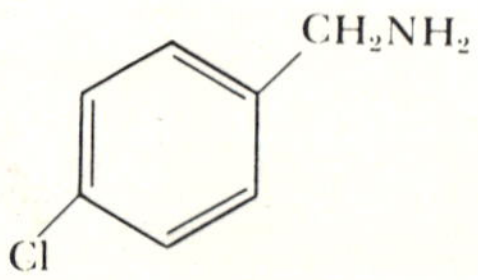

3

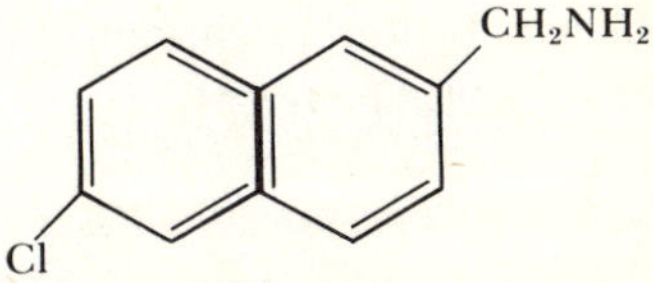

4

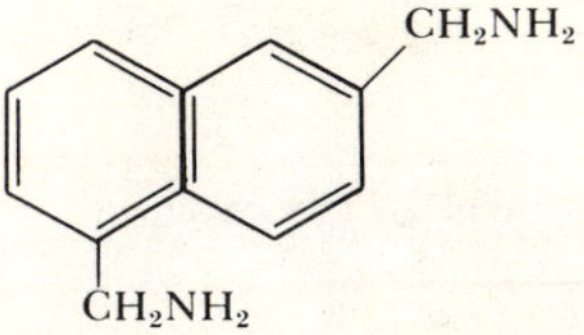

5

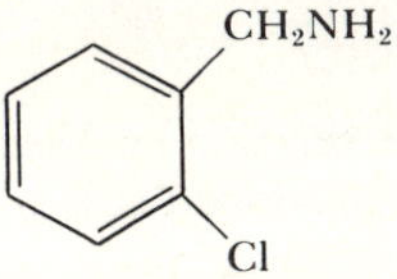

2–33 Which one of the following statements is correct?

1 w is a state function.
2 ΔH is positive for an exothermic reaction.
3 The change in the value of a state function depends on the path.
4 G must decrease for all spontaneous changes at constant T and P.
5 ΔE can be directly measured by measuring the heat of reaction at constant P.

2–34 What is ΔH^0 for the following reaction?

$$2Fe_2O_3(s) + 3C(s) \rightarrow 4Fe(s) + 3CO_2(g)$$

Substance	Standard enthalpy of formation, ΔH_f^0 (kcal mole^{-1})
$CO_2(g)$	−94
$Fe_2O_3(s)$	−197

1 −112 kcal mole^{-1}
2 −103 kcal mole^{-1}
3 +103 kcal mole^{-1}
4 +112 kcal mole^{-1}
5 Insufficient information; need ΔH_f^0 of C and Fe.

2–35 A reversible process (such as fusion, for instance) at constant T and P is characterized by which set of equations?

1 $q = 0,\ w = 0,\ \Delta E = 0$
2 $\Delta H = \Delta E$
3 $T\Delta S = \Delta H = \Delta nRT$
4 $\Delta G = \Delta E - T\Delta S$
5 $\Delta S = q_{rev}/T,\ \Delta G = 0$

2–36 The following data are known for the reaction

$Fe(s) + \frac{3}{2}Cl_2(g) \rightarrow FeCl_3(s)$

at 300°K:

$\Delta H^0 = -96.8$ kcal mole^{-1}
$\Delta S^0 = -55.3$ cal mole^{-1} deg^{-1}

What is ΔG^0 at 300°K, and will the reaction (at unit concentrations and constant pressure) spontaneously produce $FeCl_3(s)$?

1 −80.2 kcal mole^{-1}, yes.
2 −80.2 kcal mole^{-1}, no.
3 −41.5 kcal mole^{-1}, yes.
4 +41.5 kcal mole^{-1}, no.
5 +165 kcal mole^{-1}, no.

2–37 What would be the freezing point of an aqueous solution containing 17 g of C_2H_5OH (a nonelectrolyte) in 1000 g of water?
[*Note:* Regard C_2H_5OH as nonvolatile.]

[$k_f(H_2O) = 1.86$ deg/m;
at. wt: H = 1.0, C = 12, O = 16]

1 −0.69°C
2 −0.34°C
3 0.0°C
4 0.34°C
5 0.69°C

2–38 Consider the phase diagram in Figure E2–2. It is the basis of the sugar candy industry since it represents the behavior of the sucrose–water system. The regions labeled A, B, C, and D may be associated with four physical states. Which set is the correct association?

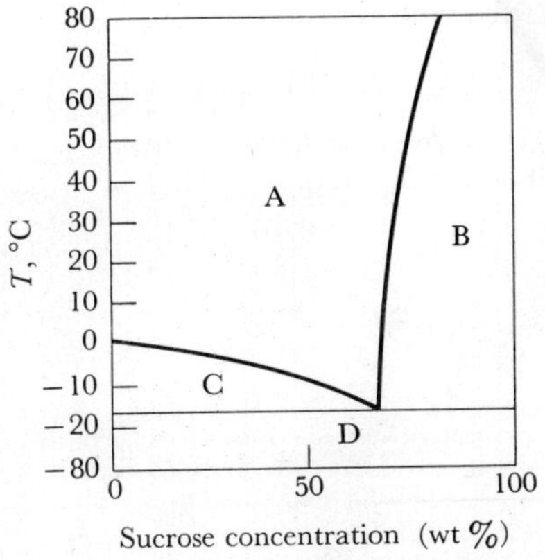

Figure E2–2

	Unsaturated solution	Solid (sucrose + ice)	Saturated solution (ice + solution)	Saturated solution (sucrose + solution)
1	A	D	C	B
2	D	A	C	B
3	D	A	B	C
4	A	D	B	C
5	A	B	C	D

2–39 The solubility of silver chloride (AgCl) in water can be increased by doing which one of the following operations? (Ignore activity coefficient effects.)

[$K_{sp}(AgCl) = 1.7 \times 10^{-10}$;
formation constant for $Ag(NH_3)_2^+$ is 1×10^8]

1 Dissolve NH_3 in the saturated solution.
2 Add $AgNO_3$ to the saturated solution.
3 Add HCl to the saturated solution.
4 Add H_2O to the saturated solution.
5 None of the above.

2–40 Which of the following statements about electromotive force are correct?

(a) $\mathscr{E} = \mathscr{E}^0$ at 0°K.
(b) $\mathscr{E} = \mathscr{E}^0$ at equilibrium.
(c) $\mathscr{E} = \mathscr{E}^0$ at unit concentration.
(d) $\mathscr{E} = \mathscr{E}^0$ at STP.

1 (a)
2 (b) and (c)
3 (a) and (c)
4 (d)
5 None of the above.

2–41 What is $\mathscr{E}^0$ for the following reaction?

$Zn(s) + Cl_2(g) \rightarrow Zn^{2+}(aq) + 2Cl^-(aq)$

Half-reaction	Standard reduction potential, $\mathscr{E}^0$ (volts)
$Zn^{2+}(aq) + 2e^- \rightarrow Zn(s)$	−0.76
$Cl_2(g) + 2e^- \rightarrow 2Cl^-(aq)$	+1.36

1 −2.12 volts
2 −0.60 volt
3 −1.36 volts
4 +0.60 volt
5 +2.12 volts

2–42 What is $\mathscr{E}^0$ for the following reaction?

$6Fe^{2+} + Cr_2O_7^{2-} + 14H^+ \rightarrow 6Fe^{3+} + 2Cr^{3+} + 7H_2O$

Reaction	Standard reduction potential, $\mathscr{E}^0$ (volts)
$Cr^{3+} + 3e^- \rightarrow Cr$	−0.74
$Fe^{2+} + 2e^- \rightarrow Fe$	−0.41
$Fe^{3+} + e^- \rightarrow Fe^{2+}$	0.77
$Cr_2O_7^{2-} + 14H^+ + 6e^- \rightarrow 2Cr^{3+} + 7H_2O$	1.33

1 −3.29 volt
2 0.56 volt
3 1.33 volts
4 2.10 volts
5 2.81 volts

2–43 What is the approximate equilibrium constant for the following reaction? (Take $2.303\ RT/\mathscr{F} = 0.0592$ volt.)

$F_2 + 2Br^- \rightarrow Br_2 + 2F^-$

Reaction	Standard reduction potential, $\mathscr{E}^0$ (volts)
$Br_2 + 2e^- \rightarrow 2Br^-$	1.06
$F_2 + 2e^- \rightarrow 2F^-$	2.87

1 $10^{0.79/0.0592} \simeq 10^{13}$
2 $10^{1.58/0.0592} \simeq 10^{27}$
3 $10^{3.62/0.0592} \simeq 10^{61}$
4 $10^{3.72/0.0592} \simeq 10^{63}$
5 $10^{7.44/0.0592} \simeq 10^{126}$

2–44 If a

$Li|LiCl|Cl_2$

cell delivered 2.1×10^2 watts for 1.0 h at a voltage of 3.4 volts, how much Li was consumed (oxidized)?

($\mathscr{F} \simeq 9.6 \times 10^4$ coulombs;
at. wt: Li = 6.9, Cl = 35.5)

1 4.4×10^{-3} g
2 16 g
3 160 g
4 1600 g
5 1500 kg

2–45 The gas reaction $A + B \rightarrow C$ is found to have a rate equal to

$k[A][B]^2$

If concentrations are expressed in mole liter^{-1}, then the units of the rate constant, k, are __________.

1 sec
2 sec^{-1}
3 liter mole^{-1} sec^{-1}
4 mole liter^{-1} sec^{-1}
5 liter^2 mole^{-2} sec^{-1}

2–46 The following data have been obtained for the homogeneous gas reaction

$2NO + 2H_2 \rightarrow N_2 + 2H_2O$

Initial rate of disappearance of NO (mole liter^{-1} sec^{-1}) $\left(\frac{-d[NO]}{dt}\right)_{t=0}$	Initial concentrations (mole liter^{-1}) [NO]	[H_2]
4.8×10^{-5}	1×10^{-2}	1×10^{-3}
43.2×10^{-5}	3×10^{-2}	1×10^{-3}
86.4×10^{-5}	3×10^{-2}	2×10^{-3}

What is the rate equation?

1 rate = $k[H_2]/[NO]$
2 rate = $k[NO][H_2]$
3 rate = $k[NO]^2[H_2]$
4 rate = $k[NO]^3[H_2]$
5 Insufficient data.

2–47 The collision theory of chemical reactions maintains that ________.

1 All collisions are reactive.
2 All energetic collisions whose energy is greater than $\frac{3}{2}RT/N$ are reactive.
3 The steric factor, p, is unity.
4 E_a(forward) − E_a(reverse) = 0.
5 None of the above.

2–48 Consider Figure E2–3. Which one of the following statements is correct?

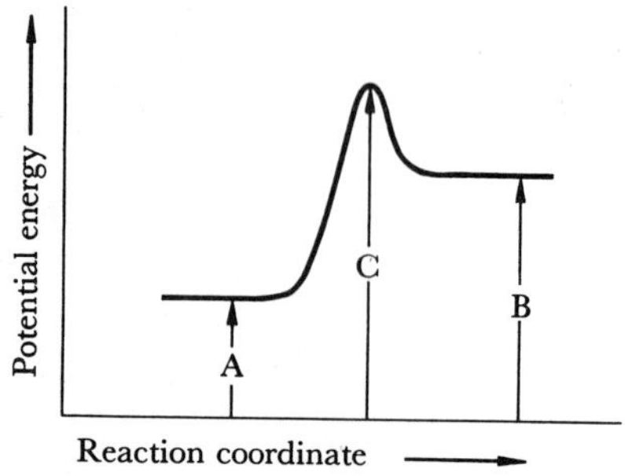

Figure E2–3

1 The activation energy of the forward reaction is less than the activation energy of the reverse reaction.
2 The activation energy for the reverse reaction is C.
3 The forward reaction is endothermic.
4 The potential energy of the activated complex is A + B.
5 The potential energy of the reactants is greater than that of the products.

2–49 For the S_N2 reaction

$$R_3CI + OH^- \rightarrow I^- + R_3COH$$

which of the following is the most reasonable choice for the activated complex?

1 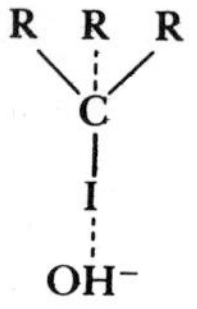

2 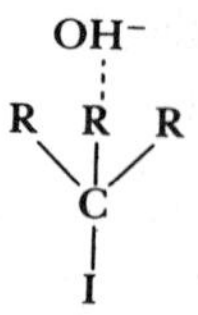

3 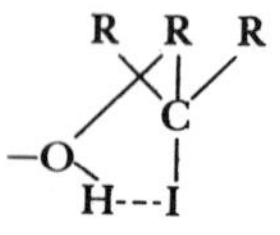

4 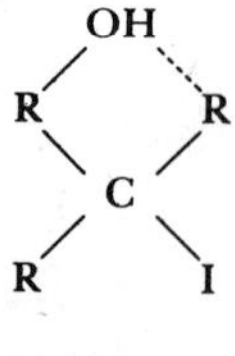

5 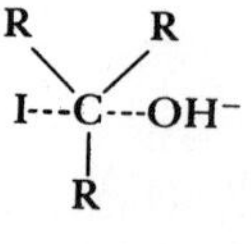

Examination II Answers

2–1 (5)
2–2 (2)
2–3 (3)
2–4 (4)
2–5 (4)
2–6 (5)
2–7 (1)
2–8 (2)
2–9 (5)
2–10 (3)
2–11 (2)
2–12 (1)
2–13 (2)
2–14 (4)
2–15 (4)
2–16 (2)
2–17 (4)
2–18 (3)
2–19 (1)
2–20 (1)
2–21 (4)
2–22 (3)
2–23 (5)
2–24 (2)
2–25 (1)
2–26 (4)
2–27 (3)
2–28 (5)
2–29 (3)
2–30 (4)
2–31 (5)
2–32 (5)
2–33 (4)
2–34 (4)
2–35 (5)
2–36 (1)
2–37 (1)
2–38 (1)
2–39 (1)
2–40 (3)
2–41 (5)
2–42 (2)
2–43 (3)
2–44 (2)
2–45 (5)
2–46 (3)
2–47 (5)
2–48 (3)
2–49 (5)

APPENDIX 2 MATHEMATICAL NECESSITIES AND DEVICES[1]

EXPONENTIAL NUMBERS

Not only does writing a number in exponential form enable us to express significant-figure information with minimal confusion, it avoids many zeros for either small or large numbers.

We use exponential numbers to express quantities as multiple powers of 10. Thus, an exponential number consists of two parts: a coefficient (chosen to be between 1 and 10) and a power of 10. For example, the Avogadro number is written 6.02×10^{23}; 6.02 is the coefficient and 10^{23} is the power of 10.

A positive exponent, n, indicates that the coefficient should be multiplied by 10 n times; that is, the decimal should be moved n places to the right of its position in the coefficient. A negative exponent, $-m$, indicates that the coefficient should be divided by 10 m times; that is, the decimal should be moved m places to the left. For example,

$$0.0000000192 = 1.92 \times 10^{-8}$$
$$10 \text{ thousand} = 1 \times 10^{4}$$
$$96500 = 9.65 \times 10^{4}$$

To add or subtract exponential numbers, we must be certain that the powers of 10 are the same. Otherwise, the operation would be like adding different things: $2x + 2y = ?$, whereas $2x + 2x = 4x$. This requirement may force you to rewrite the exponential number. This revision is easy if you remember that each time the power of 10 becomes more positive by

[1] Adapted, with the kind permission of Dr. Wilbert Hutton, from his book *A Study Guide to Chemical Principles*, W. A. Benjamin, 1974, 2nd ed.

one unit it is the same as multiplying the number by 10 or moving the decimal in the coefficient one place to the right. Similarly, if the power of 10 is made more negative, the decimal in the coefficient must be moved to the left. For example,

$$6.022 \times 10^{23} + 7.7 \times 10^{21} = ?$$

Rewrite both numbers so they have the same power of 10; for example, 21. To write 6.022×10^{23} as a multiple of 10^{21} (the exponent has been decreased by two powers of 10) requires that the coefficient be increased by two powers of 10. Therefore, its decimal point should be moved two places to the right:

$$6.022 \times 10^{23} = 602.2 \times 10^{21}$$

Now the numbers can be added:

$$\begin{array}{rl} & 602.2 \quad \times 10^{21} \\ + & 7.7 \quad \times 10^{21} \\ \hline & 609.9 \quad \times 10^{21} \quad \text{or} \quad \mathbf{6.099 \times 10^{23}} \end{array}$$

Do the following exercises to test your understanding.

(1) Add 2.46×10^{-9} cm to 2.46×10^{-8} cm.
(2) Subtract 1.625×10^{-1} cm from 2.234×10^{2} cm.
(3) Add 4.0075×10^{3} ml to 6.23×10^{2} ml.
(4) Subtract 1.725×10^{-1} g from 2.1623×10^{1} g.

Answers: (1) 2.71×10^{-8} cm; (2) 2.236×10^{2} cm; (3) 4.630×10^{3} ml; (4) 2.1450×10 g.

In *multiplication*, you need only multiply the coefficients together and then multiply the powers of 10 together (add their exponents) to obtain the coefficient and power of 10 for the answer. For example,

$$6.02 \times 10^{23} \times 1.76 \times 10^{-2} = ?$$

The product of the coefficients to the proper number of significant figures is $6.02 \times 1.76 = 10.6$. The product of the powers of 10 (exponentials) are $10^{23} \times 10^{-2} = 10^{23+(-2)} = 10^{21}$. The answer to the multiplication is 10.6×10^{21} or, writing the coefficient in the more preferred way as a number between 1 and 10, $\mathbf{1.06 \times 10^{22}}$.

In *division*, the coefficients are divided separately and then exponentials are divided. Recall that in division of exponential numbers, the exponent of the divisor (the denominator) is subtracted from the exponent of the dividend (the numerator). For example,

$$\frac{6.022 \times 10^{23}}{5.976 \times 10^{27}} = ?$$

Dividing 6.022 by 5.976 gives 1.008 to the correct number of significant figures. Dividing the exponentials gives $10^{23}/10^{27} = 10^{(23-27)} = 10^{-4}$. The answer is therefore $\mathbf{1.008 \times 10^{-4}}$.

The same general procedure is followed when raising an exponential to a power. The coefficients are done first, the exponentials next, and the results of these computations combined for the answer. Thus,

$$(6 \times 10^{3})^{3} = 6^{3} \times (10^{3})^{3} = \mathbf{2 \times 10^{11}}$$
$$(5.1 \times 10^{-2})^{2} = 5.1^{2} \times (10^{-2})^{2} = 26 \times 10^{-4} = \mathbf{2.6 \times 10^{-3}}$$

To avoid fractional exponents when extracting a root, we must adjust the power of 10 so it becomes a whole-number factor of 2 if a square root is to be extracted, a whole-number factor of 3 if a cube root is to be extracted, and so forth. Therefore, to extract the cube root of Avogadro's number, $(6.02 \times 10^{23})^{1/3}$, you must first rewrite the number so the power of 10 is a whole-number multiple of three. Since $3 \times 7 = 21$ and $3 \times 8 = 24$, either 10^{21} or 10^{24} are suitable exponentials. Let us rewrite the number as a coefficient times 10^{21} by moving the decimal in the coefficient two places to the right and decreasing the exponent of 10 by two units: $(602 \times 10^{21})^{1/3}$. The cube root of 602 is 8.45; the cube root of 10^{21} is 10^7. The answer is $\mathbf{8.45 \times 10^7}$.

As a self-test of your understanding, do the following exercises.

(1) $\dfrac{5.23 \times 10^{27}}{9.76 \times 10^{3}} = ?$

(2) $\dfrac{3.42 \times 10^{-29}}{6.704 \times 10^{5}} = ?$

(3) $\dfrac{(2.46 \times 10^{3})(1.7 \times 10^{-5})}{3.25 \times 10^{4}} = ?$

(4) $(5.2 \times 10^{-3})^3 = ?$

(5) $(7.5 \times 10^{-5})^{1/2} = ?$

Answers: (1) 5.36×10^{23}; (2) 5.10×10^{-35}; (3) 1.3×10^{-6}; (4) 1.4×10^{-7}; (5) 8.7×10^{-3}.

LOGARITHMS

There are many advantages in expressing numbers in scientific notation, that is, as $a \times 10^x$, in which a, the coefficient, is a number between 1 and 10 and x is the exponent, or power, of 10. If x is zero, the entire exponential (10^0) equals 1 (any number raised to the zero power is unity) and is not written; for example, $2.54 \times 10^0 = 2.54 \times 1 = 2.54$. Also, if the coefficient a or the exponent x is unity, it may be understood and not written as such; that is, 1×10^4 is written 10^4 and 4.2×10^1 is written 4.2×10.

Now if $10^0 = 1$ and $10^1 = 10$, it follows that any number between 1 and 10 can be expressed as a power of 10 with a coefficient of unity by using fractional exponents. Moreover, if negative powers are included, any number can be given in this way. This is the basis for expressing numbers as logarithms. The *logarithm* of a number is the exponent that must be put on 10 to give the number. For example, the number 2 can be expressed as $10^{0.301}$, in which case 0.301 is the logarithm (or "log") of 2. In Table 1 the powers of 10 needed to express the numbers 1 through 10 are listed. In other words, a table of the logarithms of the numbers 1 through 10 is given.

Table 1

$1 = 10^0$, log 1 = 0	$6 = 10^{0.778}$, log 6 = 0.778
$2 = 10^{0.301}$, log 2 = 0.301	$7 = 10^{0.845}$, log 7 = 0.845
$3 = 10^{0.477}$, log 3 = 0.477	$8 = 10^{0.903}$, log 8 = 0.903
$4 = 10^{0.602}$, log 4 = 0.602	$9 = 10^{0.954}$, log 9 = 0.954
$5 = 10^{0.699}$, log 5 = 0.699	$10 = 10^1$, log 10 = 1.000

The exponent to which 10 must be raised to give the number is referred to as a logarithm to the *base* 10. These logarithms are the most convenient for general use since, for calculations, one can follow the simple rule that 1 is added to the logarithm of a number every time the number is multiplied by 10.

However, numbers other than 10 can be used for the base. Logarithms to the base 2.718 . . . , so-called "Naperian base" or "natural base" logarithms, are used frequently in science. This base, 2.718 . . . , is designated by the letter e. In terms of e, the number 2 can be expressed as $e^{0.693}$ (i.e., e to the 0.693 power), and we would say that the logarithm of 2 to the base e is 0.693. This last statement is written $\log_e 2 = 0.693$, or $\ln 2 = 0.693$, where the abbreviation "ln" refers to the natural or Naperian base logarithm. Logarithms to the base 10 are generally indicated by the word "log," and no base is specified (e.g., $\log 2 = 0.301$). A simple relationship between base 10 and base e logs allows us to convert a logarithm from one base to the other. The relationship is $\log_e x = 2.303 \log x$.

Logarithms are useful because they make the multiplication of numbers a simple addition operation; division, a simple subtraction operation; and the computation of powers and roots, a simple multiplication or division process. Thus, much of the tedious digit-writing of arithmetic can be discarded, and operations can be performed faster and with less chance of error. The reason for this simplification is that when two powers of 10 are multiplied, their exponents are added, and when two powers of 10 are divided, their exponents are subtracted. These exponents are the logarithms. So, if we have tables that give the logarithms of numbers, we can take advantage of these simpler operations.

The logarithms of numbers expressed as powers of 10 with coefficients of unity are easily determined. They are simply the exponent. For example,

$$1 \times 10^5 = 10^0 \times 10^5 = 10^{(0+5)} = 10^5 \qquad \log 10^5 = 5$$

Similarly,

$$\log(1 \times 10^6) = \log 10^6 = 6$$
$$\log(1 \times 10^{-7}) = \log 10^{-7} = -7$$

Now let us obtain the logarithms of some numbers that have been multiplied together. Recalling that multiplication of powers of 10 is carried out by adding exponents, we can see immediately that the log of a product 10^x times 10^y is simply $x + y$.

EXAMPLE. What is the log of $(1 \times 10^7)/(1 \times 10^5)$?

SOLUTION.

$$(1 \times 10^7)(1 \times 10^5) = 1 \times 10^{12}$$
$$\log[(1 \times 10^7)(1 \times 10^5)] = \log(1 \times 10^{12}) = \mathbf{12}$$

or, if we omit the coefficients,

$$10^7 \times 10^5 = 10^{12}$$
$$\log(10^7 \times 10^5) = \log 10^{12} = \mathbf{12}$$

An alternate solution is to add the logarithms,

$$\begin{aligned}\log(10^7 \times 10^5) &= \log 10^7 + \log 10^5 \\ &= 7 + 5 \\ &= \mathbf{12}\end{aligned}$$

EXAMPLE. What is the log of $(1 \times 10^9)(1 \times 10^{-7})$?

SOLUTION.

$$\begin{aligned}\log(10^9 \times 10^{-7}) &= \log 10^9 + \log 10^{-7} \\ &= \quad 9 \quad + \quad -7 \\ &= \mathbf{2}\end{aligned}$$

Similarly, the logarithm of a quotient $10^x/10^y$ is $x-y$.

EXAMPLE. What is the log of $(1 \times 10^9)/(1 \times 10^3)$?

SOLUTION.

$$\log\left(\frac{1 \times 10^9}{1 \times 10^3}\right) = \log 10^6 = \mathbf{6}$$

An alternate solution is to subtract the logarithms.

$$\begin{aligned}\log\left(\frac{1 \times 10^9}{1 \times 10^3}\right) &= \log(1 \times 10^9) - \log(1 \times 10^3) \\ &= \quad 9 \quad - \quad 3 \\ &= \mathbf{6}\end{aligned}$$

EXAMPLE. What is the log of $(1000 \times 10)/0.0001$?

SOLUTION.

$$\begin{aligned}\log\left(\frac{10^3 \times 10}{10^{-4}}\right) &= \log 10^3 + \log 10 - \log 10^{-4} \\ &= \quad 3 \quad + \quad 1 \quad - \quad (-4) \\ &= \mathbf{8}\end{aligned}$$

In the preceding examples we have used only numbers that have coefficients of unity and whole-number powers of 10. In general, digits other than 1 will occur in the coefficients; in such cases we must look for these in a log table. Log tables always list the kind of information in Table 1, namely, the number and the exponent to which 10 must be taken to obtain the number—the number's logarithm.

Your next step should be to look up numbers expressed by coefficients larger than 1 and powers of 10 greater than zero. A few examples should illustrate how this is done.

EXAMPLE. What is the log of 2×10^3?

SOLUTION.

$$\begin{aligned}\log(2 \times 10^3) &= \log 2 + \log 10^3 \\ &= 0.301 \quad + \quad 3 \\ &= \mathbf{3.301}\end{aligned}$$

The log 2 is found in the table; the log 10^3 is merely the exponent.

EXAMPLE. Now try a number with a negative exponent. Find the log of 7×10^{-4}.

SOLUTION.

$$\begin{aligned}\log(7 \times 10^{-4}) &= \log 7 + \log 10^{-4} \\ &= 0.845 + \quad -4 \\ &= \mathbf{-3.155}\end{aligned}$$

No special operation is necessary to obtain logarithms for numbers with negative exponents.

Now you should be ready to put all of these concepts into practice. To help you to begin, here is an example worked out in detail.

EXAMPLE. What is the logarithm of

$$\frac{(2 \times 10^{-3})(6 \times 10^{23})}{5 \times 10^{4}}$$

SOLUTION.

$$\log\left[\frac{(2 \times 10^{-3})(6 \times 10^{23})}{5 \times 10^{4}}\right] = \log 2 + \log 10^{-3}$$
$$+ \log 6 + \log 10^{23}$$
$$- (\log 5 + \log 10^{4})$$

The imposing string of logarithms on the right of the equation may be a bit bewildering. Remember that the order in which a sequence of multiplications and divisions is carried out does not make any difference in the answer. You would obtain the same answer to the multiplication and division sequence if you divide the first factor in the numerator by the denominator and then multiply the result by the second factor in the numerator as you would if you multiply both factors in the numerator and then divide their product by the denominator. The same is true if you do the multiplication and division with logarithms. In this problem, the two factors in the numerator have been multiplied first, and then their product is divided by the denominator by using logarithms. Now let us find, from a four-place table of logarithms or a slide rule, the logarithms for the right side of the equation:

$$= 0.301 + (-3) + 0.778$$
$$+ 23 - 0.699 - 4$$
$$= 0.380 + 16 = \mathbf{16.380}$$

Try these exercises to test your understanding.

(1) What is the log of

$$\frac{4 \times 10^{8}}{2 \times 10^{6}}$$

(2) What is the log of

$$\frac{(6 \times 10^{-6})(3 \times 10^{14})}{(2 \times 10^{3})(1 \times 10^{6})}$$

Answers: (1) 2.301; (2) −0.046.

[*Note:* You may find the solutions to problems such as this easier if you work out the coefficients on a slide rule, since we have attempted wherever possible to simplify the arithmetic.]

You will also need to be able to reverse this procedure, so that, given the logarithm, you can express the number as an integral power of 10. To illustrate how to do this, we shall examine first some simple examples and then some more complicated ones.

EXAMPLE. What is the number whose log is 3? [*Note:* The term "antilog" indicates "the number whose log is."]

SOLUTION.

$$\begin{aligned} \log x &= 3 \\ x &= 10^3 \\ x &= 1 \times 10^3 = \mathbf{1000} \end{aligned}$$

EXAMPLE. What is the number whose log is 3.301?

SOLUTION.

$$\begin{aligned} \log x &= 3.301 \\ x &= 10^{3.301} \\ &= 10^3 \times 10^{0.301} \\ &= 10^3 \times 2 = \mathbf{2 \times 10^3} \end{aligned}$$

The method here is to take the logarithm 3.301, extract the whole number (3), and use it to derive the power of 10 (10^3), and then use the decimal part (0.301) to derive the coefficient (2). This operation requires looking in a table of logarithms for the logarithm 0.301 and then noting what number corresponds to it.

EXERCISE. What are the numbers whose logarithms are 6.903, 4.699, and 3.845?

Answers: 8×10^6; 5×10^4; 7×10^3.

Finding the antilog of a negative logarithm is complicated because the table or slide rule contains positive numbers only.

EXAMPLE. Find the number whose logarithm is −4.222.

SOLUTION.

$$\begin{aligned} \log x &= -4.222 \\ x &= 10^{-4.222} \end{aligned}$$

Note that −4.222 can be written as $-5 + 0.778$. Make this substitution and proceed as usual.

$$\begin{aligned} x &= 10^{-4.222} = 10^{-5+0.778} \\ &= 10^{-5} \times 10^{0.778} \end{aligned}$$

In effect, we have transferred all of the minus signs to the whole-number exponent of 10 and have made the decimal exponent a positive number, the antilog of which can be found in the table. The antilog of 0.778 is 6. Therefore, $x = 10^{-5} \times 10^{0.778} = \mathbf{6 \times 10^{-5}}$.

Extracting Roots and Raising Numbers to Powers

We can use logarithms to raise numbers to powers or to extract roots of numbers. For instance, the number a^x means that the number a is multiplied by itself x times. To do this with logarithms, we would write the log of a x times and add; the sum would be the log a^x. That is, $\log a^x = x \log a$.

The xth root of a number a is expressed $a^{1/x}$. Reasoning similarly to the one for raising numbers to powers, we find that $\log a^{1/x} = (1/x) \log a$.

EXAMPLE. What is the cube root of 0.008?

SOLUTION.

$$\begin{aligned}\log[(8 \times 10^{-3})^{1/3}] &= \tfrac{1}{3}(\log 8 + \log 10^{-3})\\ &= \tfrac{1}{3}[0.903 + (-3)]\\ &= \tfrac{1}{3}(-2.097)\\ &= -0.699\end{aligned}$$

Now we need to find the antilog of −0.699; this will be the cube root of 0.008.

$$10^{-0.699} = 10^{0.301-1} = 10^{0.301} \times 10^{-1}$$

We find the number equal to $10^{0.301}$; it is 2. Therefore,

$$(0.008)^{1/3} = \mathbf{2 \times 10^{-1}}$$

THE QUADRATIC FORMULA

Equations of the second order containing a single unknown that is raised to the second power have two roots. These equations can always be written as

$$ax^2 + bx + c = 0 \tag{1}$$

in which a, b, and c are numbers and x is the unknown. If $a = 0$, the equation is first order. It is also possible for b and c or both to be zero.

Once the equation is in the form of Equation 1, and a is not zero, it can be solved by substitution into the quadratic formula,

$$x = \frac{-b \pm \sqrt{b^2 - 4ac}}{2a} \tag{2}$$

In doing chemistry problems, only one of the two roots is an answer to the problem; the other is unrealistic. You can generally decide which of the two roots apply by considering the problem's physical circumstances. For instance, a common situation is when you use a quadratic equation to find the amount of a substance remaining after a chemical reaction. The solution of the equation will give two roots, each of which is a possible value for the amount of the remaining substance. If one of the roots gives a value larger than the amount of substance present when the reaction started, or if it gives a negative or imaginary number, it is obvious that this value does not apply. Hence, the other root must be the correct answer.

EXAMPLE. Consider the solution of the following equation in which x represents the amount (in moles per liter) of a substance B formed after a chemical reaction has reached equilibrium. Assume that we start with 1 mole liter^{-1} of a reacting substance and that 1 mole of it produces 1 mole of B and 1 mole of C. The equilibrium constant for the reaction is 54:

$$\frac{(x)(x)}{1 - x} = 54$$

SOLUTION. Rearrange the equation to the form of Equation 1 by multiplying both sides by $1 - x$:

$$x^2 = 54(1 - x)$$

Multiply the two quantities on the right, collect terms, and equate to zero:

$$x^2 + 54x - 54 = 0$$

Now substitute the corresponding quantities in the quadratic equation: $a = 1$, $b = 54$, and $c = -54$.

$$x = \frac{-54 \pm [54^2 - 4(1)(-54)]^{1/2}}{2(1)} = \frac{-54 \pm (2916 + 216)^{1/2}}{2}$$

$$= \frac{-54 \pm (3132)^{1/2}}{2} = \frac{-54 \pm 55.9}{2} = \frac{-54 + 55.9}{2} = \frac{1.9}{2} = 0.95$$

and

$$x = \frac{-54 - 55.9}{2} = \frac{-109.9}{2} = -55.0$$

Logic indicates which of these two solutions is correct for the problem: Since the reaction began with 1 mole of the substance, the amount left after equilibrium is reached must be less than 1.0 mole. The first root, $x = 0.95$, satisfies this requirement. The second root, -55.0, cannot possibly be correct because the concept of a negative amount of material has no physical reality. Thus, the correct answer is $x =$ **0.95 mole liter^{-1}**.

Quadratic equations and equations of higher order are often not necessary for solving problems in beginning chemistry. The reason is that frequently the conditions of the problem are such that one can foresee approximations that reduce the complexity of the calculations without introducing a significant error in the answer.

EXAMPLE. Let us look at a problem similar to the preceding one. Here, however, let us propose that the reaction reaches equilibrium when very little of the 1.0 mole liter^{-1} of starting material reacts. Thus, the amount of new substance formed, x, will be small, and the ratio of terms involving x will be smaller. The expression

$$\frac{(x)(x)}{1.0 - x} = 1.2 \times 10^{-6} \qquad (3)$$

fits these conditions and can illustrate our point.

SOLUTION. As before, to solve this expression by the quadratic formula, rearrange the equation so it is in the form

$$ax^2 + bx + c = 0$$

To do this, multiply both sides of Equation 3 by $1.0 - x$:

$$x^2 = (1.2 \times 10^{-6})(1.0 - x)$$

Now multiply the two quantities on the right, collect terms, and equate to zero:

$$x^2 + (1.2 \times 10^{-6})x - 1.2 \times 10^{-6} = 0$$

Substitute the corresponding quantities in the quadratic equation: $a = 1.0$, $b = 1.2 \times 10^{-6}$, and $c = -1.2 \times 10^{-6}$.

$$x = \frac{-b \pm \sqrt{b^2 - 4ac}}{2a}$$

$$= \frac{-1.2 \times 10^{-6} \pm [(1.2 \times 10^{-6})^2 - 4(1)(-1.2 \times 10^{-6})]^{1/2}}{2(1)}$$

$$= \frac{-1.2 \times 10^{-6} \pm (1.44 \times 10^{-12} + 4.8 \times 10^{-6})^{1/2}}{2}$$

$$= \frac{-1.2 \times 10^{-6} \pm (4.8 \times 10^{-6})^{1/2}}{2}$$

$$= \frac{-1.2 \times 10^{-6} \pm 2.2 \times 10^{-3}}{2}$$

$$= 1.1 \times 10^{-3} \quad \text{and} \quad -1.1 \times 10^{-3}$$

Since the concentration cannot be a negative number, the correct answer is $x =$ **1.1×10^{-3} mole liter^{-1}**. This is a small number; that is, only a small amount of product is formed by the time equilibrium is attained.

Consider the denominator of the original expression $(1.0 - x)$. Since we have calculated the value for x, let us now evaluate its denominator:

$$1.0 - x = 1.0 - 1.1 \times 10^{-3} = 1.0 - 0.0011 \simeq 1.0$$

Observe that if we follow the rules for subtracting significant figures, the result of subtracting the value of x from 1.0 is the same as if we had completely forgotten about the x. Its value is too small to be significant compared with 1.0.

What is the point of all this? Suppose that you know that the value for x was negligible compared with 1.0; thus, you omitted it from the $(1.0 - x)$ expression before you attempted to solve the problem. Here is the simpler equation you would have obtained:

$$\frac{x^2}{1.0} = 1.2 \times 10^{-6}$$

$$x^2 = 1.2 \times 10^{-6}$$
$$x = (1.2 \times 10^{-6})^{1/2} = 1.1 \times 10^{-3}$$

And you would have calculated the same answer as you did from your laborious calculations with the quadratic formula.

The point should now be obvious. If you had a hint at the outset that the value of x might be negligible compared with 1.0, you could have omitted it from the denominator and saved yourself effort. Generally, enough information in the problem indicates the possibility that such an omission is justified. When the opportunity presents itself, take advantage of it. Simplify the equation and solve for the answer. To ensure that your assumption was justified, place your value in the term or terms from which it was omitted. If it indeed yields the value for the term, unaltered within the precision governed by the rules for determining significant figures, you can feel confident that your answer is adequate. If this does not succeed, then the alternative is to do the problem by the more tedious, yet more exact, method.